ETHNOBOTANY
Himalayan Region

ETHNOBOTANY
Himalayan Region

(55 Colour Plates 42 Tables 25 Figures)

S. K. Sood
Anjna Kharwal
T. N. Lakhanpal
Department of Biosciences
Himachal Pradesh University
Shimla

A. K. Bhatnagar
Department of Botany
University of Delhi
Delhi

I.K. International Publishing House Pvt. Ltd.

NEW DELHI

Published by

I.K. International Publishing House Pvt. Ltd.
S-25, Green Park Extension
Uphaar Cinema Market
New Delhi – 110 016 (India)
E-mail: info@ikinternational.com
Website: www.ikbooks.com

ISBN 978-93-82332-37-4

10 9 8 7 6 5 4 3 2 1

Published by Krishan Makhijani for I.K. International Publishing House Pvt Ltd., S-25, Green Park Extension, Uphaar Cinema Market, New Delhi – 110 016 and Printed by Rekha Printers Pvt. Ltd., Okhla Industrial Area, Phase II, New Delhi – 110 020

Preface

The history and development of ethnobotany is as old as human civilization. There is practically no human activity in which plants do not play a role. Although our country, India, is renowned the world over in terms of ethnic variability, rich diversity of plants and the enormous wealth of traditional knowledge systems related to the use of plant species, unfortunately, the traditional knowledge base is fast eroding due to the advent of modern civilization and changes in sustenance economy. Evidently, need of the hour and one of the foremost challenging tasks before the world community is to inventorize and document all available ethnobotanical information among the diverse ethnic communities before it is lost forever. As a matter of fact, tribal knowledge of plants is important not just for the tribal people themselves, but for the wider world. Against this backdrop, the preparation of this invaluable compendium, the first of its type has been necessitated and is based on intensive ethnobotanical surveys undertaken by the authors in various rural pockets of Palampur Himalayas during the years 2003-2009, a region of the Himalayas which is at the confluence of the plains and blessed with remarkable natural beauty and high ranges of Dhauladhar Mountains at the backdrop with tops remains snow covered for most part of the year and significantly rich in ancient heritage, ethnic variability, floristic and cultural diversity. Main body of the textural matter comprises an exhaustive account of 336 promising plants (medicinal 224, edible 166, fodder 100, magico-religious 60, fuel 48, wood 35, household articles 16, aesthetic 12, beverages 10, fibre 10, condiments 11, hedges 9, local wine 9, dye 7, walking sticks 6, insecticidal 4, thatching 4, vermillion 4, bedding 3, insect repellent 3, nector 3, fish poison 2, miscellaneous 88) with multifaceted information on updated nomenclature; vernaculars, English, Hindi, Sanskrit and Regional Names; Distribution; Description; Flowering & Fruiting; Habitat Ecology; Part/s Used; Folk Uses; Active Constituents; Biological Activity and Uses in Literature to serve as a valuable ready-reckoner to different stakeholders like pharmacists, medical practitioners, chemists, biotechnologists, plant geneticists, new entrepreneurs and technocrats, extension workers, students, researchers, policy planners, etc. Moreover, inclusion of information on economic valuation and availability of the underutilized Himalayan plant resources, nativity, 55 photographic plates, 42 tables, 25 figures, 6 useful appendices are the other added attractions that enhance the value of this unique treatise significantly. It is hoped that the information comprehended in the present manual offers tremendous scope for its bioprospecting and as well can be utilized by policy planners in developing effective strategies for ensuring a meaningful sustainable rural development plan for hill communities of the Himalayan states. No institutional library in India and abroad should be without it.

We are indebted to Prof. A.D.N. Bajpayee, Vice Chancellor, Himachal Pradesh University, Shimla, Prof. D. Malhotra and Prof. V.K. Mattu for their encouragement, support and well-wishes; to Dr S. Kumar (Joint Director, B.S.I.) and Dr Sas Biswas (F.R.I. Dehradun) for their kind help in authentication of the plant specimens; and to Dr Kuldeep Dogra, Dr Dheeraj Rawat, Dr Suresh Kumar, Dr Romita Sharma, Mr Ved Prakash, Mr Rakesh, Ms Indu, Ms Poonam and Mr Bal Krishan for their ungrudging help in various ways. Thanks are also due to authors of books, papers and figures listed in the treatise. Finally, we thank our families for bearing the opportunity cost.

The work does not pretend to be complete; suggestions for improving it, if any, are most welcome.

S. K. Sood
Anjna Kharwal
T. N. Lakhanpal
A. K. Bhatnagar

Abbreviations

amsl	Above mean sea level
AD	*anno domini*
Alc.	Alcoholic
Alk.	Alkaline
Aq.	Aqueous
Ass.	Assamese
B.	Bengal
Bo.	Bombay
Bu	Butyl
C	Carbon
C.	Central
Ca	Calcium
cm	Centimetre
CNS	Central Nervous System
d	Dextro
E.	East
EL	Ethnobotany Laboratory
Eng.	English
et al.	(*et alia*) and other authors
etc.	(*et cetera*) and so on
Extr.(s)	Extract(s)
Fig. (s)	Figure(s)
G.	Gujarati
H.	Hindi
HCN	Hydrogen Cyanide
H.P.	Himachal Pradesh
"	Inches
i.e.	(*id est*) that is
Kan.	Kannada
Kash.	Kashmir
l	laevo
m	metre
Mal.	Malayalam
Mar.	Marathi
mg	milligram
Mg	Magnesium
M.P.	Madhya Pradesh
ml	millilitre
N.	North
(–)ve	Negative
N.W.	North West
O	Oxygen
OAc	Acetoxy
OH	Hydroxy
OMe	Oxymethyl
Ori.	Oriya
P	Phosphorus
(+) ve	Positive
(–) ve	Negative
P.	Punjab
%	Per cent
Ph	Phenyl
Pl.(s)	Plate(s)
S.	South
Sans.	Sanskrit
Sp	Specie
Spp	Species
Sq km	Square Kilometre

Syn.	Synonym
S.W.	South West
Tam.	Tamil
Tel.	Telugu
U.P.	Uttar Pradesh
Var.	Variety
Vern.	Vernacular
viz.	*Vide licet* (namely)
Vit	Vitamin
Vet	Veterinary
Vol. (s)	Volume(s)
W.	West
WTO	World Trade Organization

List of Illustrations

Figures

Coloured Photographs

Contents

Introduction

ETHNOBOTANY – A MULTIDISCIPLINARY NATURAL SCIENCE

The term ethnobotany was coined by J.W. Harshberger in 1895 to "the study of plants used by primitive and aboriginal people" (Anonymous, 1895). Since then, the subject has been variously defined and interpreted by different workers as its discipline began to follow multidisciplinary approach combining a diversity of knowledge bases and methods through the use of anthropological methods (Robbins *et al.,* 1916; Schultes & Reis, 1995). Earlier, Alcorn (1984) regarded ethnobotany as the study of contextualized plant use. As stated by Maheshwari (1983), it encompasses direct relationship between tribals/aborigines and plants during the prehistoric, historic and contemporary periods. According to Wickens (1990), the term refers to "the study of useful plants prior to their commercial exploitation and eventual domestication; it includes the use of plants by both tribal and non-tribal communities without any implication of primitive or developed societies". In wider context, it involves the reciprocal and dynamic aspect of the interaction of indigenous people with plants, and includes studies such as on food, fibre, dyes, tannins, medicinal and other useful plants, harmful plants, taboos and magico-religious beliefs about plants, material use and the acts of domestication, conservation, improvement or destruction of plants (Jain, 1987b, 1995) to unveil valuable information on unexplored or unexploited natural resources and new uses of the existing resources to the people of modern society (Schultes, 1960; Anonymous, 1984). Pushpangadan (1990) referred to it as a study of knowledge system pertaining to the multidimensional perspective of life culture, traditions as well as interaction of traditional or less advanced human communities like tribals with their local flora or fauna. Pei (2000) elaborated it as "the science which studies the relationship of a given society with its environment and, in particular, with the plant world. These relations may be social, economic, ecological, symbolic, religious, commercial or artistic". Regarding ethnobotanical knowledge system, bioscientists consider it as a first effective means for identifying as well as locating alternative food source, leads for drugs and pharmaceuticals, natural dyes, colours, gums, resins, etc. (Pushpangadan & Kumar, 2005). Nowadays, the subject of ethnobotany has been recognised as a rapidly expanding multidisciplinary natural science with many workers becoming involved in the practical application of its data in areas such as biodiversity prospecting and conservation biology (Ravishankar, 1996).

INDIGENOUS SOCIETIES

The term "indigenous peoples" (tribals/ethnic societies/rural communities) referred to those who are considered the repository of accumulated experience and knowledge of indigenous flora and fauna on which they are dependent for their sustenance and livelihood. These are indeed human conservatories which cannot be duplicated by science and technology. Harmony with nature is inherent in the culture and way of life of tribal societies. Their knowledge is not only "technical" in nature, but also incorporates integrative insights, wisdom, ideas, perceptions and innovative capabilities which pertain to ecological, biological, geographical and other physical phenomena (Mashelkar, 2001). Respect for these boundaries by different communities and social groups was the result of historically accepted formal and informal rules (Gupta & Ura, 1990). The knowledge acquired by them over a period of time is scattered and communicated orally by word of mouth from generation to generation in different parts of the world. As a matter of fact, tribal knowledge of plants is important not just for the tribal people themselves, but for the wider world (Maheshwari, 1987). There are more than 250 million tribal people all over the globe—one in every 20 on the earth, of these nearly two-thirds are living in China and India with latter having more than 550 native or tribal communities under 227 ethnic groups covering 5,000 villages and rich traditional heritage (Anonymous, 1994). Undeniably, there is an inextricable link between indigenous culture and biodiversity as areas of high biodiversity are oftenly found on indigenous community's lands and in their water bodies (Alcorn, 1996). The 15th session of the General Assembly of IUCN held in Christchurch, New Zealand, in October 1991, recognised the importance of the cultural heritage of mankind and the role of traditional cultures in conservation of nature (Mc Neely & Pit, 1995). Agenda 21 of the Rio Earth Summit (1992) stated that indigenous people have a vital role in environmental management and development because of their knowledge and traditional practices. In the words of Schultes & Swain (1976) "peoples whom we have chosen to consider members of less advanced societies have consistently looked to the Plant Kingdom ..... for the betterment of life. Should we as chemists, pharmacologists and botanists ..... with so many and varied means at our disposal.....not take a lesson from them? Unfortunately, tribal traditions and practices are now under assault everywhere in the world under the impact of industrialisation and changes in sustenance economy, and needs to be properly documented and conserved for the benefit of the present and future generations (Janaki Ammal, 1956; Maheshwari, 1987; Jha & Goel, 2004).

THE HIMACHAL HIMALAYAS – LAND OF ETHNIC COMMUNITIES

From time immemorial, the Himalayas has influenced the life and culture of the diverse ethnic communities living all along the length of its mountainous chain. Spreading over an area of 55,673 sq km (10.54% of the Himalayan landmass), the state of Himachal Pradesh including 16,997 villages under 12 districts, a veritable

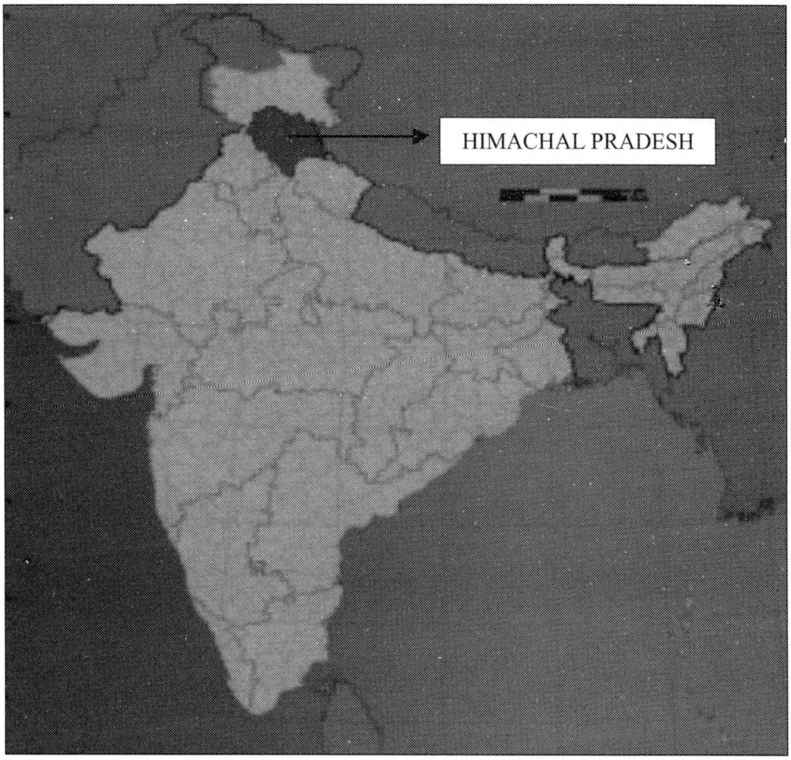

Fig. 1 Map Showing the Location of Himachal Pradesh in India.

emporium of medicinal and aromatic plants, is a green pearl nestled in the lap of N.W. Himalayan mountain ranges between latitude 30°-22′-40″ and 33°-12′-40″ north and and longitude 75°- 45′-55″ and 79°-04′-20″ east with altitudes ranging from 350-6,975 m amsl and having average rainfall of about 152 cm. Towards east it forms India's border with Tibet; Jammu and Kashmir in the north, Uttarakhand in the southeast, Haryana in the south and Punjab in the west, and its vegetation rich in Mediterranean, West Asian, Tibetan, Japanese and Burmese elements in its subtropical, temperate, alpine and cold desert regions. Different tribes such as Gaddies, Gujjars, Kinnauras, Jads, Lahoulis, Pangwalas and Swangalas inhabit its very diverse terrain. Since ages, the indigenous people of this region follow varied traditional practices of resource utilization and conservation, and the place is a very fertile area for ethnobotanical research for which the studies so far are almost negligible.

CURRENT STATUS OF WORK ON ETHNOBOTANICAL SPECTRUM OF INDIAN FLORA

Even prior to the coining of the term ethnobotany, the uses of plants by human beings found its mention in ancient Sanskrit, Greek, Chinese and Arabic literature. In India,

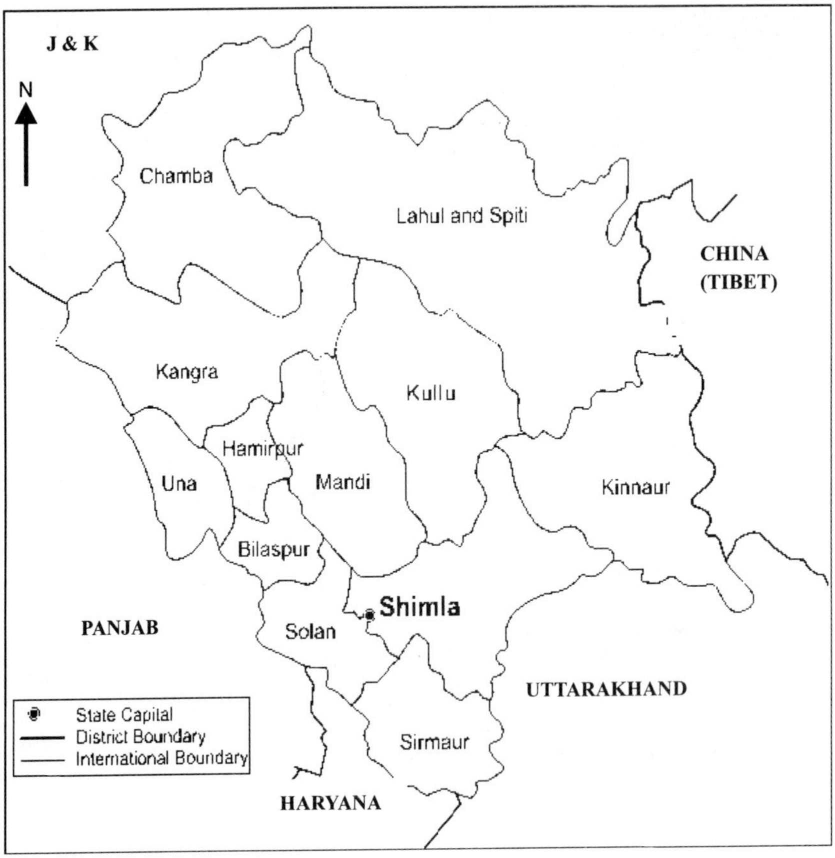

Fig. 2 Map Showing the Various Districts of Himachal Pradesh.

the earliest credit goes to the work of Watt (1889-1896, 1908) which provided the first true reflection of ethnobotany and indigenous knowledge with an index of 3,000 vernacular names of plant products and various uses from its different parts (Shah, 2005). Later, credit for efforts in this direction to pioneered ethnobotanical research had been that of Janaki Ammal (1955) who studied subsistence on food plants of certain tribals of South India particularly to explore better prospects of Dioscoreas under the official programme in the Economic Botany Section of Botanical Survey of India (BSI). Since then uses of plants by the tribals are being recorded for a variety of purposes (Jain, 1981, 1989). In the last few decades, the research and developmental activities in the area of ethnobotany has gained tremendous momentum all over the world with India not an exception to it. *Glimpses of Indian Ethnobotany* by Jain (1981) is the first book dealing with Indian ethnobotany. Subsequently, the subject matter reviewed in some of the important compilations like 'Bibliography of Ethnobotany' (Jain *et al.,* 1984). Cumulative index to ethnobotany from 1989-2000 with the analysis of regions and the themes covered (Srivastava & Jain, 2000), Indian ethnobotanical

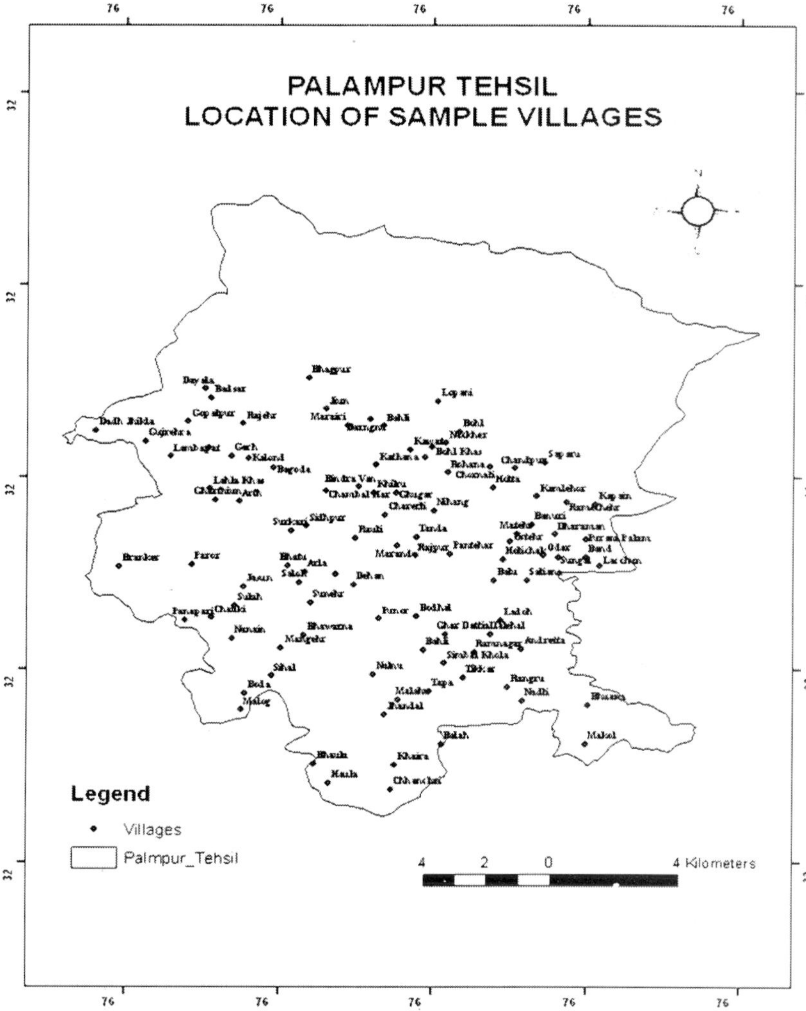

Fig. 3 Map of Tehsil Palampur of District Kangra (H.P.) Showing Places Visited for the Present Study.

literature during last two decades—a graphic view and the future directions (Jain & Srivastava, 2001), a recent detailed one listing 1,600 references of which over 1,250 relate strictly to the ethnobotanical work or very closely allied themes on India and the others on concept, global evolution of the discipline, and methodology of new areas from works of some outstanding foreign ethnobotanists (Jain, 2002) and that on ethnobotanical spectrum of Indian flora: an overview during the past 20 years in the journal ethnobotany (Goel & Tripathy, 2009) have brought together the ethnobotanical literature from disparate sources. In recent times, a number of useful treatises on the subject from Indian perspective have been those of Maheshwari

(1992), Joshi (1995), Arora & Pandey (1996), Rama Rao & Henry (1996), Kaushik (1998), Pandey (1998), Singh & Pandey (1998), Dagar & Dagar (1999), Mudgal *et al.*, (1999), Singh & Kumar (2000a), Sood *et al.* (2001, 2009a, b, 2010), Singh *et al.* (2003a, b), Sood & Thakur (2004), Trivedi (2004, 2009, 2010), Sood & Kaushal (2008), Sood & Sudersna (2008), etc., and all these contributed significantly towards our understanding on the various aspects of ethnobotanical science. Barring these, other useful works describing linkages of ethnobotany with the other sciences and disciplines (Manilal, 1989); quality control requirements of medicinal plants used in traditional medicines (Mehrotra, 1990); efforts and concerns about ethnobotanical studies on the Indian plant genetic resources (Arora, 1995); role of ethnobiologists in the traditional knowledge and the sustainable development (Rao, 1996); need to compute the relative reliability of ethnobotanical claims and ethnobotany in the 21st century in India (Khan, 2001, 2005); ethnobotanical perspective of the Sanskrit plant names (Patil, 2000); Indian plant species used in Tibetan traditional medicines (Jain, 2003; Jain & Goel, 2005); role of traditional and folklore herbals for the development of new drugs (Mehrotra & Mehrotra, 2005); CBD,WTO and Biodiversity Act of India in the ethnobotanical perspective (Pushpangadan & Kumar, 2005); ethnobotany in the new millennium—some thoughts on future direction in Indian ethnobotany (Jain, 2006); and prospects of ethnobotany in 2008 (Shah, 2008) have also been done in recent years.

Ever since the pioneering work of William Spencer in 1817, the state of Himachal Pradesh, had been extensively explored floristically by several workers (Atkinson, 1882; Hooker, 1872-1897; Collett, 1902; Nair, 1964; Chowdhery & Wadhwa, 1984; Polunin & Stainton, 1987; Stainton, 1988; Aswal & Mehrotra, 1994; Dhaliwal & Sharma, 1999; Kaur & Sharma, 2004; Retnam & Martin, 2006; Singh & Sharma, 2006). Mention of uses of plants are also there in some of the above floras. A good number of studies on diversity of medicinal and aromatic plants in different physiographic regions of the state have also been done (Ahluwalia, 1952; Rastogi, 1960; Gupta, 1961, 1964, 1971; Rau, 1961; Shabnam, 1964; Sarin, 1967; Uniyal & Chauhan, 1971-1973; Dhiman, 1976; Arora *et al.*, 1980; Sarin & Chopra, 1985; Chauhan, 1974, 1983, 1989 a, b, 1990, 1995, 1999, 2003; Singh & Aswal, 1992; Singh, 1993, 1999; Dobriyal *et al.*, 1997; Sharma & Sood, 1997; Badola, 2001; Chaurasia *et al.*, 2001). Ethnobotanically, the available information in the context of Himachal Pradesh is scarce (Goel & Tripathi, 2009). So far 42 publications including 4 books have been published and the data is available only for Gaddies, Lahaulas and Spitians. In this regard, ethnobotanical data documented for some of its districts, viz. **Chamba** (Uniyal *et al.*, 2006: Traditional use of medicinal plants among the tribal communities of Chhota Bhangal), **Hamirpur** (Prakash *et al.*, 2003: Ethnobotanical studies on some ethnobotanical plants of district Hamirpur), **Kangra** (Uniyal & Chauhan, 1973b: Traditionally important medicinal plants of Kangra valley in Dharamshala forest circle; Kapur, 1993: Ethnomedico plants of Kangra Valley; Sharma & Chauhan, 2000: Ethnobotanical studies of Gaddies—a tribal community of district Kangra; Singh & Kumar, 2000b: Observations in ethnoveterinary medicine among the Gaddi tribe of Kangra valley (H.P.) in Kangra region), **Kinnaur** (Negi & Subramani, 2002: Ethnobotanical studies in the Chhitkul of Sangla valley; Singh, 2004: Indigenous knowledge and conservation practices in tribal society of Western Himalaya: a case study of Sangla Valley), **Kullu** (Uniyal & Chauhan, 1973a: Traditionally important medicinal plants of Kullu forest division; Singh, 1990: Ethnobotanical studies of useful plants of the Kullu district; Jain & Puri, 1994: An ethno-medico-botanical survey of Parvati valley; Singh *et al.*, 1998: Ethno-cultural analysis of sacred species and ecosystems in Chhakinal watershed; Uniyal, 2003: Utilization of medicinal plants by the rural women of Kullu; Sharma *et al.*, 2004: Obeservations on the traditional phytotherapy among the inhabitants of Parvati valley), **Lahaul-Spiti** (Koelz, 1979: Notes on ethnobotany of Lahaul – a province of the Punjab; Duhoon *et al.*, 1996: Seabuckthorn (*Hippophae* spp)—a less known wonder plant of ethno-medico-botanical importance in cold desert of India; Kala, 2003: Ethno-medico-botany of Indian Trans-Himalayas—a case study from Spiti; Chauhan & Lakhanpal, 2000: Ecological and ethnobotanical studies on the flora of Spiti, Himachal Pradesh—a cold desert in the western Himalaya; Kala & Manjrekar, 2000: Ethnomedicobotanical study of Indian Trans-Himalaya—a case study from Spiti; Sood *et al.*, 2001: *Ethnobotany of Cold Desert Tribes of Lahaul-Spiti* (*N.W. Himalaya)*; Chandrasekar & Srivastava, 2003: Ethnobotanical studies in Pin Valley National Park; Srivastava & Chandrasekar, 2004; Ethnomedicine of the Pin Valley National Park; Chandrasekar & Srivastava, 2005: New reports on aphrodisiac plants from Pin Valley National Park; Singh & Chauhan, 2005: Traditional practices of herbal medicines in Lahaul valleys; Chaurasia *et al.,* 2007: Ethnobotany and plants of Trans-Himalaya), **Mandi** (Gaur & Singh, 1993: Ethnomedicinal plants of Mandi district, Himachal Pradesh; Singh, 1996b: Wild edible plants of Mandi district in North-West Himalaya; Sood & Thakur, 2004: *Ethnobotany of Rewalsar Himalaya*; Rajasekaran & Singh, 2009: Ethnobotany of Indian horse chestnut (*Aesculus indica*) in Mandi district; Sood *et al.*, 2009b: *Ethnobotanical Studies on Trees, Shrubs and Climbers of Himalaya*), **Shimla** (Chauhan, 1996: Ethno-medico-botany of Pabbar valley; Sharma *et al.,* 2009: Some important supplementary food plants and wild edible fungi of upper region of district Shimla), and **Solan** (Verma & Chauhan, 2006: Ethno-medico-botany of Kunihar forest division; Verma & Chauhan, 2007: Indigenous medicinal plants knowledge of Kunihar forest division) and the entire **state** (Chauhan, 1988: Ethnobotanical study of medicinal plants of Himachal Pradesh; Lal *et al.*, 1996: Supplementary foods of Gaddies in H.P., and Plants used as ethnomedicine and supplementary food by Gaddies of Himachal Pradesh; Sharma & Rana, 1999: Traditional medicinal uses of plants in Himachal hills; Singh & Kumar, 2000a: Ethnobotanical wisdom of Gaddie tribe of Western Himalayas; Aggarwal *et al.*, 2004: Ethnobotanical studies on some wild plants of lower hills of Himachal Pradesh; Rana, 2004: *Lesser Known Food Plants of Himachal Pradesh*; Bhalla & Bhalla, 2006: Traditional food and beverages of Himachal Pradesh) have enriched our knowledge on the subject. Besides, noteworthy contributions on the allied aspects providing useful information on the importance of *Physochaina praelta* in shaping Lahaul's economy (Sarin *et al.*, 1963), present status and future prospects of medicinal and aromatic resources of Himachal Pradesh (Sarin & Chopra, 1985; Gupta, 1988), role of forestry extension in exploitation of medicinal and aromatic plants of Himachal Pradesh (Sagwal & Gupta, 1988), endangered ayurvedic pharmacopoeial plant resources of H.P. (Chauhan, 1989c), medicinal plants of Himachal Pradesh used in Indian pharmaceutical industry (Singh & Aswal, 1992), medicinal plants of ayurvedic importance from Mandi district of H.P. (Singh, 1993), occurrence, distribution and diversity of soft fruits in N.W. and W. Himalaya and prospects of their conservation and utilization (Sharma & Chandel, 1996), endangered medicinal plant species in Himachal Pradesh (Badola & Pal, 2002), collection and marketing strategies of important medicinal and aromatic plants in H.P. (Negi & Bhalla, 2002), conservation assessment and management prioritisation for the medicinal plants of Himachal Pradesh (Ved *et al.*, 2003), baseline data on medicinal plants conservation and sustainable utilization (Himachal Pradesh) (Mishra & Gokhale, 2003) and potential application of biotechnology in medicinal plants of Himachal Pradesh (Sethi, 2003) have also appeared during the last few decades.

From the scrutiny of above literature it is evident that though some work has been done on ethnobotanical aspects of plants in many parts of Himachal Pradesh, no attempt has been made so far to document information on this aspect from tehsil Palampur of district Kangra, Himachal Pradesh (India). To bridge this gap in our understanding on the subject, the present work has, therefore, been envisaged to explore the empirical knowledge of the local inhabitants on the multifacet uses of plants from the region with a view not only to conserve it from being lost irreversibly to varied biotic pressures but also for using them as valuable clues for social forestry endeavours therapeutic agents, supplementary foods, sustainable management of species as well as their habitats, local and regional problems, and socio-economic upliftment of the mountain people and the society at large.

THE STUDY AREA

Geographical Extent

The word Palampur derives from the local word 'pulum', meaning lots of water. There are numerous streams flowing from the mountains, criss-crossing the landscape and combination of greenery and water has, thus, given tehsil a distinctive look, located at an altitude between 750-1800 m amsl at a distance of 250 km from the state headquarters between 32°07′N 76°32′E/32°12′ N 76°53′E latitudes. The nearest railway station is at Maranda at a distance of 5 km. The area is at the confluence of the plains and the hills and blessed with remarkable natural beauty and high ranges of Dhauladhar mountains at the backdrop with tops remain snow covered for most part of the year.

History

Palampur was once a part under Jalandhar kingdom before it came under the British rule. It came into being when Dr Jameson, Superintendent of Botanical Gardens, introduced the tea bush from Almora in 1849. The bush thrived and so did the town and it ultimately became a focus of the European tea estate owners.

Land and People

The natives are the Kangri people and the local language is Kangri. The majority of the people are Hindu along with Muslims, Christians, Jains and Sikhs. In recent years Tibetans practising 'Buddhism' have also settled. Traditional dresses of men are 'kurta', 'pyjamas' with a woolen jacket in winter. Women generally wear 'salwaar', 'kameez' along with 'chunni' ('chaddru' in local language). Maize and wheat are the main staple foods of the rural populace whereas those in town love to take rice at least once a day. The villagers are very fond of butter, milk, curry preparations and pickles. Economy consists of agriculture and farming with 'Kangra tea' of Palampur known internationally for its exceptional quality and medicinal value.

Climate and Agriculture

The place unfolds four broad seasons with winters spreading generally from December-February, summers from March-June, rainy season extending from July-September with landscape becoming lush green, and autumns from October- November. The temperature during winters dips to low due to lashing cold breeze of the mountain ranges of Dhauladhar. On an average the area receives annual rainfall of 2,500 mm. Agriculture is the mainstay of the inhabitants of the area. Soil varies from sandy loams to clay. The agro-climatic conditions favour the growth of food crops such as wheat, paddy, maize, potatoes, etc. Agricultural operations are carried out in two spells. Spring crops popularly called Rabi ('Harri') comprise wheat, barley, gram and oil seeds (linseed) whereas autumn crops (Kharif/'Savani') are maize, paddy, pulses, spices and potatoes. Millets are also grown in some areas. Tea cultivation is done on well drained loamy soil on the hill slopes with altitude varying from 800-1,600 m. Tea and potatoes are the two cash crops of Palampur tehsil.

Places of Interest

There is a tea factory at the very start of town, offering an insight to the processing of famous Kangra tea. One can walk through tea gardens and open fields to reach historical temple of 'Bandlamata' which was originally built about five centuries ago. Close to the temple is Neugal khad. The village of Andretta has been the home of artists Norah Richards, Sobha Singh and B. C. Sanyal. The home of Sobha Singh now houses a well known gallery of his world famous paintings. There is also a pottery and craft centre at Andretta. A mini zoo is located at Gopalpur.

Collection of Ethnobotanical Data

For documenting first hand information pertaining to ethnobotany of tehsil Palampur of district Kangra (H.P.), intensive and extensive ethnobotanical field surveys were undertaken in its various rural pockets in different seasons during the years 2003-2009. The field surveys were planned in such a way so as to collect the ethnobotanically interesting species either in flowering or fruiting stage. The desired information on ethnobotanical aspects was collected through interviews from knowledgeable people (traditional practitioners, old experienced farmers, family heads, housewives, eminent elderly persons of the community, etc.), as per the methodology outlined by Jain (1987a). The data thus collected were verified in different villages among the interviewers after showing the same plant sample, and even with the same informants on different occasions. Ethnobotanical lore was considered valid if at least three informants made similar comments about the uses.

Table I Flowering and Fruiting Season of the Ethnobotanically Important Plant Species Collected from the Study Area

S.No.	*Name of Spp.*	*Flowering & Fruiting Season*
1.	*Abelmoschus esculentus* (L.) Moench	June-December
2.	*Abrus precatorius* L.	May-August
3.	*Acacia catechu* Willd.	May-July
4.	*Achyranthes aspera* L.	May-October
5.	*Achyranthes bidentata* Blume	June-August
6.	*Acorus calamus* L.	June-July
7.	*Adiantum capillus veneris* L.	July-October
8.	*Aegle marmelos* (L.) Correa	April-July
9.	*Aerva lanata* (L.) Juss. *ex* Schult	September-March
10.	*Agave cantula* Roxb.	March-May
11.	*Agave wightii* Drumm. & Prain	July-September
12.	*Ageratum conyzoides* L.	March-April
13.	*Ajuga parviflora* Benth.	May-June
14.	*Albizia chinensis* (Osbeck) Merr.	April-May
15.	*Albizia julibrissin* Durazz.	May-August
16.	*Albizia lebbeck* Benth.	December-March
17.	*Allium cepa* L.	December-March
18.	*Allium sativum* L.	March-April
19.	*Aloe vera* (L.) Webb. & Berth.	Throughout the year
20.	*Alternanthera sessilis* (L.) R. Br. *ex* DC.	Throughout the year
21.	*Amaranthus paniculatus* L.	September-November
22.	*Amaranthus viridis* L.	May-June
23.	*Apium leptophyllum* Muell. *ex* Benth.	June-September
24.	*Araucaria columnaris* (Forst.) Hook.	September-October
25.	*Aristida cyanantha* Steud.	July-September

Contd...

Table I Contd...

S.No.	*Name of Spp.*	*Flowering & Fruiting Season*
26.	*Artemisia capillaris* Thunb.	August-September
27.	*Artemisia japonica* Thunb.	September-December
28.	*Artocarpus lacucha* Buch.-Ham.	February-September
29.	*Arundinaria falcata* Nees	May-November
30.	*Asparagus adscendens* Roxb.	October-November
31.	*Avena sativa* L.	November-May
32.	*Bambusa nutans* Wall.	Flowering sporadically
33.	*Basella alba* L.	March-December
34.	*Bauhinia vahlii* Wight & Arn.	April-June
35.	*Bauhinia variegata* L.	March-August
36.	*Berberis lycium* Royle	April-July
37.	*Bergenia ciliata* (Haw.) Sternb.	March-November
38.	*Bidens biternata* (Lour.) Merr. & Sherff.	June-September
39.	*Boehmeria macrophylla* Hornem.	July-September
40.	*Boerhaavia diffusa* L.	June-October
41.	*Bombax ceiba* L.	February-May
42.	*Brassica nigra* (L.) Koch	October-March
43.	*Brassica tournefortii* Gouan.	February-April
44.	*Butea monosperma* Kuntze	March-May
45.	*Caesalpinia decapetala* (Roth) Alston	April-July
46.	*Cajanus cajan* (L.) Millsp.	June-October
47.	*Callicarpa macrophylla* Vahl	July-September
48.	*Calotropis procera* R. Br.	February-May
49.	*Camellia sinensis* (L.) Kuntze	June-September
50.	*Cannabis sativa* L.	June-September
51.	*Capillipedium assimile* (Steud.) A. Camus.	August-November
52.	*Capsicum annuum* L.	May-September
53.	*Carica papaya* L.	Greater part of the year
54.	*Carissa congesta* Wight	April-July
55.	*Carya illinoensis* (Wangenh.) Koch	May-October
56.	*Caryopteris bicolor* (Hardw.) Mabb.	April-August
57.	*Cassia fistula* L.	March-July
58.	*Cassia floribunda* Cav.	June-December
59.	*Cassia obtusifolia* L.	July-December
60.	*Cassia occidentalis* L.	May-September
61.	*Catharanthus roseus* G. Don	Throughout the year

Contd...

Table I Contd...

S.No.	Name of Spp.	Flowering & Fruiting Season
62.	*Cedrus deodara* (Roxb.) Louden	June-July
63.	*Celastrus paniculatus* Willd.	April-December
64.	*Celtis australis* L.	April-October
65.	*Centella asiatica* (L.) Urb.	April-May
66.	*Centipeda minima* (L.) A. Br. Asch.	November-March
67.	*Centratherum anthelminticum* (L.) Kuntze	August-September
68.	*Cestrum nocturnum* L.	August-October
69.	*Cheilanthes farinosa* (Forst.) Kaulf.	July-August
70.	*Chenopodium album* L.	May-October
71.	*Chenopodium ambrosioides* L.	June-September
72.	*Cicer arietinum* L.	Throughout India
73.	*Cinnamomum camphora* (L.) Nees & Eberm	April-August
74.	*Cinnamomum tamala* Nees & Eberm	June-August
75.	*Cirsium arvense* (L.) Scop.	May-June
76.	*Cissampelos parriera* L.	April-September
77.	*Citrus grandis* (L.) Osbeck	November-March
78.	*Citrus jhambiri* Lush.	October-March
79.	*Citrus limon* (L.) Burm. f.	November-February
80.	*Citrus medica* L.	October-March
81.	*Citrus medica* var. *medica* L.	April-September
82.	*Clematis gauriana* Roxb.	October-January
83.	*Coccinia grandis* (L.) Voigt	May-August
84.	*Cocculus laurifolius* DC.	March-August
85.	*Colebrookea oppositifolia* Sm.	April-November
86.	*Colocasia esculenta* Schott	June-October
87.	*Commelina paludosa* Blume	June-August
88.	*Cordia dichotoma* Forst.	April-July
89.	*Coriandrum sativum* L.	February-April
90.	*Coronopus didymus* (L.) Sm.	December-April
91.	*Costus speciosus* Sm.	June-August
92.	*Cryptolepis buchanani* Roem. & Schult.	April-November
93.	*Cucumis melo* L. var. *Agrestis*	May-August
94.	*Cucumis sativus* L.	May-September
95.	*Cucurbita hispida* Thunb.	June-September
96.	*Cucurbita pepo* L.	June-September
97.	*Curcuma longa* Wall.	June-July

Contd...

Table I Contd...

S.No.	*Name of Spp.*	*Flowering & Fruiting Season*
98.	*Cuscuta reflexa* Roxb.	July-October
99.	*Cyclanthera pedata* (L.) Schrad.	August-September
100.	*Cymbopogon citratus* (DC.) Stapf	August-March
101.	*Cynodon dactylon* (L.) Pers	July-January
102.	*Cyphomandra betacea* (Cav.) Sendtn.	May-September
103.	*Dalbergia sissoo* Roxb.	May-August
104.	*Datura innoxia* Mill.	June-August
105.	*Datura stramonium* L.	June-October
106.	*Debregeasia hypoleuca* Wedd.	May-August
107.	*Deeringia amaranthoides* Merr.	July-October
108.	*Dendrocalamus hamiltonii* Nees & Arn. *ex* Munro	Flowering sporadically
109.	*Didymocarpus pedicellata* R. Br.	July-March
110.	*Dioscorea bellophylla* Voigt	September-November
111.	*Dioscorea bulbifera* L.	June-August
112.	*Dioscorea oppositifolia* L.	August-September
113.	*Diospyros kaki* L.	May-June
114.	*Diplazium esculentum* (Retz.) Sw.	April-July
115.	*Diplocyclos palmatus* (L.) C. Jeffrey	September-October
116.	*Dodonaea viscosa* Jacq.	February-April
117.	*Dolichos biflorus* L.	June-October
118.	*Drimia indica* (Roxb.) Jessop	May-September
119.	*Duranta repens* L.	April-August
120.	*Ehretia acuminata* R. Br.	May-July
121.	*Ehretia laevis* Roxb.	January-June
122.	*Eleusine coracana* Gaertn.	June-September
123.	*Emblica officinalis* Gaertn.	April-September
124.	*Engelhardia colebrookeana* Lindl.	April-October
125.	*Equisetum debile* Roxb.	July-September
126.	*Eriobotrya japonica* Lindl.	July-September
127.	*Eriophorum comosum* Wall.	August-October
128.	*Erythrina suberosa* Roxb.	March-June
129.	*Eucalyptus tereticornis* Sm.	March-April
130.	*Eulaliopsis binata* (Ritz.) C. E. Hubb.	July-November
131.	*Euonymus hamiltonianus* Wall.	August-September
132.	*Euphorbia royleana* Boiss.	June-July
133.	*Fagopyrum cymosum* Meisn.	June-October

Contd...

Table I Contd...

S.No.	*Name of Spp.*	*Flowering & Fruiting Season*
134.	*Fagopyrum esculentum* (L.) Moench	August-October
135.	*Ficus benghalensis* L.	February-June
136.	*Ficus carica* L.	February-June
137.	*Ficus clavata* Wall.	June-September
138.	*Ficus foveolata* Wall.	March-July
139.	*Ficus glomerata* Roxb.	April-July
140.	*Ficus hispida* L.	June-August
141.	*Ficus infectoria* Roxb.	February-June
142.	*Ficus palmata* Forsk.	April-September
143.	*Ficus religiosa* L.	March-October
144.	*Ficus roxburghii* Wall. *ex* Miq.	April-August
145.	*Ficus rumphii* Blume	April-July
146.	*Ficus semicordata* Buch.-Ham. *ex* Sm.	May-September
147.	*Flacourtia indica* (Burm. f.) Merr.	March-April
148.	*Foeniculum vulgare* Gaertn.	August-September
149.	*Fragaria indica* Andrz.	June-August
150.	*Fumaria indica* (Hausskn.) Pugsley	October-March
151.	*Galium aparine* L.	July-August
152.	*Geranium nepalense* Sweet	March-May
153.	*Glycine max* Merr.	July-September
154.	*Gnaphalium affine* L.	April-September
155.	*Gonatanthus pumilus* (D. Don) Engl. & Krause	June-August
156.	*Gossypium arboreum* L.	July-September
157.	*Grevillea robusta* A. Cunn. *ex* R. Br.	March-July
158.	*Grewia disperma* Rottl. ex Spreng	September-November
159.	*Grewia elastica* Royle	April-July
160.	*Grewia optiva* Drumm. *ex* Burret	April-June
161.	*Hamiltonia suaveolens* Roxb.	September-November
162.	*Helicteres isora* L.	September-November
163.	*Heteropogon contortus* (L.) Beaub. *ex* Roem. & Schult.	July-October
164.	*Hibiscus rosa sinensis* L.	February-October
165.	*Holarrhena antidysentrica* Wall.	June-February
166.	*Hypericum oblongifolium* Choisy	April-October
167.	*Impatiens balsamina* L.	July-September
168.	*Indigofera cassioides* Rottl. *ex* DC.	March-May
169.	*Ipomoea batatas* (L.) Lam.	June-March

Contd...

Table I Contd...

S.No.	Name of Spp.	Flowering & Fruiting Season
170.	*Ipomoea turbinata* Lag.	July-September
171.	*Jasminum dispermum* Wall.	April-November
172.	*Jasminum grandiflorum* L.	April-September
173.	*Jasminum humile* L.	May-September
174.	*Jasminum multiflorum* (Burm. f.) Andrews	November-April
175.	*Jatropha curcas* L.	February-April
176.	*Juglans regia* L.	April-October
177.	*Justicia adhatoda* L.	April-November
178.	*Kalanchoe pinnata* (Lam.) Pers.	December-March
179.	*Lannea coromandelica* (Houtt.) Merr.	March-September
180.	*Lantana camara* L.	April-June
181.	*Lathyrus aphaca* L.	April-May
182.	*Lathyrus sativus* L.	May-August
183.	*Lawsonia inermis* L.	July-October
184.	*Lepidagathis cuspidata* Nees	December-June
185.	*Lepidium sativum* L.	March-May
186.	*Leucaena glauca* Benth.	September-November
187.	*Leucas cephalotes* Spreng.	July-October
188.	*Linum usitatissimum* L.	November-April
189.	*Litchi chinensis* Sonn.	April-June
190.	*Litsea monopetala* (Roxb.) Pers.	March-July
191.	*Litsea sebifera* Pers.	May-September
192.	*Litsea semicarpifolia* Wall.	April-September
193.	*Loranthus falcatus* L.	July-December
194.	*Lotus corniculatus* L.	April-September
195.	*Luffa acutangula* (L.) Roxb.	May-September
196.	*Luffa cylindrica* (L.) Roem.	May-September
197.	*Mallotus philippinensis* (Lam.) Muell.-Arg.	April-October
198.	*Mangifera indica* L.	March-July
199.	*Medicago polymorpha* Roxb.	February-April
200.	*Melia azedarach* L.	May-October
201.	*Mentha longifolia* (L.) Huds.	November-February
202.	*Mentha piperita* L.	July-September
203.	*Michelia champaca* L.	April-October
204.	*Micromeria biflora* (Buch.- Ham.) Benth.	January-July
205.	*Milletia extensa* Benth. *ex* Baker	July-September

Contd...

Table I Contd...

S.No.	Name of Spp.	Flowering & Fruiting Season
206.	*Mirabilis jalapa* L.	June-September
207.	*Momordica charantia* L.	July-October
208.	*Momordica dioica* Roxb.	June-August
209.	*Moringa oleifera* Lam.	February-April
210.	*Morus alba* L.	March-May
211.	*Mucuna pruriens* (L.) DC.	July-September
212.	*Murraya koenigii* Spreng.	March-June
213.	*Murraya paniculata* (L.) Jack.	March-June
214.	*Musa sapientum* L.	Throughout the year
215.	*Neolitsea umbrosa* (Nees) Gamble	March-May
216.	*Nerium indicum* Mill.	April-November
217.	*Ocimum basilicum* L.	June-September
218.	*Ocimum gratissimum* L.	June-August
219.	*Ocimum kilimandscharicum* Guerke	August-September
220.	*Ocimum sanctum* L.	August-September
221.	*Opuntia dillenii* Haw.	April-June
222.	*Origanum vulgare* L.	July-September
223.	*Oroxylum indicum* Vent.	May-August
224.	*Ougeinia dalbergioides* Benth.	March-May
225.	*Oxalis corniculata* L.	April-November
226.	*Panicum antidontale* Retz.	July-October
227.	*Pennisetum americanum* (L.) Leeke	June-July
228.	*Pentanema indicum* L.	Throughout year
229.	*Perilla frutescens* (L.) Britt.	May-August
230.	*Persea odoratissima* (Nees) Kosterm.	March-April
231.	*Phoenix acaulis* Buch.- Ham. *ex* Roxb.	April-June
232.	*Phoenix sylvestris* Roxb.	May-June
233.	*Phyllanthus urinaria* L.	August-September
234.	*Pinus roxburghii* Sarg.	March-May
235.	*Pistacia khinjuk* Stocks	March-May
236.	*Plantago lanceolata* L.	January-April
237.	*Plumbago zeylanica* L.	April-September
238.	*Pogostemon benghalense* (Burm. f.) Kuntze	December-April
239	*Polygala chinensis* L.	July-September
240.	*Polygonum barbatum* L.	April-July
241.	*Populus ciliata* Wall.	April-July

Contd...

Table I Contd...

S.No.	*Name of Spp.*	*Flowering & Fruiting Season*
242.	*Populus nigra* L.	April-July
243.	*Porana paniculata* Roxb.	October-February
244.	*Portulaca grandiflora* Hook.	July-August
245.	*Prinsepia utilis* Royle	June-October
246.	*Prunus cerasoides* D. Don	November-May
247.	*Prunus domestica* L.	April-August
248.	*Prunus persica* Benth. & Hook. f.	February-May
249.	*Psidium guajava* L.	July-September
250.	*Punica granatum* L.	May-September
251.	*Pyrus communis* L.	June-August
252.	*Pyrus pashia* Ham.	May-August
253.	*Pyrus serotina* Rehder	March-May
254.	*Quercus leucotrichophora* A. Camus	April-October
255.	*Randia dumetorum* (Retz.) Lam.	April-September
256.	*Ranunculus arvensis* L.	April-June
257.	*Ranunculus laetus* Wall. *ex* Don	March-April
258.	*Ranunculus muricatus* L.	March-June
259.	*Raphanus sativus* L.	April-May
260.	*Reinwardtia indica* Dumort.	November-May
261.	*Rhamnus triquetra* Wall.	May-August
262.	*Rhododendron arboreum* Sm.	March-May
263.	*Ricinus communis* L.	March-May
264.	*Rorippa nasturtium-aquaticum* (L.) Hayek	April-September
265.	*Rosa brunonii* Lindl.	April-October
266.	*Rosa chinensis* Jacq.	Throughout year
267.	*Rosa damascena* Mill.	April-June
268.	*Roylea cinerea* (D. Don) Baill.	March-November
269.	*Rubus ellipticus* Sm.	January-May
270.	*Rubus niveus* Thunb.	March-May
271.	*Rumex hastatus* Don	November-April
272.	*Rumex nepalensis* Spreng.	May-October
273.	*Saccharum officinarum* L.	January-April
274.	*Saccharum spontaneum* L.	June-October
275.	*Salix karelini* Turcz. *ex* Stsch.	May-July
276.	*Salix tetrasperma* Roxb.	April-July
277.	*Salvia coccinea* L.	June-August

Contd...

Table I Contd...

S.No.	Name of Spp.	Flowering & Fruiting Season
278.	*Santalum album* L.	September-January
279.	*Sapindus mukorosii* Gaertn.	April-September
280.	*Sapium insigne* (Royle) Benth.	May-August
281.	*Sapium sebiferum* (L.) Roxb.	June-August
282.	*Scandix pecten veneris* L.	May-September
283.	*Sechium edule* Sw.	June-September
284.	*Selaginella chrysocaulos* (Hook. & Grev.) Spring	July-September
285.	*Sesamum orientale* L.	July-September
286.	*Sida cordifolia* L.	July-September
287.	*Silene conoidea* L.	March-May
288.	*Sisymbrium irio* L.	December-May
289.	*Solanum erianthum* Don	July-October
290.	*Solanum melongena* L.	June-September
291.	*Solanum nigrum* L.	January-July
292.	*Solanum torvum* Sw.	April-August
293.	*Solanum viarum* Dunal	May-August
294.	*Sorghum bicolor* (L.) Moench	July-August
295.	*Spilanthes oleracea* L.	May-October
296.	*Spondias pinnata* (L.f.) Kurz	April-June
297.	*Sporobolus indicus* R. Br.	July-September
298.	*Stellaria media* (L.) Vill.	March-April
299.	*Stephania glabra* (Roxb.) Miers	June-August
300.	*Syzygium cumini* (L.) Skeels	April-August
301.	*Syzygium operculatum* (Roxb.) Nied.	April-July
302.	*Tagetes minuta* L.	September-October
303.	*Terminalia arjuna* Wight & Arn.	May-September
304.	*Terminalia bellirica* (Gaertn.) Roxb.	May-September
305.	*Terminalia chebula* Retz.	April-June
306.	*Terminalia tomentosa* (DC.) Wight & Arn.	August-March
307.	*Thuja orientalis* L.	July-October
308.	*Thysanolaena maxima* (Roxb.) Kuntze	July-October
309.	*Tinospora cordifolia* Miers	May-August
310.	*Toona ciliata* Roem.	May-June
311.	*Trichosanthes cucumerina* L.	May-August
312.	*Trifolium resupinatum* L.	May-June
313.	*Trigonella foenum-graecum* L.	March-May

Contd...

Table I Contd...

S.No.	Name of Spp.	Flowering & Fruiting Season
314.	*Triticum aestivum* L.	October- May
315.	*Tulipa stellata* Hook.	March-May
316.	*Urtica ardens* Link	February-May
317.	*Urtica dioica* L.	June-October
318.	*Vanda tessellata* (Roxb.) G. Don	January-December
319.	*Verbascum thapsus* L.	January-April
320.	*Vicia faba* L.	December-March
321.	*Vicia hirsuta* Koch	March-April
322.	*Vicia sativa* L.	January-February
323.	*Vigna radiata* (L.) Wilez.	June-October
324.	*Viola patrinii* Ging.	April-May
325.	*Viola pilosa* Blume	February-April
326.	*Vitex negundo* L.	March-August
327.	*Vitis repanda* Wight & Arn.	April-June
328.	*Vitis vinifera* L.	March-April
329.	*Woodfordia fruticosa* (L.) Kurz	March-June
330.	*Wrightia arborea* (Dennst.) Mabb.	May-January
331.	*Zanthoxylum armatum* DC.	April-June
332.	*Zea mays* L.	June-September
333.	*Zephyranthes carinata* Herb.	April-July
334.	*Zizyphus jujuba* Mill.	July-August
335.	*Zizyphus nummularia* (Burm.f.) Wight & Arn.	June-September

ENUMERATION

The collected plant specimens were dried, preserved and mounted as per the known herborizing practices outlined by Jain & Rao (1977). Botanical identification of the selected species was first done with the help of regional floras (Chauhan, 1999; Chowdhery & Wadhwa, 1984; Collett, 1902; Dhiman, 1976; Hooker, 1872-1897; Nair, 1977; Polunin & Stainton, 1984; Stainton, 1988) and later carefully matched with the authenticated specimens at the herbaria of Botanical Survey of India, Northern Circle and Forest Research Institute, Dehradun. Nomenclature of these taxa was confirmed from Bennet (1986) and Wielgorskaya (1995). Excepting for the minor modifications to conform to the present-day circumscription, families and genera are delimited mainly after *The Flora of British India* (Hooker, 1872-1897). Economic valuation of all the presently recorded ethnobotanical species was also carried out to calculate the total importance values (TIV) as per detailed methodology outlined by Belal & Springuel (1996). The nativity of the species has been determined on the basis of Matthew (1969), Maheshwari & Paul (1975), Nayar (1977), Sharma (1984), Hajra & Das (1982), Saxena (1991), Pandey & Parmar (1994), Reddy *et al.* (2000), Reddy & Raju (2002), Reddy & Reddy (2004), Murthy *et al.* (2007), Negi & Hajra (2007) and Reddy (2008). For future reference and studies voucher specimens of aforementioned taxa (HPU-EL-SKS-5401-5745) have been deposited in the Ethnobotanical herbarium of Department of Biosciences, Himachal Pradesh University, Shimla. Besides, colour photographs of the plant species in their natural habitat were also clicked.

Systematic enumeration of the plants is in alphabetical order of their botanical names, synonym of the species followed by their respective families. Besides this, information on English, Sanskrit and regional names, distribution, description, reproductive cycle, habitat ecology, material examined, part/s used, folk use/s, active constituents, biological activity and uses in literature has also been furnished and presented, if available. The data gathered were screened with the help of available literature (Ambasta, 1986; Anonymous, 1948-1976; Chopra *et al.*, 1956, 1969; Jain, 1991; Kirtikar & Basu, 1984; Maheshwari & Singh 1984; Singh *et al.*, 1987; Singh, 1988), besides many other books and articles published in different journals.

Promising Himalayan Plants for Sustenance of Mankind

Abelmoschus esculentus **(L.) Moench** (Pl. 1A)

Syn.: *Hibiscus esculentus* L.; *H. longifolius* Roxb.

Family: Malvaceae.

Vern. : Ghandoli.

English, Sanskrit & Regional Names:

Eng. : Edible Hibiscus, Gobba, Gumbo, Lady's finger

Sans. : Asrapatraka, Bhenda, Chatupunda, Darivka, Sushaka

Ass. : Bhenda, Darivka, Ghandhamorla, Tindisa

B. : Bhindi, Katavandai, Ramturai

G. : Bhinda, Bhindu

H. : Bhindi, Bhindi Tori, Katavandai, Ramturai, Ranturi

Kan. : Bendakainaru, Bendu

Mal. : Vanta, Venda

Mar. : Bhenda, Bhendi

P. : Bhindatori, Bhindi, Ramturai

Tam. : Vendai, Vendi

Tel. : Benda, Bendi, Penda, Venda.

Distribution : Cultivated throughout India and all other tropical countries.

Description : Erect annual herbs. Leaves elliptic, base cordate, palmately lobed. Flowers axillary, yellow with a crimson centre. Fruits (capsule) elongated, creamy-green. Seeds many rounded.

Reproductive Cycle : June-December.

Habitat Ecology : Cultivated.

Material Examined : HPU-EL-SKS-5548; Bhaura (940 m), Chimbalhar (1,150 m), Dadh (1,100 m), Nanaun (900 m), Tanda (1,120 m).

Parts Used : Fruit. Stem.

Folk Uses : Fruits used for preparing **soups**, **pickle**, and also relished as **vegetable**. Stem yields fibre for **ropes**.

Active Constituents : Rhamnose, galactose, galacturonic acid and glucuronic acid; malvanic, stereulic and veronic acids (*seed oil*); quercetin, hyperin and β-sitosterol identified in fruits.

Uses in Literature : Reported earlier to be used as antispasmodic, cordial, demulcent, diuretic, emollient, stimulant, vegetable; and for burning sensation during urination **(Ambasta, 1986; Khanna, 2002).**

Abrus precatorius **L.** (Pl. 1B)

Syn.: *A. minor* Desv.; *A. pauciflorus* Desv.

Family: Fabaceae.

Vern. : Chadannu, Rati.

English, Sanskrit & Regional Names:

Eng. : Indian Liquorice, Jeryquity, Papernoster pea, Rosary pea, Weather plant

Sans. : Bhilabhushana, Gunja

Ass. : Gunja, Krishanala, Kakachinchi

B. : Chunhati, Gunj, Kawet

Bo. : Ghungchi

G. : Chanoti, Gunja

H. : Chirmiti, Gnunghachi

Kan. : Guengungi

Kash. : Shangir

Mal. : Irattimadhuram, Shekkunni

Mar. : Chanoti, Kunch

Ori. : Kainsho, Kotibepolo

P. : Labri, Ratak

Tam. : Adisamiyai, Atti, Ghurie, Ghenza

Tel. : Kukkutamu, Sinnaguruginja

U.P. : Ujarki ghumchi.

Distribution : Throughout India upto 1,200 m in forest tract. China, Pacific Island, Tropical Africa, The West Indies, below 1,200 m elevation.

Description : Copiously branched woody climbers having 20-40 membranous leaflets, reddish corolla and bright scarlet seeds with a dark spot in hilum.

Reproductive Cycle : May-August.

Habitat Ecology : Forests. Shrubberies.

Material Examined : HPU-EL-SKS-5551; Balah (900 m), Boda (1,000 m), Panapari (1,100 m), Rangru (910 m), Tikkar (910 m).

Parts Used : Roots. Seeds.

Folk Uses : Seed paste applied for **skin diseases** and **paralysis**. Poultice of seeds with *Piper nigrum* (black pepper) applied over head and covered with leaves of *Justicia adhatoda* to relieve **headache**. Seed and root paste applied externally to treat **goitre**. Seeds used for **chanting mantras** and also to **weigh** gold. Also, they are put in eyes to check **dust**.

Active Constituents : Active ingredients are precol, abrol, glycyrrhizin, abrasine, precasine and isoflavon-quinones I, II, III (*Roots*), abrin A & B, abricin, abridin, abric acid, hypaphorine, alainine, serine, abralin, abrine and abranin (*seed*).

Biological Activity : Antitumourous, anthelmintic, antispermatogenic and spasmogenic activities (+)ve.

Uses in Literature : Known to be used as an abortifacient, alexiteric, analgesic, anthelmintic, antidote to poison, antiperiodic in preventing malaria, antiseptic, antitumourous, aphrodisiac, blood purifier, CNS depressant, colic, diuretic, emetic, febrifuge, laxative (vet.), masticatory, measure of weights for goldsmiths, purgative, refrigerant, suppository for causing abortion, symbolic (to ward off evil spirits), tonic; and for affections of hair and nervous system, anthrax (vet.), appetite, boils, bone fracture (vet.), bronchial problem (vet.), bronchitis, convulsions, cough, cure of galls on the neck, decoration of baskets and bracelets, diarrhoea, dropsy, eczema, eye complaints, facilitating delivery, flatulence, gall neck (vet.), gonorrhoea, haemorrhoids, hastening labour, headache, hemicrania, contraceptive, impotency, indigestion, inflammation, inhibiting conception, insect bite (vet.), itching, killing lice, leprosy, leucorrhoea, liver (vet.), menstrual complaints, muscular pain, nervous disorders (vet.), paralysis, parturition, peptic ulcers, pnuemonia, retaining placenta (vet.), rheumatism, scabies, sciatica, scorpion bite, skin eruptions, snakebite (vet.), sore throat, stiffness of shoulder joints, paralysis, stomach complaints, strengthening and toning stomach, swelling, syphilis, ulcer (vet.), wounds and ulcers, wounds (vet.) (Ambasta, 1986; Chauhan, 1999; Dagar & Dagar, 1996; Datt, 1996; Gill & Nyawuame, 1994; Diwakar & Ansari, 2003; Dwivedi, 2003; Gogoi & Das, 2003; Goud & Pullaiah, 1996b; Jain, 1991, 1996; Jha & Kumar, 2003; Kothary & Moorthy, 1996; Kumar *et al.,* 2004; Kurian, 1995; Malik, 1996; Maheshwari, 1996; Mandal & Basu, 1996; Mishra *et al.,* 1996b; Nadanakunjidam & Abirai, 2003; Nayak *et al.,* 2003; Nazarudeen *et al.,* 1996; Painuli & Meheshwari, 1996; Pawar & Patil, 2004; Prasad *et al.,* 1996; Pullaiah & Kumar, 1996; Rajendran *et al.,* 2003; Rama Rao & Henry, 1996; Ranjan, 2003; Ravishankar & Henry,

1992; Rawat & Chowdhary, 1998; Saini, 1996a; Satapathy & Brahmam, 1996; Sharma, 1996; Singh & Pandey, 1998; Singh & Prakash, 1994; Singh *et al.*, 1996b; Siwakoti & Varma, 1996; Srivastava et al., 2003a; Thomas & Britto, 2003; Tirkey, 2004; Upadhyay, 2003).

***Acacia catechu* Willd.** (Pl. 1C)

Syn.: *A. polyacantha* Willd.; *A. wallichiana* DC.; *A. catechuoides* DC.; *A. sundra* Bedd.; *Mimosa catechu* Roxb.; *M. catechuoides* Roxb.

Family: Fabaceae.

Vern. : Khair.

English, Sanskrit & Regional Names:

Eng. : Black Catechu, Catechu Nigraum, Cutuch, Pegu Catechu
Sans. : Dantadhavana, Jivumangalaya, Vakrakantaka
Ass. : Kat, Koir
B. : Kahyer, Kuth
Bo. : Khaderi, Lalkhair
G. : Kher, Kheribayal
H. : Babul, Katha, Khair
Mal. : Kadaram, Senkainnali
Mar. : Khaderi, Lalkheir
Ori. : Bimbu, Khodira
Tam. : Karangalli, Kasukkalti
Tel. : Khadiramu, Mallasandra.

Distribution : Throughout Peninsula and from Indus eastwards to Assam: 500-1,350 m.

Description : Trees possessing recurved prickles, pale yellow flowers and glabrous, oblong stalked pods.

Reproductive Cycle : May-July.

Habitat Ecology : Planted along hilly slopes.

Material Examined : HPU-EL-SKS-5421; Andretta (980 m), Bhauda (1,050 m), Malahu (1,050 m), Thandol (950 m), Tikkar (910 m).

Parts Used : Heartwood ('katha'). Leaves. Pods. Stem.

Folk Uses : Stem decoction taken for **cough**. Wood used as **timber wood**. Stem employed for making '**mool**' (pestle) for removing husk from paddy. Leaves and pods fed to goats for enhancing **milk**. 'Katha' from heartwood applied locally for **mouth ulcers**.

Active Constituents : These are catechin, catechu, tannic acid, tannin (*bark*) and α, β, γ,-catechin and 1-epicatechin (*wood*).

Biological Activity : Antiviral, antiinflammatory hepatoprotective, hypoglycaemic and spasmolytic activities confirmed.

Uses in Literature : Reported in India as an abortifacient, anticancerous, antifungal, astringent, aphrodisiac, blood purifier, diabetes, cutch used in paan, digestive, dye, expectorant, febrifuge, fibre, galactagogue, gum arabic, host for lac insects, plywood adhesives, vasoconstrictor, tonic; and for affection of gum, asthma, bleeding from nose, blood dysentery, bronchial affections, burns, causing impotency if in excess, chest pain, childbirth, colic pain, congestion of faeces, conjunctivitis, constipation (vet.), consumption, cough, cracked palate (vet.), diarrhoea, eczema, gonorrhoea, gravel phthisis, haemoptysis, hoarseness, increasing the secretion of milk after delivery, inflammation, intermittent fevers, itching, leprosy, leucorrhoea, menorrhagia after confinement, menstrual complaints, Mithila paintings, muscular growth (vet.), obstinate skin diseases, otorrhoea and tonsils, pain in chest, piles, printing, rheumatism, scurvy, skin diseases, snakebite, soothening throat, sore gums, sore in mouth, sores of nails and soles (vet.), strangulation of intestine, suppressing sexual desire, swelling (vet.), swelling of foot, syphilis, tanning, toothache, ulcers, and wounds (vet.) (Agarwal, 2003; Ahuja, 1965; Ambasta, 1986; Asolkar *et al.,* 1992; Beg *et al.,* 2005; Bhatt *et al.,* 2003a; Bose, 1984; Chauhan, 1999; Chevallier, 1996; Dey & Bahadur, 1973; Duke *et al.,* 2002; Dwivedi, 2003; Harborne & Baxter, 2001; Islam & Jha, 2003; Jain, 1991; Jain *et al.,* 1991; Jha & Basak, 1994; Karnick, 1994; Khare, 2004; Kothary & Moorthy, 1996; Mahato & Chaudhary, 2005; Masih, 2003; Mishra *et al.,* 1996b; Panthi & Chaudhary, 2003; Prajapati *et al.,* 2006; Rai & Upadhyay, 1996; Ranjan, 2003; Said, 1997; Saini, 1996b; Sanyal, 1994; Sharma, 2003; Shukla & Verma, 1996; Shukla *et al.,* 2003; Singh & Pandey, 1998; Sinha & Sinha, 2001; Siwakoti & Varma, 1996; Watt, 1972).

Achyranthes aspera **L.** (Pl. 1D)

Syn.: *Digera muricata* Mart.

Family: Amaranthaceae.

Vern. : Puthkanda.

English, Sanskrit & Regional Names:

Eng. : Prickly Chaff Flower

Sans. : Adhoganta, Adhvashalya, Aghamargava, Aghata, Apamarga, Apangaka, Dhamargava, Malakanta, Mayuraka, Sthalamangari

Ass. : Apang

B. : Apang

Bo. : Aghada

G. : Agheda, Aghedi

H. : Apang, Chichra, Chirchira, Latjira
Kan. : Uttarane
M.P. : Apamar
Mal. : Katalati
Mar. : Aghada, Aghara
P. : Kutri
Tam. : Nayuviri
Tel. : Antisha, Apamaragamu
U.P. : Latzira, Lahchichra.

Distribution : Throughout India.

Description : Herbs with leathery leaves, four-sided branches and greenish-white, recurved flowers in rigid spikes. Achenes enclosed in hardened perianth. Seeds oblong and inverted.

Reproductive Cycle : May-October.

Habitat Ecology : Wasteplaces. Roadsides. Forest edges.

Material Examined : HPU-EL-SKS-5544; Banuri (1,150 m), Bodhal (970 m), Dharaman (1,100 m), Paror (940 m), Sulah (950 m).

Parts Used : Aerial plant parts. Seeds.

Folk Uses : Seed powder considered **diuretic**. Seed ash taken along with honey twice daily for **cough**. Aerial plant parts considered an important ingredient of '**havan**'(religious fire).

Active Constituents : Constituents isolated are linoleic (49.4), oleic (22.6), palmitic (18.6), stearic (4.4), behenic (1.8), arachidic (1.6), myristic (1.2) and lauric (0.4%) acids (*seed oil*), two saponins; saponin A and saponin B, saponin β-methyl ester (*seeds*).

Biological Activity : Antibacterial, anthelmintic, anti-inflammatory, cytotoxic, diuretic, cardiotonic and abortifacient activities confirmed.

Uses in Literature : Known to be used as abortifacient, antidote to snakebite, antiseptic, appetiser (vet.), asthma, astringent, cardiotonic, cataract, cereals (seeds), in times of scarcity, colic, diuretic, dye, emetic, febrifuge, general tonic, germicide, laxative, magico-theraupetic in labour to achieve easy delivery, menstruation, manure, oligogalactia, psychoactive plant, spinach, tonic after delivery, tooth brush, toothpowder, vegetable (leaves); and for abortion, anasarca, antifertilty in women, anti-inflammatory activity, ascites, asthma, biliousness, bleeding in delivery, blood dysentery, bronchitis, bodyache, boils, bowel complaints, bronchitis (vet.), chest pain, chicken pox, cold, cough, cracks (vet.), cutaneous diseases, cuts, dandruff, debility, delivery (vet.), dental caries, dog bites, dropsy, ear complaints, eruptions of skin, expelling placenta, expelling worms, external cuts, eye (vet.), eye complaints, feeding newborn babies, fever (vet.), gastric (vet.), general dysentery, goitre, gonorrhoea, gout, gum troubles, headache, hydrophobia, improving potency of man during coition, indigestion, injuries (vet.), insect bite (vet.), intestinal problems, jaundice, keeping of evil's spirits, leucoderma, liver complaints, night blindness, malarial fever, menorrhagia, migraine, muscle sprains, opacity of cornea (vet.), pain in ribs, piles, poisonous insect bites, pneumonia (vet.), pyorrhoea, reducing hunger, regaining consciousness, relieving fever, removing dead foetus (vet.), renal complaints, respiratory problems, rheumatism, scorpion bite, sore, stomachache, swelling of eyelid, syphilis, throat irritations, toothache, ulcers, warts of body, and whooping cough (Ambasta, 1986; Aminuddin & Girach, 1991, 1996; Arora & Pandey, 1996; Bennet *et al.,* 1991; Bhatt *et al.,* 1999, 2001, 2003a; Beg *et al.,* 2005; Bhandary *et al.,* 1996; Billore *et al.,* 1998; Borthakur, 1993; Chandra, 1997; Chandrasekar & Srivastava, 2003; Chauhan, 1999; Chowdhery, 1996; D'Rozario *et al.,* 1999; Das & Agarwal, 1991; Datt, 1996; Deshmukh & Rothe, 2003; Dey & Bahadur, 1973; Dwarakan & Ansari, 1996; Gill *et al.,* 1997; Girach, 1992, Girach & Aminuddin, 1995; Gogoi & Borthakur, 2001; Gogoi & Das, 2003; Goud & Pullaiah, 1996b; Hema *et al.,* 2006; Hembrom, 1996; Henry *et al.,* 1996; Hosagoudar & Henry, 1996c; Jadeja, 1999; Jain & Patole, 2001; Jain *et al.,* 1994; Jha *et al.,* 1991, 1996; Katewa *et al.,* 2001; Khan *et al.,* 2003; Khanna & Kumar, 2000; Khanna *et al.,* 1996b; Khare, 2004; Kirtikar & Basu, 1984; Kumar & Lal, 2003; Kumar *et al.,* 2003, 2004; Kurian, 1995; Lal *et al.,* 1996; Mahato & Chaudhary, 2005; Maliya, 2004, 2005; Maliya & Singh, 2003; Manandhar, 1989a, b, 1990, 1996; Mishra *et al.,* 1996b; Mitaliya *et al.,* 2003; Mohanty & Padhy, 1996; Murthy *et al.,* 2003; Nath & Maiti, 2003; Nayak *et al.,* 2003; Panthi & Chaudhary, 2003; Parrotta, 2001; Pawar & Patil, 2004; Prajapati *et al.,* 2006; Prasad *et al.,* 1996; Pullaiah & Kumar, 1996; Raghupathy & Mahadevan, 1991; Rai & Upadhyay, 1996; Rajendran *et al.,* 1996, 2003; Rama Rao & Henry, 1996; Ranjan, 2003; Rao *et al.,* 1996; Reddy et al., 1996; Roy *et al.,* 1998; Said, 1997; Saini, 1996b; Saklani & Jain, 1994; Samwatsar & Diwanji, 1996a; Sanyal, 1994; Sarin, 1996; Seetharam *et al.,* 1999; Sen & Behera, 2003; Sharma, 2003; Sharma & Borthakur, 2003; Sharma & Sood, 1997; Shome *et al.,* 1996; Shukla & Verma, 1996; Sikarwar, 1996; Singh, 2000, 2001; Singh & Kumar, 2000a; Singh & Maheshwari, 1992; Singh & Pandey, 1998; Singh & Prakash, 1996; Singh & Rao, 2003; Singh *et al.,* 2003a, b; Sinha & Sinha, 2001; Sivakumar *et al.,* 2003; Siwakoti & Siwakoti, 2000; Sood & Thakur, 2004; Srivastava *et al.,* 2003a, c; Subramani & Goraya, 2003; Subramaniam, 2000; Sundaravelu *et al.,* 2003; Swami & Gupta, 1996; Tirkey *et al.,* 2003b; Upadhyay & Singh, 2005; Verma & Chauhan, 2006; Vihari 1995).

***Achyranthes bidentata* Blume** (Pl. 1E)

Syn.: *A. wightiana* Wall.; *A. lanceolata* Wall.; *A. aspera* Wall.

Family: Amaranthaceae.

Vern. : Puthkanda.

Regional Names :

- **Ass.** : Apamarga, Bankhat
- **H.** : Adhajhara, Oga
- **Kash.** : Puthkandiari
- **U. P.** : Latjeera.

Distribution : Temperate & Sub-Tropical Himalaya.

Description : An undershrub with thin membranous leaves and slender spikes which are rarely more than 10-12.5 cm long. Flowers small, numerous. Utricle with small, oblong seed.

Reproductive Cycle : June-August.

Habitat Ecology : Ravine slopes, as a weed of wasteplaces.

Material Examined : HPU-EL-SKS-5640; Bhagpur (1,600 m), Jiun (1,550 m), Kathena (1,290 m), Paror (940 m), Sungal (1,050 m).

Parts Used : Whole Plant. Leaves.

Folk Uses : Decoction of plant useful in reducing accumulation of **uric acid** in body, and also considered useful in the preparation of local wine, '**sur**'. Paste of leaves applied as an antidote to **scorpion sting**.

Active Constituents : *Roots* yield ecdysterone, ecdysone, emodins and physcion.

Biological Activity : Antimicrobial (root extr.) and abortifacient (stem-bark) activities (+) ve.

Uses in Literature : Used in India as an astringent, diuretic, galactagogue (vet.), toothbrush, tooth powder; and for blisters in mouth, cholera, dental caries, edible purposes (leaves and seeds), hemicrania, indigestion, jaundice, pyorrhea, scorpion sting, scrofula, skin diseases, swelling, toothache and whooping cough (Chandra, 1997; Chhetri, 2005; Chevallier, 1996; Chowdhery, 1996; Hema *et al.*, 2006; Jain, 1991; Jain *et al.*, 1991; Kapur & Singh, 1996; Kirtikar & Basu, 1984; Lalramnghinglova, 1996; Manandhar, 1990, 1996; Molla & Roy, 1996; Prajapati *et al.*, 2006; Roy *et al.*, 1998; Saini, 1996b; Singh & Kumar, 2000a; Singh *et al.*, 1980).

Acorus calamus L. (Pl. 1F)
Family: Araceae.

Vern.: Barain, Brein.

English, Sanskrit & Regional Names:

Eng. : Calamus, Sweet Flag
Sans. : Shadgrantha, Ugragandhaha, Vacha
B. : Bach
G. : Gandhilovaj, Godavaj, Vekhand
H. : Bach, Ghor, Gorbach
P. : Bariboj, Warch
Tam. : Vashambu
Tel. : Vadaja, Vasa, Wasa.

Distribution : Throughout India.

Description : An erect aromatic herb with branched, creeping rootstock. Spadix sessile, leaf-like, long, narrow, dense-flowered with yellow-green flowers. Berries 1-3 seeded.

Reproductive Cycle : June - July.

Habitat Ecology : Frequent along banks of lakes, water streams and in marshy places.

Material Examined : HPU-EL-SKS-5418; Arla (1,100 m), Chachian (1,300 m), Dharaman (1,100 m), Gopalpur (1,350 m), Panchrukhi (990 m).

Parts Used : Rhizome. Leaves.

Folk Uses : Rhizome chewed as such or with honey for **sore throat** and **cough**. Its paste with mustard oil good for massage in children during winters due to its **warming effect**. Leaves put in grainries to keep away **insects**, and also used for **lining roofs** as resistant to **insect attack**.

Active Constituents : *Leaves* and *rhizomes* contain Vit.-C, luteolin-6 and 7-diglucoside. Besides, fresh *leaves* contain oxalic acid 0.078 and calcium 0.006%. Oil extracted from *rhizome* showed asarone, α-pinene, d-camphene, calamene, calamenol and sesquiterpene alcohols.

Biological Activity : Rhizome CNS depressant and spasmogenic.

Uses in Literature : Recorded in India for the treatment of abdominal pain, asthma, bodyache, bronchitis, cold, cough, cuts, dysentery, kidney troubles, leucoderma, improving memory, insect bite, epilepsy, fever, headache, heart and lung trouble, hysteria, indigestion, dislocation of bones, snakebite, inflammation, malaria, pain in neck, phthisis, skin diseases, sore throat, stomachache, teething in children, wounds (cattle); and as an appetizer, insecticidal and tonic (Bhuyan, 2003; Borthakur *et al.,* 2004; Dwivedi, 2003;

Gogoi & Das, 2003; Gopal & Chandra, 2003; Jain, 1991; Jha & Kumar, 2003; Mahato & Chaudhary, 2005; Nath & Bordoloi, 1989; Panthi & Chaudhary, 2003; Rawat & Chowdhury, 1998; Sharma *et al.,* 2003a; Upadhyay, 2003; Upadhye & Kulkarni, 2004).

***Adiantum capillus veneris* L.** (Pl. 2A)
Family: Polypodiaceae.

Vern. : Dum-Tuli

English & Regional Names:

Eng. : Maidenhair fern, Maria's fern, Our Lady's Hair
G. : Hanspadi
H. : Hansraj, Mubarka, Pursha
Kan. : Hansraj, Kashmir
Kash. : Dumtuli.

Distribution : Tamil Nadu, westside upto 1,700 m on the mountains, Sri Lanka, N. India, Africa, America and Australia.

Description : Stipes suberect, slender, blackish, naked; fronds bipinnate; sori rounded or obreniform.

Reproductive Cycle : July-October.

Habitat Ecology : Frequent along nallahs, moist rocky slopes.

Material Examined : HPU-EL-SKS-5497; Banuri (1,150 m), Bhawarna (1,100 m), Lolani (1,600 m), Saparu (1,250 m), Tapa (945 m).

Parts Used : Aerial plant parts.

Folk Use : Decoction of Aerial plant parts recommended as **hair dye**.

Active Constituents : Mixture of esters, ketone, triterpine epoxide, astragalin, isoquercetin, rutins, querciturone (*whole plant*) characterized from the plants.

Biological Activity : Antiviral activity against vesicular stomatitis.

Uses in Literature : Used so far as colic, demulcent, expectorant, diuretic, emmanagogue, febrifuge, hair tonic, substitute of tea; and for cough, dandruff, throat affections, amenorrhoea, bronchial disordess, catarrhal affections; cough, demulcent, heart & liver ailments, infertility, killing lice, phlegmatic, throat affections, vaginal discharge, and women disorders (Ambasta, 1986; Chandra, 1997; Chatterjee & Pakrashi, 1991; Kirtikar & Basu, 1984; Nadanakunjidam & Abirai, 2003; Shah, 2005; Sharma, 2003; Singh & Pandey, 1998).

Aegle marmelos **(L.) Correa** (Pl. 2 B,C)

Syn.: *Feronia pellucida* Roth; *Crateava marmelos* L.

Family: Rutaceae.

Vern. : Bil, Bael.

English, Sanskrit & Regional Names:

Eng.	: Bael Fruit Tree, Bal Quince, Holy Fruit, Indian Quince, Stone Apple
Sans.	: Atimangaliya, Mahakapithakya, Satyadharma, Tripatra
Ass.	: Bel
B.	: Bela, Vilva
Bo.	: Bila
G.	: Bil, Billy
H.	: Bel, Siriphal
Kan.	: Bilvapatra
M.P.	: Bael
Mal.	: Kuvulam, Mavilamu
Mar.	: Bel
Ori.	: Bilwa, Sripholo
Tam.	: Aluvigam, Iyalbudi
Tel.	: Bilvamu, Sailushamu.

Distribution : Throughout India.

Description : A small deciduous spiny tree. Leaves glabrous, leaflets 3, ovate-lanceolate. Flowers white, sweet-scented. Berries grey or yellow, globose.

Reproductive Cycle : April-July.

Habitat Ecology : Planted; wastelands or near houses.

Material Examined : HPU-EL-SKS-5540; Balah (900 m), Bhauda (1,050 m), Boda (1,000 m), Malahu (1,050 m), Thandol (950 m).

Parts Used : Fruit. Leaves. Gum.

Folk Uses : Fruits form one of the constituents of '**havan**', ripe ones **edible** and considered **refrigerant**; and also prescribed for **gastrointestinal ailments**. Fruit pulp taken as refreshing drink to relieve **burning sensation** of hands and soles, and also as an **astringent** to mucous membrane of intestine. Gum from fruits employed as an **adhesive.** Ash from pericarp taken with honey for **constipation** and **piles**. Its powder used for making '**rangoli**' (colourful decorative patterns) on **religious ceremonies**. Leaves offered to **appease** Lord 'Shiva'. Fruit pulp taken with equal quantity of ginger to check **blood dysentery**.

Active Constituents : Main constituents are aegeline, γ-sitosterol (*leaves*) and umbelliferone, dictamnine, xanthotoxol, xanthotoxin, scoparone, isopimpinellin, isoimperation, N-2-methoxy-2-(4-Methoxy-phenyl) ethyl cinnamamide, marmelin and its methyl ether, bergapten, marmesin, osthol and auraptin (*fruit pericarp*).

Biological Activity : Activities exhibited are hypoglycaemic.

Uses in Literature : Used earlier as an adhesive, appetizer, aromatic, astringent, beads, candy, cardiac depressant, cementing mixtures, fish poison, funeral pyre of rich, ingredient of 'chayawanprash', laxative, plaster for walls, post-natal, psychoactive plant, purgative, refrigerant, sacred, scorbutic, soaps, soothing agents for intestines, stomachic, sweet drinks, fuel, tree motif on 'Shiva' temples, substitute for tea, tannin and varnish, wish granting tree; and for abdominal accumulation of liquid in the cellular tissue accompanied with constipation and jaundice, abdominal pain for abscess, allaying urinary troubles, anthrax (vet.), asthma, backache, bite of mad dog, blisters (vet.), boil (vet.), bowel complaints, breast pain, bronchitis, burning sensation of the soles of the feet, catarrhs, chest pain, cholera, chronic dysentery, cold, conjunctiva, convulsions, cough, cramp, dandruff, darkening and shining hairs, diabetes, diarrhoea (vet.), dyeing, dysuria, eczema, edible fruits, sherbet and juice, eye complaints, fever (vet.), feverishness, gastric trouble, heart palpitations, hyperacidity, hypochondriasis, indigestion, inflammation of the mucous membrane having a free discharge, injuries, intermittent fevers, intestinal disorders, irritation of alimentary canal, joint pain (vet.), keeping evil spirits away (planted near stones with snake sculptures on it), menstrual disorders, mouth sores, nausea, night fever, opthalmia, paintings (Mithila), piles, relieving headache, scurvy, skin diseases, snakebite, sore (vet.), strengthening and giving tone to the stomach, sunstroke, toys, tympany, ulcer (vet.), urinogenital disorders, vomiting and wounds (Agarwal, 2003; Ambasta, 1986; Aminuddin & Girach, 1993; Arora & Pandey, 1996; Bajpayee & Dixit, 1996; Bhattacharya *et al.,* 2005; Bennet *et al.* 1991; Bhatt *et al.,* 2001; Bhandary & Chandrashekar, 2002; Borthakur, 1996b; Bose, 1984; Chauhan, 1999; Choudhary, 1996; Chowdhury *et al.,* 1996; D'Rozario *et al.,* 1999; Das & Sharma, 2003; Datt, 1996; Devi, 2003; Dey & Bahadur, 1973; Diwakar & Ansari, 2003; Dixit & Kumar, 2003; Duke *et al.,* 2002; Dwarakan & Ansari, 1996; Girach, 2001; Goud & Pullaiah, 1996a; Hebber *et al.,* 2003; Hembrom & Goel, 2005; Huidrom, 1996; Jain *et al.,* 1994; Jha & Basak, 1994; Kalakoti, 2001; Katewa & Arora, 1997; Katewa *et al.,* 2000; Khanna *et al.,* 1996b; Khare, 2004; Kirtikar & Basu, 1984; Kothari & Moorthy, 1996; Kothari *et al.,* 2003; Kshirsagar & Singh, 2000; Kumar, 2003; Kumar *et al.,* 2003; Kurian, 1995; Lalramnghinglova, 1996; Mahato & Chaudhary, 2005; Manandhar, 1990; Masih, 2003; Mehrotra *et al.,* 1995; Mishra et al., 1996a; Mitaliya & Bhatt, 2003; Nadanakunjidam & Abirai, 2003; Nath & Maiti, 2003; Nayak *et al.,* 2003; Negi & Gaur, 1991;

Painuli & Maheshwari, 1996; Pal *et al.,* 2000; Pandey & Joshi, 2001; Pandey & Verma, 2003; Panthi & Chaudhary, 2003; Parmar & Kaushal, 1982; Patole & Jain, 2002; Prakash, 2001; Prajapati *et al.,* 2006; Prasad *et al.,* 1996; Pullaiah & Kumar, 1996; Rai & Upadhyay, 1996; Rajendran *et al.,* 2003; Rama Rao & Henry, 1996; Ranjan, 2000; Ravishankar & Henry, 1992; Roy *et al.,* 1998; Sahni, 1998; Said, 1997; Saikia & Nath, 2003; Sanyal, 1994; Sarin, 1996; Sen & Batra, 1997; Sen & Behera, 2003; Seetharam *et al.,* 1999; Sharma, 2003; Sharma & Singh, 2001; Sharma & Sood, 1997; Shome *et al.,* 1996; Singh, 1991, 1996b, 1999; Singh & Kumar, 2000a; Singh & Maheshwari, 1992; Singh & Pandey, 1998; Singh & Rao, 2003; Singh *et al.,* 1997; Sinha & Sinha, 2001; Siwakoti & Siwakoti, 2000; Siwakoti & Varma, 1996; Srivastava *et al,,* 2003a; Sudhakar & Vedavathy, 2000; Thomas & Britto, 2003; Tirkey *et al.,* 2003b; Upadhyay & Chauhan, 2000; Vihari 1995; Watt, 1972).

***Aerva lanata* (L.) Juss. *ex* Schult** (Pl. 2D)

Syn.: *Achyranthus lanata* L.; *Aerva elegans* Moq.; *Illecebrum lanatum* (L.) Muer.

Family: Amaranthaceae.

Vern. : Aung.

Sanskrit & Regional Names:

Sans. : Astamabayota; Bhadra

B. : Chaya

G. : Bur, Gorakha, Ganja, Kapurimadhurai

H. : Chaya, Kapurijadi, Gorkhabundi,

Kan. : Bilihandi, Soppu

Mal. : Cherula, Cherupula

Mar. : Kapurimadhurai, Kapurmadhura, Kapurapindi

Ori. : Paunsia

P. : Buikallan

Tam. : Pullai, Sirupulai

Tel. : Pindichettu, Pindiconda, Pindikunda.

Distribution : Weed throughout India upto 900 m.

Description : Prostrate herbs with woolly branches and simple, alternate leaves. Flowers minute, greenish-white. Fruits ovoid, acute; seeds black, bean-shaped.

Reproductive Cycle : September-March.

Habitat Ecology : Borders of agricultural fields. Drier slopes.

Material Examined : HPU-EL-SKS-5430; Ladoh (980 m), Nadli (1,000 m), Paror (940 m), Rajpur (1,100 m), Saliana (1,000 m).

Parts Used : Aerial plant parts. Root.

Folk Uses : Garland of aerial parts hung on main entrance of houses for warding off **evil spirits**. Dried aerial parts are kept under pillow of patient suffering from **herpes** and **measals**; also form one of the ingredients of '**havan**'. Plant paste applied to cure **cutaneous** diseases. Decoction of plant parts considered **diuretic.** Root-paste with curd good for **jaundice** whereas decoction of inflorescence (twice daily) prescribed to dissolve **kidney stone** and for inducing **rumination** (vet.).

Active Constituents : Flowering and fruiting parts contain hemicellulose, starch, an acid soluble polysaccharide and water-soluble polysaccharide

Uses in Literature : So far reported as antidote to snakebite, diuretic, demulcent, vermifuge; and for lithiasis, cephalalgia, stranguary, headache, jaundice, urinary disorders, diarrhoea, religious rituals, cough, diabetes, and easy discharge after childbirth (Ambasta, 1986; Bhatt et al., 2003a, b; Chatterjee & Pakrashi, 1991; Goud & Pullaiah, 1996a; Jadeja, 1999; Kirtikar & Basu, 1984; Maya et al., 2003; Mohanty & Padhy, 1996; Nayak et al., 2003; Parrotta, 2001; Singh & Pandey, 1998; Srivastava et al., 2003a; Thomas & Britto, 2003).

Agave cantula **Roxb.** (Pl. 2E)

Syn.: *A. americana* auct. (non L.) Hook. f.

Family: Amaryllidaceae.

Vern.: Ramban, Ranhbaan.

English, Sanskrit & Regional Names:

Eng. : Amercan Aloe, Carata, Century plant

Sans. : Kantala

B. : Banskeora, Bilati-ananash, Bilatipat, Carata

G. : Fangli-kunvara

H. : Banskeora, Barakanwar, Hatisengar, Kantala, Rakas-pattah, Ramkanta

Kan. : Bhuttale, Budukattalenaru

Mal. : Panam, Katrazha

Mar. : Parkand

Tam. : Anaik-kat-razhai, Pithakalabuntha

Tel. : Rakshimatalu.

Distribution : Throughout India.

Description : A xerophyte with rhizome and rosette of thick, fleshy leaves which are spiny, and with curved tips. Flowers dull-green. Berries possess black seeds.

Reproductive Cycle : March-May.

Habitat Ecology : Open slopes. Roadsides. Wastelands.

Material Examined : HPU-EL-SKS-5544; Malahu (1,050 m), Panapari (1,100 m) Sungal (1,050 m), Thandol (950 m).

Parts Used : Whole Plant. Leaves. Peduncle.

Folk Uses : Plant grown as **hedges**. Fibres from leaves employed for making **baskets**, **ropes** and **mats**. Leaf strips also used as **ropes**. Leaf juice applied over maggot infested wounds as **antiseptic**. Peduncle consumed as **vegetable**, and also **pickled**.

Active Constituents : Hecogenin, chlorogenin, 10 steroidal saponins-agavoside A, B, C, C′, D, E, F, G, H and I- isolated from *leaves*.

Biological Activity : Antibacterial and anticancerous activities found to be positive.

Uses in Literature : Used earlier as an anthelmintic, antisyphilitic, diaphoretic, diuretic, emmenagogue, blood purifier, fire wood, fish poison, fodder, laxative, paper material, poultice, resolvent, substitute for soap; and for anasarca, ascites, boils, bruishes and contusions, brushes, burns, cancer, conjuctivitis, controlling exparasite (vet.), cordage, cut, diarrhoea, dropsy, dysentery, falling hairs, fever, fever (vet.), goitre, gonorrhoea, liver and kidney problems mats, paralysis, proof against ravages of white ants, scrofula, scurvy and venereal sores, skin diseases (vet.), gastric and liver disorders, skin disorders, thatching, tuberculosis, warts (vet.) and wounds (Ambasta, 1986; Dey & Bahadur, 1973; Harborne & Baxter, 2001; Jain, 1991; Jain *et al.,* 1991; Kapur, 1996a; Khare, 2004; Kirtikar & Basu, 1984; Kurian, 1995; Lalramnghinglova, 1996; Prajapati *et al.,* 2006; Rana *et al.,* 1996; Rao, 1996; Said, 1997; Saxena & Vyas, 1983; Singh, 1996b, 2000; Watt, 1972).

Agave wightii **Drumm. & Prain** (Pl. 2F)

Syn.: *A. angustifolia* Haw.; *A. ektliodes* Hook. f.

Family: Amaryllidaceae.

Vern. : Kyor.

English, Sanskrit & Regional Names:

Eng. : Dwarf aloe

Sans. : Kantala

H. : Khetki

Mar. : Chota ghayal
Tam. : Chinna erumaikkattalai, Malaikkattalai
Tel. : Balurakkasi, Kittanara, Peddakalabanda.

Distribution : Sub-Himalayan hills, outer Himalayas & peninsular India.

Description : A short stemmed perennial. Leaves in rosette, greyish-green, and dagger-shaped. Prickles slender-cusped. Flowers greenish, borne in inflorescence. Capsule beaked and many seeded.

Reproductive Cycle : July-September.

Habitat Ecology : Wasteplaces. Fieldsides.

Material Examined : HPU-EL-SKS-5714; Bhaura (940 m), Bhuana (1400 m), Chhanchri (930 m), Molichak (1020 m), Rangru (910 m).

Part Used : Peduncle.

Folk Uses : Peduncle consumed as a **vegetable** and **pickle**.

Uses in Literature : Known earlier as a diuretic, diaphoretic, and edible (peduncle); good against bruises (Agarwal, 2003; Ambasta, 1986; Kirtikar & Basu, 1984; Parrotta, 2001).

Ageratum conyzoides L. (Pl. 3A)

Syn.: *A. cordifolium* Roxb.

Family: Asteraceae.

Vern. : Jaihar, Phulanu.

English & Regional Names:

Eng. : Goat Weed, White Weed
B. : Dochunty
Bo. : Osari, Sahadevi
G. : Ajgandha, Gandhari sedardi, Mankdamari
M.P. : Koobhi
Mal. : Appa, Muryanpacha
Mar. : Ghaneraosadi
Ori. : Boksunga, Poksunga.

Distribution : Throughout India.

Description : Erect, hairy annuals having opposite, cordate, serrated leaves. Flowers bluish, borne in dense small terminal corymbose heads. Achenes black, 5-angled. Pappus awned.

Reproductive Cycle : July - August.

Habitat Ecology : Edges of cultivated areas. Meadows. Wasteplaces.

Material Examined : HPU-EL-SKS-5647; Banuri (1,150 m), Banuri (1,150 m), Chauki (900 m), Gharthun (900 m), Lahla (930 m).

Parts Used : Leaves. Flowers.

Folk Uses : Leaf-paste applied over **wounds** and injury to check **bleeding**. Flower paste good for killing **lice**.

Active Constituents : These are phenols (eugenol) 5% and a phenol ester (*leaves, flower*), friedelin, β-sitosterol and stigmasterol; hydrocarbons, 2 unidentified esters, stigmasterol and α-spinasterol, coumarin and stigmasterol; flavonoids and antigonadotropic hormones (*plant*).

Biological Activity : Antibacterial, anthelmintic and anticancerous activities confirmed.

Uses in Literature : Known to be used in India as antidysentric, antilithic, antiseptic, edible, styptic, haemostatic, opthalmia, insecticidal, shampoo, symbolic; and considered useful for body swellings, boils, burns, cancer, cuts, diabetes, diarrhoea, fever, flatulence, headache, leprosy, muscular pain, piles, infection between toes, prolapsed anus, ringworm, scabies, skin diseases, snake-bite, sores, tumour, uterine disorders and wounds (Augustine & Sivadasan, 2005; Baruah & Sarma, 1984; Bhuyan, 2003; Chhetri, 2005; Gogoi & Das, 2003; Islam & Hasin, 2003; Jain, 1991; Jha & Kumar, 2003; Mahato & Chaudhary, 2005; Maliya, 2005; Nayak *et al.,* 2003; Padhy & Dash, 2003; Ranjan, 2003; Rawat & Chowdhury, 1998; Sen & Behera, 2003; Sinha, 1986; Srivastava *et al.,* 2003a, c; Upadhyay & Singh, 2005).

Ajuga parviflora Benth. (Pl. 3B)

Syn.: *Teucrium hispidum* Wall.

Family: Lamiaceae.

Vern. : Neelkanthi.

Distribution : W. Himalayas, Kashmir- Nepal.

Description : Softly villous, perennial herbs. Branches ascending. Whorls crowded, in bracteate spikes. Flowers blue. Ovary 4-lobed. Nutlets pitted.

Reproductive Cycle : March - April.

Habitat Ecology : Rock-crevices. Moist places. Rice fields.

Material Examined : HPU-EL-SKS-5721; Balu (1,030 m), Chandpur (1,290 m), Dadh (1,100 m), Purani-palam (1,100 m).

Parts Used : Aerial Plant Parts. Leaves.

Folk Uses : Leaves (1-2 daily) chewed as such to check **diabetes**. Leaf-paste with jaggery applied to treat **sores**, **boils** and other **cutaneous diseases** both in **humans** and **cattle**. Leaf extr. applied as **hair tonic**. Juice from aerial parts good against **killing lice**.

Albizia chinensis (Osbeck) Merr. (Pl. 3C)

Syn.: *A. stipulata* Boiv.; *Mimosa chinensis* Osbeck; *M. stipulata* Roxb.; *Acacia stipulata* (Roxb.) DC.; *A. smithiana* Wall.; *A. marginata* Ham.; *Arthrosprion stipulatum* Hassk.

Family : Fabaceae.

Vern. : Oyee.

Regional Names:

Ass. : Sau, Selcha
B. : Amluki, Chakua
Bo. : Udala
H. : Kanujera, Sam Sundra
Kan. : Hottabage, Kalbage
Mal. : Mottavaka, Pottuvaka
Mar. : Udala
Ori. : Gudanandia, Reyi
P. : Kasir, Oe, Sirin
Tam. : Katturmjil, Silaivogai
Tel. : Chinduga, Kondachiguru.

Distribution : India, China, S.E. Asia: 300-1,300 m.

Description : Trees. Leaves pubescent or hairy. Stipules pink and large-sized. Flowers yellow. Pods glabrous.

Reproductive Cycle : May-June.

Habitat Ecology : Riversides. Cultivated areas.

Material Examined : HPU-EL-SKS-5449; Chornali (1,280 m), Dehan (1,130 m), Pahra (950 m), Jasun (1,000 m), Rajeher (1,050 m).

Parts Used : Bark. Leaves. Root. Wood.

Folk Uses : Bark paste applied for **suppuration** of **boils**. Roots form an important constituent of local wine, **'sur'**. Leaves serve as **forage**. Wood employed for **house-construction**, **agricultural implements** and as **fuel**.

Biological Activity : 50% EtOH extr. of aerial parts spasmogenic and diuretic. Seeds showed marked hypoglycaemic activity in normal rats but not in alloxan-diabetic rats. Aq. extr. of bark acts as uterine consrictor.

Uses in Literature : Used as a gum, fodder, fuel; and for sizing paper, agricultural implements, tea boxes, boils, blisters and swellings (Ambasta, 1986; Singh & Kumar, 2000a; Watt, 1972).

Albizia julibrissin Durazz. (Pl. 3D)

Syn.: *A. mollis* Boiv.; *Acacia mollis* Wall.; *A. nemu* Willd.; *Mimosa kalkora* Roxb.; *M. arborea* Thunb.

Family: Fabaceae.

Vern. : Karmarua

Regional Names:

- **B.** : Kankora
- **H.** : Barau, Kurmura, Lalsiris
- **Ori.** : Tentuliya
- **P.** : Brind, Mathirshi, Surangru
- **Tam.** : Selaivagai
- **Tel.** : Nallasinduga.

Distribution : Pakistan-Bhutan, Asia, Africa: 1,000-2,100 m.

Description : Small deciduous trees having membranous, pubescent leaflets. Flowers pink, fragrant. Pods glabrous, membranous, grey-pale brown.

Reproductive Cycle : May-August.

Habitat Ecology : Forests. Riversides. Cultivated areas, often planted as a shade tree.

Material Examined : HPU-EL-SKS-5431; Chimbalhar (1,150 m), Dehan (1,130 m), Khaira (940 m), Makol (900 m), Rajeher (1,050 m).

Parts Used : Leaves. Wood. Gum.

Folk Uses : Leaves given as an excellent **fodder**. Gum oozing out of stem prescribed for **backache**. Wood used as **timber** and **fuelwood**.

Active Constituents : *Flowers* rich in cis- and trans-linalool oxide, β-ocimene, isopentanol and 4-penten-2-one.

Uses in Literature : Reported in India as a psychoactive plant, forage; and for furniture (Chowdhery, 1996; Jain *et al.,* 1994; Sood & Kaushal, 2008; Watt, 1972).

***Albizia lebbeck* Benth.** (Pl. 3E)

Syn.: *A. latifolia* Boiv.; *Acacia speciosa* Willd.; *A. sirisa* Ham.; *A. lebbeck* Willd.; *Mimosa speciosa* Jacq.; *M. sirisa* Roxb.; *M. lebbeck* L.

Family: Fabaceae.

Vern. : Sirin, Sarin.

English, Sanskrit & Regional Names:

Eng. : Parrot Tree, Siris Tree, Sizzling Tree

Sans. : Barhapushpa, Bhandi, Bhandika, Kapitaka, Kapitana, Krishnasiris, Lomashapushpaka, Murdhapushpa, Plavaga, Shankiniphala, Vishaghati, Vishahanta

B. : Siris, Sirisha

Bo. : Garso, Harreri, Siris

G. : Kaliosavas, Kalokansakio, Kalosadasadu, Kaloshirish, Pilosarshio

H. : Garso, Kalshish, Kalsir, Lasrin, Mathirshi, Shiris, Shirish, Sirai, Sirar, Sirin, Tantia

Kan. : Baggera

M. : Vagu

Mal. : Kattuvaka, Nenmani, Vaka

Mar. : Chichola, Kalashiras, Mothasiras, Shirish

Ori. : Vaghe, Kotvaghe

Tam. : Adukkavajai, Kalindi, Kattuvagai, Pandil, Perunvagai, Piram, Undugam

Tel. : Dirisanamu, Girisamu, Kapitanamu, Peddadirisanamu, Sirishamu, Siriso.

Distribution : Throughout India, ascending to 1,650 m.

Description : A large deciduous tree with fissured bark and young pubescent branches. Leaves evenly bipinnate of 3-9 pairs, petiole 8-12 cm long. Flowers in white heads. Pods thin, flat, glabrous. Seeds sub-orbicular-oblong.

Reproductive Cycle : April-May.

Habitat Ecology : Frequently planted.

Material Examined : HPU-EL-SKS-5630; Chhanchri (930 m), Paror (940 m), Sulah (950 m), Tapa (945 m), Thandol (950 m).

Parts Used : Leaves. Bark. Gum. Seeds. Wood.

Folk Uses : Leaves used as good **fodder**. Bark chewed to strengthen gums and for relieving **toothache**. Gum from bark taken with milk to check **gastric trouble**. Decoction of seeds used as eye drops to heal **eye injuries** in livestock. Wood used for **house construction** and as **fuel**.

Active Constituents : Chief constituents are flavandiols-melacacidin and lebbeacacidin (*heartwood*), d-catechin, leuco-cyanidin, condensed tannins (*bark*), vicenin-2, rcynoutrin, rutin, myricitin, robibin, albegenin (*leaves*).

Biological Activity : Anti-asthmatic and anti-anaphylactic activities (+)ve.

Uses in Literature : So far used as an alterative, antidote to poisons, aphrodisiac, astringent, blood purifier, camel fodder, detergent, expectorant, firewood, poultice for swelling of cervical glands, psychoactive plant, restorative, symbolic (god of love uses flowers on arrow of desire), tannin, tonic; and for agricultural implements, arthritis, anaemia, asthma, boats, boils, canoes, carbuncle, carriages, cold, cooling, cough, diarrhoea, dyeing, dysentery, edible purposes, eruptions, eye troubles, fever (vet.), furniture, gonorrhoea, gum ulcers, house building, infected teeth, interior decoration, leather, leprosy, leucoderma, night blindness, oil-mills, opthalmia, picture frames, piles, railway carriage work, reducing enlargement of cervical glands, kidney stones, retaining seminal fluid, rheumatism (vet.), snakebite, sugar cane-crushers, swelling, toys manufacturing, tympany (vet.), ulcers, well curbs, and wounds (Agarwal, 2003; Ambasta, 1986; Chauhan, 1999; Ganesan & Kesavan, 2003; Gill & Nyawuame, 1994; Goud & Pullaiah, 1996b; Islam & Jha, 2003; Jain, 1991; Jain & Patole, 2001; Jain *et al.,* 1994; Khanna *et al.,* 1996b; Khare, 2004; Kirtikar & Basu, 1984; Kurian, 1995; Mahato & Chaudhary, 2005; Murthy *et al.,* 2003; Parrotta, 2001; Raghupathy & Mahadevan, 1991; Rajendran *et al.,* 2003; Rama Rao & Henry, 1996; Said, 1997; Shah, 2005; Sharma, 2003; Singh & Kumar, 2000a; Siwakoti & Varma, 1996; Srivastava *et al.,* 2003b; Watt, 1972).

***Allium cepa* L.** (Pl. 3F)

Family: Liliaceae.

Vern. : Payaj.

English, Sanskrit & Regional Names:

Eng. : The Onion

Sans. : Dirghapatra, Mahakanda, Palandu, Rajapriya

Ass. : Piyaz

Bo. : Kanda, Puyaj

Mal. : Bawang

Mar. : Kanda

P. : Piyaj

Tam. : Iravengayam, Irulli

U.P. : Nirulli, Vulligandalu.

Distribution : Cultivated throughout India.

Description : A bulbous herbaceous plant. Leaves cylindrical. Umbels large; flowers capitate, and pedicelled. Seeds black.

Reproductive Cycle : December-March.

Habitat Ecology : Moist places.

Material Examined : HPU-EL-SKS-5614; Banuri (1,150 m), Ladoh (980 m), Malnu (1,000 m), Sidhpur (1,100 m).

Parts Used : Bulbs. Leaves.

Folk Uses : Bulbs and leaves used as **vegetable**, **chutney**, and **salad**. Bulb juice good as **ear-drops** for relieving **itching**. Half-roasted bulbs applied to hasten suppuration of **boils**. Extr. of bulb and leaves of *Mentha longifolia* given with sugar to relieve **vomiting**.

Active Constituents : Isolated constituents are cycloallin, 1-β-fructosylsucrose, 6-β-fructosylsucrose, arabinose, rhamnose, xylose, N-2, 4-dinitrophenyl, propenyl sulfuric acid (*bulbs*) (*bulbs*); carbohydrates (*green leaves, bulbs*).

Biological Activity : Hypoglycaemic effect in rats and rabbits confirmed.

Uses in Literature : Reported as acrid, anthelmintic, appetiser, aphrodisiac, colic, diuretic, edible, emmanagogue, expectorant, pungent, rubefacient, stimulant, stomachic, tonic asthma, biliousness, bleeding piles, blood dysentery, body pains, boils, caries of teeth, catarrh, chronic bronchitis, cold, condiment, cough, diseases of spleen, dropsy, earache, emmanagogue, eye pains, fainting, fattening; fever, foot sores, haemorrhoids, hydrophobic, hypertension, improves taste; impotency, inflammations, insect-bite, itch, jaundice, liver complaints, malaria, migraine, night-blindness, piles, pimples, poultice, prolapsed ani, pthisis, pyorrhoea, scabies, scurvy; sedative, skin-diseases, sore-throat, tumours, ulcer, urinary discharges, vomiting, worm infection, and wounds (Ambasta, 1986; Dixit & Kumar, 2003; Katewa *et al.,* 2001; Khanna & Kumar, 2000; Kirtikar & Basu, 1984; Mohanty & Padhy, 1996; Pal & Paul, 2003; Rajan *et al.,* 2003; Sen & Batra, 1997; Sharma, 2003; Sharma & Singh, 2001; Sharma *et al.,* 2003b; Singh & Pandey, 1998; Singh & Singh, 2003; Sivakumar *et al.,* 2003; Vishwanathan et. al., 2001).

***Allium sativum* L.** (Pl. 4A)
Family: Liliaceae.

Vern. : Lahsun, Lassan.

English, Sanskrit & Regional Names:

Eng. : Churl's Treacle, Garlic, Poor Man's Treacle
Sans. : Arishtha, Bhutabhna, Grinjana, Lashuna, RasonaVatari
B. : Lashan, Lasan, Rasem
Bo. : Lusoon
H. : Lahsan, Lasan
Tam. : Vallaippundu
Tel. : Vellaitelagada.

Distribution : Throughout India.

Description : Bulbous plants. Leaves long, flat, narrow. Flowers white. Bulbils in a globose mass on a solid scape.

Reproductive Cycle : December-March.

Habitat Ecology : Moist places.

Material Examined : HPU-EL-SKS-5548; Bindraban (1,270 m), Chachian (1,300 m), Chimbalhar (1,150 m), Gharthun (900 m), Nanaun (900 m).

Parts Used : Bulbs. Leaves.

Folk Uses : Bulbs used for flavouring **vegetable** and **pulses**, and also consumed as **pickle**. Bulb juice prescribed for lowering serum **cholesterol**, **earache**, **gas trouble**, **skin diseases**, **rheumatic pains**, and as **colic**. Half roasted bulb taken to check **stomachache**, **gastric trouble** and **indigestion**. Leaves consumed as **chutney**.

Active Constituents : Garlic extr. contains ajoene. A new furostanol glycoside-protoeruboside B-isolated from *bulbs*.

Biological Activity : Prophylactic, bactericidal, stimulating effect on uterus +ve.

Uses in Literature : So far reported as alexipharmic, analgesic anthelmintic, antipyretic, antiseptic, aphrodisiac, carminative, condiments, digestive, diuretic, emmanagogue, expectorant, stimulant, pungent, tonic; and for arthritis, ascites, asthma, bronchitis, complexion, diseases of spleen, liver and lungs; dyspepsia, earache, epileptic fits, fattening, fevers, fishing, gum infection, heating, improving appetite, inflammations, leucoderma, lumbago, malaria, nervous diseaeses, paralysis, piles, removing maggots from wounds, rheumatism, scorpion-bite, sexual potency, swelling, thinning blood, thirst, tumours, voice and diseases of the eye and the heart (Ambasta, 1986; Borthakur *et al.,* 2004;

Dixit & Kumar, 2003; Ganai & Nawachoo, 2003; Gogoi & Borthakur, 2001; Kirtikar & Basu, 1984; Panthy & Choudhary, 2003; Sen & Bahera, 2003; Sen & Batra, 1997; Sharma, 2003; Sharma & Singh, 2001; Sharma *et al.*, 2003b; Singh & Pandey, 1998; Singh & Singh, 2003; Sivakumar *et al.*, 2003).

***Aloe vera* (L.) Webb. & Berthel.** (Pl. 4B)

Syn.: *A. barbadensis* Mill.; *A. perfoliata* var. *vera* L.

Family: Liliaceae.

Vern. : Ghi kanwar, Kuaarein.

English, Sanskrit & Regional Names:

Eng. : Aloe, Barbados Aloe, Common Indian Aloe, Curcago Aloe, Jaffarabad Aloe

Sans. : Adala, Ajara, Amara, Ambudhisrava, Aphala, Atipichhila, Bahupatri, Brahmaghni, Grihakanya, Kantakapravrita, Sthuladala, Sukantaka,Tarani, Vipulsrava

Ass. : Chai-kuwari

B. : Ghritakumari

G. : Kadvikunwar, Kunvar, Kuvara

H. : Ghiguvara, Ghikumari, Gvarapatha, Kumari, Kuvarapatha

Kan. : Loli-sara

Mal. : Chenninayakam, Cherukattala, Ghritakumari, Kaltaavala, Kaltala, Kaltavaya, Kumari

Mar. : Koraphada, Koraphanta, Korkand, Kunvarpatha

Ori. : Ghikunri, Ghritokumari, Kumari

Tam. : Angani, Angini, Ettakalavanda, Kalavanda, Kattalai, Kodiyan, Manjaikattalai, Musamvaramu, Sirukattalai, Sivappusottukkattalai, Sirappukkattalai, Sottukk-attalai.

Distribution : Kerala, Tamilnadu and other warm regions of India, below 1,500 m.

Description : Acaulescent, stoloniferous shrubs. Leaves large, spiny, succulant and in rosettes. Flowers red coloured, in racemes. Fruits a 3-seeded capsule.

Reproductive Cycle : March-April.

Habitat Ecology : Wild in drier areas.

Material Examined : HPU-EL-SKS-5681; Bodhal (970 m), Chandpur (1,290 m), Darugnu (1,400 m), Jasun (1,000 m), Saparu (1,250 m).

Parts Used : Leaves. Flowers.

Folk Uses : Leaf-slices applied as such on head, chest and soles to check **burning sensation**. Leaf pulp washed with ash and water, and cooked as **vegetable**. One teaspoonful of leaf juice taken daily for **jaundice** and relieving pain during **menstruation**; also applied locally for **suppuration of boils** and **pimples**, **burning sensation** of soles and for healing **burns**. Flowers used in culinary preparations like '**pakoras**'.

Active Constituents : Chief ingredients are barbaloin aloesin, aloesone β-sitosterol (*leaves*) and 4 partially acetylated glucomannans, β-barbaloin and aloe-anodin (*jelly*).

Biological Activity : Barbaloin inhibits the growth of *Mycobacterium tuberculosis*. Plant extr. (100 mg/kg) prevents implantation in 50 and 100% rats at oral dose of 100 and 200 mg/kg. Supernatant of gel cytotoxic against C3HIOTI/2 cells at initiation stage of transformation.

Uses in Literature : Used earlier as an aperient in apoplexy, aphrodisiac, aromatic, blood purifier, carminative, cαthartic, colic, coolant in fever, diphoretic, diuretic, edible (pickled leaves), emmenagogue, fibre, flavouring agent, refrigerant, stimulant, stomachic, tonic, vermicidal; and for acute indigestion, asthma, backache, blisters, boil, boil (vet.), breast tissue hardening, bronchitis, burns, burns (vet.), checking growth of tumours, chest pain, combining with tonics, curing bone sprain and in ostisis of horses, cutaneous leishmaniasis, delivery (vet.), dermatitis, digestion (vet.), dyeing, easy shedding of placenta during delivery (vet.), enlarged lymphatic glands, eye diseases, falling hair, flatulence, giddiness, gin preparation, gonorrhoea, headache, healing wounds and fresh cuts, hepatomegaly, indigsetion (vet.), injuries, itching (vet.), leprosy, liver and spleen ailments, mastitis (vet.), menstrual complaints, miscarriage (vet.), *Mycobacterium tuberculosis,* obstinate constipation, pickling, piles, prevention of miscarriage, relieving pains in muscles, rheumatism, ringworm (vet.), sexual vigour and vitality, skin diseases, softening swollen parts, soothing burns, sore (vet.), sores of bullocks, swelling (vet.), swelling of foot, typhoid, washing hairs (vet.), wounds, and X-ray burns (Agarwal, 2003; Ambasta, 1986; Arinathan *et al.,* 2003; Arora & Pandey, 1996; Bhatt *et al.,* 1999, 2003b; Borthakur & Sarma, 1996; Borthakur *et al.,* 2004; Chauhan, 1999; Chevallier, 1996; Choudhary, 1996; Datt, 1996; Dey & Bahadur, 1973; Diwakar & Ansari, 2003; Duke *et al.,* 2002; Dwivedi, 2003; Ganesan & Kesavan, 2003; Geetha *et al.,* 1996a, b; Girach & Aminuddin, 1995; Gogoi *et al.,* 2003a; Gopal & Chandra, 2003; Harborne & Baxter, 2001; Hosagoudar & Henry, 1996a; Jain, 1991; Jain & Patole, 2001; Jamir, 1997; Katewa *et al.,* 2001; Khare, 2004; Kirtikar & Basu, 1984; Kurian, 1995; Mishra *et al.,* 1996b; Mitaliya, 2003; Panthi & Chaudhary, 2003; Parrotta, 2001; Prasad *et al.,* 2003; Rajendran *et al.,* 2003; Rama Rao & Henry, 1996; Said, 1997; Saini, 1996a; Samwatsar & Diwanji, 1996a; Satapathy & Brahmam, 1996; Sen & Batra, 1997; Sen *et al.,*

2002; Sharma, 2003; Sharma & Borthakur, 2003; Shome *et al.,* 1996; Singh & Pandey, 1996, 1998; Sinha & Sinha, 2001; Sivakumar *et al.,* 2003; Thomas & Britto, 2003; Verma & Chauhan, 2006; Watt, 1972).

***Alternanthera sessilis* (L.) R. Br. *ex* DC.** (Pl. 4C)
Syn.: *A. nodiflora* R. Br.; *A. triandra* Lam., *Gamphrena sessilis* L.
Family: Amaranthaceae.

Vern. : Rubadi-Khubdi.

English, Sanskrit & Regional Names:

Eng. : The light galangal
Sans. : Giojihra, Matsyaksi
B. : Sanchi
G. : Jalajamboo, Paninibhaji
H. : Kanchari
Kan. : Honagonne
Mal. : Ponnagannikeeray
Mar. : Kanchari
Tam. : Ponnagannikkirai
Tel. : Ponnagantakura, Ponnagantikura.

Distribution : Throughout the warmer parts of India up to 1,200 m.

Description : Prostrate herbs. Leaves opposite, fleshy. Flowers inconspicuous, white; in axillary heads. Utricle cordiform, compressed. Seeds orbicular.

Reproductive Cycle : Throughout the year.

Habitat Ecology : Moist-shady places.

Material Examined : HPU-EL-SKS-5613; Chandpur (1,290 m), Chornali (1,280 m), Gharthun (900 m), Molichak (1,020 m), Purani-palam (1,100 m).

Parts Used : Aerial Plant Parts.

Folk Uses : Tender aerial plant parts consumed as **vegetable** and **fodder**. Poultice of aerial plant parts good for suppuration of **boils**.

Active Constituents : *Leaves* rich in a saponin having oleanolic acids as aglycone and glucose and rhamnose as sugars.

Uses in Literature : So far reported as antidote to poisons, cataract of eye, condiments, galactagogue, edible, refrigerant; and for dyspepsia, enhance lactation (vet.), eye complaints, fever, fodder, leprosy, night blindness, pruritus, spermatorrhoea, snakebite, stomach disorders, tuberculosis and vegetable good for breast milk (Agarwal, 2003; Ambasta, 1986; Arora & Pandey, 1996; Balasubramaniam & Prasad, 1996; Banerjee & Ghora, 1996; Bennet *et al.*, 1991; Bondya *et al.*, 2006; Borthakur, 1996b; Borthakur *et al.*, 2004; Chatterjee & Pakrashi, 1991; Das, 2000; Goud & Pullaiah, 1996a; Gogoi & Das, 2003; Hema *et al.*, 2006; Jain & Saklani, 1992; Jain *et al.*, 1994; Jha *et al.*, 1996; Khare, 2004; Kirtikar & Basu, 1984; Kumar & Goel, 1998; Nayak *et al.*, 2003; Ranjan, 2000, 2003; Roy *et al.*, 1998; Mitaliya *et al.*, 2003; Murthy *et al.*, 2003; Nath & Maiti, 2003; Prajapati *et al.*, 2006; Saini, 1996a; Samant *et al.*, 2001; Singh & Maheshwari, 1992; Singh *et al.*, 1999, 2003b; Sinha, 1996; Sivakumar *et al.*, 2003; Siwakoti & Varma, 1996; Srivastava *et al.*, 2003a, c; Tiwari & Tiwari, 1996).

Amaranthus paniculatus L. (Pl. 4D)

Syn.: *A. frumentaceus* Ham.

Family: Amaranthaceae.

Vern. : Seul.

Sanskrit & Regional Names:

- **Sans.** : Rahadri, Rajagiri, Rajashakini
- **B.** : Bathu, Chuko, Natya
- **G.** : Chuko, Rajgaro
- **H.** : Chaulai, Chuamarsa, Ganhar
- **Kan.** : Kire soppu
- **Kas.** : Bustanafroz
- **Mar.** : Rajagira
- **Tam.** : Pungikirai.

Distribution : E. & W. Asia (throughout India). Africa.

Description : Annual tall herbs upto 180 cm with stout, grooved stem, long bracteoles, shortly-awned sepals and numerous flowers in dense spikes of magenta colour. Seeds black.

Reproductive Cycle : September-November.

Habitat Ecology : Quite common as a weed in maize fields, moist places.

Material Examined : HPU-EL-SKS-5505; Arla (1,100 m), Bhagpur (1,600 m), Dehan (1,130 m), Kathena (1,290 m), Nanaun (900 m).

Parts Used : Leaves. Seeds.

Folk Uses : Leaves used as a **pot-herb**. Seeds relished as '**kheer**' and '**ladoo**' especially during **fasting**.

Active Constituents : Leaves contain oxalic acid.

Uses in Literature : Known in India as antidote to snakebite, cereals, diuretic, gonorrhoea, laxative, vegetable, and for boils, piles, purifying blood, and sores (Agarwal, 2003; Ambasta, 1986; Arora & Pandey, 1996; Chopra *et al.*, 1956; Kirtikar & Basu, 1984; Mao, 1993; Roy *et al.*, 1998; Samant *et al.*, 2001; Sharma & Rana, 2005; Sood & Thakur, 2004; Sood *et al.*, 2001).

Amaranthus viridis L. (Pl. 4E)
Syn.: *Euxolus viridis* Moq.
Family: Amaranthaceae.

Vern. : Chalairi, Chaulai.

English, Sanskrit & Regional Names:

Eng. : Green or Wild Amaranth, Green Pigweed
Sans. : Tanduliya, Vishaghna
B. : Ban Note
G. : Dhimdo
Mal. : Bayam munyit
Mar. : Lhanamat
Tam. : Kuppai Keerai, Sinna Keerai
Tel. : Chailaka Thotakura.

Distribution : Throughout India. All tropical and warm countries.

Description : Glabrous much-branched herbs, 30-60 cm high with purplish branches, long petioled leaves and pale green flowers in panicled slender spikes. Sepals membranous, keeled. Utricle indehiscent. Seeds shining black.

Reproductive Cycle : May-June.

Habitat Ecology : Frequent on wastelands. Weed in cultivated fields. Roadsides.

Material Examined : HPU-EL-SKS-5416; Balu (1,030 m), Bindraban (1,270 m), Malnu (1,000 m), Pahra (950 m), Rajeher (1,050 m).

Parts Used : Aerial Plant Parts. Seeds.

Folk Uses : The plant used as a **pot-herb**, **purgative** and **diuretic**. Seed powder roasted in 'desi ghee' taken to check excessive bleeding during **menses**.

Uses in Literature : Used as appetizer, cattle fodder, pot-herb, vermifuge; and for abortion, constipation, diabetes, diarrhoea, excessive bleeding during menses, kidney stones, lessening effect of poison, scorpion sting, snakebite, and urinary problems (Agarwal, 2003; Ambasta, 1986; Arora & Pandey, 1996; Banerjee & Ghora, 1996; Bennet *et al.*, 1991; Bhatt *et al.*, 2003a; Borthakur,

1996b; Chakraborty *et al.*, 2003; Chhetri & Shrestha, 2004; Chopra *et al.*, 1956; Das, 2000; Das & Choudhary, 2003; Hema *et al.*, 2006; Jain, 1991; Jha *et al.*, 1996; Kshirsagar *et al.*, 2003; Lal *et al.*, 1996; Lalramnghinglova, 1996; Manandhar, 1990; Murthy *et al.*, 2003; Negi *et al.*, 1999; Panthi & Chaudhary, 2003; Patole & Jain, 2002; Roy *et al.*, 1998; Saini, 1996a; Saklani & Jain, 1994; Shah, 2005; Sharma & Rana, 2005; Singh, 1996b; Singh & Kumar, 2000a; Singh & Pandey, 1998; Sood & Thakur, 2004; Srivastava *et al.*, 2003a; Upadhyay & Chauhan, 2000).

***Apium leptophyllum* Muell. *ex* Benth.**

Family: Apiaceae.

Vern. : Jungli ajwain, Pahadi ajwain Juanhe.

Distribution : Throughout the greater parts of India.

Description : Glabrous herbs. Leaves dark-green. Flowers white in compound heads. Fruits flattened.

Reproductive Cycle : June-September.

Habitat Ecology : Edges of cultivated areas. Wastelands. Damp places.

Material Examined : HPU-EL-SKS-5517; Bindraban (1,270 m).

Parts Used : Leaves. Seeds.

Folk Uses : Leaves **chewed** raw. Leaves and seed decoction prescribed for **stomach disorders** both in case of human beings as well as livestock. Its leaves along with that of 'gandhela' (*Murraya koenigii),* 'banna' (*Vitex nigundo*), onion tuber (*Allium cepa)* and 'kali-mirch' (*Piper nigrum*) prescribed for **fever** in livestock.

Active Constituents : Anhydrorutaretin, dihydrofurocoumarins, leptophyllidin, leptophyllin, its glucoside and umbelliferone, isopimpinellin and mannitol, marmesin, marmesinin, sitosterol glucoside, skimmin, trans-khellactone, 9-OH-4-OMe-psoralen (*seeds*); bergapten, seselin (*fruits*); and Me-ethers of thymol, carvacrol and thymoquinol (*plant essential oil*) have been isolated.

Uses in Literature : Used in India as a carminative, flavouring plant, nervine tonic, sedative, stimulant, and to scent hair-oil (Anonymous, 1985; Jha & Kumar, 2003; Kayal *et al.,* 2003; Sood & Thakur, 2004).

Araucaria columnaris (Forst.) Hook. (Pl. 4F)

Syn.: *A. cookii* R. Br. *ex* Lindl.

Family: Araucariaceae.

Vern. : Araucaria

English Name:

Eng. : Monkey Puzzle Tree.

Description : Trees with lower branches usually abscising and short shoots arising from the adventitious buds at the base. Male inflorescences 2.5-6.0 cm long. Scales about 2-4 cm wide. Seeds winged with a terminal thorny tip.

Reproductive Cycle : September-October.

Habitat Ecology : Gardens.

Material Examined : HPU-EL-SKS-5596; Andretta (980 m), Bandla (1,280 m), Holta (1,200 m), Sungal (1,050 m), Thakur dwara (1,150 m).

Parts Used : Whole Plant. Wood.

Folk Uses : Plant grown for **ornamental** purposes. Wood used for **furniture**.

Uses in Literature : Used in general carpentry work (Ambasta, 1986).

Aristida cyanantha Steud.

Family: Poaceae.

Vern. : Kaduli-Tuli.

Distribution : W. Himalayas, 1,000-1,750 m.

Description : Perennials with stems upto 2 m. Panicle purple, branches very long, droopimg. Spikelets 4 cm long. Empty glumes shortly-awned, the upper one much longer than flowering glumes. Awn 5-6 cm long.

Reproductive Cycle : July-September.

Habitat Ecology : Open slopes.

Material Examined : HPU-EL-SKS-5654; Balah (900 m), Bhauda (1,050 m), Bhuana (1,400 m), Malahu (1,050 m), Nadli (1,000 m).

Part Used : Stem.

Folk Uses : Stem employed for making **winnowing-pans**; and for stitching **leaf-plates** ('pattalu') and **leaf-bowls** ('dunnu').

Uses in Literature : Reported to be used as fodder when tender (Ambasta, 1986).

Artemisia capillaris Thunb. (Pl. 5A)

Syn.: *A. elegans* Roxb.; *A. scoparia* Waldst. & Kit.; *A. trichophylla* Wall.

Family: Asteraceae.

Vern. : Jhauncha.

Regional Names:

Bo. : Churisaroj, Danti

P. : Biur, Dona, Jhan, Jhau, Lasaj, Marua, Pila.

Distribution : India.

Description : Faintly scented shrub-like herbs with slender stem and purplish branches. Leaves radical, long-petioled and finely dissected. Heads minute in panicled racemes. Achenes very minute.

Reproductive Cycle : August - September.

Habitat Ecology : Drier areas. Crevices of rocks of dry slopes.

Material Examined : HPU-EL-SKS-5424; Chhanchri (930 m), Jasun (1,000 m), Panapari (1,100 m), Thandol (950 m), Rangru (910 m).

Parts Used : Leaves. Inflorescence.

Folk Uses : Leaves and inflorescence used as **spice** and for **rheumatic** pain; also taken with curd to check **diarrhoea**. Inflorescence paste applied to check **swellings**.

Active Constituents : Reported to contain β-sitosterol, oxalic acid and a lactone (*roots*), capillin, 1-Ph-2,4-hexadiyne-1-ol, vanillin and scoparone (*leaves and flowers*), capillarin, coumarin, lactone and scoparone (*flower heads*), and essential oil with characteristic odour of eugenol, dillapiol and carvone (*air-dried flowering tops*).

Biological Activity : Hypotensive, tranquilizing and sleep-inducing (6-7-di-MeO-coumarin); and increased secretion of bile and bile salts (p-OH-acetophenone) activities chracterized.

Uses in Literature : Recorded earlier for curing burns, earache and constipation (Ambasta, 1986; Chopra *et al.*, 1956; Kirtikar & Basu, 1984; Nautiyal, 1981; Sharma *et al.*, 1979; Watt, 1972).

***Artemisia japonica* Thunb.** (Pl. 5B)
Syn.: *A. parviflora* Don
Family: Asteraceae.

Vern. : Chharmara.

Regional Names:

H. : Pati
Ladak : Burmar, Basbna tashang
P. : Kanyurts.

Distribution : Temperate Himalaya.

Description : Tall aromatic undershurbs with lower wedge-shaped leaves. Heads small, discoid, greenish-yellow. Pappus none.

Reproductive Cycle : September-December.

Habitat Ecology : Meadows. Open slopes. Wastelands.

Material Examined : HPU-EL-SKS-5465; Banuri (1,150 m), Chandpur (1,290 m), Gopalpur (1,350 m), Jasun (1,000 m), Makol (900 m).

Parts Used : Aerial Plant Parts. Leaves.

Folk Uses : Fomentation of hot water containing its aerial plant parts and that of *Vitex negundo*, *Murraya koenigii* and *Artemisia japonica* checks **sprain** and **fracture**. Moreover, its leaf-paste alongwith that of *Trachyspermum ammi* applied externally for same purpose. Leaves rubbed over body of livestock to **keep away flies** during rainy season.

Active Constituents : *Plant* shows presence of coumarins 6-7-di-Me-aesculetin, aesculentin and scopoletin.

Biological Activity : Antiviral and diuretic activity +ve.

Uses in Literature : So far used in India for convulsion in children, skin diseases and throat infections (Jain, 1991; Megoneitso & Rao, 1983).

Artocarpus lacucha Buch.-Ham. (Pl. 5C)

Syn.: *Artocarpus lacucha* Roxb.; *A. lakoocha* Roxb.

Family: Urticaceae.

Vern. : Dehu.

English, Sanskrit & Regional Names:

Eng.	: Monkey jack
Sans.	: Lakudi
Ass.	: Chama, Dewa-Dewasali
B.	: Dahu, Dehua, Dehu
G.	: Lakoocha
H.	: Barhal, Dahu, Dheu, Lakooch
Kan.	: Esuluhuli, Lakoocha, Wonta
Mal.	: Chimpa, Lakucham, Pulinjakka
Mar.	: Bhadhar, Vatara, Wotomba
Ori.	: Jeuto
Tam.	: Hagusam
Tel.	: Dhum, Kammaregu, Tiun
Urdu	: Barhal.

Distribution : Moist deciduous forests in Northern India upto 1,800 m.

Description : Deciduous trees with large spreading crown. Leaves leathery, oblong, elliptic. Flowers born in globose heads. Fruits subglobose, smooth.

Reproductive Cycle : Feburary-September.

Habitat Ecology : Wastelands. Borders of agricultural fields.

Material Examined : HPU-EL-SKS-5515; Balah (900 m), Khaira (940 m), Boda (1,000 m), Tapa (945 m), Malahu (1,050 m),

Parts Used : Fruits. Flower buds. Leaves.

Folk Uses : Fruits and flower-buds relished as **vegetable** and **pickle**. Leaves used as **fodder** and tender ones as '**pakoras**'.

Uses in Literature : Known earlier for biliousness, boat-building, blood complaints, constipation, colouring, cracked skin, dental carries, eye troubles, fever due to cough, fodder, pimples, impotence, loss of appetite, and toothache; and also as edible and tonic to the liver (Ambasta, 1986; Arora & Pandey, 1996; Bennet *et al.*, 1991; Brahmam *et al.*, 1996; Chatterjee & Pakrashi, 1991; Chopra *et al.*, 1956; Chun-Lin & Jieru, 1995; Das & Agarwal, 1991; Khare, 2004; Kirtikar & Basu, 1984; Kumar & Goel, 1998; Kumar & Rao, 2003; Kumar *et al.*, 2003; Negi *et al.*, 1993; Prajapati *et al.*, 2006; Saini, 1996a; Saklani & Jain, 1994; Sharma & Rana, 2005; Siwakoti & Varma, 1996; Sood & Kaushal, 2008; Tigga *et al.*, 2003).

Arundinaria falcata Nees

Syn.: *A. utilis* Cugh.

Family: Poaceae.

Vern. : Banji.

English & Regional Names:

Eng. : Himalayan Bamboo

H. : Gurwa, Spiag, Spikso.

Distribution : W. Himalayas: 1,200-2,100 m.

Description : Shrub-like bamboos. Leaves not net-veined and have glabrous sheath. Flowers occurring on leafless stem. Glumes ribbed.

Reproductive Cycle : May-November; flowering sporadically.

Habitat Ecology : Wastelands. Nallahs.

Material Examined : HPU-EL-SKS-5533; Darugnu (1,400 m), Gopalpur (1,350 m), Nadli (1,000 m), Rajeher (1,050 m), Sulah (950 m).

Parts Used : Leaves. Stem.

Folk Uses : Leaves given as **fodder**. Stem employed for **agricultural implements**, **house-construction** and **walking sticks**.

Uses in Literature : Reported to be used as base material for house, cattle shed, containers for compost fertilizer, fibre, fodder; and for agricultural implements, baskets, brooms, grains, roofing, storage wings, mats, fishing rods and hookahs (Ambasta, 1986; Negi *et al.*, 1993; Watt, 1972).

Asparagus adscendens **Roxb.** (Pl. 5D)

Syn.: *A. volubilis* Wall.; *Asparagopsis adscendens* Kunth

Family: Liliaceae.

Vern. : Sanspain.

Regional Names:

Bo. : Dholimusali, Saphetamusali

G. : Dholimusali, Saphedmusli, Ujlimusli

H. : Satavar, Sufedmusli

Mar. : Safedmusli.

Distribution : W. Himalayas. Punjab-Kumaon.

Description : Much branched, tall undershrubs with terete grooved stems. Spines stout, straight. Cladodes densely crowded.

Flowers white, bisexual in racemes. Berries globose, 1-seeded.

Reproductive Cycle : October-November.

Habitat Ecology : Scrub forests. Damp-shady places.

Material Examined : HPU-EL-SKS-5548; Balu (1,030 m), Bhaura (940 m), Chauki (900 m), Rangru (910 m), Saliana (1,000 m).

Parts Used : The Whole Plant. Roots.

Folk Uses : Plant worshipped while soleminizing **marriages** and performing other **religious rites**. Root-juice **astringent** to bowels and good for dysentery and enhancing **lactation**.

Active Constituents : Isolated constituents are asparagine and saponin steroidal (*roots*); β-sitosterol, diosgenin and sarsasapogenin (*fruits*).

Biological Activity : Antimicrobial activity found to be +ve.

Uses in Literature : Used in India as an aphrodisiac, appetizer, astringent, coolant, demulcent in veterinary medicine, diaphoretic, diuretic, edible (young shoots and roots), galactagogue, sacred, and tonic; and for bodyache, cough, diarrhoea, debility, dog bite, dysentery, dyspepsia, infertility, lumbago, leucorrhoea, nerve tonic, obesity, rheumatism, scorpion sting, swelling, throat complaints, urinary stones, urinary tract inflammation and substitute for 'salep misree' (Agarwal, 2003; Ambasta, 1986; Arinathan *et al.*, 2003; Arora & Pandey, 1996; Beg *et al.*, 2005; Bhatt *et al.*, 2003b; Chauhan, 1999; Devi, 2003; Dhiman, 1976; Diwakar & Ansari, 2003; Dwivedi, 2003; Ganesan & Kesavan, 2003; Gopal & Chandra, 2003; Jha & Kumar, 2003; Khare, 2004; Kirtikar & Basu, 1984; Kumar & Lal, 2003; Manandhar, 1980b; Nayak *et al.*, 2003; Negi *et al.*, 1993; Prajapati *et al.*, 2006; Roy *et al.*, 1998; Sen & Batra, 1997; Sharma, 2003; Sharma *et al.*, 2003b; Singh, 1999, 2003; Singh & Kumar, 2000a; Singh & Pandey, 1998; Singh *et al.*, 2003a, b; Sivakumar *et al.*, 2003; Sood & Prakash, 2007; Sood & Thakur, 2004; Srivastava *et al.*, 2003c; Subramani & Goraya, 2003; Thomas & Britto, 2003).

***Avena sativa* L.** (Pl. 5E)

Family: Poaceae.

Vern. : Luanhe.

English & Regional Name:

Eng. : Groats, Oat, Oats

H. : Jai.

Distribution : Himalayas upto 4,000 m.

Description : An annual herb with erect cylindrical stem; linear, sessile leaves with parallel venation and flowers in pendulous spikes. Glumes very long acuminate. Fruits caryopsis. Seeds albuminous.

Reproductive Cycle : November-May.

Habitat Ecology : Weed in agricultural fields, moist shady places.

Material Examined : HPU-EL-SKS-5635; Bandla (1,280 m), Bodhal (970 m), Gopalpur (1,350 m), Panchrukhi (990 m), Saparu (1,250 m).

Parts Used : Aerial plant parts.

Folk Use : Aerial plant parts employed as a **fodder** for milching cattle.

Active Constituents : Flavanoids of apigenin, luteolin and tricin types isolated from *leaves*, *stem* and *flowers*.

Uses in Literature : Considered antispasmodic, breakfast foods, cakes, fodder, nerve tonic, porridge, stimulant, antispasmodic, tonic in spermatorrhoea and insomnia; good for febrile diseases, heart muscles and urinary organs, relieving spasmodic conditions of the bladder and ureter, itch, leprosy (Ambasta, 1986; Caius, 2003; Kirtikar & Basu, 1984; Sood & Kaushal, 2008).

***Bambusa nutans* Wall.** (Pl. 5F).

Syn.: *Bambusa falconeri* Munro

Family: Poaceae.

Vern. : Dang.

Regional Name:

B. : Mahlbans.

Distribution : H.P. – Sikkim; Commonly cultivated upto 1,700 m.

Description : Solitary stems with spiny branches arising from a creeping rhizome. Internodes 38-45 cm long. Leaves linear-lanceolate, pubescent beneath.

Reproductive Cycle : Flowering sporadically.

Habitat Ecology : Shady areas. Scrub forests.

Material Examined : HPU-EL-SKS-5674; Chimbalhar (1,150 m), Dadh (1,100 m), Malnu (1,000 m), Sidhpur (1,100 m), Tapa (945 m).

Parts Used : Stem. Leaves.

Folk Uses : Leaves used as **fodder**. Stem good for **building works**, **walking sticks** and **wooden utensils**.

Uses in Literature : Reported earlier to be used as base material for house, cattleshed, containers for compost fertilizer, fibre, fodder; and for agricultural implements, asthma baskets, brooms, grains, cough, roofing, storage wings, mats, fishing rods, hookahs (Ambasta, 1986; Arora & Pandey, 1996; Negi *et al*., 1993; Saklani & Jain, 1994; Samant *et al*., 2001; Sharma & Rana, 2005; Thomas & Britto, 2003; Watt, 1972).

***Basella alba* L.** (Pl. 6A)

Syn.: *Basella rubra* L.

Family: Chenopodiaceae.

Vern. : Poe.

English, Sanskrit & Regional Names:

- **Eng.** : Indian spinach
- **Sans.** : Apodike, Kalambi, Madushaka, Mohini, Potaki, Potaki, Putica, Upoti, Vishala, Vishavatulsi
- **B.** : Puisak, Ruktopuri
- **Bo.** : Mayakbhaji, Velgond
- **G.** : Pothi, Pethinivel, Valchibhagi
- **H.** : Bonpoe, Lalbachlu, Poe, Poikivel, Sufedbachla
- **Mal.** : Basellakkira, Lakkira
- **Mar.** : Mayalu, Wahlea
- **Tam.** : Shivappuvaslakkirai, Vaslakkirai
- **Tel.** : Allabatsalla, Alubabachali.

Distribution : Throughout India wild or cultivated, Sri Lanka—Tropical Asia and Africa.

Description : A fleshy, glabrous, perennial twining herb. Leaves alternate, broad, entire. Flowers sessile, white or red, few, and in spike. Fruits of the size of pea, red, white or black. Seeds erect, subglobose.

Reproductive Cycle : March-December.

Habitat Ecology : Moist-shady places. Borders of agricultural fields.

Material Examined : HPU-EL-SKS-5649; Andretta (980 m), Bodhal (970 m), Chornali (1,280 m), Ghugar (1,250 m), Lolani (1,600 m).

Part Used : Leaves.

Folk Uses : Leaves cooked as '**saag**' and '**pakodas**;' and also prescribed for '**Vaata**'.

Uses in Literature : Considered antipyretic, aphrodisiac, appetizer, demulcent, diuretic, fattening, laxative, narcotic, red dye, tonic; and for allaying heat, balanites, bioliousness, burns, constipation, cooling effect, diarrhoea, dysentery, gonorrhoea, heating, increase semen content, itching of urticaria, labour pain, leprosy, making curry, piles, sweetish and ulcer (Agarwal, 2003; Ambasta, 1986; Arora & Pandey, 1996; Bennet *et al*., 1991; Bondya *et al*., 2006; Chatterjee & Pakrashi, 1991; Jha & Basak, 1994; Jha *et al*., 1996; Khan *et al*., 2003; Khanna *et al*., 1996b; Kirtikar & Basu, 1984; Kumar & Rao, 2003; Kumar *et al*., 2003; Kurian, 1995; Mitaliya & Bhatt, 2003; Mohanty & Padhy, 1996; Murthy *et al*., 2003; Nadanakunjidam & Abirai, 2003; Prajapati *et al*., 2006; Ranjan, 2000; Roy *et al*., 1998; Saini, 1996a; Saklani & Jain, 1994; Singh, 2000; Singh & Singh, 2003; Subramani & Goraya, 2003; Sudhakar & Vedavathy, 2000; Upadhyay & Chauhan, 2000; Vishwanathan, 1997).

Bauhinia vahlii **Wight & Arn.** (Pl. 6B).

Syn.: *B. racemosa* Vahl; *Phanera vahlii* Benth.

Family: Fabaceae.

Vern. : Tor, Taur.

English & Regional Names:

Eng. : Camel's Foot Climber
B. : Chehur
G. : Chambelli
H. : Jallur, Maljhan
Mal. : Motanvally
Mar. : Chambal, Chambura
Ori. : Shiali
P. : Taur
Tam. : Mandarai
Tel. : Adattige, Madapu.

Distribution : Kashmir-Sikkim: upto 1,500 m.

Description : A large climber having branches terminating in a pair of opposite tendrils, cordate-orbicular leaves, whitish yellow petals, and 8-12 seeded woody pods with brown hairs.

Reproductive Cycle : April-June.

Habitat Ecology : Forests. Shady-rocky steep slopes.

Material Examined : HPU-EL-SKS-5545; Khaira (940 m), Jasun (1,000 m), Makol (900 m), Malahu (1,050 m), Nadli (1,000 m), Rangru (910 m).

Parts Used : Leaves. Stem. Seeds.

Folk Uses : Stem-juice used as nasal drops to kill **leeches** in livestock. Moreover, crushed stems also added in water reservoirs infested with **leeches**. Fibres obtained after retting of stem used to make **ropes**. Leaves used for making 'pattal' **(meal-plates)** and 'duna' **(meal-bowls)**, and also as **fodder**. Roasted seeds **relished** by local people.

Active Constituents : Chief constituents are quercitroside, iso-quercitroside, rutoside (*plant*); tannin (17%) (*bark*); and campesterol, stigmasterol, β-sitosterol, agathisflavone and its mono- and d-mrthyl derivatives, (kaempferol, quercitin) its 3-glucoside and betulinic acid (*leaves*).

Uses in Literature : Used so far as adhesive, antifertility, aphrodisiac, demulcent, edible (fruits and roasted seeds, eaten raw), finger rings, stomachic, tonic, vemifuge; and for amenorrhoea, bone fracture (vet.), cancer, caps, dropsy, dysentery, fibres, food serving plates, local cigarretes, malaria, meal plates, mucilaginous remedies, rain hats, ropes and cordage, rough table clothes, snakebite, syphilis, thatching, and umbrellas (Agarwal, 2003; Ambasta, 1986; Arora & Pandey, 1996; Banerjee, 1996; Chauhan, 1999; Chhetri, 2005; Das & Agarwal, 1991; Goud & Pullaiah, 1996a; Jain, 1991, 1996, 2003a; Jha & Kumar, 2003; Kirtikar & Basu, 1984; Kumar, 2003; Kumar & Pullaiah, 2003; Mahato & Chaudhary, 2005; Manandhar, 2003; Murthy *et al.*, 2003; Nayak *et al.*, 2003; Parrotta, 2001; Patole & Jain, 2002; Prakash, 2001; Prasad *et al.*, 1999; Rai & Upadhyay, 1996; Rama Rao & Henry, 1996; Rao, 1996; Roy *et al.*, 1998; Sahu, 1996; Saini, 1996a, b; Samant *et al.*, 2001; Sharma, 2003; Sharma & Rana, 2005; Singh, 1996b, 2000; Singh & Kumar, 2000a; Singh & Prakash, 1994; Sood & Thakur, 2004; Sudhakar & Vedavathy, 2000; Tirkey, 2004; Tosh, 1996; Varma, 1997; Watt, 1972).

***Bauhinia variegata* L.** (Pl. 6C)

Syn.: *B. purpurea* Wall.; *B. candida* Roxb.; *Phanera variegata* Benth.

Family: Fabaceae.

Vern. : Karale, Karaalein.

English, Sanskrit & Regional Names:

Eng.	: Buddhist Bauhinia, Mountain Ebony, Orchid Tree
Sans.	: Kovidara
B.	: Raktanchan
Bo.	: Segapa-Munthari
G.	: Kovindara
H.	: Kachnar
Kan.	: Kanchavala
Mal.	: Chuvannamandaram
Mar.	: Raktakanchan
Ori.	: Boroda
Tam.	: Sagapumanchori
Tel.	: Devakanchanamo.

Distribution : Common in India, Myanmar, China upto 1,800 m.

Description : Trees with deeply bilobed glabrous leaves. Petals with purple veins and fragrant. Pods glabrous hard, flat, 12-15 seeded.

Reproductive Cycle : March-August.

Habitat Ecology : Forests, planted in villages.

Material Examined : HPU-EL-SKS-5639; Balah (900 m), Boda (1,000 m), Chhanchri (930 m), Malahu (1,050 m), Panapari (1,100 m).

Parts Used : Flowers. Leaves. Bark. Wood.

Folk Uses : Petals and flower buds cooked as **vegetable**, and also **pickled**. Boiled petals mashed in curd given to check **dysentery**; and as an **astringent** to bowels both in human beings as an well livestock. Leaves considered excellant **fodder** for milching cattle for enhancing fat content in milk. Bark employed for **dyeing** leather and as ropes. Wood used as **fuelwood**.

Active Constituents : Active properties attributed to isoquercitin, astragalin quercitin, hentriacontene octacosanol reducing sugars, stigmasterol, β-sitosterol and β-p-glucopyranoside (*flowers*).

Uses in Literature : Reported to be used as an alterative, anthelmintic, antidote to snake poison, antiinflammatory, astringent, carminative (roots), culminative, fodder, gentle laxative, pot herbs (flower bud, leaves), sacred, tonic, vermifuge (vet.); and for agricultural implements, beedi manufacturing (leaves), body/muscular swelling, cuts (vet.), cutaneous troubles, diarrhoea, dyeing, dysentery, dyspepsia, enhancing lactation, epistaxis, expelling placenta (vet.), fatness, flatulence, gastropathics (vet.), goitre, gripping pains from the stomach and bowels, indigestion, leprosy, malaria, mumps, oilseeds, paintings (Mithila), curd preparation, pickling, piles, preventing obesity and decomposition of the blood, reducing corpulence, piles, scrofula, skin diseases, snakebites, suppuration, syphilis, tanning, tuberculosis, tumours, ulcer (vet.), uterine bleeding, ulcers, and wounds (vet.) (Agarwal, 2003; Ambasta, 1986; Arora & Pandey, 1996; Banerjee, 1996; Bhattacharya *et al.*, 2005; Chauhan, 1999; Chhetri, 1994; Chun-Lin & Jieru, 1995; Devi, 2003; Dey & Bahadur, 1973; Dwivedi, 2003; Gill & Nyawuame, 1994; Hosagoudar & Henry, 1996a; Jain, 1991; Jain *et al.*, 1991; Jha & Basak, 1994; Jha *et al.*, 1996; Khanna *et al.*, 1996a, c; Kirtikar & Basu, 1984; Kumar, 2003; Kumar & Lal, 2003; Kumar *et al.*, 2003; Kurian, 1995; Lal *et al.*, 1996; Lalramnghinglova, 1996; Manandhar, 1995; Masih, 2003; Murthy *et al.*, 2003; Nadanakunjidam & Abirai, 2003; Negi *et al.*, 1985; Pandey & Joshi, 2001; Parrotta, 2001; Purohit & Samant, 1995; Rai & Upadhyay, 1996; Rao, 1996; Roy *et al.*, 1998; Saini, 1996a, b; Shah, 2005; Sharma, 2003; Shukla *et al.*, 2003; Singh, 1991, 1999; Singh & Kumar, 2000a; Sinha, 1996; Srivastava *et al.*, 2003a; Verma & Chauhan, 2006; Watt, 1972).

***Berberis lycium* Royle** (Pl. 6D)

Syn.: *B. insignis* Hook.f.

Family: Berberidaceae.

Vern. : Kashmal, Kasmalein.

English, Sanskrit & Regional Names:

Eng. : Raisin Berberry

Sans. : Daruharidra

Bo. : Kasmal Rasout

H. : Kushmal.

Distribution : Pakistan-E. Nepal. India:1,500-3,000 m.

Description : Shrubs with angular branches. Leaves narrowly lanceolate and have a few, small teeth. Flowers pale-yellow. Berries ovoid and violet, and covered with blume.

Reproductive Cycle : April-July.

Habitat Ecology : Drier hilly slopes.

Material Examined : HPU-EL-SKS-5450; Bhagpur (1,600 m), Chauki (900 m), Rajeher (1,050 m), Saparu (1,250 m).

Parts Used : Roots. Stem. Leaves. Fruit.

Folk Uses : Root decoction taken orally as **blood purifier**, and also used to wash **sores** and **boils** due to **antiseptic** properties. Root decoction also employed as eye drop for **opthalmic ailments**. Ripe fruits **edible** and considered **blood purifier**. Twigs coloured and worshipped on the day of '**Holi**' festival. Leaves excellent **fodder** for goat.

Active Constituents : Properties of roots due to umbellatine, berberine, berbericine hydrochloride, berbericinine hydroiodide, oxyberberine and berbamine.

Biological Activity : Antianotic, antisecretory, anticonvulscent and spasmolytic activities confirmed.

Uses in Literature : Used earlier in India as an alterative, antidote for poisoning, antimalarial, antiseptic, antispasmodic, aperient, blood purifier, carminative, coagulant, febrifuge, fodder, mild laxative, sedative, tonic; and for allaying irritability and thirst, bleeding piles, boils, chest and throat troubles, cholera, cooling effect, chronic diarrhoea, cough, disorders of urinary tract, dysentery, ear problems, edible purposes, enlargement of liver and spleen, epigastrium, eye diseases (chronic opthalmia), increasing appetite, jaundice, leprosy, malaria, menorrhagia, oral ulcers, oriental sores, periodic neuralgia, piles, promoting digestion, skin diseases, sores, spleen troubles, stomach disorders, throat pain, and urinogenital disorders (Agarwal, 2003; Ambasta, 1986; Arora & Pandey, 1996; Bennet *et al.*, 1991; Chandra, 1997; Chauhan, 1999; Chopra *et al.*, 1956; Jain, 1991; Jha & Kumar, 2003; Kant & Dutt, 2005; Kapur *et al.*, 1996; Kaul, 1997; Kirtikar & Basu, 1984; Lal *et al.*, 1996; Nautiyal *et al.*, 2003; Negi & Gaur, 1991; Rana *et al.*, 1996; Rao, 1990; Roy *et al.*, 1998; Said, 1997; Samant *et al.*, 1996; Singh, 1996b; Singh & Kumar, 2000a; Sood & Thakur, 2004; Sood & Kaushal, 2008; Watt, 1972).

***Bergenia ciliata* (Haw.) Sternb.** (Pl. 6E)

Syn.: *Megasea ciliata* Haw.; *Saxifraga ciliata* (Haw.) Royle; *B. ligulata* Wall.

Family: Saxifragaceae.

Vern. : Saprotri.

Distribution : Afganistan-S.E. Tibet.

Description : Stunted creeping herbs with stout cylindrical rootstock. Leaves stalked, circular in circumference; ciliate with stiff hairs. Flowers pinkish, in spreading panicles.

Reproductive Cycle : March-November.

Habitat Ecology : Forests. Moist-rock edges.

Material Examined : HPU-EL-SKS-5705; Bhagpur (1,600 m), Jiun (1,550 m), Lolani (1,600 m).

Parts Used : Flowers. Rootstock.

Folk Uses : Flowers used for making '**pakodas**'. Rootstock chewed for **kidney stones**. Also handful of it mixed with wheat flour given for 2-3 days to check **dysentery** (vet.).

Uses in Literature : Earlier reported as an abortifacient, acrid, alexiteric, analgesic, aphrodisiac, astringent, bitter, carminative, diuretic, edible, laxative, tonic; and for asthma, biliousness, bodyache, bone fracture, cooling, diseases of bladder, dysentery, edible, eye sores, fever, heart diseases, high fever, hydrophobia, kidney stones, liver ailments, lung ailments, menorrhagia, muscular pains, piles, pulmonary affections, tumours, sickness (vet.), spleen enlargement, ulcers, urinary discharges, urinary complaints, vital diet of breast feeding mothers, weakness (Ambasta, 1986; Arora & Pandey, 1996; Chauhan, 1999; Chhetri, 2005; Kumar & Rao, 2001; Lal *et al.*, 1996; Manandhar, 2002; Panthi & Chaudhary, 2003; Rana *et al.*, 1996; Roy *et al.*, 1998; Saklani & Jain, 1994; Sarin, 1990; Sharma, 2003; Sharma *et al.*, 2003a; Singh, 2001; Singh & Kumar, 2000a).

Bidens biternata (Lour.) Merr. & Sherff. (Pl. 6F)

Syn.: *Bidens pilosa* auct. (Non L.) Hook. f.

Family: Asteraceae.

Vern. : Lumb.

Regional Names:

- **G.** : Phutium, Samarakokadi
- **H.P.** : Chor pushi
- **H.** : Chirchitta
- **P.** : Chirchitta.

Distribution : Common throughout India.

Description : Herbs upto 60-90 cm tall with quadrangular grooved stems and pinnate leaves. Branches opposite. Heads white and on long peduncles. Achenes barbed.

Reproductive Cycle : June-September.

Habitat Ecology : Wastelands. Moist and shady slopes.

Material Examined : HPU-EL-SKS-5583; Banuri (1,150 m), Dadh (1,100 m), Jasun (1,000 m), Ladoh (980 m), Lahla (930 m).

Parts Used : Aerial Plants Parts.

Folk Use : Paste of aerial plant parts good for **mouth ulcers**.

Uses in Literature : Used as an appetizer, galactagogue (vet.) medicine, salad, styptic, substitute for tea, symbolic to ward off evil spirits, vegetable; and for cough, cuts, diarrhoea, ear complaints, eye sores, fistulae, headache, leprosy, phthisis, pustules, skin diseases, snakebite, sores, toothache, tumours, ulcers and wounds (Arora & Pandey, 1996; Bennet *et al.*, 1991; Chandra, 1997; Chhetri, 2005; Chopra *et al.*, 1956; Chowdhery, 1996; Devi, 2003; Islam & Hasin, 2003; Jain, 1991; Kirtikar & Basu, 1984; Manandhar, 1990; Roy *et al.*, 1998; Singh & Kumar, 2000a, b; Sood & Thakur, 2004; Subramaniam, 2000).

Boehmeria macrophylla **Hornem.** (Pl. 7A)

Syn.: *B. macrostachya* Wedd.; *B. platyphylla* D. Don; *Urtica macrostachya* Wall.

Family: Urticaceae.

Vern. : Handa.

Regional Name:

H. : Gargela.

Distribution : India. S.W.China. Burma: 800-2700 m.

Description : Shrubs having opposite leaves and small, nearly white flowers. Fruits compressed and enclosed in dry perianth.

Reproductive Cycle : July-September.

Habitat Ecology : Common in Wastelands. Secondary forests.

Material Examined : HPU-
EL-SKS-5491; Balu (1,030 m), Bhawarna (1,100 m), Darugnu (1,400 m), Makol (900 m), Tikkar (910 m).

Parts Used : Leaves. Seeds.

Folk Uses : Leaves used for **scouring** teeth. Seed powder (one teaspoonful) roasted with wheat flour and desi-ghee taken twice or thrice to check excessive bleeding during **menses**.

Active Constituents : Cryptopleurine, 5 α - stigmastan - 3, 6 - dione, pomolic acid, oleanolic acid, lupeol and β-sitosterol, secophenanthroquinolizidine isolated from *leaves* and *twigs*.

Biological Activity : Hypotensive and spasmolytic activities and effect on CNS found +ve.

Uses in Literature : Known to be used as fibre for cordages and fishing lines, fodder, galactagogue; and for cuts, diarrhoea (vet.), dress making, dysentery, eczema, stomach disorders, and wounds (Ambasta, 1986; Jain, 1991; Kharkongor & Joseph, 1981; Mahato & Chaudhary, 2005; Manandhar, 1987a, 1989a; Purohit & Samant, 1995; Rama Rao & Henry, 1996; Rao, 1996; Uniyal & Chauhan, 1973b).

Boerhaavia diffusa L. (Pl. 7B)

Syn.: *B. procumbens* (Roxb.); *B. repens* L.

Family: Nyctaginaceae.

Vern. : Eat-seat.

English, Sanskrit & Regional Names:

Eng. : Hogweed, Pigweed
Sans. : Bhauma, Kathillaka, Karura, Lohita, Nava, Navya, Nila, Nilini
B. : Gadhapurna, Punarnaba
Bo. : Glhetuli, Khapra, Punarnava;
G. : Dholisalurdi, Motosatodo,Vashakapara;
H. : Sant,Thikri;
Kan. : Balavadikae, Gajjeru;
Mal. : Thirmizhazhura, Titudamma;
Mar. : Kharapraya, Raktavasu, Vasu;
Tam. : Mukkurattai, Keerai;
Tel. : Atikamamidi, Punarnava.
Urdu : Bashkhira.

Distribution : Throughout India.

Description : Dffuse herbs with large, fusiform root. Stem prostrate or ascending, slender, cylindrical. Leaves in unequal pairs, broadly ovate-suborbicular, green and glabrous above. Flowers small, nearly sessile; in umbels. Fruits rounded, bluntly 5-ribbed.

Reproductive Cycle : June-October.

Habitat Ecology : Along fields borders.

Material Examined : HPU-EL-SKS-5512; Bindraban (1,270 m), Molichak (1,020 m), Rajpur (1,100 m), Saparu (1,250 m), Sulah (950 m).

Parts Used : Leaves.

Folk Uses : Leaves in combination with that of *Cannabis sativa* (1:1 ratio) and wheat flour cooked as '**pakodas**' and taken to relieve **rheumatic pains** and checking **dysentery**. Leaf-juice used as **eye drop** to check redness. Also, leaves form one of the ingredients of '**saag**'.

Active Constituents : *Aerial plant parts* contain alkaloids, free and combined amino acids.

Biological Activity : Glycoprotein reported to be active against spherical and tubular viruses in hypersensitive and systemic hosts and responsible for inducing systemic resistance in plants.

Uses in Literature : Useful as alexeteric, anthelmintic, appetiser, ascites, astringent, bitter blood purifier, carminative, cooling, diuretic, edible, emetic, expectorant, febrifuge, laxative, stomachic, tonic; and for anaemia, anasarca, asthma, anti-inflammatory, biliousness, blood pressure, boils, calculi, cataract of eye, colitis, ear and nasal drops; cough, diabetes, gonorrhoea, hasten delivery, heart diseases, internal inflammations, insect-bite, jaundice, kidney ailments, leucorrhoea, obesity, oedema, opthalmic diseases, piles, post-natal care, redness of eyes, scanty urine, rheumatism, snakebite, and scorpion-sting; swellings, urticaria (Ambasta, 1986; Arora & Pandey, 1996; Balasubramaniam & Prasad, 1996; Bhatt *et al.*, 1999, 2003b; Bondya *et al.*, 2006; Borthakur, 1996a; Borthakur *et al.*, 2004; Bose, 1984; Bhattacharya, 1996; Binu *et al.*, 2003; Chandra, 1997; Chakraborty *et al.*, 2003; Chatterjee & Pakrashi, 1991; Chopra *et al.*, 1956; Choudhary, 1996; Das, 2000; Deshmukh & Rothe, 2003; Diwakar & Ansari, 2003; Dixit & Kumar, 2003; Dwivedi, 2003; Gill *et al.*, 1993; Gogoi & Borthakur, 2001; Gogoi & Das, 2003; Henry *et al.*, 1996; Jain *et al.*, 1991, 1994; Jha & Varma, 1996; Jha *et al.*, 1991, 1996; Khanna & Kumar, 2000; Khanna *et al.*, 1996b; Kirtikar & Basu, 1984; Kumar, 2003; Kumar & Lal, 2003; Mehrotra *et al.*, 1995; Mitaliya *et al.*, 2003; Nayak *et al.*, 2003; Panthi & Chaudhary, 2003; Prakash, 1999, 2001; Prasad *et al.*, 2003; Rai & Upadhyay, 1996; Rana *et al.*, 1996; Ranjan, 1996, 2003; Roy *et al.*, 1998; Saini, 1996b; Samant *et al.*, 2001; Saklani & Jain, 1994; Sanyal, 1994; Sarin, 1996; Shah, 2005; Shanavashan *et al.*, 2003; Sharma, 2003; Singh, 2000, 2003; Singh & Kumar, 2000a; Singh & Maheshwari, 1992; Singh & Prakash, 1996; Singh *et al.*, 2003b; Siwakoti & Varma, 1996; Srivastava *et al.*, 2003a; Subramani & Goraya, 2003; Sudhakar & Vedavathy, 2000; Thomas & Britto, 2003; Verma & Chauhan, 2006).

***Bombax ceiba* L.** (Pl. 7C, D)

Syn.: *B. heptaphylla* Cav. Diss.; *B. heptaphyllum* Roxb.; *B. malabaricum* DC.; *Salmalia malabarica* (DC.) Schott & Endl. Melet.; *Gossampinus malabarica* (DC.) Merr.; *G. rubra* Ham.

Family: Malvaceae.

Vern. : Simbal.

English, Sanskrit & Regional Names:

Eng. : Cotton Tree, Red Silk Cotton Tree;
Sans. : Bahuvirya, Nirgandhapuspi;
B. : Roktosimul,Tula;
Bo. : Saer, Shembal;
G. : Ratoshemalo, Shimul;
H. : Kantishembal, Simal;

Mal. : Mullilavu, Unnamurrika;
Mar. : Kanterisamar, Shevari;
Ori. : Buro, Mochorso;
Tam. : Ilavam, Pougar, Selavagu;
Tel. : Kondaburaga, Pinnaburaga.

Distribution : India, Bhutan, S. China. S.E. Asia to 1,400 m.

Description : Tall trees with prickly stems. Leaflets 5-7, glabrous shortly-stalked. Flowers whitish red. Capsules ovoid, downy.

Reproductive Cycle : February-May.

Habitat Ecology : Common in lower valleys and sub-Himalayan regions.

Material Examined : HPU-EL-SKS-5568; Chauki (900 m), Malahu (1,050 m), Sungal (1,050 m), Thandol (950 m), Tikkar (910 m).

Parts Used : Whole Plant. Leaves. Flowers. Fruits. Gum. Thorns. Wood. Bark. Seeds.

Folk Uses : Plant considered '**abode**' of evil spirits. Flower buds and pods cooked as **vegetable**. Gum from bark **edible** and 1-teaspoonful of it taken with water for 2-3 times a day to check **dysentery**. Mucous oozing out after soaking bark in water used for **smoothening mud floor**. Poultice of thorns good for **suppuration** of **sores**. Epidermal fibres from seed coat mixed with butter given to check **dysentery** in livestock. Cotton used for stuffing **pillows** and **mattresses**. Wood good for **house construction**. Bark decoction (1-teaspoonful thrice a day for one week) taken to treat **cough**.

Active Constituents : Hydrolysis of *gum* yields arabinose, galactose, galacturonic acid, catechu-tannic acid and traces of rhamnose.

Biological Activity : Antiviral, hypoglycaemic and oxytocic activities found (+)ve.

Uses in Literature : Used so far as an alterative, aphrodisiac, astringent, contraceptive, cooling tonic, filling material for jackets, cushions and pillows, demulcent, diuretic, edible (calyces, leaves, buds root) emetic, expectorant, fibre, firewood, fodder, gynaecological disorders, haemostatic, laxative, restorative in excessive loss of vital fluid, stimulant, stomachic, styptic, tonic, tubercular, upholstery; and for abdominal pain, acne pimples, anaemia, asthma, atrophy, bacillary dysentery, bleeding gums, blisters, boils, bone fracture, calculus infections, canoes, catarrhal affections, chickenpox, cholera, chronic cystitis, coffins, cordage, cough, delivery (vet.), diarrhoea (vet.), dislocation of bone (vet.), doors, emaciation, enlarged spleen, facilitating delivery, fever, fishing float, fistula, flatulence (vet.), glandular swellings, gonorrhoea, haemorrhoids, hematuria, impotency, indigestion (vet.), inflammation, intercostal neuralgia, leprosy and rinder pest diseases, leucorrhoea, lining of wells, mattresses, menorrhagia, night-pollution, oilseeds, packing fragile material, pain in legs, painful menstruations, pickling, planking, pleurisy, pneumonia, relieving giddiness, rheumatism, rutting (vet.), scab boards, scabies, scorpion bite, shingles, skin disorders, sleeping bags, smallpox, snakebite, sore in mouth, spermatorrhoea, spleen complaints, styptic, syphilis, tea boxes, tonsillitis, toothache, toys, ulceration of bladder, ulcers, urinary troubles, spermatorrhoea, vitality, water troughs, and window-shutters (Agarwal, 2003; Ambasta, 1986; Aminuddin & Girach, 1991; Arora & Pandey, 1996; Aswal, 1996; Bajpayee & Dixit, 1996; Banerjee, 1996; Brahmam & Saxena, 1990; Chauhan, 1999; Chun-Lin & Jieru, 1995; Das & Sharma, 2003; Deshmukh & Rothe, 2003; Diwakar & Ansari, 2003; Dwivedi, 2003; Girach, 1992; Girach & Aminuddin, 1995; Gogoi & Borthakur, 1991; Gopal & Chandra, 2003; Harborne & Baxter, 2001; Katewa & Arora, 1997; Katewa *et al.*, 2001; Kshirsagar *et al.*, 2003; Kumar *et al.*, 2003; Lalramnghinglova, 1996; Maliya, 2004; Manandhar 1989a, b; Mishra *et al.*, 1996b; Mukherjee & Namhata, 1990; Nadanakunjidam & Abirai, 2003; Panthi & Chaudhary, 2003; Patil, 2004; Rai & Upadhyay, 1996; Rama Rao & Henry, 1996; Ranjan, 2003; Rao, 1996; Roy *et al.*, 1998; Sahni, 1998; Saini, 1996b; Sarangi & Sahu, 2005; Sen & Behera, 2003; Sharma, 2003; Shome *et al.*, 1996; Sikarwar, 1996; Sikarwar & Kaushik, 1992; Singh, 1996b, 2001, 2003; Singh & Kumar, 2000a; Singh & Maheshwari, 1992; Singh & Pandey, 1998; Singh & Zaheer, 1996; Sinha & Sinha, 2001; Siwakoti & Varma, 1996; Siwakoti & Siwakoti, 2000; Upadhyay & Chauhan, 2003; Watt, 1972).

Brassica nigra **(L.) Koch** (Pl. 7E)

Syn.: *Sinapis erysimoides* Roxb.; *S. nigra* L.

Family: Brassicaceae.

Vern. : Kali-rai.

English, Sanskrit & Regional Names:

Eng. : Black Mustard, True Mustard

Sans. : Asuri, Jwalanti, Krimika, Madhurika, Raji

H. : Aslrai, Banarasi-rai, Ghorrai, Jagrai, Kalirai, Lahi, Makrarai, Rai, Taramira,Tira.

Distribution : Cultivated in various parts of India; chiefly on the hills.

Description : Herbaceous, biennials with tap root; cylindrical stem and sessile leaves. Inflorescence corymbose raceme. Flowers complete, pedicellate. Siliqua having numerous, minute, exalbuminous seeds.

Reproductive Cycle : October-March.

Habitat Ecology : Cultivated fields, but often met as an escape in wasteplaces.

Material Examined : HPU-EL-SKS-5572; Banuri (1,150 m), Chachian (1,300 m), Dehan (1,130 m), Ladoh (980 m), Panapari (1,100 m).

Parts Used : Leaves. Seeds.

Folk Uses : Leaves cooked as '**patrodas**'. Seeds yield **cooking oil**. Leaves used for **flavouring** 'curries'.

Uses in Literature : So far considered anthelmintic, digestive condiment, laxative, stimulant, stomachic; and for appetite, boils, nose, ear, eye troubles, and dyspepsia, enlargement of spleen, fever; increasing bile, rheumatism, skin diseases, repelling flies and mosquitoes, snakebite, toothache (Ambasta, 1986; Caius, 2003; Kirtikar & Basu, 1984).

Brassica tournefortii **Gouan.** (Pl. 7F)

Syn.: *B. stocksii* Hook. f.

Family: Brassicaceae.

Vern. : Rai.

Distribution : Upper Gangetic valley.

Description : Erect hispid herbs upto 30 cm tall with radical leaves petioled. Flowers pale yellow. Pods compressed.

Reproductive Cycle : February - April.

Habitat Ecology : Moist places.

Material Examined : HPU-EL-SKS-5582; Bindraban (1,270 m), Dadh (1,100 m), Darugnu (1,400 m), Pahra (950 m), Sungal (1,050 m).

Parts Used : Leaves. Seeds.

Folk Uses : Leaves consumed as **vegetable** and **chutney**. Seed oil used for **cooking**, **body massage**, and as **blood purifier**. Seeds along with *Brassica compestris* var. *toria* and 'Kakkad' (*Muntiacus muntjak*) skin made to paste and given orally to check **hair falling** in young ones of livestock.

Active Constituents : *Seed* cake contains N, 4.9%, P_2O_5, 2.5%, K_2O, 1.5%.

Uses in Literature : Seed oil edible and seed cake as cattle feed (Anonymous, 1948).

Butea monosperma Kuntze (Pl. 8A)

Syn.: *B. frondosa* Koen. *ex* Roxb.; *Erythrina monosperma* Lam.; *Plaso monosperma* Kuntze

Family: Fabaceae.

Vern. : Palah, Palash.

English, Sanskrit & Regional Names:

Eng.	: Bastard Teak, Bal Kino, Flame of the Forest, Butea Gum
Sans.	: Bijasneha, Ksharashreshtha, Vatapotha
B.	: Palas, Polasi
Bo.	: Khakra, Palasa
G.	: Khakra, Palasso
H.	: Chichra, Kankrei, Palas
Kan.	: Muttuga, Muttuga mara, Thoras
Mal.	: Brahmavriksham, Palasinjamata, Palasin-samatha
Mar.	: Kakradha, Phalas
Ori.	: Kinjuko, Polaso
Tam.	: Kutumurukku
Tel.	: Tellamoduga, Vatapodhamu
U.P.	: Dhak.

Distribution : Common throughout greater parts of India:1,300-1,400 m.

Description : Trees with pinnately trifoliate leaves; brown silky pubescence on shoots, and bright, orange red coloured petals. Pods argenteo-canescent which narrowed suddenly into a stalk.

Reproductive Cycle : March-May.

Habitat Ecology : Common throughout drier areas. Borders of agricultural fields. Forests.

Material Examined : HPU-EL-SKS-5581; Balah (900 m), Bhauda (1,050 m), Jasun (1,000 m), Panapari (1,100 m), Tikkar (910 m).

Parts Used : Whole plant. Leaves. Twigs. Flowers. Roots. Seeds. Bark. Wood.

Folk Uses : Whole plant considered **sacred**. Leaves used as **fodder**. Leaves and wood employed for performing various **religious** ceremonies. Twigs used for **scouring** teeth and **healthy** gums. Flower decoction (100 ml daily for one month) taken for **urinary troubles** and **piles;** also used for washing **pimples**. Poultice of petals considered good for **skin ailments** and skin glow. Seeds **eaten** raw and also considered **anthelmintic**. Toasted flowers tied around belly to regulate **menstrual cycle** and associated **pain**. Root-juice used as eye drops to cure **opthalmic** ailments.

Active Constituents : These are butrin, isobutrin-5, coreopsin, isocoreopsi, sulphurein, monospermoside andho isomonospermoside (*flower*), tetramers of leucocyanidin (*bark*), a nitrogenous acidic compound along with palasonin; α-amyrin, β-sitosterol, its glucoside and sucrose and a new lactone-n-hemicosanoic acid-δ-lactone (*seeds*).

Biological Activity : Anthelmintic activity +ve due to palasonin.

Uses in Literature : Known to be used as abortifacient, alterative, anthelmintic (vet.), antiseptic, antispasmodic, aperient, aphrodisiac, astringent, beedi wrapper, colic, colouring material, contraceptive, cooling, depurative, diabetic, diuretic, dye (flowers), edible (fruits, flowers), febrifuge, fibre for paper making and cordage, fish poison, fodder, fuel, haemostatic, insecticidal, lac host, laxative, ornament (ear top), psychoactive plant, rubiefacient (vet.), sacred, sore throat, steriliser, stomachic (vet.), stomatitis, tonic, vermifuge (vet.); and for amenorrhoea, ascaris, asthma, bladder, blood pressure, boils, bone fracture, bowels, butea gums, catarrh, cold, controlling parasite (vet.), conspitation, cough, delirium, diarrhoea, diuresis, dog bite, snakebite, dysentery, dyspepsia, elephantiasis, eye complaints, flatulence, gastric problems, gout, gum, haemorrhage affections, heart-burn, hydrocele, inflammation in the inner side of the mouth, jaundice, kidney trouble, leprosy, leucorrhoea, meal plates for serving food, menorrhagia, menstrual disorders, newsprint manufacture, night blindness, offerings, orchitis, paintings as colouration (Mithila), phthisis, piles, pimples, promoting menstrual flow, muscle pain, pterygium, relaxed sore throat, removing dhobie's itch, saline soils, septic and congested throat, skin diseases, sore (vet.), structural works, sun stroke, swellings, tanning, tubercular, tumours, ulcers, urine complaints, water scoops, well curbs, worshipping, wounds, and dummy corpse (Ambasta, 1986; Arora & Pandey, 1996; Bajpayee & Dixit, 1996; Banerjee, 1996; Bhandary *et al.*, 1996; Bhatt *et al.*, 2001, 2003a, c; Bose, 1984; Chandra, 1997; Chevallier, 1996; Chopra *et al.*, 1956; D'Rozario *et al.*, 1999; Das & Agarwal, 1991; Dey, 1980; Diwakar & Ansari, 2003; Gill & Nyawuame, 1994; Girach & Aminuddin, 1995; Goel *et al.*, 1984; Jain, 1968, 1991, 1992, 1996, 2003a; Jain *et al.*, 1994; Jha & Basak, 1994; Jha & Kumar, 2003; Jha & Varma, 1996; Kalakoti, 2001; Katewa *et al.*, 2001; Khanna & Kumar, 2000; Khare, 2004; Kirtikar & Basu, 1984; Kothari & Moorthy, 1996; Kumar & Lal, 2003; Kumar *et al.*, 2003; Kurian, 1995; Lalramnghinglova, 1996; Masih, 2003; Meena *et al.*, 2003; Mohanty & Padhy, 1996; Nayak *et al.*, 2003; Pal *et al.*, 2000; Parrotta, 2001; Pawar & Patil, 2004; Rai & Upadhyay, 1996; Rama Rao & Henry, 1996; Ranjan, 1996; Rao, 1996; Roy *et al.*, 1998; Ravishankar & Henry, 1992; Reddy *et al.*, 1996; Sahni, 1998; Said, 1997; Saini, 1996a; Sarangi & Sahu, 2005; Satapathy &

Brahmam, 1996; Seetharam *et al.*, 1999; Shah & Singh, 1990; Sharma, 1996, 2003; Sikarwar & Kaushik, 1992; Singh 1991, 1999, 2000, 2003; Singh & Kumar, 2000a; Singh & Prakash, 1994; Singh & Zaheer, 1996; Sinha & Sinha, 2001; Sivakumar *et al.*, 2003; Siwakoti & Siwakoti, 2000; Siwakoti & Varma, 1996; Srivastava *et al.*, 2003a; Sood & Prakash, 2007; Sood & Thakur, 2004; Thomas & Britto, 2003; Upadhye & Kumbhojkar, 2003; Varma, 1997; Prakash & Mehrotra, 1987; Watt, 1972).

Caesalpinia decapetala (Roth) Alston (Pl. 8B)

Syn.: *C. crista* Thunb.; *C. sepiaria* Roxb.; *C. ferox* Hassk.; *C. japonica* Sieber & Zucc.; *C. bonduc* (L.) Roxb.; *C. bonducella* (L.) Flem.; *Reinchardia decaptela* Roth; *Bianacaea scadens* Tod.

Family: Fabaceae.

Vern. : Drogaadi, Taranghadi.

English, Sanskrit & Regional Names:

Eng.	: Bonduc:Nut, Fevernut, Indian Filbert, Niekar, Physic Nut, The Mysore Thorn
Sans.	: Kantakini, Putikaranja, Tinagachhika
B.	: Natakanja, Notukoranja
Bo.	: Sagurghota
G.	: Galga, Kachhki
H.	: Kanja, Kanthe, Karanja
Kan.	: Gajagakayi
Kash.	: Kando
Mal.	: Kazanchik-kuru
Mar.	: Kanchaki, Karbath
Ori.	: Kokokolenja
P.	: Chillur
Tam.	: Kachukkai, Shriyinda
Tel.	: Gacha, Sukajambuka
U.P.	: Kuchaitha, Karjee.

Distribution : India. Pakistan - Bhutan. China. Japan. S.E. Asia to 2,200 m.

Description : Climbing prickly shrubs having pinnate leaves; yellow coloured flowers and glabrous, flat pods.

Reproductive Cycle : April-July.

Habitat Ecology : Scrub forests. Ravines.

Material Examined : HPU-EL-SKS-5513; Bhaura (940 m), Bhuana (1,400 m), Chauki (900 m), Malahu (1,050 m), Tapa (945 m).

Parts Used : Leaves. Seeds.

Folk Uses : Powdered seeds as such or with curd applied against **burns**. Leaves employed as forage for goat with **cooling** effect.

Active Constituents : *Seeds* rich in α-caesalpin, β-caesalpin, γ-caesalpin, δ-caesalpin, aspartic acid (675.0 μg/g), arginine (224.0 μg/g), citrulline (3.64%), a new diterpine-ξ-caesalpin and bonducellin.

Biological Activity : Plant extr. (500 mg/kg p.o.) administered on days 1-8 of post-coitum exhibits significant contraceptive activity in female hamsters.

Uses in Literature : Reported earlier as an anthelmintic, antiperiodic (vet.), antipyretic, colic, emmenagogue, febrifuge, insecticidal (vet.), laxative, ornamental bracelets, pulmonary tubercular, soap nuts, symbolic (seeds strung to red silkworm as an abortion amulet collar), uterine tonic, yellow dye obtained from flowers; and for asthma, boils, burns, cholera (vet.), cosmetics, diarrhoea (vet.), discharges from ears, epilepsy, essential oil, expelling intestinal worms (vet.), fencing, fever (vet.), filariasis, general debility, hydrocele, indigestion and liver trouble, inflammation of umblicis (vet.), inflammed piles, inflammed swellings, lac, malarial fever, menstrual complaints, nervous complaints, remittent fever, removal of placenta (vet.), rheumatism (vet.), scrofula, skin diseases, snakebite, tanning, ulcer, and wounds (Ambasta, 1986; Gill & Nyawuame, 1994; Girach *et al.*, 1996, 1997; Henry *et al.*, 1996; Jain, 1991; Kothari & Moorthy, 1996; Ragupathy & Mahadevan, 1991; Rama Rao & Henry, 1996; Reddy *et al.*, 1996; Saini, 1996b; Singh & Maheshwari, 1992; Singh *et al.*, 1996b, 2003b; Watt, 1972).

Cajanus cajan (L.) Millsp. (Pl. 8C)

Syn.: *Cajanus indicus* Spreng.; *C. obcordifolia* Singh; *Cystisus cajan* L.

Family: Fabaceae.

Vern. : Dhinger.

English, Sanskrit & Regional Names:

Eng. : Pegeon-pea

Sans. : Adhaki, Tuvarica, Tuwari

B. : Arhar, Tur, Tuver

G. : Dangri, Turdal, Tuver

H. : Arhar, Tur, Tuver

Kan. : Thogari, Bele, Thuvarg
Mal. : Adhaki, Kaksi, Thuvara
Mar. : Arhar, Tur, Turi, Tuver
Ori. : Horodo, Kanetulo
P. : Arhar, Dinger, Tohar
Tam. : Adagam, Paruppu, Thovary, Tuvari
Tel. : Adhaki, Kandula, Kandulu.

Distribution : Throughout India.

Description : Erect hairy herbaceous annuals. Leaves trifoliate; leaflets lanceolate. Flowers yellow, densely hairy. Fruits (pods) 3-7 seeded; seeds red or pale yellow.

Reproductive Cycle : June-October.

Habitat Ecology : Cultivated.

Material Examined : HPU-EL-SKS-5744; Chhanchri (930 m), Dharaman (1,100 m), Jasun (1,000 m), Nadli (1,000 m), Sidhpur (1,100 m).

Parts Used : Leaves. Stem. Seeds.

Folk Uses : Leaves chewed to check **jaundice** and **mouth ulcers**; also fed to livestock to relieve **dysentery**. Leaf decoction (1 cup) prescribed to facilitate **delivery**. Seeds consumed as **pulses**. Stem used as **fuelwood**.

Active Constituents : Vitexin, salicylic acid, lacarol, longistylin A, longistylin C and B-sitosterol have been isolated.

Uses in Literature : So far used for checking secretion of milk, toothache, jaundice, smallpox, bad breath, chickenpox, jaundice, cholera, mouth ulcers, house fencing, easy delivery, liver problem, piles, making huts; and as fodder, green manure, and pain reliever (Ambasta, 1986; Bhatt *et al.*, 2003c; Borthakur, 1993; Borthakur *et al.*, 2004; Das & Saikia, 2001; Ghosh, 1997; Islam, 2003; Khanna, 2002; Kshirsagar, 2003; Kshirsagar *et al.*, 2003; Kumar *et al.*, 2003; Manandhar, 2003; Pandey & Verma, 2002; Ranjan, 2003; Rastogi & Mehrotra, 1999; Sikarwar, 2002).

***Callicarpa macrophylla* Vahl** (Pl. 8D)
Syn.: *C. incana* Roxb.; *C. tomentosa* Koenig *ex* Vahl; *C. roxburghii* Wall.
Family: Verbenaceae.

Vern. : Danther.

Regional Names:

Ass. : Priyangu
B. : Mathara, Mattranja
G. : Lata-priyangu
H. : Daya, Shiwali
Kan. : Gandhapriyangu, Priyangu
Mal. : Chimpompil
Mar. : Gavhala, Priyangu
Ori. : Priyangu
P. : Priyangu
Tam. : Gnazalpoo, Gnazhal
Tel. : Prenkanamu, Prkhanam
U.P. : Daia.

Distribution : Throughout India; ascending to 2,000 m.

Description : Shrubs having branches and inflorescence densely clothed with wool-like tomentum. Leaves crenate, shortly stalked. Flowers pink coloured, succulent, globose, in white coloured droop.

Reproductive Cycle : July-September.

Habitat Ecology : Commonly occurring in shady localities of ravines and scrub jungles.

Material Examined : HPU-EL-SKS-5474; Boda (1,000 m), Khaira (940 m), Malnu (1,000 m), Panapari (1,100 m), Tikkar (910 m).

Parts Used : Inflorescence. Leaves. Fruits.

Folk Uses : Leaf-paste applied for healing **wounds**. Ripe fruits **edible**. Gargle with decoction of its fruits good for **mouth ulcers**. Inflorescence forms one of the constituents of local wine, '**sur**'.

Active Constituents : Main ingredients of *leaves* are luteolin, apigenin, luteeolin-7-O-glucusonide and apigenin-7-O-glucuronide.

Uses in Literature : Used earlier as an astringent, blood purifier, edible (fruits), emetic, refrigerant; and for abdominal trouble, agricultural implements, boils, cuts, diarrhoea, dysentery, fever, gout, fever, headache, indigestion, intestinal infections, rashes on tongue, rheumatism, skin diseases, stomach disorders, throat, ulcers, and wound (Ambasta, 1986; Arora & Pandey, 1996; Chandra, 1997; Khare, 2004; Jain, 1991; Maliya, 2005; Manandhar, 1980a,b, 1987a,

1989b, 1990; Megoneitso & Rao, 1983; Negi & Gaur, 1991; Negi *et al.*, 1999; Rama Rao & Henry, 1996; Roy *et al.*, 1998; Samant *et al.*, 2001; Singh, 1991, 1999, 2000, 2001; Sood & Thakur, 2004; Watt, 1972).

Calotropis procera **R. Br.** (Pl. 8E)

Syn.: *C. hamiltonii* Wall.; *C. wallichii* Wight; *C. heterophylla* Wall.; *Asclepias procera* Willd.

Family: Asclepiadaceae.

Vern. : Aak, Ak.

Sanskrit & Regional Names:

Sans. : Adyitapushpika, Alarka, Arka, Bimbora, Dirghpuspa, Ganarupaka, Kashthila, Mandara, Shamblau, Sharakarapushpa, Shitakarka, Shuklaphala, Vasuka, Vritamallika

Ass. : Ag, Ak, Akada, Madar, Safedak

Mar. : Mandara

Ori. : Orko

P. : Ak, Shakarallighal, Shakarulusar

Tam. : Vellerukku.

Distribution : Throughout India, ascending to 1,350 m.

Description : Erect, white-downy shrubs with milky latex, pink coloured purple spotted flowers and thick, wrinkled follicles covered with woolly pubescence.

Reproductive Cycle : February-May.

Habitat Ecology : Wastelands. Ravines. Along rivers. Roadsides. Forest clearings.

Material Examined : HPU-EL-SKS-5631; Balah (900 m), Bhauda (1,050 m), Pahra (950 m), Rangru (910 m).

Parts Used : Aerial Plant Parts. Latex. Seeds.

Folk Uses : Aerial plant parts form one of the ingredients of '**Havan**'. Latex applied locally for **toothache**. Seed-oil massaged for **rheumatic** pains.

Active Constituents : Active principles isolated from *latex* are uzarigenin, syriogenin, proceroside, calotropin, calotoxin, uscharin, uscharidin, α- and β-amyrin, β-sitosterol, calotropagenin; uzarigenin and syriogenin.

Biological Activity : Calotropin exhibits digitalis like action on the heart.

Uses in Literature : Used earlier as an abortifacient, alterative, antidote in scorpion sting, antiseptic, bow strings, camel fodder, cardiac, diaphoretic, emetic, gun powder charcoal, infanticide, laxative, magico-therapeutic (put on forehead of women in labour to achieve easy delivery), masticatory, psychoactive plant, symbolic (abode of god and taboo to offer to godesses), tonic, toothbrush; and for asthma, boils, bronchitis, bronchitis (vet.), cholera, chronic rheumatism, cold, constitutional syphilis, corn, cuts (vet.), dental caries, diarrhoea, dropsy, dysentery, eczema, elephantiasis, epilepsy, expelling worms, fever, fibre, gastric troubles of children, guinea worm, idiopathic ulceration, insect bite, leprosy, malarial fever, menstrual disorders, mercurial cachexia, mouth diseases (vet.), neck injury (vet.), pain, piles, pyrexia (vet.), rheumatoid arthritis, ringworm, scorpion bite, skin diseases (vet.), smallpox, sores, spleen complaints, stomach disorders, swellings, syphilis, toothache, whooping cough, and wounds (vet.) (Ambasta, 1986; Bhattacharya, 1996; Datt, 1996; Huidrom, 1996; Jain, 1991, 1996; Jain *et al.*, 1994; Kapur *et al.*, 1996; Khanna *et al.*, 1996a; Maheshwari & Singh, 1984; Masih, 2003; Mishra *et al.*, 1997; Pandey & Verma, 2003; Saini, 1996a, b, Sharma & Borthakur, 2003; Sharma & Sood, 1997; Shome *et al.*, 1996; Sikarwar, 1996; Singh & Maheshwari, 1983; Srivastava *et al.*, 2003a-c; Watt, 1972).

Camellia sinensis (L.) Kuntze (Pl. 8F)

Syn.: *C. theifera* Griff.; *C. thea* Link; *Thea sinensis* L.; *T. assamica* Mast.

Family: Theaceae.

Vern. : Cha, Chai.

English & Regional Names:

Eng. : Assam Tea, China Tea, Indian Tea, Tea
Ass. : Hilkat
B. : Chha
Mal. : Chaya
Mar. : Chaha
Ori. : Cha
P. : Cha
Tam. : Karupputeyilai, Pachai Heyilai, Teyilai
Tel. : Nallateyaku, Teyaku, Tiyaku.

Distribution : Extensively cultivated in Darjeeling, Assam, in and around Palampur in W. Himalayas to 2,300 m.

Description : An evergreen shrub with white flowers and light green coloured pods.

Reproductive Cycle : June-September.

Habitat Ecology : Cultivated.

Material Examined : HPU-EL-SKS-5712; Bandla (1,280 m), Bhatu (1,040 m), Bhawarna (1,100 m), Bindraban (1,270 m), Saliana (1,000 m).

Part Used : Leaves.

Folk Uses : Tea leaves chewed for removing **foul smell** from mouth; its beverage considered **refreshing**, **stimulant** and **anti-hypertensive**. Also, these are used as a **colouring material**.

Active Constituents : *Leaves* possess a new proanthocyanidin gallate epigallocatechin (4β→8)-3-O-galloylepicatechin (I) along with two hydrolyzable tannine-1,4,6-tri-O-galloyl-β-D-glucose and 1-O-galloyol-4,6-(-) hexahydroxydiphenoyl-β-D-glucose, procya-nidins B-2, B-4, C-1, three new red pigments-epitheaflagallin, epitheaflagallin-3-O-gallate and theaflagallin together with theaflavin and its 3-O-gallate, 3′-O-gallate and 3,3′-di-O-gallate; a growth inhibitor identified as ABA; camellianin A and camellianin B, epigenin and triacontanol.

Biological Activity : Anticoagulant and antithrombic properties of green tea polysaccharides demonstrated *in vitro* and *in vivo* in mice, rabbit and guinea pig.

Uses in Literature : Useful so far in India as an abortifacient, antihypertensive, anti-inflammatory, astringent, as colouring material in Mithila paintings, diuretic, folk beverage, nervine tonic, psychoactive plant, sedative, stimulant; and for conjuctivitis, cuts and injuries, diarrhoea, dysentery, removing bad smell from mouth, and skin burns (Ambasta, 1986; Jain, 1991; Jain *et al.*, 1994; Jha & Basak, 1994; Lalramnghinglova, 1996; Shome *et al.*, 1996).

Cannabis sativa L. (Pl. 9A)

Syn.: *C. indica* ssp. *indica* var. *indica* (Lam.) Wehmer

Family: Urticaceae.

Vern. : Bhang.

English, Sanskrit & Regional Names:

Eng. : Hemp, Indian Hemp;

Sans. : Bhanga, Chapola, Ganja, Ganjika, Hursini, Indrasana, Jaya, Ujaya, Viyaya, Vrijpatta;

B. : Bhang, Ganja, Sidhi;

H. : Bhang, Charas, Ganja, Gur, Kinnab, Siddhi;

Kan. : Bhangi-gida;

Mal.	: Ginjilachilachi, Kanchavachetto;
Mar.	: Bhangachajhada;
Tam.	: Ganja-chedi, Kalpam, Korkkarmuli;
Tel.	: Ganjari-chettu.

Distribution : Throughout India.

Description : Aromatic pubescent annuals with palmately divided leaves and yellowish-green, unisexual flowers on separate plants. Achenes small, rounded.

Reproductive Cycle : June-September.

Habitat Ecology : Common in Wastelands.

Material Examined : HPU-EL-SKS-5458; Bhaura (940 m), Chornali (1,280 m), Ghugar (1,250 m), Malnu (1,000 m), Thakurdwara (1,150 m).

Parts Used : Leaves. Seeds.

Folk Uses : Roasted seeds (1 teaspoonful thrice a day) taken to relieve **diarrhoea** and **flatulance**. Leaf-juice applied for **healing cuts**, **wounds** and **bee-sting**. Handful of leaves in combination with *Boerhaavia diffusa* eaten as 'pakodas' to check **rheumatic pains**.

Active Constituents : Active constituents are glucocapparins (seeds), phthalic acid and l-stachydrine (*flowers, fruits*).

Biological Activity : Antibacterial, hypotensive and spasmolytic activity and effect on CNS found +ve.

Uses in Literature : Useful in India as an abortifacient, analgesic, anthelmintic, antispasmodic, appetizer, CNS stimulant, edible (roasted and fried seeds), laxative, narcotic, nervine stimulant and sleep pills; and for bowel complaints, bronchitis, cold, cough, convulsions, cramps, delaying ejaculation in human beings, delirium, dysentery, epilepsy, cuts, dyspepsia, ear complaints, eye diseases, gonorrhoea, headache, hydrocele, labour pains, paralysis of tongue, piles, rough clothes, skin diseases, skin eruptions, stomach and intestine disorders, sores, tetanus, wasp sting and wounds (Agarwal, 2003; Ambasta, 1986; Bose, 1984; Chandra, 1997; Chhetri & Shrestha, 2004; Chevallier, 1996; Chatterjee & Pakrashi, 1991; Chauhan, 1999; Chopra *et al.*, 1956; Das & Sharma, 2003; Dash *et al.*, 2003; Dey & Bahadur, 1973; Duke *et al.*, 2002; Ganai & Nawachoo, 2003; Harborne & Baxter, 2001; Jain, 1991; Kaul, 1997; Khanna & Kumar, 2000; Khare, 2004; Kirtikar & Basu, 1984; Kumar *et al.*, 2003; Negi *et al.*, 1999; Pandey & Joshi, 2001; Panthi & Chaudhary, 2003; Ranjan, 2003; Roy *et al.*, 1998; Saikia & Nath, 2003; Samant *et al.*, 1996, 2001; Sarin, 1990; Shah, 1997; Sharma & Rana, 2005; Singh, 2001; Singh & Kumar, 2000a; Sinha & Sinha, 2001; Sood & Thakur, 2004; Srivastava *et al.*, 2003b; Watt, 1972).

***Capillipedium assimile* (Steud.) A. Camus.** (Pl. 9B)
Syn.: *Andropogon assimile* Steud. *ex* Zoll.; *A. subrepens* Steud.
Family: Poaceae.

Vern. : Dholu.
Distribution : Common in India.
Description : A grass; leaves conspicuously narrowed at base. Inflorescence spike of spikelet. Lower glume flat. Panicle often open.
Reproductive Cycle : August-November.
Habitat Ecology : Grasslands. Field edges.
Material Examined : HPU-EL-SKS-5466; Boda (1,000 m), Lahla (930 m), Rangru (910 m), Sulah (950 m).
Parts Used : Aerial Plant Parts. Stem.
Folk Uses : Aerial plant parts considered an excellent fodder. Stems after removing leaves used as brooms, and stem-juice (1/2 teaspoonful regularly for one week) prescribed to check leucorrhoea.
Uses in Literature : Known to be a forage grass (Sharma & Rana, 2005).

***Capsicum annuum* L.** (Pl. 9C)
Syn.: *C. frutescens* Cl.
Family: Solanaceae.

Vern. : Mirach, Pipali.
English, Sanskrit and Regional Names:
Eng. : Red Pepper, Red Chili
Sans. : Marichi-phalam
B. : Gach-marich, Lal-marich, Lanka-marich
G. : Lal-mirich, Marchu
H. : Gachmirch, Lal-mirch, Marcha, Mattisa, Mirch, Wangru
Kan. : Menasina-kayi
Mar. : Mirsinga
P. : Gachmirch, Lal-mirch, Mattisa, Mirch
Tam. : Milagay, Mollaghai, Mollagu, Mulagay
Tel. : Merapu-kai, Mirapakaya.
Distribution : Throughout India.
Description : Annual herbs. Leaves ovate, elliptic. Flowers white. Berries red, many seeded.

Reproductive Cycle : May-September.

Habitat Ecology : Sunny places.

Material Examined : HPU-EL-SKS-5460; Chauki (900 m), Darugnu (1,400 m), Dehan (1,130 m), Saparu (1,250 m), Thandol (950 m).

Part Used : Fruits.

Folk Uses : Fruits eaten raw and **pickled**, commonly used as **condiment**, and also hung on door to ward off **evil spirits**. Fruit powder applied externally for **dog bite**.

Active Constituents : *Seeds* contain capsicidin and fruits contain capsaicin, and capsanthin carotenoid.

Biological Activity : Capsaicin caused transient rise in blood pressure, caused a sustained increase in tension of spiral strips of proximal and distal mesenteric arteries.

Uses in Literature : So far considered analgesic, expectorant, condiment, carminative, earache, magico-religious, pickles, stomachic, lumbago, neuralgia, heart trouble, stimulant, rubefacient, sore throat, tonsillitis, neuralgic affections, antidote to snakebite, cholera, hoarseness, dyspepsia, yellow fever, piles, against snakebite; enriches the blood, lesions inflammation and pain (Ambasta, 1986; Bennet *et al.*, 1991; Bhatt *et al.*, 2001; Chauhan, 1999; Das & Saikia, 2001; Gogoi & Borthakur, 2001; Kumar *et al.*, 2004; Pal & Paul, 2003; Reddy *et al.*, 2003a; Sen *et al.*, 2002; Singh & Pandey, 1998; Thakur, 2005; Thomas & Britto, 2003; Vishwanathan *et. al.*, 2001).

Carica papaya L. (Pl. 9D)

Family: Caricaceae.

Vern. : Kharbuja, Papita.

English, Sanskrit & Regional Names:

Eng. : Melon Tree, Papaw, Papaya, Paw-paw

Sans. : Chirbita, Eranda, Naikadala

B. : Papai

Bo. : Chibhado, Katha, Paputa

G. : Chibda, Eranda Kakdi, Thadchibadi

H. : Anda Kharbuja, Papita

Mal. : Kappalam, Nalikadala

Mar. : Papaya

Ori. : Omrytobhondo, Popoya

P. : Popai

Tam. : Pappli, Parangiyananakki

Tel. : Madananaba, Madhurnakamu.

Distribution : Cultivated mostly throughout India.

Description : Rapidly growing sub-herbaceous, branchless trees. Trunk succulent with milky sap. Leaves palmately lobed. Flowers creamy yellow.

Reproductive Cycle : Greater part of the year.

Habitat Ecology : Cultivated; often met as an escape.

Material Examined : HPU-EL-SKS-5444; Ghugar (1,250 m), Gopalpur (1,350 m), Nanaun (900 m), Sungal (1,050 m), Tikkar (910 m),

Parts Used : Milky Latex. Fruit.

Folk Uses : Ripe fruits **edible** and prescribed for **digestive** and **liver disorders**. Unripe ones cooked as a **vegetable** and '**pakodas**'. They are taken empty stomach to check **diabetes** and also fed to cattle for inducing **rumination**. Latex applied for **cutaneous** diseases.

Active Constituents : These are myristoleic, palmitoleic arachidic acids, pectin and free sugars (*fruits*).

Biological Activity : Carpaine exhibits antitumour activity *in vitro* against mouse lymphoid leukamia (1-210), lymphocytic leukamia (P-338) and *Ehlrich ascites* tumour cells, and also as anti-microbial against Gram-positive and Gram-negative bacteria.

Uses in Literature : Reported to be used as an abortifacient, adulterant, anthelmintic, antidote for cancer, blood anticoagulant, carminative, diuretic, emmenagogue, face creams, febrifuge, galactagogue, gastric, hair shampoos, purgative, stomachic, source of vitamins and papain, substitute of soap, tonic, vegetable; and for amoebic dysentery, asthma, beri-beri, bleeding piles, boils, bone fracture, candies, cardiac diseases, chest problems, cooling, corns and warts, removing freckles and other blemishes, cough, cuts and wounds, dental caries, diabetes, diptheria, dressing of foul wounds, dysentery, dyspepsia, eczema, excessive bleeding of the kidneys, expelling intestinal worms, flatulence, flavouring meat, jaundice, gonorrhoea, haemoptysis, heart depressions, hoarseness, inflammation, itching, jaundice, kidney stone, malaria, marmalades, pain in gums, pickling, promoting menstrual flow, reducing elephantoid growth, rheumatism, ringworm, skin diseases, soft drinks, spleenic and hepatic enlargements, stomach and bowel disease, stomach and duedonal ulcers, syphilis, tonsils, tooth and gumache, tuberculosis, ulcers, urinary troubles, warts, and wounds (Ambasta, 1986; Bhuyan, 2003; Borthakur, 1993; Borthakur *et al.*, 2004; Dagar & Dagar, 1996; Ghosh *et al.*, 1996; Gill *et al.*, 1993, 1997; Gogoi *et al.*, 2003b; Hosagoudar & Henry, 1996a; Jain, 1991; Jain *et al.*, 1997; Kayal *et al.*, 2003; Kumar *et al.*, 2003; Kurian, 1995; Lalramnghinglova, 1996; Nadanakunjidam & Abirai, 2003; Rajan *et al.*, 2003;

Rama Rao & Henry, 1996; Ranjan, 2003; Rao *et al.*, 1996; Saini, 1996b; Sharma, 2003; Singh & Zaheer, 1996; Siwakoti & Varma, 1996; Thomas & Britto, 2003; Watt, 1972).

Carissa congesta Wight (Pl. 9E)

Syn.: *C. carandus* L.

Family: Apocynaceae.

Vern. : Garna, Garnae.

English, Sanskrit & Regional Names:

Eng.	: Bal Currents
Sans.	: Bokkarambuka, Jalipushpa, Krishna Pakaphala, Vanalaya
B.	: Karamcha, Kurumia
Bo.	: Hartundi, Karinda
G.	: Karamadan, Karamarda
H.	: Karaunda, Karrona
Kan.	: Karekayi
Mal.	: Kalavu, Karakka
Mar.	: Boronda, Haradundi, Karavandi
Ori.	: Khyirokoli, Korondo
Tam.	: Kalakkay, Perungala
Tel.	: Paddavaka, Peddakalive, Vaka.

Distribution : Throughout India, ascending to 1,750 m.

Description : Erect spinous shrubs with leathery leaves. Flowers white. Berries nearly black when ripe.

Reproductive Cycle : April-July.

Habitat Ecology : Open slopes.

Material Examined : HPU-EL-SKS-5523; Balah (900 m), Khaira (940 m), Makol (900 m), Panapari (1,100 m), Tikkar (910 m).

Parts Used : Whole Plant. Leaves. Stem. Fruits. Wood.

Folk Uses : Plant employed extensively for **fencing** grasslands and agricultural fields. Its leaves considered good **fodder** for sheep and goats. Fruits **edible** and regarded good for **anaemia**, and as **blood purifier**. Wood used as fuel and for making **coal, combs**; also its ash applied externally on joints to cure **pain**. Poultice of tender stem good for **skin diseases**.

Active Constituents : A new triterpine carissol isolated from *fruits* along with glucose, galactose, serine, glutamine, alamine, valine, phenylalanine, glycine and carissic acid.

Biological Activity : Hypotensive effect and cardiotonic activity (+)ve.

Uses in Literature : So far used as an anthelmintic, antiscorbutic, astringent, contraceptive, edible, expectorant, laxative, antidote for snakebite, stomachic; and for anasarca, bite of rabbit dog, blood dysentery (vet.), checking bleeding of internal organs, chutney, combs, convulsions, cooling in bilious complaints, curry, diarrhoea (vet.), dropsy, dyeing and tanning, epilepsy, fencing, fever, fodder, indigestion, itching, jellies, pickling, madness, postnatal care, puddings, remittent fevers, rheumatism, animal sores, scurvy, skin infections, sore (vet.), spoons, stimulating appetite, strengthening and giving tone to the stomach, tarts and puddings, tussar silkworms and wounds (vet.) (Ambasta, 1986; Arora & Pandey, 1996; Bennet *et al.*, 1991; Bhandary *et al.*, 1996; Chandra, 1997; Chopra *et al.*, 1956; Das & Agarwal, 1991; Diwakar & Ansari, 2003; Jain, 1991; Khanna *et al.*, 1996a; Kirtikar & Basu, 1984; Kurian, 1995; Mitaliya & Bhatt, 2003; Murthy *et al.*, 2003; Prajapati *et al.*, 2006; Roy *et al.*, 1998; Sahu, 1996; Samant *et al.*, 2001; Sharma & Rana, 2005; Singh & Kumar, 2000a; Singh & Prakash, 1996; Sood & Thakur, 2004; Sood & Kaushal, 2008; Srivastava *et al.*, 2003a; Watt, 1972).

Carya illinoensis **(Wangenh.) Koch** (Pl. 9F).

Syn.: *C. pecan* Engl. & Graebn.

Family: Juglandaceae.

Vern. : Kagzi khod.

English Names:

Eng. : Hickory Nut, Pecan Nut Tree.

Distribution : Native of South-East USA; introduced in India.

Description : A medium-sized tree. Leaves lanceolate with serrated margin. Drupes ovoid with thick, fleshy rind enclosing a woody, wrinkled two-valved nut.

Reproductive Cycle : May-October.

Habitat Ecology : Cultivated, often an escape.

Material Examined : HPU-EL-SKS-5602; Bhagpur (1,600 m), Chornali (1,280 m), Jiun (1,550 m), Lolani (1,600 m), Molichak (1,020 m).

Parts Used : Fruit. Stem-bark. Leaves.

Folk Uses : Fruits **edible**, also 1-2 fruits recommended daily for one month to check **bed-wetting** in children. A **black dye** from its rind useful for colouring clothes. Bark employed for **scouring** teeth; also used as a **hair dye** after boiling it in mustard oil along with fruits of *Terminalia chebula*. Leaf paste applied for **sore toes**.

Active Constituents : Quercitin-3-glucoside, 3-galactoside, 3- rhamnoside and 3-arabinoside isolated from *leaves*.

***Caryopteris bicolor* (Hardw.) Mabb.** (Pl. 10A).

Syn.: *C. odorata* (D. Don) B. L. Rob.; *C. wallichiana* Schauer; *Volkameria odorata* Roxb.; *Clerodendrum odoratum* Don

Family: Verbenaceae.

Vern. : Ban-dhavein.

Distribution : Sub-Tropical Himalayas: 300-2,100 m.

Description : Tall glabrous shrubs having shortly-stalked, toothed lanceolate leaves and white, fragrant flowers tinged with blue. Fruits dark blue.

Reproductive Cycle : April-August.

Habitat Ecology : Open slopes. Foothills.

Material Examined : HPU-EL-SKS-5613; Bhuana (1,400 m), Darugnu (1,400 m), Jasun (1,000 m), Rajeher (1,050 m), Tikkar (910 m).

Parts Used : Stem. Flowers.

Folk Uses : Flowers regarded as a source of **nectar** for rearing honey-bees. Stem used as **fuelwood**.

Uses in Literature : Reported in India for walking sticks (Ambasta, 1986).

***Cassia fistula* L.** (Pl. 10B).

Syn.: *C. rhombifolia* Roxb; *Cathartocarpus fistula* Pers.

Family: Fabaceae.

Vern. : Alia, Amaltas.

English, Sanskrit & Regional Names:

Eng. : Drum Stick, Golden Shower, Indian Laburnum, Purging Cassia

Sans. : Aragvadha, Chaturangula, Kritamalluka

Ass. : Soanru

B. : Amultas, Sonali

G. : Balla, Garmala

H. : Amaltas, Bandarlauri

Kan. : Kakke

Mal. : Saturangulam, Svarnnakam
Mar. : Bhawabaya, Chimkani
Ori. : Soturongulo, Sunari
P. : Alash, Karangal
Tam. : Arogaram, Irayavinnadagam
Tel. : Aragvadhamu, Sampakamu.

Distribution : Common in Himalayas upto 1,400 m.

Description : Small trees. Leaves petiolate, stipulate with entire margins; leaflets in 3-5 pairs. Flowers yellow, in racemose. Pods long, cylindrical. Seeds flat embedded in soft pulp.

Reproductive Cycle : March-July.

Habitat Ecology : Forests. Shrubberies. Open hill sides.

Material Examined : HPU-EL-SKS-5411; Chachian (1,300 m), Dharaman (1,100 m), Holta (1,200 m), Malahu (1,050 m), Tapa (945 m).

Parts Used : Whole Plant. Leaves. Flowers. Wood.

Folk Uses : Plant considered **sacred**; also grown in courtyards to ward off **evil spirits**. Paste from tender leaves applied for healing **wounds, sore toes** and **skin ailments**. Flower decoction (1 teaspoonful) taken daily for **constipation**. Wood used as **fuel** and for **agricultural implements**.

Active Constituents : Various constituents are fistucacidin (*bark*), leucoanthocyanidin- 5-4′-dihydroxy flavon-3, 4-diol (I) (*sap wood*), rhein, glucose, sucrose and fructose (*fruit pulp*), galactomannan (seed), rhein glucoside, fistulic acid (*pods*), vernolic (6.1), sterculic (2.0) and malvalic (1.5%) acids (*seed oil*), seven new biflavinoids, β-sitosterol (I-VII) and two triflavonoids (VIII-IX), (-) epicatechin, procyanidin B-2, (-) epiafzelechin and its 3-O-glucoside (*leaves*).

Biological Activity : Aq. extr. of fruit pulp exhibits antibacterial activity.

Uses in Literature : Recorded earlier as an antibacterial (vet.), antinarcotic to revive nervous system paralysed under the influence of liquor, antiseptic, blood purifier, colic, cooling agent, emetic, febrifuge, insecticide (vet.), laxative (vet.), mild cathartic, ornamental, psychoactive plant, purgative (vet.), symbolic abode of god, tonic (roots), vegetable (buds, leaves & flowers), vermicide; and for abdominal pain, amenorrhoea, anthrax (vet.), antifertilty, asthma, biliousness, black water fever, blindness, body swellings, boils, breast diseases, burns, cancer, cardiac diseases, carts, cataract, chest infection, chest pain, cold (vet.), confectionery, cough, diabetes, diarrhoea (vet.), dog bite, dropsical swellings, dyeing, dysentery, dyspepsia, dysuria, eczema, epilepsy, expelling hookworms, flatulence, flavouring coffee and tobacco, gout, heart diseases, house construction, intestinal worms (vet.), jaundice, leprosy, liver complaints, malarial fever, paralysis, pimples, ploughs, pustules, rheumatic pains, ringworm, scorpion bite, skin diseases, snakebite, stomach disease (vet.), swelling, swelling in throat (vet.), syphilis, tanning, thoracic obstructions, tongue sore (vet.), tool handles and wheels, toothache, urinary troubles, whooping cough, and wounds (vet.) (Agarwal, 2003; Ambasta, 1986; Arora & Pandey, 1996; Balasubramaniam & Prasad, 1996; Bhakta *et al.*, 1997; Bhandary & Chandrashekar, 2002; Bhandary *et al.*, 1996; Bhattacharya *et al.*, 2005; Bondya & Sharma, 2004; Bose, 1984; Chandra, 1997; Chauhan, 1999; Das & Agarwal, 1991; Dey & Bahadur, 1973; Diwakar & Ansari, 2003; Dixit & Pal, 1996; Dwivedi, 2003; Gill & Nyawuame, 1994; Girach & Aminuddin, 1995; Gogoi & Borthakur, 1991; Gopal & Chandra, 2003; Goud & Pullaiah, 1996b; Jain, 1968, 1991, 1996; Jain *et al.*, 1994; Joshi & Joshi, 2005; Khare, 2004; Kirtikar & Basu, 1984; Kothari & Moorthy, 1996; Kumar, 1996, 2003; Kumar & Lal, 2003; Kumar *et al.*, 2003; Kurian, 1995; Lalramnghinglova, 1996; Mahato & Chaudhary, 2005; Manandhar, 1989a; Masih, 2003; Meena *et al.*, 2003; Molla & Roy, 1996; Nadanakunjidam & Abirai, 2003; Nayak *et al.*, 2003; Pal *et al.*, 2000; Panthi & Chaudhary, 2003; Patole & Jain, 2002; Prajapati *et al.*, 2006; Pullaiah & Kumar, 1996; Rai & Upadhyay, 1996; Rajendran *et al.*, 2003; Rama Rao & Henry, 1996; Ranjan, 2003; Roy *et al.*, 1998; Sahni, 1998; Sahoo & Mudgal, 1995; Said, 1997; Saini, 1996a; Samant *et al.*, 2001; Sanyal, 1994; Sarin, 1996; Sharma, 1996, 2003; Sharma & Sood, 1997; Sikarwar, 1996; Sikarwar & Kaushik, 1992; Singh, 1991, 1999, 2000, 2003; Singh & Kumar, 2000a; Sinha & Sinha, 2001; Sivakumar *et al.*, 2003; Siwakoti & Siwakoti, 2000; Siwakoti & Varma, 1996; Srivastava *et al.*, 2003b; Thomas & Britto, 2003; Tiwari & Tiwari, 1996; Tosh, 1996; Watt, 1972).

***Cassia floribunda* Cav.** (Pl. 10C).
Syn.: *C. laevigata* Willd.
Family: Fabaceae.

Vern.	: Bada-rellu.
Distribution	: Throughout India—naturalised in Himalayas to 2,200 m.
Description	: Erect shrubs having ovate-elliptic, long-pointed leaflets. Flowers yellow and in cylindric pods with a thick stripe.
Reproductive Cycle	: June-December.
Habitat Ecology	: Scrub forests. Wastegrounds.
Material Examined	: HPU-EL-SKS-5420; Bhatu (1,040 m), Dadh (1,100 m), Gopalpur (1,350 m), Paror (940 m), Sungal (1,050 m).
Parts Used	: Leaves. Fruits.

Folk Uses : Pods consumed as a **vegetable** and **purgative**. Poultice of leaf applied to relieve **sprain**.

Active Constituents : Main constituents are chrysophenol, physcion and quercetin-3,7-dirhmnopyranoside (*seeds*) and floribundone 1 and floribundone along with N1, N8-di-benzoyl spermidine and physicion (*leaves*).

Uses in Literature : Known for its edible pods as a vegetable, substitute of coffee & tea, (roasted seeds); and for sprain with leaf paste of *Vitex negundo* (Chandra, 1997; Singh & Kumar, 2000a; Singh *et al*., 1999).

Cassia obtusifolia L. (Pl. 10D)

Syn.: *C. foetida* Salisb.; *C. gallinaria* Collad.; *C. numilis* Collad.; *C. tora* L.; *C. toroides* Roxb.

Family: Fabaceae.

Vern. : Aellu, Rellu.

English, Sanskrit and Regional Names:

Eng. : The Foetid Cassia

Sans. : Dadamardan, Dadamari, Prabunatha, Prapunnala, Prinsnaparni, Sakaramardaka, Taga, Uranakkhyaka, Vimardaka, Vyavartaka

B. : Chakunda, Panwar

Bo. : Kovariya, Kowaria, Tankala

G. : Kawaria, Kovariya

H. : Chakunda, Panevar

Mar. : Takala, Takla, Tanku, Tarota

P. : Chakunda, Pawar, Pawas, Panwar

Tam. : Toratah, Ushit, Tagarai

Tel. : Tagarisha-chettu.

Distribution : Throughout India, ascending to 1,750 m.

Description : Annual herbs having long leaves, grooved rachis; and 3-pairs of opposite, oblong-lanceolate leaflets. Flowers subsessile, pale-yellow. Pods subteragonous, obliquely septate. Seeds rhombohedral.

Reproductive Cycle : July-December.

Habitat Ecology : Barren areas, shady-slopes.

Material Examined : HPU-EL-SKS-5514; Balah (900 m), Chhanchri (930 m), Jasun (1,000 m), Nadli (1,000 m), Tikkar (910 m).

Parts Used : Aerial parts. Seeds.

Folk Uses : Aerial parts used as **bedding** for cattle. '**Chutney**' of seeds considered **delicious**; also prescribed to treat

cough. One teaspoonful of seed powder taken every morning to cure **diabetes**.

Active Constituents : Obtusin, chrysoobtucin, aurantioobtusin (*seeds*).

Biological Activity : Chrysophanic acid-9-anthrone showed fungicidal activity.

Uses in Literature : Known to be used for cardiac disorders, dysentery, eye ailments, intestinal troubles, itch, liver complaints, ringworm, skin diseases, wounds; and as anthelmintic, antidote to snakebite, antiperiodic, aperient, colic, dye, edible, galactagouge, laxative, stomachic (Ambasta, 1986; Aminuddin & Girach, 1991; Bennet *et al.*, 1991; Bhandary & Chandrashekar, 2002; Bhatt *et al.*, 1999, 2003a; Bhattacharya, 1996; Bhattacharya, *et al.*, 2005; Bondya *et al.*, 2006; Bose, 1984; Chandra, 1997; Chauhan, 1999; Chopra *et al.*, 1956; Devi, 2003; Dey & Bahadur, 1973; Dwivedi, 2003; Gill & Nyawuame, 1994; Girach, 1992; Gogoi & Das, 2003; Gogoi *et al.*, 2003a; Jain *et al.*, 1991; Kalakoti, 2001; Katewa & Arora, 1997; Katewa *et al.*, 2000; Kirtikar & Basu, 1984; Kshirsagar *et al.*, 2003; Kumar *et al.*, 2003; Kurian, 1995; Lalramnghinglova, 2002; Manandhar, 2003; Meena *et al.*, 2003; Murthy *et al.*, 2003; Nath & Maiti, 2003; Nayak *et al.*, 2003; Pal *et al.*, 2000; Patole & Jain, 2002; Prajapati *et al.*, 2006; Punjani, 2002; Radhakrishnan *et al.*, 1996; Ramdas *et al.*, 2000; Ranjan, 1996, 2003; Roy *et al.*, 1998; Saklani & Jain, 1994; Said, 1997; Saini, 1996b; Samant *et al.*, 2001; Sanyal, 1994; Sarin, 1996; Sharma, 2003; Shukla *et al.*, 2003; Singh, 1996b, 2000; Singh & Pandey, 1998; Singh & Singh, 2003; Sinha, 1996; Sinha & Sinha, 2001; Srivastava *et al.*, 2003c; Swami & Gupta, 1996; Tosh, 1996; Upadhyay & Singh, 2005).

***Cassia occidentalis* L.** (Pl. 10E)

Syn.: *Senna occidentalis* Roxb.; *C. foetida* Pers.; *C. sophera* Wall.

Family: Fabaceae.

Vern. : Chhota-rellu.

English, Sanskrit & Regional Names:

Eng. : Fetid Cassia, Negro Coffee, Rubbish Cassia, Stinking weed
Sans. : Arimard, Kasari, Kasamara
B. : Kalkashunda
Bo. : Hikal
H. : Barikasondi, Chakunda, Kasunda
Mal. : Karintakara
Ori. : Kasundri
Tam. : Nattandagarai, Peyavirai
Tel. : Kasindha
Urdu : Kasonji.

Distribution : Common throughout India, naturalized.

Description : Undershrubs, 60-150 cm tall. Branches purplish. Flowers yellow, in short peduncled racemes. Pods recurved, tortulose, 15-30 seeded.

Reproductive Cycle : May-September.

Habitat Ecology : Wastelands. Edges of cultivated areas.

Material Examined : HPU-EL-SKS-5471; Balah (900 m), Bodhal (970 m), Ghugar (1,250 m), Jasun (1,000 m), Tikkar (910 m).

Parts Used : Aerial Plant Parts. Leaves. Pods.

Folk Uses : Decoction of aerial plant parts taken as **expectorant**, **diuretic** and **purgative**. Tender pods consumed as **vegetable**, and also **pickled**. Leaves used for **scouring** teeth and healing **mouth ulcers**.

Active Constituents : Possesses 1,8-dihydroxy- and oxymethyl-anthraquinones, emodol, luteolin, dimethylether-rhamnoside, quercetin and polysaccharides (*plant*); c-flavonosides of apigenin (*pericarp*); emodin, phyocion and its β-glucopyranoside (*flowers*); aloe-emodin, a bianthraquinone, chrysophenol, dianthronic heterosides, Jaceidin-7-rhamnoside, matteucino 1-7-rhamnoside rhein and potassium chelidonate (*leaves*); and chrysolein, linoleic, linolenic and oleic acids, campesterol, 8-carbomethoxy - 1,7-dihydroxy – 3 - methylxanthone, 1, 8 – dihydroxy – 2-methylanthraquinone, 1, 4, 5 - trihydroxy anthroquinone derivatives, tannic acid and toxalbumin (*seeds*).

Uses in Literature : Known earlier as abortifacient, alexiteric, anti-periodic flowers decoction, aphrodisiac, diuretic, edible (pods, roasted seeds, leaves as a vegetable), febrifuge, hepatic stimulant, indigestion (vet.), laxative; and good for anaemia, asthma, boils, burns, cold, cough, diabetes, diarrhoea, dysentery, dog-bite, eczema, elephantiasis, epilepsy, fever, heart disorders, hydrophobia, leucorrhoea, liver, mental disorders, paralysis, saliva secretion, scabies, stomachache, syphilis, wheezing, whooping headache, and kidney, skin, stomach, and urinary disorders (Ambasta, 1986; Balasubramaniam & Prasad, 1996; Bhatt *et al.*, 2003a; Chandra, 1997; Chauhan, 1999; Das & Agarwal, 1991; Dey & Bahadur, 1973; Gill & Nyawuame, 1994; Gill *et al.*, 1993; Gogoi & Das, 2003; Gogoi *et al.*, 2003b; Hosagoudar & Henry, 1996b, c; Jamir & Rao, 1990; Jha & Kumar, 2003; Jha & Varma, 1996; Katewa & Arora, 1997; Khanna, 2002; Kirtikar & Basu, 1984; Manandhar, 2003; Mohanty & Padhy, 1996; Murthy *et al.*, 2003; Nayak *et al.*, 2003; Panthi & Chaudhary, 2003; Patole & Jain, 2002; Prajapati *et al.*, 2006; Ranjan, 2000; Reddy *et al.*, 2003b; Samant *et al.*, 2001; Samwatsar & Diwanji, 1996a; Shah, 2005; Sharma, 2003; Singh, 2000; Singh & Kumar, 2000a; Sinha & Sinha, 2001; Sivakumar *et al.*,

2003; Sood & Thakur, 2004; Subramani & Goraya, 2003; Swami & Gupta, 1996; Upadhyay & Chauhan, 2000; Vishwanathan, 1997).

***Catharanthus roseus* G. Don** (Pl. 10F)

Syn.: *Vinca rosea* L.; *Lochnera rosea* (L.) Reichb.

Family: Apocynaceae.

Vern. : Sadabahar.

English & Regional Names:

Eng. : The Periwinkle
B. : Nayantara
H. : Sadabahar
Mal. : Ushamalari
Mar. : Sad-Phul
Ori. : Ainskati
Tel. : Billaganneru.

Distribution : Cultivated throughout India, also met as an escape.

Description : Perennial shrubs with simple, opposite, glossy, obovate leaves and axillary flowers pink in cymose clusters. Follicles many-seeded.

Reproductive Cycle : Throughout the year.

Habitat Ecology : Shady Places.

Material Examined : HPU-EL-SKS-5727; Bandla (1,280 m), Bodhal (970 m), Chauki (900 m), Chornali (1,280 m), Jiun (1,550 m).

Parts Used : Whole Plant. Leaves.

Folk Uses : Plant grown for **ornamental** purposes. 1 teaspoonful leaf-extr. taken in morning to check **diabetes** and **kidney ailments**. Leaf-paste applied against **wasp stings**.

Uses in Literature : Recorded earlier to cure anthrax (vet.), boils, cold (vet.), cough (vet.), cut (vet.), diabetes, diarrhoea, fever (vet.), high blood pressure, madness (vet.), mastitis (vet.), menorrhagia, worm, wounds, malignant lymphoma, paralysis (vet.), rabies (vet.), swelling of body (vet.), toothache, wasp stings, wounds; and as antileukemic, anticancerous, haemostat (vet.), hypotensive, sedative, mosquito repellent (vet.), eye complaints, purgative, stomachic (Ambasta, 1986; Hosagoudar & Henry, 1996c; Jain, 1991; Jha & Kumar, 2003; Kurian, 1995; Lalramnghinglova, 1996; Nadanakunjidam & Abirai, 2003; Prasad *et al.*, 1996; Sharma, 2003; Sharma & Sood, 1997; Siwakoti & Varma, 1996; Srivastava *et al.*, 2003c; Thomas & Britto, 2003; Watt 1972).

***Cedrus deodara* (Roxb.) Louden** (Pl. 11A).
Syn.: *C. libani* Barrel. var. *deodara* Hook. f.; *Abies deodara* Lindl.; *Pinus deodara* Roxb.
Family: Pinaceae.

Vern. : Deyar, Devdar.

English, Sanskrit & Regional Names:

Eng. : Himalayan Cedar
Sans. : Bhadravata, Indravriksha, Svehavriksha
B. : Gunduguragi, Pitadaru
G. : Vanseodeodar
H. : Deodar, Kilankaper
Kash. : Dadar, Dar, Deowar
Mal. : Devataram
Mar. : Devadaru
P. : Geyar, Kairval, Keolikelmang
Tam. : Tevadaram, Vandugolli
Tel. : Devadaravu.

Distribution : Himalayas: 1,800-3,000 m.

Description : Robust lofty trees. Cones oblong. Seeds winged, broadly triangular and with rounded sides.

Reproductive Cycle : June-July.

Habitat Ecology : Forming forests.

Material Examined : HPU-EL-SKS-5593; Bandla (1,280 m), Chandpur (1,290 m), Jiun (1,550 m), Kathena (1,290 m), Lolani (1,600 m).

Part Used : Wood.

Folk Uses : Wood excellent for **furniture** and **building works**.

Active Constituents : Active constituents are cedeodarin, cedrin, dihydromyricetin, cedrinoside, taxifolin meso-secoisolariciresinol, cedrusinin (*wood*).

Biological Activity : Stem-bark extr. exhibits significant anti-inflammatory activity in rat.

Uses in Literature : Reported to be used as an alterative with turmeric and guggul, anthelmintic, carminative, diaphoretic, diuretic, stomachic, symbolic (abode of gods); and for ascites, bilious fevers, bridges, chronic skin diseases, construction, costiveness, diarrhoea, doors, dropsy, dysentery, indolent swellings, intermittent fevers, flatulence, furniture, gonorrhoea, gout, headache, heart palpitations, inflammation, leprosy, material requirements, packing cases, paralysis, piles, pulmonary complaints, railway sleepers, rheumatism, ring worm (vet.), shingles, skin diseases, syphilis, ulcers, urinary disorders, and window frames (Ambasta, 1986; Chauhan, 1999; Jain, 1991; Kapur & Singh, 1996; Nautiyal *et al.*, 2003; Singh, 1991; Watt, 1972).

***Celastrus paniculatus* Willd.** (Pl. 11B).

Syn.: *Ceanothus paniculatus* Roth; *Celastrus alnifolia* Don; *C. dependens* Wall.; *C. multiflora* Roxb.; *C. nutans* Roxb.; *C. rothiana* Roem.

Family: Celastraceae.

Vern. : Sankheeru.

Sanskrit and Regional Names:

Sans. : Jiotishmati
B. : Mal-kangni
Bo. : Kanguni, Mal-kangni
G. : Malkangana
H. : Mal-kangni, Mal-kungi
Kan. : Kariganne
Mar. : Kangani, Mal-kanganitela, Malkarigoni, Pigavi
P. : Sankhil, Sankhu
Tam. : Atriparich-cham, Valulval
Tel. : Bavungie, Gundumedia, Malkanguni-maneru; Moierikata, Vittulu.

Distribution : Tropical & Sub-Tropical Himalayas: 330-1,200 m.

Description : Woody climbing shrubs upto 10 m tall. Leaves elliptic-ovate with acute apex and elevated midrib on both surfaces. Flowers in terminal panicles. Fruits a subglobose, bright yellow capsule. Seeds ellipsoid.

Reproductive Cycle : April-December.

Habitat Ecology : Borders of agricultural fields. Dry deciduous forests.

Material Examined : HPU-EL-SKS-5661; Chhanchri (930 m), Chauki (900 m), Malahu (1,050 m), Rajeher (1,050 m), Tapa (945 m).

Parts Used : Fruits. Seeds.

Folk Uses : Ripe fruits **edible**. 'halwa' (**sweet pudding**) of seeds with wheat flour and 'desi ghee' considered **nutritious**, and also prescribed for **backache**. Fruits also form one of the ingredients of '**dahli**' used for making local **wine**.

Active Constituents : *Seed oil* contains triterpene diol-paniculatadiol, β-amyrin, β-sitosterol and fatty acids.

Uses in Literature : Earlier considered abortifacient, aphrodisiac, appetiser, brain tonic, edible, emetic, laxative, rheumatic, stomachic; and for abdominal complaints, arthritis, asthma, blood purification, body pains, cancer, cough, diabetes, headache, hiccups, impotency, jaundice, joint pains, leprosy, leucoderma, leucorrhoea, paralysis, promoting menstrual cycle, skin diseases, stimulant, ulcers of intestine, weakness (Agarwal, 2003; Ambasta, 1986; Beg *et al.*, 2005; Bennet *et al.*, 1991; Brahmam & Saxena, 1990; Deshmukh & Rothe, 2003; Dey & Bahadur, 1973; Duke *et al.*, 2002; Girach & Aminuddin, 1995; Hosagoudar & Henry, 1996a; Jadeja, 1999; Jain, 2003b; Jain & Patole, 2001; Jain *et al.*, 1994; Jha & Kumar, 2003; Kalakoti, 2001; Khanna & Kumar, 2000; Khare, 2004; Kirtikar & Basu, 1984; Kothari *et al.*, 2003; Kumar & Lal, 2003; Kurian, 1995; Murthy *et al.*, 2003; Nadanakunjidam, 2003; Nayak *et al.*, 2003; Pal *et al.*, 2000; Panthi & Chaudhary, 2003; Parrotta, 2001; Patole & Jain, 2002; Paul & Pal, 2003; Rai & Upadhyay, 1996; Rajan *et al.*, 2005; Roy *et al.*, 1998; Sahoo & Mudgal, 1995; Said, 1997; Saini, 1996b; Sarangi & Sahu, 2005; Shanavashan *et al.*, 2003; Sharma, 2003; Singh, 1996a, 1999; Singh & Prakash, 1994; Siwakoti & Siwakoti, 2000; Siwakoti & Verma, 1996; Tirkey, 2004; Watt, 1972).

***Celtis australis* L.** (Pl. 11 C,D).
Syn.: *C. caucasia* Willd.; *C. acata* Ham.
Family: Urticaceae.

Vern. : Khirak, Khidak.

English & Regional Names:

Eng. : European Nettle Tree, The Honey Berry Tree
Kash. : Brimji
P. : Khark, Khirk, Ku, Batkar.

Distribution : Temperate Himalayas, 1,350-2,650 m.

Description : Trees. Leaves 2-nerved, alternate. Flowers pale yellow. Drupes long-stalked.

Reproductive Cycle : April-October.

Habitat Ecology : Borders of agricultural fields. Forests.

Material Examined : HPU-EL-SKS-5442; Gopalpur (1,350 m), Pahra (950 m), Paror (940 m), Saparu (1,250 m), Thandol (950 m).

Parts Used : Leaves. Wood.

Folk Uses : Leaves lopped for **fodder**. Wood used for **agricultural implements**, **house construction** and as **fuel**.

Uses in Literature : Known to be used as cattle fodder, colic, seeds for fatty oil, vegetable; and as agricultural implements, ammenorrhoea, blow instruments, coach whips, cups, edible fruits, hay-forks, leprosy, oars, ram-rods, spoons, timber, turnery, wagons, walking sticks, and whip handles (Ambasta, 1986; Arora & Pandey, 1996; Chandra, 1997; Das & Agarwal, 1991; Jain, 1991; Jain *et al.*, 1991; Kapur & Nanda, 1996; Kapur *et al.*, 1996; Lalramnghinglova, 2002; Maheshwari *et al.*, 1981; Purohit & Samant, 1995; Roy *et al.*, 1998; Sharma *et al.*, 2000; Singh & Kumar, 2000a; Sood & Thakur, 2004; Watt, 1972).

***Centella asiatica* (L.) Urb.** (Pl. 11E)

Syn.: *Hydrocotyle asiatica* L.

Family: Apiaceae.

Vern. : Brahmi, Meenaki.

English, Sanskrit & Regional Names:

Eng. : Indian Pennywort
Sans. : Bhekaparni, Divya, Manduki
Ass. : Manimuni
B. : Brahmamanduki
Bo. : Karinga, Karivana
H. : Brahmamanduki, Khulakhudi
Kan. : Von-de-laga
Mal. : Kodagam, Muthal
Mar. : Brahmi
Tam. : Babassa, Vallari
Tel. : Bokkudu.

Distribution : Tropical and sub-tropical regions; common throughout India

Description : Small trailing herbs with kidney-shaped leaves and purplish flowers in simple, axillary a few-flowered umbels. Fruits orbicular, reticulated.

Reproductive Cycle : April-May.

Habitat Ecology : Prominent along water channels, shady-damp areas.

Material Examined : HPU-EL-SKS-5459; Darugnu (1,400 m), Lahla (930 m), Malnu (1,000 m), Sidhpur (1,100 m), Sulah (950 m).

Parts Used : Aerial Plant Parts. Leaves.

Folk Uses : Drink prepared from aerial parts taken to relieve **body heat**, **urinary inflammation**, **improve memory**, **liver disorders**, **digestive ailments**, and as **brain tonic**. Leaves chewed as **blood purifier**, and its paste applied to hasten **suppuration** of **boils**.

Active Constituents : Therapeutic properties attributed to centoic, centellic, centic acids, brahmoside, brahminoside, two triterpenic acids—brahmic acid and iso-brahmic acid, betulic acid and stigmasterol.

Biological Activity : EtOH (50%) extr. of the plant antiprotozoal and spasmolytic.

Uses in Literature : Known in India for boils, cholera, colic pain, cough, curing blood disorder, diarrhoea, dysentery, eczema, eye complaints, fever, gastric disorders, headache, insanity, leprosy, leucorrhoea, liver complaints, memory, mouth sores, nervine disorder, respiratory disorder, scabies, skin diseases, syphilis, tubercles, tumour, worms, urinary complaints and wounds; and as an anthelmintic, brain tonic, coolant, 'chutney', diuretic, emmenagogue, galactagogue (vet.), insecticidal, psychoactive drug, postnatal tonic, stomachic, scorpion bite, tonic, and local drink, vegetable (leaves) (Agarwal, 2003; Ambasta, 1986; Arora & Pandey, 1996; Augustine & Sivadasan, 2005; Banerjee & Ghora, 1996; Bondya *et al.*, 2006; Borthakur *et al.*, 2004; Chakraborty, 2003; Chauhan, 1999; Chevallier, 1996; Chowdhery, 1996; Chowdhury *et al.*, 1996; Das, 2000; Devi, 2003; Dwivedi, 2003; Gogoi & Borthakur, 2001; Gogoi *et al.*, 2003a, b; Goud & Pullaiah, 1996a; Harborne & Baxter, 2001; Hembrom & Goel, 2005; Henry *et al.*, 1996; Jain, 1968, 1991, 1996; Jain & Patole, 2001; Jain *et al.*, 1994; Jamir, 1997; Jha & Varma, 1996; Karnick, 1994; Katewa & Arora, 1997; Kayal *et al.*, 2003; Khare, 2004; Kirtikar & Basu, 1984; Kshirsagar & Singh, 2000; Kurian, 1995; Lalramnghinglova, 1996, 2002; Mao, 1993; Maya *et al.*, 2003; Murthy *et al.*, 2003; Nadanakunjidam & Abirai, 2003; Negi *et al.*, 1999; Panda, 1996; Panthi & Chaudhary, 2003; Prajapati *et al.*, 2006; Prasad *et al.*, 1996; Rai & Upadhyay, 1996; Rajendran & Sikarwar, 2003; Rajendran *et al.*, 2003; Ranjan, 2000, 2003; Rawat & Chowdhary, 1998; Said, 1997; Saklani & Jain, 1994; Samant *et al.*, 2001; Sarma *et al.*, 2001; Satapathy & Brahmam, 1996; Sharma, 2003; Sharma & Rana, 2000; Sharma & Sood, 1997; Singh, 1999, 2000, 2003; Singh & Kumar, 2000a; Singh & Singh, 2003; Sinha, 1996; Sood & Thakur, 2004; Srivastava *et al.*, 2003b; Subramaniam, 2000; Thomas & Britto, 2003; Tiwari & Tiwari, 1996; Upadhyay & Singh, 2005; Verma *et al.*, 1995; Vishwanathan, 1997).

Centipeda minima (L.) A. Br. Asch. (Pl. 11F)

Syn.: *C. orbicularis* Lour.

Family: Asteraceae.

Vern. : Nakk-chhikannu.

Sanskrit & Regional Names:

Sans. : Chhikani, Chhikika
B. : Mechilla
H. : Nakk-chhikannu, Nagdowana, Pachittie
Mar. : Nakashikani.

Distribution : Throughtout India.

Description : A prostrate herb with small, serrated leaves. Flowers in greenish-yellow head. Achenes upto 2mm long without pappus.

Reproductive Cycle : November-March.

Habitat Ecology : Moist-shady places, often rice fields.

Material Examined : HPU-EL-SKS-5486; Bhatu (1,040 m); Dadh (1,100 m), Dharaman (1,100 m), Ladoh, (980 m), Panapari (1,100 m).

Parts Used : Flower heads. Leaves.

Folk Uses : Crushed flower heads used as snuff to relieve **headache** due to cold. Leaves chewed for **toothache**.

Active Constituents : Taraxasteryl palmitate, taraxasteryl acetate, taraxasterol, stigmasterol and anidiole isolated from *leaves*.

Uses in Literature : Employed so far as snuff, vermifuge; and for cold, ophthalmia, swollen gums, toothache (Ambasta, 1986; Islam & Hasin, 2003; Kumar & Lal, 2003; Srivastava *et al.*, 2003c).

Centratherum anthelminticum (L.) Kuntze (Pl.12A)

Syn.: *Conyza anthelmintica* L.; *Serratula anthelmintica* Roxb.; *Vernonia anthelmintica* (L.) Willd.

Family: Asteraceae.

Vern. : Brahmjeeri.

English, Sanskrit & Regional Names:

Eng. : Purple Fleabane
Sans. : Agnibija, Tiktajiraka, Vanajiraka
B. : Babchi, Kaliziri, Somraj
Bo. : Kalenjiri
G. : Kadvojiri

	H.	: Bakshi, Kalijhiri, Vapchi
	Kan.	: Kadu-jirage
	Mal.	: Kalajirakam
	Mar.	: Kalajira
	P.	: Bukoki, Kalazira
	Ori.	: Somraj
	Tam.	: Kattu-chiragam
	Tel.	: Adavijilakatta
	Urdu	: Janglijiri.

Distribution : Throughout India, common.

Description : Stout, leafy aromatic herbs. Leaves obovate with serrated margins. Flower heads small, purple. Achenes 10-ribbed. Pappus reddish.

Reproductive Cycle : August-September.

Habitat Ecology : Waste places. Agricultural fields.

Material Examined : HPU-EL-SKS-5405; Bodhal (970 m), Dadh (1,100 m), Jiun (1,550 m), Panchrukhi (990 m), Sungal (1,050 m).

Part Used : Seeds.

Folk Uses : Seeds taken along with jaggery to purify blood to check **cutaneous diseases** and **digestive ailments**; also considered **anthelmintic**. 1 teaspoonful of powdered seeds taken with cold water on empty stomach every morning for one month to treat **digestive disorders**.

Active Constituents : *Seeds* possess elemanolide lactone. *Seed oil* contains vernosterol whereas *leaves* and *stem* yield centratherin and germacranolide.

Biological Activity : Seeds antiviral and spermicidal.

Uses in Literature : Known in India as an abortifacient, anthelmintic, antiemetic, antiseptic, diuretic, febrifuge, stimulant, stomachic and vermifuge; and for acidity, asthma, cholera, constipation, cough, leucoderma, pain, paralysis, scorpion sting, skin diseases, swelling, syphilis, urinary complaints and veterinary diseases (Chopra *et al.*, 1956; Pal, 1981; Shah, 1984; Sharma & Sood, 1997; Uphof, 1968).

Cestrum nocturnum L. (Pl. 12B)

Family: Solanaceae.

Vern. : Rat-Ki-Rani.

English & Regional Names:

Eng. : Lady of Night, Night Blooming Jasmine

B. : Hasna

H. : Rat-Ki-Rani.

Distribution : Throughout India; commonly cultivated upto 1,500 m.

Description : Evergreen shrubs with ovate-lanceolate, long pointed, shining leaves. Flowers greenish yellow; scented at night. Fruit a berry.

Reproductive Cycle : August-October.

Habitat Ecology : Wastelands. Borders of agricultural fields.

Material Examined : HPU-EL-SKS-5537; Arla (1,100 m), Bhagpur (1,600 m), Makol (900 m), Panapari (1,100 m), Pahra (950 m).

Parts Used : Whole Plant. Flowers.

Folk Uses : Plant grown in courtyards for **scented flowers**. Flowers considered **sacred** for offering to deities.

Active Constituents : Active principles of *flowers* are benzaldehyde, benzyl alcohol, benzyl acetate, methyl benzoate, phenyl acetaldehyde, phenyl ethyl alcohol, linalool and α-farnesene.

Biological Activity : Aerial parts spasmolytic, diuretic and hypotensive.

Uses in Literature : Recorded in India for cosmetics, epilepsy and perfumes (Sharma, 2003; Uphof, 1968).

Cheilanthes farinosa (Forst.) Kaulf. (Pl. 12C)

Family: Cheilanthaceae.

Vern. : Duman-tuli.

Distribution : Throughtout India, ascending to 2400 m in hills.

Description : Small tufted ferns with elongated black glossy stipes. Fronds with black outline and underside covered with indumentum of usually 2-pinnate papillae. Sori globose.

Reproductive Cycle : July-August.

Habitat Ecology : Damp shady areas.

Material Examined : HPU-EL-SKS-5710; Bhaura (940 m), Chornali (1,280 m), Jiun (1,550 m), Palampur (1,280 m), Saparu (1,250 m).

Parts Used : Whole Plant. Petiole.

Folk Uses : Petiole employed as **nose-pin**. Plant juice prescribed for **ear ailments**.

Active Constituents : *Plant* contains cheilanthatriole and cheilarinosin.

Uses in Literature : Earlier used for earache, eczema, wounds, stomachache, antibacillus activity, menstrual disorders, seasonal cold and fever, stomachache (Singh & Kumar, 2000a; Rastogi & Mehrotra, 1999).

***Chenopodium album* L.** (Pl.12 D)

Syn.: *C. giganteum* Don; *C. nepalense* Hort.; *C. viride* L.; *C. laciniatium* L.; *C. purpurascens* Ham.

Family: Chenopodiaceae.

Vern. : Kanah.

English, Sanskrit & Regional Names:

Eng. : Bacon Weed, Dirty Dick, White Goose-foot, Wild Spinach
Sans. : Agralohila, Chillika, Kankella, Mridapatri, Shakarata, Vastuki
B. : Bathu sag, Chandan betu
Bo. : Chakwit
H. : Bathua, Khartua sag
P. : Bathu
Tam. : Parupukire
Tel. : Pappukura.

Distribution : Throughout India.

Description : Glabrous herbs upto 3 m tall with striped stems and clustered flowers. Leaves stalked, variable. Flowers and underside of the leaves mealy white. Utricles included in the perianth.

Reproductive Cycle : May-October.

Habitat Ecology : Cultivated fields. Wasteplaces.

Material Examined : HPU-EL-SKS-5573; Bodhal (970 m), Dehan (1,130 m), Dharaman (1,100 m), Jasun (1,000 m), Saliana (1,000 m).

Parts Used : Aerial Plant Parts. Leaves.

Folk Uses : Aerial plant parts eaten as a **pot-herb**, and also prescribed once a week as **diuretic**, **laxative** and **anthelmintic**. Leaf-juice applied to relieve **burns**, and its poultice applied externally as an **antiseptic** for skin problems.

Active Constituents : These are ecdysteroids, β-ecdysone and polypodine B isolated (*roots*), ascorbic acid, β-carotene, catechins, gallocatechin, caffeic acid, p-coumaric acid, ferulic acid, β-sitosterol, campesterol, xanthotoxin, stigmasterol, n-tri-acontanol, scopoletin, imperatorin, spinasterol and 7-stigmasterol (*plant*).

Biological Activity : Plant EtOH (50%) extr. CNS depressant.

Uses in Literature : Known to be used as an aphrodisiac, anthelmintic, antiseptic, blood purifier, diuretic, edible (vegetable; seeds as flour), laxative, purgative, tonic; and for abdominal pain, anaemia, diarrhoea, eye troubles, intestinal ulcers, local drinks, piles, reviving taste, scabies, throat, skin diseases, spleen and bilious disorders, wounds and urinary complaints (Ambasta, 1986; Arora & Pandey, 1996; Asolkar *et al.*, 1992; Bajpayee & Dixit, 1996; Banerjee & Ghora, 1996; Chakraborty *et al.*, 2003; Chatterjee & Pakrashi, 1991; Chopra *et al.*, 1956; Dey, 1973; Dwivedi, 2003; Jain, 1991; Jain & Goel, 2005; Kirtikar & Basu, 1984; Lal *et al.*, 1996; Mahato & Chaudhary, 2005; Mahato & Mahato, 1996; Murthy *et al.*, 2003; Panthi & Chaudhary, 2003; Parrotta, 2001; Prajapati *et al.*, 2006; Ranjan, 2000, 2003; Roy *et al.*, 1998; Said, 1997; Saini, 1996a; Saklani & Jain, 1994; Samant *et al.*, 2001; Sen & Batra, 1997; Shah, 2005; Shanavashan *et al.*, 2003; Sharma, 2003; Sharma & Rana, 2005; Singh, 1996b; Singh & Kumar, 2000a; Singh *et al.*, 2001; Sood & Thakur, 2004; Sood *et al.*, 2001; Srivastava *et al.*, 2003a, b; Subramani & Goraya, 2003; Viswanathan, 2000).

***Chenopodium ambrosioides* L.** (Pl. 12 E)

Syn.: *C. vulpinum* Wall.; *Ambrina ambrosioides* Moq.

Family: Chenopodiaceae.

Vern. : Bathu.

English & Regional Names:

Eng. : The Sweet Pigweed, Mexican tea

Mal. : Katu ayamodakam.

Distribution : Common in many parts of India.

Description : Much-branched aromatic herbs upto 100 cm. Flowers clustered in long slender spikes. Utricles included in the sepals.

Reproductive Cycle : June-September.

Habitat Ecology : Field borders. Roadsides. Wastelands.

Material Examined : HPU-EL-SKS-5564; Balah (900 m), Bindraban (1,270 m), Darugnu (1,400 m), Lolani (1,600 m), Thakurdwara (1,150 m).

Parts Used : Aerial Parts.

Folk Use : Aerial parts consumed as **pot-herb**.

Active Constituents : Therapeutic properties attributed to ascaridole, quercetin, kaempferol-7-rhamnoside and ambroside (*leaves*).

Biological Activity : Essential oil from leaves antifungal (against human pathogenic fungi) and antimicrobial. Leafy extr. anthelmintic whereas shoot extr. nematicidal.

Uses in Literature : Used so far for amoebic dysentery, appendicitis, asthma, cough, cramps, diarrhoea, essential seed oil, opthalmia, tuberculosis; and as anthelmintic, cholera, colic pain, edible aerial parts (vegetable), and wounds (Ambasta, 1986; Chatterjee & Pakrashi, 1991; Chevallier, 1996; Dey & Bahadur, 1973; Duke *et al.*, 2002; Kapur & Singh, 1996; Mahato & Chaudhary, 2005; Nunez & Castro, 1995; Prajapati *et al.*, 2006; Saklani & Jain, 1994; Sharma, 2003; Sharma & Rana, 2005; Singh & Kumar, 2000a; Sood & Thakur, 2004; Sood *et al.*, 2001).

***Cicer arietinum* L. (Pl. 12F)**

Family: Fabaceae.

Vern. : Chana.

English, Sanskrit & Regional Names:

- **Eng.** : Bal gram, Chickpea
- **Sans.** : Canaka, Canakah, Chanaka
- **B.** : But, Butkalai, Chhola
- **H.** : But, Chana, Chanaka
- **Kan.** : Karikadake, Kempukadali, Kudoly
- **Mar.** : Harbara
- **P.** : Channa, Chola
- **Tam.** : Kadalai
- **Tel.** : Harimandhakam, Sanagulu
- **Urdu** : But, Chana.

Distribution : Throughout India.

Description : Erect annual herbs. Leaves pinnately compressed. Flowers white-pink. Pods pubescent. Seeds obovate or subglobose, beaked.

Reproductive Cycle : February-March.

Habitat Ecology : Cultivated fields, met as an escape in open areas.

Material Examined : HPU-EL-SKS-5616; Banuri (1,150 m), Chauki (900 m), Gopalpur (1,350 m), Malnu (1,000 m), Tikkar (910 m).

Parts Used : Fruits. Seeds.

Folk Uses : Fruits (pods) **eaten** raw when fresh. Seeds **edible** as pulses, and 'chapati'.

Active Constituents : Vanillic and β-hydroxybenzoic acids isolated from *leaves.*

Uses in Literature : Used in India as abortifacient, adhesive, anthelmintic, aphrodisiac, appetiser, astringent, coolant, diuretic, refrigerant, stimulant, tonic; and for biliousness, blood troubles, bronchitis, cooling bath, constipation, curing thirst and burning, dyspepsia, jaundice, leprosy, skin diseases, sprains of cattle, sunstroke, and throat complaints (Ambasta, 1986; Aswal & Mehrotra, 1994; Kirtikar & Basu, 1984; Kshirsagar, 2003; Kshirsagar *et al.*, 2003; Pandey, 2003; Parrotta, 2001; Sen & Batra, 1997).

***Cinnamomum camphora* (L.) Nees & Eberm** (Pl. 13A)

Syn.: *Laurus camphora* L.

Family: Lauraceae.

Vern. : Kapur.

English & Regional Names:

Eng. : Camphor, Camphor Cinnamomum, Official Camphor

Mal. : Ghanarasam, Himavaluka

Ori. : Chondrongyon, Romokorpuro

Tam. : Indu, Kirasemiran, Pugalvisurandiram

Tel. : Chandramu, Ghanasaramu, Karpuramu.

Distribution : Cultivated in Himalaya upto 1,500m.

Description : Trees with alternate leaves. Buds enclosed in silky, caducous, imbricating scales. Flowers hairy within. Fruits globular, with thick club-shaped base.

Reproductive Cycle : April-August.

Habitat Ecology : Open Grasslands.

Material Examined : HPU-EL-SKS-5443; Bhaura (940 m), Bhuana (1,400 m), Chachian (1,300 m), Chimbalhar (1,150 m), Molichak (1,020 m).

Parts Used : Seeds. Leaves. Wood.

Folk Uses : Seed-oil massaged to relieve **rheumatic pains**. Vapours from boiling leaves inhaled to treat **cough**. Wood used for performing **'havan'**.

Active Constituents : Isolated constituents are 3,7-dimethylocta-1,7-dien-3,6-diol and 3,7-dimethylocta-1,5-dien-3,7-diol, (-) trans-2-(3,4-dimethoxybenzyl)-butrolactone (I)- and (-) trans

3-hydroxy-5,7-dimethoxy –3′,4′-methylenedioxyflavan along with dimethyl matairesinol and kusunokinin (*heartwood*) and sesquiterpene hydrocarpone-α-ylangene, β-santalene, δ-guanine, δ-cadinene, calamenene, calacorrone γ-patchoulene and 1,6-dimethyl-4-isopropyl-7,8-dihydronaphthalene (*camphor oil*).

Uses in Literature : Reported earlier as acronarcotic poison, analeptic in cardiac depressions, anthelmintic, antidote to strychnia, antileprotic, anti-phlegmatic, antiseptic, antispasmodic, aphrodisiac, carminative, counter-irritant in muscular strains, diaphoretic, natural camphor, psychoactive plant, sedative, stimulant; and for affections of the genito-urinary system, angina pectoris, anodyne properties, asthma, bed-sores, camphor, cancer, curing wounds in hoofs, delirium, dysmenorrhoea, fever, gangrene, gout, hysteria, impotency, inflammation, inflammatory conditions of the eye, insanity, malaria, muscular strains, muttering, nervous headaches, nervousness, nymphomania, painful affections of the uterus, perfumery, rheumatic conditions, sleeplessness, small pox, spermatorrhoea, whooping-cough, and wounds (vet.) (Ambasta, 1986; Chatterjee & Pakrashi, 1991; Jain *et al.*, 1994; Mishra *et al.*, 1996b; Sharma & Sood, 1997; Vihari, 1995; Watt, 1972).

Cinnamomum tamala **Nees & Eberm** (Pl. 13B).

Syn.: *C. albiflorum* Nees; *C. cassia* Don; *Persea tamala* Spreng.; *Laurus cassia* Roxb.; *L. albiflora* Wall.; *L. tamala* Ham.; *L. tazia* Ham.; *L. somcaurium* Ham.; *L. sailyana* Ham.

Family: Lauraceae.

Vern. : Gudpatraj, Mithpatra, Tejpatta.

English, Sanskrit & Regional Names:

Eng. : Cassia cinnamomum, Indian Cassia Lignea
Sans. : Dalavhaya, Patrakhya, Vasanavhaya
Ass. : Dopati
B. : Tejpat, Chota sinkoli
Bo. : Darchini, Tamala
G. : Tamalpatra
H. : Dalchini, Tezpat
Kan. : Thit-kya-bo
Kash. : Dalchini-Pan, Tajpatra
Mal. : Karuvapatta Patram
Mar. : Dalchinitiki, Sambharpana
Ori. : Tejapatra

P. : Tajpater
Tam. : Talishappaalttiri
Tel. : Aakupatri, Tallishapatri.

Distribution : Tropical & sub-tropical Himalayas : 450-2,100 m.

Description : Trees with leathery leaves and white flowers in pubescent panicles. Drupes black when ripe.

Reproductive Cycle : June-August.

Habitat Ecology : Forests.

Material Examined : HPU-EL-SKS-5632; Balah (900 m), Gharthun (900 m), Makol (900 m), Purani-palam (1,100 m), Tikkar (910 m).

Part Used : Leaves.

Folk Uses : Leaves used for **flavouring** tea and curries. Its decoction taken to check **lung ailments**, **cholesterol** and as **carminative**. Also, its powder good for **scouring teeth**.

Active Constituents : Essential oil from *leaves* contains cinnamic aldehyde, phellandrene, eugenol, linalool, cinnamaldenyde, 3,4′,5,7-tetrahydroxy-flavone, 3,3′,4′,5,7-pentahydroxy-flavone, kaempferol-3-O-glucopyranoside,kaempferol-3-0-sophoroside, kaempferol-3,7-di-O-rhamnopyranoside and quercetin-3-O-rutinoside.

Uses in Literature : Known so far as an adulterant of true cinnamon, carminative, colic, condiment (leaves), expectorant, ingredient of 'chywanprash', narcotic, 'paan' stimulant, stomachic, sudorific; and for affection of the nerves or heart, asthma, bites of serpents, bronchitis, calico printing, cold, cough, debility of stomach, diarrhoea, dropsy, dyeing, enlargement of spleen, essential oil, fever, flatulence, gonorrhoea, headache, heart and throat complaints, indigestion, manufacturing of soap, mumps, nausea, oil in perfumery, pain in the womb, piles, poisoning by opium, retention of urine and catamenia, rheumatism, and scorpion sting (Agarwal, 2003; Ambasta, 1986; Arora & Pandey, 1996; Bose, 1984; Chatterjee & Pakrashi, 1991; Chauhan, 1999; Chakraborty *et al.*, 2003; Chandra, 1997; Dey & Bahadur, 1973; Duke *et al.*, 2002; Ghosh *et al.*, 1996; Jain, 1991; Lalramnghinglova, 1996, 2000; Mahato & Chaudhary, 2005; Mehrotra *et al.*, 1995; Mehtani & Ingle, 1999; Nath & Maiti, 2003; Negi & Pant, 1990; Prajapati *et al.*, 2006; Pruthi, 1976; Rana *et al.*, 1996; Ranjan, 2000; Rawat & Chowdhary, 1998; Roy *et al.*, 1998; Sahni, 1998; Saikia & Nath, 2003; Saklani & Jain, 1994; Samant *et al.*, 2001; Sharma, 2003; Sharma & Sood, 1997; Singh, 1991, 1996b, 1999, 2001; Sinha, 1996; Watt, 1972).

Cirsium arvense (L.) Scop. (Pl. 13C)

Syn.: *Sarratula arvensis* L.; *Breea arvensis* (L.) Less.; *Cnicus arvensis* (L.) Roth; *C. candicans* Wall.; *Carduus lanatus* Roxb.; *Aplotaxis cirsiodes* DC.

Family: Asteraceae.

Vern. : Bhruce.

English Name: Creeping Thistle.

Distribution : N.W. Himalayas.

Description : Erect thistles with sessile leaves and cottony stems. Heads purple. Outer involucral bracts foliaceous. Pappus brownish-white. Achenes black.

Reproductive Cycle : May-June.

Habitat Ecology : Damp-shady areas. Wastelands.

Material Examined : HPU-EL-SKS-5603; Balu (1,030 m), Bhawarna (1,100 m), Lolani (1,600 m), Nanaun (900 m), Sidhpur (1,100 m).

Part Used : Rhizome.

Folk Uses : Crushed rhizome as such or its decoction taken orally as **refrigerant**; also given orally to milching cattle for enhancing **lactation**.

Active Constituents : These are glucuronides of apigenin and acacetin, taraxasterol, its -OAc and stigmasterol (*plant*) and phytotoxic compound (*root*).

Biological Activity : Whole plant spasmolytic and hypotensive. Root diuretic.

Uses in Literature : Used in India as an emetic, tonic, diaphoretic, edible (seeds, stem, leaves), emmenagogue and for coagulating milk (Ambasta, 1986; Aswal, 1996; Bennet *et al.*, 1991; Chopra *et al.*, 1956; Maheshwari *et al.*, 1981; Roy *et al.*, 1998; Singh *et al.*, 1996a; Sood & Thakur, 2004).

***Cissampelos parriera* L.** (Pl. 13D).

Syn.: *Cissampelos pareira* L. var. *hirsuta* (Buch.-Ham. *ex* DC.) Forman; *C. caapeba* L.; *C. convolvulacea* Willd.; *C. hirsuta* Buch.-Ham. *ex* DC.; *C. orbiculata* Buch. DC.; *C. heradiflora* Wall.; *C. obectata* Wall.; *C. discolor* Buch.; *C. diversa* Miers; *C. elata* Miers; *C. grallatoria* Miers; *C. eriantha* Miers; *C. subpeltata* Thwaites; *C. delicatula* Miers; *C. nepalensis* Rhode; *Cocculus orbiculatus* DC.; *C. villous* Wall.; *Menispermum orbiculatum* L.

Family: Menispermaceae.

Vern. : Bhatindu.

English, Sanskrit & Regional Names:

Eng. : False Periera, Periera Brava, Velvet Leaf
Sans. : Ambashthika, Ruchaishya, Vidhakarni
Ass. : Saruthubukilata
B. : Akanadi, Nimuka
Bo. : Pahadmul, Pahadvel
G. : Venivel
H. : Akauadi, Dekhnirbissi
Kan. : Padavali
M.P. : Karepar
Mal. : Kattuvalli, Patuvalli
Mar. : Padavalli, Padavel
Ori. : Okanibindhi
P. : Bat, Batindu Path
Tam. : Appatta, Puttutiruppi, Vatattirupi
Tel. : Pata.

Distribution : Tropical & sub-tropical parts of India upto 2,000 m.

Description : A climber with orbicular-ovate leaves. Flowers pale-green in racemose. Drupes red.

Reproductive Cycle : April-September.

Habitat Ecology : Rocky slopes. Screes.

Material Examined : HPU-EL-SKS-5488; Bhauda (1,050 m), Jasun (1,000 m), Jiun (1,550 m), Kathena (1,290 m), Malahu (1,050 m).

Parts Used : Leaves. Roots.

Folk Uses : Poultice of leaf-paste applied to heal **bone fracture**; also the paste mixed with lukewarm water given to livestock for **dysentery**. A handful of its crushed roots along with 7 fruits of black pepper (*Piper nigrum*) made to tablets and prescribed one daily for seven days to check **leucorrhoea**.

Active Constituents : Active principles are menismine, cissamine, parierine, 1-bebeerine, hayatinin, hayanin, d-isochendrodendrine, 4″-0-methyleurine (*root bark*), isoguinoline, pelosine, berbenine, reserpine, cissampetive, hyatin, hyatinin, quercitel and sterol (*roots*).

Biological Activity : CNS depressant activity confirmed.

Uses in Literature : Reported to be used in different parts of India as an anthelmintic, antidote against poisons, antilithic, antiperiodic, astringent, carminative, colic, country wine, diuretic, expectorant, febrifuge, fibre yielding, fish poison, health tonic, psychoactive plant, purgative used in dyspepsia, sedative, stomachic, suppurative, symbolic (wearing in neck cures headache), tonic; and for abscess, antifertility, asthma, bodyache, boils, bone fracture, bowel complaints, bronchitis, burns, carbuncle, catarrhal affections of the bladder, chest pain, chronic liver disorders associated with alcoholism, cold, cough, country wine, cystitis, diabetes, diarrhoea, diphtheria, dropsy, dysentery, enlarged spleen, epilepsy, eye complaints, fermentation, fibre, gastric complaint, headache, heart burn, inflammation of kidney, insect bites, itching, jaundice, labour pain, leprosy, leucoderma, leucorrhoea, malaria, pains, peptic ulcer, piles, pimples, prolapsed uterus, recovering energy after childbirth, regularizing menstruation, rheumatism, scorpion bite, sedative action on the mucous membranes of the genito-urinary organs, sinuses, skin diseases, small pox, stomach ulcer, stone, swelling of gum, toothache, ulcer, unhealthy sores, urinary troubles, venereal complaints, and weakness of stomach (Ambasta, 1986; Aminuddin & Girach, 1991; Borthakur, 1993; Borthakur *et al.*, 2004; Brahmam & Saxena, 1990; Brahmam *et al.*, 1996; Chauhan, 1999; Datt, 1996; Diwakar & Ansari, 2003; Dwarakan & Ansari, 1996; Gogoi & Boissya, 1984; Gogoi *et al.*, 2003a; Hosagoudar & Henry, 1996 a, c; Islam, 1996; Jain, 1991, 1996, 2003a; Jain *et al.*, 1991, 1994, 1997; Jha & Varma, 1996; Kumar *et al.*, 1987; Manandhar, 1980a, 1996; Masih, 2003; Painuli & Maheshwari, 1996; Nayak *et al.*, 2003; Rajan *et al.*, 2003; Rama Rao & Henry, 1996; Rana *et al.*, 1996; Rao, 1996; Reddy *et al.*, 1996; Saini, 1996a; Saxena *et al.*, 1981; Shah & Joshi, 1971; Sharma & Sood, 1997; Singh, 1991, 2003; Singh & Rao, 2003; Sinha, 1996; Siwakoti & Varma, 1996; Srivastava *et al.*, 2003a, c; Thomas & Britto, 2003; Upadhyay, 2003; Varma, 1997; Watt, 1972).

Citrus grandis (L.) Osbeck (Pl. 13E)

Syn.: *C. maxima* (Burm.) Merr.; *C. decumana* L.; *Aurantium maximum* Burm.

Family: Rutaceae.

Vern. : Chakotra.

English, Sanskrit & Regional Names:

Eng.	: Forbidden Fruit, Paradies Apple, Pommel-Moose Shadock Pummelo
Sans.	: Madhukarkati
B.	: Batavinebu, Botornebu
Bo.	: Papanass
G.	: Obakotru
H.	: Batavinebu, Chakotra, Sadaphal
Kan.	: Chaoktre, Sakkota
Mal.	: Bambitinarakam
Mar.	: Panis
Ori.	: Batabilembu
P.	: Chakotra
Tam.	: Pambalimasu
Tel.	: Pampanpanasa.

Distribution : Foothills of Himalayas.

Description : Trees with large, ovate-oblong leaflets. Flowers white. Fruits large-sized, pale globose with thick rind.

Reproductive Cycle : November-March.

Habitat Ecology : Planted near houses.

Material Examined : HPU-EL-SKS-5658; Andretta (980 m), Balah (900 m), Makol (900 m), Panapari (1,100 m), Rangru (910 m).

Parts Used : Fruits.

Folk Uses : Fruits **edible**, **refreshing** and check **vomiting**.

Active Constituents : Main ingredients isolated are 5-geranyloxy-7-methoxycoumarin, isoimperatorin, bergapten, and 5-geranyloxypsoralen (bergamottin) (*fruits*).

Uses in Literature : Used as edible, nutritive, refrigerant; and for cholera, convulsive coughs, constipation and inflamed nerves, deserts, diabetes, diarrhoea (vet.), epilepsy, flatulent pains, heart disease, indigestion (vet.), jams and marmalades, insanity, morbid cancer, nervous disorders, offerings, painful swellings, rheumatism, scurvy, swellings, ulcer, vomiting, waist and back pains (Agarwal, 2003; Ambasta, 1986; Borthakur, 1996a; Chopra *et al.*, 1956; Chun-Lin & Jieru, 1995; Das & Agarwal, 1991; Gogoi & Borthakur, 1991; Kurian, 1995; Prajapati *et al.*, 2006; Saklani & Jain, 1994; Sharma & Rana, 2005; Sharma *et al.*, 2000).

Citrus jambhiri **Lush.** (Pl. 13F)
Family: Rutaceae.

Vern. : Ghambirdi, Jhamirdi, Khatirali.

English Name:

Eng. : Rough Lemon.

Distribution : Hot valleys of Himalayas.

Description : Shrubs or small trees. Leaves alternate. Flowers white in racemes. Fruits berry. Seeds numerous.

Reproductive Cycle : October-March.

Habitat Ecology : Border of the fields.Wastelands.

Material Examined : HPU-EL-SKS-5677; Boda (1,000 m), Chauki (900 m), Jasun (1,000 m), Pahra (950 m), Tikkar (910 m).

Part Used : Fruit.

Folk Uses : Fruits **edible** and considered a rich source of vitamin C. Sacred vermillion, '**kungu**' is made by boiling its fruit juice with 'turmeric' (*Curcuma longa* L.) and 'suhaaga' (borax) for performing **religious rites**.

Citrus limon **(L.) Burm.f.** (Pl. 14A)
Syn.: *C. medica* var. *lomonum* L. (Risso) Hook.f.; *C. limonum* Risso; *C. medica* var. *limon* L.
Family: Rutaceae.

Vern. : Nimbu.

English & Regional Names:

Eng. : Lemon
B. : Bara Nebu, Goranebu
G. : Motalimbu
H. : Bara Nibu, Jambhira, Paharinembu, Paharikagzi
Kan. : Bijapura, Bijuri
Mar. : Idalimbu, Thoralimbu
Ori. : Periya Yelumichai
Tel. : Bijapuram.

Distribution : Valley of the Himalayas.

Description : Small-sized thorny trees having pinkish young foliage and flower buds. Leaves ovate; petioles marginate or winged. Fruits medium-sized, ovoid-yellow with thin rind.

Reproductive Cycle : November-February.

Habitat Ecology : Cultivated in gardens.

Material Examined : HPU-EL-SKS-5574; Banuri (1,150 m), Chachian (1,300 m), Nanaun (900 m), Rangru (910 m), Sungal (1,050 m).

Parts Used : Leaves. Fruit.

Folk Uses : Leaves used for **scouring** teeth and checking **foul** breath. Toasted leaves tied on chest of kids to check **cough**. Fruits relished as such or as **pickle**; also used for **offering** to Goddess 'Tulsi'. Fruits roasted by burrowing in hot ash of fireplace eaten for **rheumatic** pains.

Active Constituents : *Fruit peel* yields apigenin, luteolin, chrysoseriol, quercitin, isorhamnetin, new flavonols- limocitrin and limocitrol-sinapic acid, p-coumaric acid, scopoletin, umbelliferone, hesperidin, citrusins A, B, C and D, coniferin syringin and dehydro-diconeferyl alcohol-4β-glucoside.

Biological Activity : Coniferin and syringin (1.0 mg/100 g, *i. v.*) found to raise blood pressure of SHR-Sp rats whereas dehydrodiconiferyl alcohol -4β-glucoside (*i. v.*) lowers it in spontaneously hypertensive rats.

Uses in Literature : Useful in India as an anthelmintic (vet.), arteriosclerosis, antiseptic, cardiotonic, carminative, disinfectant, juice (vet.), febrifuge, nervine tonic; and for acidity, burns, citric acid, colds, corns, coughs, culinary purposes and in the preparation of beverages, dandruff, flavouring liqueurs, flu, headache, hypertrophy of spleen, indigestion, lemon oil (obtained as by-products), lowering body heat, pectin, pickles, preventing dandruff in hairs, ceremonial sacrifice, rheumatism, scabies, scurvy, stomachache, throat diseases, vitality to the body, and warts (Agarwal, 2003; Ambasta, 1986; Arora & Pandey, 1996; Chevallier, 1996; Chopra *et al.*, 1956; Dagar & Dagar, 1996; Harborne & Baxter, 2001; Khare, 2004; Kurian, 1995; Mishra *et al.*, 1996a; Prajapati *et al.*, 2006; Roy *et al.*, 1998; Saikia & Nath, 2003; Sharma *et al.*, 2000; Shome *et al.*, 1996; Tirkey *et al.*, 2003a; Viswanathan & Singh, 1996).

Citrus medica **L.** (Pl. 14B)
Family: Rutaceae.

Vern. : Baranimbu

English, Sanskrit & Regional Names:

Eng. : Adam's Apple, Cedrat, Citron, Melon Lime
Sans. : Amlakshera, Bijaphalaka, Rochanaphala
B. : Baranimbu, Nimbu
Bo. : Bijapura, Limbu
H. : Baranimbu, Nimbu
Kan. : Madala, Mahaphala, Rusaka
Mal. : Gilam, Matalanarakam
Mar. : Mahalunga
P. : Bajauri
Tam. : Kogilachan, Kommattimadalai
Tel. : Lungamu.

Distribution : Foothills of Himalayas.

Description : A spiny tree with pinkish white flowers. Fruit oblong or ovoid and usually mammilate at the apex.

Reproductive Cycle : April-September.

Habitat Ecology : Cultivated.

Material Examined : HPU-EL-SKS-5570; Balah (900 m), Bhauda (1,050 m), Jasun (1,000 m), Thandol (950 m), Tikkar (910 m).

Parts Used : Leaves. Fruits.

Folk Uses : Leaves useful for **scouring teeth** and **against foul breath**. Fruits **edible** and also **pickled**.

Active Constituents : A new triterpenoid-citrusin-isolated along with palmitic and succinic acids, citropten, β-sitosterol and daulosterol (*fruits*).

Biological Activity : Anthelmintic activity found positive.

Uses in Literature : So far reported as an anthelmintic, antidote to acro-narcotic and animal poisons, antiscorbutic, carminative, febrifuge, flavouring agent, refreshing drinks, stomachic, vermifuge (vet.); and for abdominal pain, acnes, acute rheumatism, agricultural implements, biliousness, bleeding gums, cardiac diseases, checking constipation, diarrhoea, dysentery, dyspepsia, essential oil, fatigue, feckles, foul breath, haemorrhage, hypertrophy of the spleen, improving general weakness and trembling of the limbs, indigestion, inflammatory affections, jaundice, kidney diseases, lemon syrup, malaria, marmalades, mental disorders, mucous and pus from mouth (vet.), nausea, pickling, scabies, sciatica, scurvy, stimulating appetite, stomach ulcer, strengthening and giving tone to the stomach, urinary calculus, urinary diseases, veterinary-mucus and vomiting (Ambasta, 1986; Borthakur & Sarma, 1996; Chowdhery, 1996; Jain, 1991; Kurian, 1995; Lalramnghinglova, 1996; Mohanty & Padhy, 1996; Negi & Gaur, 1991; Singh *et al.*, 1997; Tiwari & Tiwari, 1996; Watt, 1972).

Citrus medica var. medica L. (Pl. 14C)

Family: Rutaceae.

Vern. : Bajoorein.

Distribution : Valley of the Himalayas.

Description : A small thorny tree having pink flower buds. Leaflets oblong. Fruits oblong or obovoid with warted thick rind.

Reproductive Cycle : October-March.

Habitat Ecology : Along fields edges.

Material Examined : HPU-EL-SKS-5633; Banuri (1,150 m), Bhaura (940 m), Malnu (1,000 m), Nadli (1,000 m), Sulah (950 m).

Parts Used : Fruit.

Folk Uses : Fruits **eaten** as such or **pickled**, also considered **sacred** and form one of the ingredients of '**havan**'.

Uses in Literature : Known for its edible rind, pulp, pickle; good for perfumery (Ambasta, 1986; Watt, 1972).

Clematis gauriai *Roxb.* (Pl. 14D)

Family: Ranunculaceae.

Vern. : Baaker-bel.

Regional Names:

- **Eng.** : Indian Travellar's Joy, Old Man's Beard
- **Bo.** : Morial, Morvel, Ranjal
- **Kan.** : Telejadari
- **Ori.** : Boromojhanti, Idiya.

Distribution : Wild in hills.

Description : Climbing shrubs. Leaves opposite, compound. Flowers greenish-white, fragrant. Styles hairy. Calyx 4, petaloid. Corolla absent. Fruitlets feathery.

Reproductive Cycle : October-January.

Habitat Ecology : Hedges.

Material Examined : HPU-EL-SKS-5660; Boda (1,000 m), Chandpur (1,290 m), Saparu (1,250 m), Thandol (950 m).

Part Used : Leaves.

Folk Uses : Leaves chewed to check **diabetes**, and also employed for **scouring** teeth.

Uses in Literature : Reported as vesicant; and for boils, skin diseases, spleen trouble (Ambasta, 1986; Gopal & Chandra, 2003; Singh & Kumar, 2000a).

Coccinia grandis (L.) Voigt (Pl. 14E)

Syn.: *C. indica* Wight & Arn.; *Cephalandra indica* Naudin

Family: Cucurbitaceae.

Vern. : Kalakadu.

Sanskrit & Regional Names:

- **Sans.** : Bimba, Bimbaka, Bimbi, Chhardim, Dantachhadipame, Govhi, Jhundikeshi, Kamboja, Karmakari, Kaluka, Katutundika, Oshtni, Ostopamaphala, Piluparni, Raktaphala, Tiklabumbi, Tiktatundi, Tundi, Tundika, Tundikeri, Vimba
- **B.** : Bimbu
- **Bo.** : Bhimb, Ramtondala, Tendli, Tenduli
- **G.** : Galedu, Galluda
- **H.** : Bhimb, Kanduri, Kunduri
- **Mal.** : Gwel, Kova, Kwel
- **Mar.** : Bimbi, Tenduri, Kundru
- **P.** : Ghol, Kanduri
- **Tam.** : Kovai, Kwai
- **Tel.** : Bimbika, Kaidopnda, Kakidonda
- **Urdu** : Kundaru.

Distribution : Throughout India and Sri Lanka.

Description : Much branched perennials with grooved stem and slender tendrils. Leaves long, palmately 5-nerved. Flowers creamish, glabrous. Fruits fusiform ellipsoid, slightly beaked. Seeds compressed, yellowish grey.

Reproductive Cycle : May-August.

Habitat Ecology : Cultivated fields. Wastelands.

Material Examined : HPU–EL–SKS-5461; Banuri (1,150 m), Chornali (1,280 m), Gharthun (900 m), Lahla (930 m), Nanaun (900 m).

Part Used : Fruits.

Folk Use : Unripe fruits cooked as **vegetable** and considered good for **weak** people.

Active Constituents : *Fruit* contains B-amyrin, lupeol, a bitter glycoside containing cucurbitacin B, B-sitosterol and tarexerol.

Biological Activity : Hypoglycaemic (fruits).

Uses in Literature : Used so far as antipyretic, aphrodisiac, astringent to bowels, edible, galactagogue, purgative; and for asthma, biliousness, blood diseases, burning sensation of hand and feet, cooling, dysentery, flatulence, flavour, fracture of limb, gonorrhoea, headache, itching, jaundice, liver and skin diseases, migraine, ringworm, snakebite and scorpion sting, sprains, wounds, urinary stress and discharges, urinogenital diseases (Agarwal, 2003; Ambasta, 1986; Bennet *et al.*, 1991; Bhandary & Chandrashekar, 2002; Bhattacharya *et al.*, 2005; Dey & Bahadur, 1973; Ghose, 1997; Jha & Kumar, 2003; Khanna *et al.*, 1996a; Khare, 2004; Kirtikar & Basu, 1984; Ramdas *et al.*, 2000; Ranjan, 2003; Rajendran *et al.*, 2003; Reddy & Raju, 2000; Roy *et al.*, 1998; Sarin, 1996; Sharma & Rana, 2005; Shukla *et al.*, 2003).

***Cocculus laurifolius* DC.** (Pl. 14F)

Syn.: *C. leaeba* DC.; *C. pendulus* (Forsk.) Diels

Family: Menispermaceae.

Vern. : Parauda.

Regional Names:

- **G.** : Parwatti
- **P.** : Illarbillar, Parwatti
- **Tel.** : Dusaratige.

Distribution : Outer Himalayas upto 1,700 m.

Description : Erect shrubs or small trees. Leaves glabrous, shortly-stalked, lanceolate, entire. Flowers numerous; panicles shorter than the leaves. Drupes black.

Reproductive Cycle : March-August.

Habitat Ecology : Shady places along nallahas.

Material Examined : HPU-EL-SKS-5611; Boda (1,000 m), Bhuana (1,400 m), Khaira (940 m), Malahu (1,050 m), Tapa (945 m).

Parts Used : Aerial Parts.

Folk Uses : Aerial parts good for making **huts**, and for **roofing** purposes.

Active Constituents : Laurifoline, magnoflorine chlorides, o-methylisocorydine, isocorydine and boldine methochlorides isolated.

Biological Activity : Isocorydine, o-methylisocorydine and boldine methochlorides exhibited neuromuscular blocking and hypertensive activities.

Uses in Literature : So far reported as febrifuge, tonic, fish-poison, beverage; and for making poles (Ambasta, 1986; Singh & Kumar, 2000a).

Colebrookea oppositifolia Sm. (Pl. 15A)

Syn.: *C. ternifolia* Roxb.; *Elsholtzia oppositifolia* Poir.

Family: Lamiaceae.

Vern. : Doos.

Regional Names:

- **Bo.** : Bahmani, Bhamini, Darai, Dasari, Dussarica
- **H.** : Binda, Bindu, Pansra
- **Mar.** : Darigopa, Merota
- **Ori.** : Bosili, Dangoti
- **P.** : Barmera, Briali, Dashane, Duss, Phisbekkar, Sampru, Shakardana, Sulai.

Distribution : Tropical & sub-tropical parts of India : ascending to 1,350 m.

Description : Small shrubs. Leaves oblong-lanceolate, leathery, crenate. Flowers white, in terminal panicles. Nutlets usually one with hairy tip.

Reproductive Cycle : April-November.

Habitat Ecology : Waste places.

Material Examined : HPU-EL-SKS-5562; Bhaura (940 m), Bodhal (970 m), Nadli (1,000 m), Panapari (1,100 m), Rangru (910 m).

Parts Used : Leaves. Stem.

Folk Uses : Leaf paste good against **bleeding** from cuts. Stem used as **fuel**.

Active Constituents : Active properties due to flavones-5-hydroxy-4′, 6, 7, 8,-tetramethoxy flavone and hydroxy-4′, 6, 7,-trimethoxy flavone (*aerial parts*).

Uses in Literature : Known earlier for asthma, burns, cough, cuts, eye complaints, hysteria, epilepsy, fuel, indigestion, sneezing, mouth ulcer, sores, toothache, veterinary diseases, wounds; and as nasal drops (Agarwal, 2003; Ambasta, 1986; Girach, 2001; Jain, 1991; Jha & Varma, 1996; Mahato

& Chaudhary, 2005; Manandhar, 1990; Nayak *et al.*, 2003; Prakash, 1999; Rajendran & Sikarwar, 2003; Sood & Thakur, 2004; Srivastava *et al.*, 2003c; Verma & Chauhan, 2006; Watt, 1972).

***Colocasia esculenta* Schott** (Pl. 15B)

Syn.: *C. antiquorum* (L.) Schott

Family: Araceae.

Vern. : Janglee kachalu.

English & Regional Names:

- **Eng.** : Cocoyam, Dashun, Eddo, Taro
- **Sans.** : Kachu, Kachwi
- **B.** : Kachu
- **BO.** : Kachaalu
- **H.** : Arvi, Kachalu, Ghuiya
- **Kan.** : Kachchi, Shamagadde
- **Mal.** : Kaladi, Shembu
- **Mar.** : Alu
- **Ori.** : Saru
- **Tam.** : Seppankizhangu, Shamakkilangu
- **Tel.** : Chamadumpa, Chemagadda.

Distribution : Wild as well as cultivated throughtout the hotter parts of India.

Description : Stem swollen at the base of the leaf-sheaths, arising from a hard tapering rhizome. Petiole erect, lamina thinly coriaceous. Peduncle much shorter than the petiole. Spathe pale-yellow, lamina narrowly lanceolate. Female inflorescence as long as the sterile male inflorescence.

Reproductive Cycle : June-October.

Habitat Ecology : Shady localities.

Material Examined : HPU-EL-SKS-5663; Bandla (1,280 m), Dehan (1,130 m), Lolani (1,600 m), Purani-palam (1,100 m), Tanda (1,120 m).

Parts Used : Leaves. Corms.

Folk Uses : Leaves and corms used as a **vegetable**, and **pickled** after boiling. Petiole slices mixed with paste of 'urd' pulse (*Phaseolus mungo*) and dried as nuggets for addition to various **vegetable** preparations.

Active Constituents : Phytoalexin 9, 12, 13 – trihydroxy – 10 (E) – octadecenoic acid, dihydrosterols, nonacosane, cyaniding – 3 – glucoside, β – sitosterol and stigmasterol isolated from *corms*.

Uses in Literature : Known as antidote to wasp stings and other insects, astringent, good food for infants, laxative, patrodu, rubefacient, stimulant, styptic, vegetable; and for alopaecia, bodyaches, congestion of portal system, earache, edible, flatulence, haemorrhage, industrial alcohol, inflamed glands and buboes, mucilage as a size for impermeable paper, otorrhoea (Agarwal, 2003; Ambasta, 1986; Arora, 1981; Augustine & Sivadasan, 2005; Banerjee & Ghora, 1996; Bennet *et al.*, 1991; Bondya *et al.*, 2006; Chopra *et al.*, 1956; Chowdhery, 1996; Chun-Lin & Jieru, 1995; Das, 2000; Devi, 2003; Dey & Bahadur, 1973; Gogoi & Das, 2003; Huidrom, 1996; Islam, 1996; Kulkarni & Kumbhojkar, 1996; Lal *et al.*, 1996; Murthy *et al.*, 2003; Prajapati *et al.*, 2006; Saini, 1996b; Samwatsar & Diwanji, 1996b; Sharma *et al.*, 2003b; Singh & Kumar, 2000a; Singh & Singh, 2003; Sood & Prakash, 2007).

Commelina paludosa Blume (Pl.15C)

Syn.: *C. obliqua* auct. (non Vahl) Ham. *ex* Don; *C. polyspatha* Wight; *C. maculata* Edgew.; *C. striata* Edgew.; *C. semi-ovata* Ham.; *C. communis* Roxb.; *C. donii* Dietr.

Family: Commelinaceae.

Vern. : Chhura.

Distribution : Common throughout India.

Description : Erect branched herbs upto 90 cm with sessile leaves, blue flowers and trigonous capsules. Seeds ellipsoid, compressed.

Reproductive Cycle : June-August.

Habitat Ecology : Edges of cultivation. Moist slopes. Wastelands.

Material Examined : HPU-EL-SKS-5511; Khaira (940 m), Gharthun (900 m), Pahra (950 m), Jasun (1,000 m), Tikkar (910 m).

Parts Used : Aerial Parts, Leaves.

Folk Uses : Aerial parts used as a **pot-herb**. Leaf-juice good for healing **wounds**.

Biological Activity : Plant extr. hypotensive, CNS depressant and diuretic.

Uses in Literature : Considered useful against barrenness in women, cuts and wounds, eyelid diseases, fungal infection, herpes, insect-bite, leprosy, leucorrhoea, skin diseases, urinary troubles; and as demulcent, emollient, laxative, pot herb (edible boiled rhizomes), refrigerant (Ambasta, 1986;

Arinathan *et al.*, 2003; Arora & Pandey, 1996; Bajpayee & Dixit, 1996; Bennet *et al.*, 1991; Bondya *et al.*, 2006; Chandra, 1997; Chopra *et al.*, 1956; Devi, 2003; Dwivedi, 2003; Gogoi & Das, 2003; Kapur & Singh, 1996; Kharkongor & Joseph, 1981; Kshirsagar *et al.*, 2003; Kumar & Goel, 1998; Mitaliya & Bhatt, 2003; Murthy *et al.*, 2003; Painuli & Maheshwari, 1996; Ranjan, 2003; Roy *et al.*, 1998; Saklani & Jain, 1994; Sen & Behera, 2003; Sen *et al.*, 2002; Singh & Kumar, 2000a; Sood & Thakur, 2004; Srivastava *et al.*, 2003a, c).

Cordia dichotoma **Forst.** (Pl. 15D)

Syn.: *C. obliqua* Willd.; *C. latifolia* Roxb.; *C. myxa* Thwaites; *C. lowriana* Brandis; *C. indica* Lam.

Family: Boraginaceae.

Vern. : Lasura, Lasyada.

English, Sanskrit & Regional Names:

Eng. : Sebastan Plum
Sans. : Kshudrashleshmataka, Madhubhutadruma
B. : Bahubara, Chhotabohnaari
Bo. : Bargund
G. : Rayagundo, Vadagundo
H. : Chhotalasura, Lasura
Kan. : Chikka Challe
Mal. : Madaviriyasam, Vidi
Mar. : Chokri, Goden
Ori. : Domolo, Kophokoturi
P. : Laswara
Tam. : Naruvilli, Viriyam
Tel. : Chinnanakkera, Nekkara.

Distribution : Throughout India upto 650 m.

Description : A medium-sized tree having ovate or orbicular, tomentose leaves when young and glabrous at adult stage. Corolla white. Drupes a yellowish-green with viscid pulp.

Reproductive Cycle : April-July.

Habitat Ecology : Common in moist forests.

Material Examined : HPU-EL-SKS-5438; Balah (900 m), Bhauda (1,050 m), Jasun (1,000 m), Malahu (1,050 m), Rangru (910 m).

Parts Used : Inflorescence. Leaves. Stem. Fruit.

Folk Uses : Flower-buds and fruits consumed as **vegetable**, and also **pickled**; and considered **anthelmintic**, **astringent** to bowels and **expectorant**. Leaves lopped for **fodder**. Stem as **fuelwood**.

Uses in Literature : Used in India as an anthelmintic, antidysenteric, astringent, blood purifier, colic, demulcent, diuretic, expectorant, fibre yielding, snakebite, vegetable; and for agricultural implements, anasarca, boats, bone fractures (vet.), bleeding, burning of throat, chest diseases, cholera, chronic fever, cold, constipation, cough, diseases of chest, dropsy, dysentery, dyspepsia, edible purposes (fruits), expelling retained placenta on childbirth, expelling worms (vet.), fodder, gonorrhoea, headache, indigestion, jaundice, joint pains, lung disease, pickling, soothing effect on skin, spleen diseases, stomachache, syphilis, ulcers, urinary diseases, urticaria, and well-curbs (Agarwal, 2003; Ambasta, 1986; Arora & Pandey, 1996; Beg *et al.*, 2005; Bennet *et al.*, 1991; Chopra *et al.*, 1956; Dagar & Dagar, 1996; Das & Agarwal, 1991; Dey & Bahadur, 1973; Diwakar & Ansari, 2003; Girach *et al.*, 1997; Hosagoudar & Henry, 1996a; Jain, 1991; Khare, 2004; Kirtikar & Basu, 1984; Kumar & Pullaiah, 2003; Lalramnghinglova, 1996, 2002; Mitaliya *et al.*, 2003; Nayak *et al.*, 2003; Negi & Gaur, 1991; Parmar & Kaushal, 1982; Patole & Jain, 2002; Pawar & Patil, 2004; Prajapati *et al.*, 2006; Rama Rao & Henry, 1996; Rao, 1996; Roy *et al.*, 1998; Saini, 1996 a; Singh, 1996b, 2000, 2001; Singh & Prakash, 1996; Sood & Kaushal, 2008; Srivastava *et al.*, 2003b; Sudhakar & Vedavathy, 2000; Tirkey, 2004; Watt, 1972).

***Coriandrum sativum* L.** (Pl. 15E)

Family: Apiaceae.

Vern. : Beenh, Dhania.

English, Sanskrit & Regional Names:

Eng.	: Coriander Fruit
Sans.	: Allaka, Dhanya, Janapriya, Kustumburi, Nisara
B.	: Dhane
G.	: Dhana, Dhania, Konphir
H.	: Dhania, Dhaniya
Kan.	: Havija, Kotambari, Kothambri
Mal.	: Kothumpalari, Kotampalari
Mar.	: Dhane, Dhanya, Khobir, Khotimbir, Kothmir
Tam.	: Kotamalli
Tel.	: Dhaniyalu, Kotimiri.

Distribution : All over India.

Description : Glabrous annual herbs; upper leaves short-petioled. Flowers in compound umbels. Fruits cremocarp, orbicular, ribbed.

Reproductive Cycle : February-April.

Habitat Ecology : Cultivated land. Often met as an escape in moist places.

Material Examined : HPU-EL-SKS-5558; Bindraban (1,270 m), Chauki (900 m), Jiun (1,550 m), Kathena (1,290 m), Paror (940 m).

Parts Used : Leaves. Fruits.

Folk Uses : Leaves employed for flavouring **curries**, and in preparation of **chutney**; its juice applied to remove skin **blemishes**. Leaves eaten for improving **eyesight**. Fruits employed as **condiment**, and considered **refreshing**.

Active Constituents : These are β-sitosterol, D-mannitol and flavonoid, glycosides (*seeds*), α-pinine, limonene, β-phelliandrene, linalool, borneol, β-caryophyllene, citronellol, geraniol, elemol and methylheptenone (*seed oil*).

Uses in Literature : Considered so far acrid, anthelmintic, antipyretic, aphrodisiac, appetiser, diuretic, laxative, stimulant, stomachic, tonic to the brain, heart and liver; and for analgesic, biliousness, bleeding of the gums, bodyache, bronchitis, condiment, cooling, dysentery, dyspepsia, flavouring, headache, hypnotic, mouth ulcers, nose-bleeding, perfumery, piles, scabies, scrofula, sweet, thirst, vomiting and applied to ulcer on the penis (Ambasta, 1986; Aswal & Mehrotra, 1994; Chauhan, 1999; Kirtikar & Basu, 1984; Pandey & Verma, 2002; Parrotta, 2001; Kashirsagar & Singh, 2000; Sharma & Singh, 2001).

***Coronopus didymus* (L.) Sm.** (Pl. 15F)
Family: Brassicaceae.

Vern. : Garhbeenhi.

Distribution : A widely distributed weed in Asia, Europe and N. America; occuring as nationalized weed in India.

Description : Prostrate or ascending herbs, often forming a rosette. Leaves pinnatifed. Flowers minute, pale green. Pods separating into 2. Seeds brown or brown-black.

Reproductive Cycle : December-April.

Habitat Ecology : Shady places.

Material Examined : HPU-EL-SKS-5554; Balu (1,030 m), Bhagpur (1,600 m), Ladoh (980 m), Sidhpur (1,100 m), Sungal (1,050 m).

Parts Used : Aerial Parts.

Folk Use : Aerial parts cooked as **vegetable**.

Uses in Literature : Described earlier as a vegetable; and for earache (Singh & Kumar, 2000a; Gopal & Chandra, 2003; Srivastava *et al.*, 2003b).

Costus speciosus Sm. (Pl. 16A, B)

Syn.: *C. arabicus* Jacq.; *Hellenia grandiflora* Retz.; *Banksia speciosa* Koenig *ex* Retz.

Family: Zingiberaceae.

Vern. : Jangli-haldi.

Sanskrit & Regional Names:

Sans.	: Bramhatirtha, Kashmira, Kushtha, Kusthabeda, Padmakarna, Padmapatramula, Padmapunya, Paushkara, Pushkarajata, Pushkarshifa, Subandhu, Vira, Virapushkaravhaya, Vriksharuha
Ass.	: Jamlakhuti
B.	: Keu, Kust
Bo.	: Gudurichakanda, Kemuka
G.	: Pakaramula
H.	: Keu, Kust
Kan.	: Chengalvakoshtu
Mal.	: Anakkuva, Anappu, Channa, Channakuva, Marujanna, Narikkurampu, Patimukam, Pushkarmulam
Mar.	: Penva, Pinnaga, Pushkarmula
Ori.	: Chittorokudho, Ghanapohora, Kudho
Tam.	: Kottam, Kudaram, Kugaimanjal, Kuravam, Kuttaiyidukki, Kuttam, Malaiva-sambhu
Tel.	: Chengalvakostu.

Distribution : U.P.- Sikkim upto 1,500 m.

Description : Undershrubs. Leaves spirally arranged with funnel-shaped white flowers having contrasting large, bright red bracts. Capsule red. Seeds arillate.

Reproductive Cycle : June-August.

Habitat Ecology : Forests. Shrubberies.

Material Examined : HPU-EL-SKS-5510; Bhauda (1,050 m), Bodhal (970 m), Dehan (1,130 m), Jasun (1,000 m), Saparu (1,250 m).

Parts Used : Rhizome.

Folk Uses : Poultice of rhizome hastens suppuration of **boils**; its juice considered good for **earache**.

Active Constituents : Active ingredients of *rhizome* are tetradecyl-11-methyl-tridecanoate, 14-oxotricosanoic acid, 14-oxohepta-cosanoic acid, 15-oxo-octacosanoil acid, methyl 3-(4-hydroxyphenyl)-2E-propeonate, tigo-genin, and diosgenin.

Biological Activity : Antifertility, antispasmodic, anti-inflammatory, antiarthritic, and CNS depressant activities of alkaloidal fraction from rhizome have been found to be +ve.

Uses in Literature : Known earlier as anthelmintic, antidote to snakebite, aphrodisiac, coolant, diuretic, edible (leafy stems), febrifuge, opthalmic, purgative, preservative, tonic and urinary disorders, and for anthrax, asthma, burning sensation, blisters, catarrhal fever, cuts, dog bite, dyspepsia, earache, headache, inflammation, internal injury (vet.), jaundice, mumps, nausea, pinworm trouble, reducing the size of uterus, rheumatism, sexual debility, skin diseases and tuberculosis (Agarwal, 2003; Ambasta, 1986; Aminuddin & Girach, 1991, 1993; Arora & Pandey, 1996; Augustine & Sivadasan, 2005; Bhandary & Chandrashekar, 2002; Borthakur *et al.*, 2004; Brahmam & Saxena, 1990; Chauhan, 1999; Chopra *et al.*, 1956; Dash *et al.*, 2003; Dey & Bahadur, 1973; Diwakar & Ansari, 2003; Gogoi & Das, 2003; Goud & Pullaiah, 1996a, b; Islam, 1996; Jain, 1968; Jain & Saklani, 1992; Jain *et al.*, 1991; Jamir & Rao, 1990; Kirtikar & Basu, 1984; Kshirsagar & Singh, 2000; Lalramnghinglova, 2002; Mahato & Mahato, 1996; Mandal & Basu, 1996; Nadanakunjidam, 2003; Nath & Maiti, 2003; Nayak *et al.*, 2003; Prajapati *et al.*, 2006; Prasad *et al.*, 1999; Radhakrishnan *et al.*, 1996; Rai & Upadhyay, 1996; Rajan *et al.*, 2005; Rajendran & Sikarwar, 2003; Rawat & Chowdhury, 1998; Reddy *et al.*, 1997; Roy *et al.*, 1998; Saklani & Jain, 1994; Shah & Singh, 1990; Saini, 1996a; Samant *et al.*, 2001; Sharma, 2003; Sharma *et al.*, 2003b; Siwakoti & Siwakoti, 2000; Sood & Prakash, 2007; Thomas & Britto, 2003; Tirkey, 2004; Upadhyay, 2003; Varma, 1997).

Cryptolepis buchanani **Roem. & Schult.** (Pl. 16C)

Syn.: *C. reticulata* Wall.; *Nerium reticulum* Roxb.; *Echites reticulata* Roth; *E. cuspidata* Heyne

Family: Asclepiadaceae.

Vern. : Dudhi-bel.

Sanskrit & Regional Names:

- **Sans.** : Kashmari
- **Ass.** : Gamar
- **H.** : Dudhi, Karanta
- **Mal.** : Kalipalvalu, Kattupalvalli
- **Ori.** : Gopikonioro, Maloti
- **Tam.** : Umi-Thekku
- **Tel.** : Adavipalatige, Gurgupaltige, Rokalliphala.

Distribution : Throughout India, ascending to 1,350 m.

Description : Twining shrubs with milky latex and yellow-green flowers in racemes. Seeds oblong-ovate, hairy which are contracted below the tip.

Reproductive Cycle : April-November.

Habitat Ecology : Wastelands. Forests.

Material Examined : HPU-EL-SKS-5645; Balah (900 m), Bhaura (940 m), Makol (900 m), Tapa (945 m), Tikkar (910 m).

Parts Used : Leaves. Stem.

Folk Uses : Leaves given as **fodder** for sheep and goats, for enhancing **lactation**. Stem used for **cordage** purposes.

Active Constituents : Active properties of *leaves* due to cardenolide-cryptosin and crytanoside B.

Uses in Literature : Recorded earlier as fibre yielding, galacatagogue (vet.); and for abdominal pain, anasarca, ascites, babesiosis, bleeding, bodyache, toothache, increase lactation, jaundice, bone fractures, cholera, cuts, dropsy, dysentery, fever, fissures on sole, fits, making ropes and cordage, paralysis, rheumatism, rickets in children, skin diseases, snakebites, sores, stomachache, syphilis, venereal diseases, worshipping (twigs worshipped in the harvesting place to get good yield) and yoke gall (Ambasta, 1986; Brahmam *et al.*, 1996; Henry *et al.*, 1996; Hosagoudar & Henry, 1996a-c; Jain, 1991; Nadanakunjidam, 2003; Rama Rao & Henry, 1996; Rana *et al.*, 1996; Rao, 1996; Ravishankar & Henry, 1992; Reddy *et al.*, 2003b; Singh, 1996a; Singh & Prakash, 1996; Singh & Rao, 2003; Watt, 1972).

Cucumis melo L. *var. agrestis* (Pl. 16D)

Syn.: *C. trigonus* Roxb.; *C. pubescence* Willd.

Family: Cucurbitaceae.

Vern. : Phaphanu.

Sanskrit & Regional Names:

Sans. : Bahaphala, Chitra, Chitraphala, Chitravalli, Deis, Garakshi-vrishka, Godumba, Katphala, Kumbhari, Laghuchirbhita, Maruya, Mrigadani, Mrigakshi, Mrigeshana, Pathya, Vichitra, Vishala

B. : Bangumak, Gomuk

H. : Bhakwia, Bislambhi, Bislombi, Gorakhakakadi, Janglindrayan;

Tam. : Adavipuchcha, Kodibudana, Kodibudinga, Nallapudama.

Distribution : Throughout India, Sri Lanka, Afghanistan, Persia, Malaya, N. Australia.

Description : Annual climbing or trailing, hispid or scabrous herbs. Leaves deeply cut into 5-7 lobes. Flowers monoecious, yellow. Fruits fleshy, globose or cylindrical.

Reproductive Cycle : May-August.

Habitat Ecology : Grasslands. Maize fields.

Material Examined : HPU-EL-SKS-5578; Bhaura (940 m), Dadh (1,100 m), Dehan (1,130 m), Rangru (910 m), Saparu (1,250 m).

Part Used : Fruits.

Folk Uses : Ripe fruits **edible**.

Uses in Literature : Known so far as astringent; and for bilious disorders, edible, fever, snakebite, purging (Agarwal, 2003; Ambasta, 1986; Chopra *et al.*, 1956; Das & Agarwal, 1991; Kirtikar & Basu, 1984; Roy *et al.*, 1998; Sharma & Rana, 2005; Vartak, 1981).

***Cucumis sativus* L.** (Pl. 16E)

Syn.: *C. muricatus* Wall.; *C. hardwickii* Royle

Family: Cucurbitaceae.

Vern. : Kheera.

English, Sanskrit & Regional Names:

Eng. : Cucumber, Musk Melon, Sweet Melon

Sans. : Bahpuspha, Kandalu, Kantakilaka, Kantakiphala, Pitapushpa, Sudhavasa, Sushitala, Trapukarkati, Trapusha, Tundilaphala

B. : Khira, Sasa

Bo. : Kakri, Kankari

G. : Khira

Mar. : Kakdi, Khira, Tavase

Ori. : Kaknai

P. : Khira, Khiyar

Tam. : Piping Kay

Tel. : Dozakaya.

Distribution : Cultivated throughout India.

Description : Climbing herbs with hirsute stem and simple tendrils. Leaves hispid on both sides. Flowers yellow. Fruits oblong, cylindrical.

Reproductive Cycle : May-September.

Habitat Ecology : Grown locally.

Material Examined : HPU-EL-SKS-5561; Banuri (1,150 m), Bhawarna (1,100 m), Chachian (1,300 m), Chornali (1,280 m), Sulah (950 m).

Part Used : Fruit.

Folk Uses : Fruits **edible** and its juice considered **refreshing**; also prescribed to check **kidney stones**, **blemishes** and **black circles** around eye.

Active Constituents : Active properties due to palmitic, stearic and oleic acids, their glycerides, sterols, squalene (*fruits*).

Uses in Literature : Useful earlier as diuretic, facial cosmetics, fruit salads, gastric (vet.), refrigerant, tonic; and for confectionery, fever, headache, locust swarm on nagali, perfumery, pnuemonia, sunstroke, sweets, throat affections, urinary infections, and whooping cough (Ambasta, 1986; Jain, 1991; Kulkarni & Kumbhojkar, 1993, 1996; Negi & Gaur, 1991; Satapathy & Brahmam, 1996; Vihari, 1995; Watt, 1972).

Cucurbita hispida Thunb. (Pl. 16F)

Syn.: *Benincasa cerifera* Savi

Family: Cucurbitaceae.

Vern. : Dhuda-petha.

English, Sanskrit & Regional Names:

Eng. : Ash Pumpkin, Tallow Gourd, White Gourd Melon
Sans. : Gramya Karkati, Kushmanda, Nagapuspa Phala
B. : Chalkumra
Bo. : Kohala
H. : Golkaddu, Petha
Kan. : Budagumbala Kayi
Mal. : Kumpalanna
Mar. : Kohala
P. : Chalkumra, Golkaddu
Tam. : Kalytanppushinikkay
Tel. : Burdagumuda, Pendligummadikaya.

Distribution : Throughout India.

Description : Climbers with rigid petiole having pungent hairs and large yellow flowers. Fruits flesh white berry. Seeds flat.

Reproductive Cycle : June-September.

Habitat Ecology : Grown in house gardens.

Material Examined : HPU-EL-SKS-5484; Chhanchri (930 m), Darugnu (1,400 m), Gopalpur (1,350 m), Malnu (1,000 m), Paror (940 m).

Parts Used : Twigs. Fruit.

Folk Uses : Tender twigs consumed as **vegetable**. Fruits used for performing **sacrificial rites** and hung on the main entrance of house to keep away **evil spirits**; also find extensive use in confectionery in preparation of a **sweet** called 'petha'.

Active Constituents : *Fruits* possess β-sitosterol, lupeol, n-tricontanol, mannitol, arginine, aspartic acid, glutamic acid, asparagine, glutamine, proline, hydroxy proline, isoleucine, cysteine, L-leucine, glucose and rhamnose.

Uses in Literature : Reported earlier as an anthelmintic, aphrodisiac, diuretic, general tonic for horses, tonic in insanity, vegetable (flowers); and for asthma, epilepsy, haemoptysis, intestinal worms, preserving corpse, and tuberculosis (Ghosh *et al.*, 1996; Jha *et al.*, 1996; Lalramnghinglova, 1996; Mahato & Mahato, 1996; Rama Rao & Henry, 1996; Sharma & Borthakur, 2003; Siwakoti & Varma, 1996).

***Cucurbita pepo* L.** (Pl. 17A)
Syn.: *C. pepo* Roxb.
Family: Cucurbitaceae.

Vern. : Kaddu.

English, Sanskrit & Regional Names:

Eng. : Pumpkin, Vegetable Marrow
Sans. : Karkaru, Kurkaru
B. : Kadimah, Kumra
Bo. : Kaula
G. : Dudia
H. : Kadimah, Safed Kaddu
Mar. : Kohala
Ori. : Panikakhara
Tam. : Suraikayi
Tel. : Budadegummedi.

Distribution : Throughout India.

Description : Climbers with 5-lobed leaves, large fruit and strongly grooved woody, peduncle.

Reproductive Cycle : June-September.

Habitat Ecology : Planted in house gardens.

Material Examined : HPU-EL-SKS-5716; Chornali (1,280 m), Holta (1,200 m), Rajeher (1,050 m), Sidhpur (1,100 m), Thandol (950 m).

Parts Used : Twigs. Fruit. Rind. Seeds.

Folk Uses : Tender twigs and fruits cooked as **vegetable**. Seeds **eaten** as such or after roasting. Dried pericarp used as a **utensil**.

Active Constituents : These are cholesterol, ergosterol, campesterol, β-sitosterol (*seeds*).

Biological Activity : Seeds exhibit anthelmintic activity *in vitro* and *in vivo* against *Hymenolepis nana* and *Dicrocrelium denddriticum.*

Uses in Literature : Known so far as an anthelmintic, poison (vet.), psycoactive plant, taenicide (seeds); and for burns, diarrhoea, edible purposes (leaf, fruit), utensils, worshipping (Ambasta, 1986; Beigh *et al.*, 2003; Jain, 1991; Jain *et al.*, 1994; Jha *et al.*, 1996; Watt, 1972).

Curcuma longa Wall. (Pl. 17B,C)

Syn.: *C. angustifolia* Roxb.

Family: Zingiberaceae.

Vern. : Haldee.

English, Sanskrit & Regional Names:

Eng. : Wild or East Indian Arrowroot, Narrow-leaved Turmeric, Travancore Starch

Sans. : Gavayodbhava, Payakshira, Yavaja

Bo. : Tickar

G. : Tavakhara

H. : Tavakhira, Tikhur

Mar. : Tavakila, Tavakhira

Tam. : Ararutkilangu

Tel. : Ararutgaddalu.

Distribution : Tropical Himalayas.

Description : Herbs with small, tuberous rootstocks. Flowers yellow, in a spike. Bracts pink. Capsule ovoid. Seeds small.

Reproductive Cycle : June-July.

Habitat Ecology : Moist shady slopes.

Material Examined : HPU-EL-SKS-5477; Andretta (980 m), Bhauda (1,050 m), Chhanchri (930 m), Pahra (950 m), Malnu (1,000 m).

Part Used : Rhizome.

Folk Uses : Young rhizome (150 g) fed to livestock for 2-3 days after conception to check **abortion**; also 25 g of its powder alongwith powdered seeds (50 g) of *Centratherum anthelminticum* roasted in 'desi ghee'(100 g) and prescribed for '**bashahar**' (boils all over the body) both for **human beings** and **cattle** every morning for 15 days. Product of rhizome paste soaked in citrus fruit-juice and covered with leaves of *Justicia adhatoda* for 2-3 days for preparing '**kungu**' and used as **vermillion** in various **religious ceremonies**. Paste of rhizome roasted in mustard oil applied for healing **cuts** and suppuration of **boils**. Its powder blended with milk and taken frequently to check **cough** and **internal injuries**.

Uses in Literature : Reported earlier as antiseptic, blood purifier, condiments, curative, stomachic, symbolic (as sacred food), tonic, yellow dye; and for asthma, bone fractures, cold, cough, diarrhoea, fever, insect bite, internal injuries, jaundice, pickles, soups, sprain and swelling of body (Bennet *et al.*, 1991; Borthakur *et al.*, 2004; Dey, 1973; Gangwar & Ramkrishnan, 1990; Jain, 1991; Kumar *et al.*, 2003; Mahato & Chaudhary, 2005; Mokat & Deokule, 2004; Nadanakunjidam & Abirai, 2003; Nath & Maiti, 2003; Nayak *et al.*, 2003; Pal & Paul, 2003; Pandey, 2003; Panthi & Chaudhary, 2003; Rao & Jamir, 1982; Saikia & Nath, 2003; Sharma *et al.*, 2003b; Srivastava *et al.*, 2003a; Thomas & Britto, 2003).

Cuscuta reflexa Roxb. (Pl. 17D)

Syn.: *C. grandiflora* Wall.; *C. verrucosa* Sweet; *C. hookeri* Sweet; *C. macrantha* G. Don; *C. elatior* Choisy

Family: Convolvulaceae.

Vern. : Akashbel, Amarbel,Gaasbel.

English, Sanskrit & Regional Names:

Eng. : Dodder
Sans. : Akshspavana, Amaravallika
Ass. : Amarlati
B. : Hadialgusilutta
G. : Amarbel
H. : Akashabel, Amarabela
Mar. : Nirmulia, Kashavela
Ori. : Kolanirmuli
P. : Amil-nilathari, Niradhar, Zarbuti

Tam. : Kodiya Gundal
Tel. : Savarapukada
U.P. : Ka-banwar, Banwar.

Distribution : Afganistan - S.W. China: 600-3,300 m.

Description : Climbers with succulent stem, densely overlacing small trees or shrubs. Flowers waxy white and fragrant. Inflorescence racemose. Capsules fleshy, black with 2-4 seeds.

Reproductive Cycle : July-October.

Habitat Ecology : Forests. Waste places.

Material Examined : HPU-EL-SKS-5542; Arla (1,100 m), Boda (1,000 m), Chachian (1,300 m), Nadli (1,000 m), Thakurdwara (1,150 m).

Parts Used : Whole Plant.

Folk Uses : Bath of the water boiled with crushed plant parts prescribed for relief from **post-delivery pains** and **rheumatism**. One teaspoonful of stem extract taken orally for seven days to check **jaundice**. Paste of aerial plant parts applied to relieve **swelling of testicle**.

Active Constituents : Main ingredients are dulcitol, luteolin, quercitin, a glucoside of luteolin (*stem*), and luteolin, quercetin, a glycoside of luteolin, amerbelin, cuscutin and wax (*plant*).

Biological Activity : Aq. and Alc. extrs. have relaxant, cardiotonic, spasmolytic and acetylcholine-like action in atropinised rabbit auricles and frog's heart.

Uses in Literature : Reported to be used as alterative, anodyne, anthelmintic, astringent, blood purifier, cough, cardiac, carminative, demulcent, depurative, diaphoretic, fodder, psychoactive plant, purgative, symbolic (sorcery), wounds, redding, of eyes, tonic; and for antifertility, bilious disorders, bodyache (vet.), bone fracture (vet.), bronchitis, burns, cold, cough, cuts, dermatitis, diarrhoea, diptheria, displaced kneecap, dyeing, epilepsy, eye diseases, fall of hairs, haematuria, headache, hydrocele, indigestion, inflammation, itching, jaundice, killing lice (vet.), melancholy madness, nervine weakness, oedema, poultry disease (vet.), protracted fevers, rheumatism, sores, spermatorrhoea, sprain, stomachache, swelling of legs and testicles, tongue ulcers, warts, and wounds (Ambasta, 1986; Chauhan, 1999; Dash *et al.*, 2003; Datt, 1996; Dixit & Pal, 1996; Dwivedi, 2003; Girach, 1992; Girach & Aminuddin, 1995; Girach *et al.*, 1996; Gogoi & Das, 2003; Hembrom, 1991; Jain, 1991; Jain *et al.*, 1991; Katewa & Arora, 1997; Kumar *et al.*, 2003; Mahato & Mahato, 1996; Maliya, 2005; Manandhar, 1985, 1987a; Nayak *et al.*, 2003; Patil, 2004; Sahoo & Mudgal, 1995; Saini, 1996a; Satapathy & Brahmam, 1996; Singh, 1996a;

Singh & Prakash, 1996; Singh *et al.*, 1996a, b, 1997; Siwakoti & Varma, 1996; Srivastava *et al.*, 2003b; Tirkey *et al.*, 2003b; Tiwari & Tiwari, 1996; Watt, 1972).

Cyclanthera pedata (L.) Schrad. (Pl. 17E)

Family: Cucurbitaceae.

Vern. : Meetha karela.

English Name:

Eng. : Wild cucumber.

Distribution : H.P.- S.W. China.

Description : Annual vines. Stem branched, glabrous. Leaves ovate-orbicular, margins serrulate-dentate. Flowers whitish. Fruits greenish-white, striate. Seeds black, turtle-shaped.

Reproductive Cycle : August-September.

Habitat Ecology : Along field sides.

Material Examined : HPU-EL-SKS-5532; Bhagpur (1,600 m), Bhuana (1,400 m), Darugnu (1,400 m), Lolani (1,600 m), Panchrukhi (990 m).

Part Used : Fruits.

Folk Uses : Fruits cooked as a **vegetable**.

Uses in Literature : Known as a vegetable (Sharma & Rana, 2005).

Cymbopogon citratus (DC.) Stapf (Pl. 17F)

Syn.: *Andropogon citratus* DC.

Family: Poaceae.

Vern. : Makoda.

English, Sanskrit & Regional Names:

Eng. : Melissa Grass, West Indian Lemon Grass

Sans. : Atigandha, Badhira, Bhustrina, Chhatra

B. : Gandhabena

G. : Lilicha, Lolacha

H. : Agin Ghas, Gandhatrina

Kan. : Maijigehullu, Purhalihulla

Mal. : Shambharapulla, Vasanappula, Vasana pulls, Vasanappula

Mar. : Mirvacha, Olaucha

	P.	: Khawi
	Tam.	: Karpurappillu, Vasanappillu
	Tel.	: Chippagaddi, Nimmagaddi, Vasanagaddi.

Distribution : Widely distributed over the tropics.

Description : Perennial grasses. Leaves in fascicles, aromatic. Flowers reddish, in panicles.

Reproductive Cycle : August-March.

Habitat Ecology : Dry slopes. Meadows.

Material Examined : HPU-EL-SKS-5498; Bhuana (1,400 m), Chornali (1,280 m), Khaira (940 m), Sidhpur (1,100 m), Tikkar (910 m).

Part Used : Whole Plant.

Folk Uses : Plant considered a good **fodder** for cattle. Root juice possesses **antiseptic** properties and good for washing **sores** and **boils**. Stem chewed as such to treat **leucorrhoea**; its decoction taken to relieve **body heat** and for making local wine (**'sur'**).

Active Constituents : *Aerial parts* contain cymbopogone, cymbopogonol, α-and β-citrals, myrcene, methylheptanone, linalool, linalyl acetate, geranyl acetate, citronellol, nerol, luteolin and 6-C-glucosides.

Biological Activity : Antimicrobial activity confirmed.

Uses in Literature : Recorded as alexipharmac, aphrodisiac, anthelmintic, appetiser, bitter, carminative, diphoretic, diuretic, febrifuge, hot in spasmodic affections of the bowels, laxative, pungent; stimulant, sudorific and tonic; and for cold, cough, digestive disorders, fever, gastric irritability, headache, malaria, measles, neuralgia, painful affections, pains, perfumery rheumatism, ringworm and sprains (Ambasta, 1986; Chevallier, 1996; Chopra *et al*., 1956; Chun-Lin & Jieru, 1995; Gill *et al*., 1993, 1997; Harborne & Baxter, 2001; Jain & Sharma, 2000; Jain *et al*., 1995; Kirtikar & Basu, 1984; Kurian, 1995; Nunez & Castro, 1995; Parrotta, 2001; Prajapati *et al*., 2006; Srivastava *et al*., 2003b).

***Cynodon dactylon* (L.) Pers.** (Pl. 18A)

Syn.: *C. linearis* Willd.; *C. maritimus* H. Bock; *C. occidentalis* Willd.; *C. portoricensis* Willd.; *C. radiatus* Roth; *C. sarmentosus* Gray; *C. stellatus* Willd.; *C. virgatus* Nees; *Chloris cynodon* Trin.; *Dactylon officinale* Vill.; *Fibichia umbellata* Koel.

Family: Poaceae.

Vern. : Dharub, Daroov.

English, Sanskrit and Regional Names:

Eng. : Couch Grass, Creeping Panic Grass, Doorwa
sans. : Dhurba, Durva; Durva, Dhruva, Druva, Haritali
B. : Dhub, Dubla, Durba
G. : Dhro, Dhrokad, Durva, Gharo
H. : Dhoboghas, dhub, Dub, Hariali
Mal. : Karuka, karukappullu
Mar. : Durva, Haryeli, Karala
P. : Daurva, Dub, Dubra, Kabbar, Khakkal, Talla
Tam. : Arugam-pilla, Hariali
Tel. : Gaddi, Gurka, Ghericha, Haryali.

Distribution : Throughout India.

Description : Perennial creeping grasses with narrow, flat, glabrous leaves and hairy ligules. Spikes radiating, green or purplish and rachis very slender. Spikelets minute ovate. Keels scabrid. Flowers purplish, borne on spikes. Grains laterally compressed.

Reproductive Cycle : July-January.

Habitat Ecology : Fields borders. Grasslands. Open spaces.

Material Examined : HPU-EL-SKS-5575; Andretta (980 m), Bhagpur (1,600 m), Chandpur (1,290 m), Chornali (1,280 m), Gopalpur (1,350 m), Rangru (910 m).

Parts Used : Aerial plant parts. Leaves.

Folk Uses : Leaf-juice **haemostatic**, **refrigerant**, **diuretic** and also taken to check **nose-bleeding**. Aerial parts used for **worshipping** especially 'Ganesha'.

Active Constituents : These are 3-methoxy-4-hydroxyl-benzoic acid, 4-hydroxyl benzoic acid, phytol, sitosterol and stigmasterol acetate (*aerial parts*).

Uses in Literature : Considered earlier acrid, anasarca, antidote to snake-bite, astringent, diuretic, expectorant, haemostatic, sweet, vulnary; and for biliousness, blood diseases, bruises, burning sensation, chest burning, cobra-bite, cooling, cuts, diarrhoea, dropsy, dysentery, dysuria, epileptic fits, fatigue, fever;

for retained placenta (vet.), goat and rheumatism, heartburn, improves digetion, leprosy, menorrhagia, nose-bleeding, piles, prickly heat, scabies, scorpion sting, skin diseases, sores, stomatitis, swelling, tuberculosis, ulcers, urticaria, thirst, vomiting and wounds (Ambasta, 1986; Bhandary & Chandrashekar, 2002; Bhatt *et al.*, 2001; Bhuyan, 2003; Bose, 1984; Chauhan, 1999; Chopra *et al.*, 1956; Diwakar & Ansari, 2003; Dwivedi, 2003; Ghose, 1997; Gogoi & Borthakur, 2001; Gogoi & Das, 2003; Kapur & Singh, 1996; Kirtikar & Basu, 1984; Jain *et al.*, 1994; Maya *et al.*, 2003; Nautiyal *et al.*, 2003; Nayak *et al.*, 2003; Panthi & Chaudhary, 2003; Parrotta, 2001; Raghupathy & Mahadevan, 1991; Ramdas *et al.*, 2000; Ranjan, 2003; Rawat & Chowdhury, 1998; Reddy *et al.*, 2003b; Roy *et al.*, 1998; Said, 1997; Saikia & Nath, 2003; Sharma *et al.*, 2001; Sen & Batra, 1997; Sharma, 2003; Sharma & Rana, 2005; Sharma *et al.*, 2003b; Singh & Pandey, 1998; Singh & Singh, 2003; Srivastava *et al.*, 2003b; Subramani & Goraya, 2003; Swami & Gupta, 1996; Upadhyay & Singh, 2005).

***Cyphomandra betacea* (Cav.) Sendtn.** (Pl. 18B)

Syn.: *Solanum betaceum* Cav.

Family: Solanaceae.

Vern. : Desi Tamatar, Bada-tamater.

English Name:

Eng. : Tree Tomato.

Distribution : Cultivated.

Description : Small trees. Leaves ovate, alternate with entire margins. Fruits green berry when young and reddish-black on ripening.

Reproductive Cycle : May-September.

Habitat Ecology : Planted near houses.

Material Examined : HPU-EL-SKS-5623; Bodhal (970 m), Darugnu (1,400 m), Panapari (1,100 m), Paror (940 m), Thandol (950 m).

Part Used : Fruits.

Folk Uses : Fruits **eaten** raw or cooked as **vegetable**.

Active Constituents : Isolated constituents are pelarigonidin-3-glucosy-lglucoside (*peel of fruits*) peonidin-3-glucosyl glucoside and malvidin-3-glucosylglucoside (*seeds*).

Uses in Literature : Used in India for edible purposes (fruit), jams and preserves (Ambasta, 1986).

***Dalbergia sissoo* Roxb.** (Pl. 18C)
Syn.: *D. pendula* Ten.
Family: Fabaceae.

Vern. : Shisham, Tahli.

English, Sanskrit & Regional Names:

Eng. : South Indian Red Wood
Sans. : Agurushinshupa, Mandalapatri
Ass. : Sissu
B. : Shishu
Bo. : Sissu,Tanach
G. : Sissom, Tanach
H. : Shisham, Sissai
Kan. : Agaru, Biridi
Mal. : Iruvil
Ori. : Simsapa, Sisu
P. : Nelkar, Shewa, Shishai, Tali
Tel. : Etta Sissu, Sinsupa.

Distribution : Throughout India ascending to 1,650 m.

Description : Trees with pale-green, odd pinnate leaves and 3-5, broadly ovate leaflets. Flowers yellow-white, complete. Pods glabrous, flat, indehiscent. Seeds with hard seed coat, cotyledons 2.

Reproductive Cycle : May-August.

Habitat Ecology : Grassy slopes.

Material Examined : HPU-EL-SKS-5427; Boda (1,000 m), Khaira (940 m), Jasun (1,000 m), Thandol (950 m), Tikkar (910 m).

Parts Used : Leaves. Wood.

Folk Uses : Leaf-paste mixed with jaggery and ash of tobacco leaves applied to hasten **suppuration** of **boils**. Poultice of leaves useful to check **inflammation** of toes. 1 teaspoonful of leaf-extr. taken daily for seven days to check **leucorrhoea**. Wood good for **furniture**, **agricultural implements**, **building works** and as **fuel**.

Active Constituents : Principal ingredients are biochanin A-7-glucoside (sissortin), isoflavone-0-glycoside (*leaves*), and a new allylphenol of latifolin type–dalbergiphenol along with dalbergenone, dalbergin, methyldalbergin (*heartwood*).

Biological Activity : Spasmolytic.

Uses in Literature : So far recorded as an alterative, antiemetic, astringent, excellent fuelwood, fodder, ornamental (ear pendent), sacred, spasmolytic (vet.); and for agricultural implements, allaying vomiting, building works, diarrhoea, boils, cabinet wood, carved work, charcoal, charpai legs, diarrhoea (vet.), dysentery, eruptions, framing of carriages, gonorrhoea, headache, high class furniture, leprosy, musical instruments, naves of wheel, railway sleepers, scurf, skin diseases, and sleepers (Ambasta, 1986; Borthakur & Sharma, 1996; Diwakar & Ansari, 2003; Dwivedi, 2003; Jain, 1991; Kumar *et al.*, 2003; Mandal & Basu, 1996; Nayak *et al.*, 2003; Shukla *et al.*, 2003; Siwakoti & Varma, 1996).

Datura innoxia Mill. (Pl. 18D)

Family: Solanaceae.

Vern. : Dhatura.

Sanskrit & Regional Names:

Sans. : Dhustura
Kan. : Ummatta
Tam. : Vellummattai
Tel. : Datturamu.

Distribution : N.W. Himalayas.

Description : Undershrubs with broadly-ovate leaves and creamish flowers. Capsule armed with spines.

Reproductive Cycle : June-August.

Habitat Ecology : Wastelands. Shady areas.

Material Examined : HPU-EL-SKS-5409; Bhaura (940 m), Chimbalhar (1,150 m), Gopalpur (1,350 m), Lolani (1,600 m), Kathena (1,290 m).

Parts Used : Flowers. Seeds.

Folk Uses : Flowers emnployed **offered** to appease Lord 'Shiva'. Seed paste mixed with hot mustard oil applied locally for **rheumatic** pain.

Uses in Literature : Known to be used as an anodyne, antihydrophobic, antiseptic, sedative and cerebral depressant; and for asthma, boils, country liquor, diarrhoea, fever, pyorrhoea, rheumatism, skin diseases and offering to Lord Shiva (Agrawal, 1981; Hajra, 1987; Jain *et al.*, 1991; Kumar *et al.*, 2003; Saxena & Vyas, 1981; Shah & Joshi, 1971; Shekhawat & Anand, 1984).

***Datura stramonium* L.** (Pl. 18E)

Syn.: *D. stramonium* var. *tatula* (L.) Cl.; *D. tatula* L.; *D. ferox* Nees; *D. wallichii* Dunal; *D. inermis* Jacq.; *Stramonium vulgatum* Gaertn.

Family: Solanaceae.

Vern. : Dhatura.

English, Sanskrit & Regional Names:

Eng. : Apple of Peru, Devil's Apple, Devil's Trumphet, Dewtry, Jamestown Weed, Mad Apple, Stink Weed, Thorn Apple

Sans. : Devika, Dhattura, Dhurta, Dhurtakrit, Dhustura, Dhultura, Ghantapushpa, Kalama, Kanaka, Kanakaohaya, Kantaphala, Madanaka, Mahamohi, Mahashatha, Matulaka, Savisha, Shivapriya, Shivasekhara, Shyama, Turi, Unmatta, Unmattaka

B. : Sadadhutura

G. : Dholo Dhaturo

H. : Dhatura

Kan. : Ummatta

Mal. : Matulam, Ummam, Ummata

Ori. : Dhutura, Sukladhutura

P. : Dattura, Tattur

Tam. : Emanamam, Simalyumattai, Turutturam, Umattai, Vellumattai

Tel. : Dutturamu, Tellavummetta, Ummetta.

Distribution : Temperate Himalayas (Kashmir-Sikkim): 200-2,200 m.

Description : Shrubs. Leaves stalked, irregularly lobed and toothed. Flowers white. Capsule ovoid with rigid, sharp prickles.

Reproductive Cycle : June-October.

Habitat Ecology : Wastegrounds. Cultivated areas.

Material Examined : HPU-EL-SKS-5455; Banuri (1,150 m), Chimbalhar (1,150 m), Dharaman (1,100 m), Malahu (1,050 m), Saparu (1,250 m).

Parts Used : Leaves. Flowers.

Folk Uses : Flowers employed as **offering** to appease Lord Shiva. Smoke of burning leaves inhaled to cure **asthma**.

Active Constituents : Various constituents are vitanolide-vitastra-monolide along with daturalactone and its keto derivative characterised as 5α,12α,27-trihydroxy-1-oxo-6α, 7α-epoxy-22R-vita 2,24-dienolide (*leaves*), hyoscine, 3α,6β-ditigloycoxytropane, 3α, 6β-ditigloyloxytropane-7β-OL, apoatropine, tigcoidine, tropine, meteliodine (Nigerian plant), 0.3 to 0.5% of alkaloids chiefly as hyoslyamine with smaller amount of hyoscine and atropine (*dried leaves, flowering tops*).

Uses in Literature : Recorded in literature as anodyne, antirheumatic, antispasmodic, cigarettes (leaves), mydriatic, narcotic in treatment of bronchitis, psychoactive plant; and for asthma, boils, breast inflammation due to excessive formation of milk, controlling salivation and exparasite, dental caries, diarrhoea, dislocation of joints, dysmenorrhoea, enlarged testicles, epilepsy, haemorrhoids, headache, inflammatory pains, itch, jaundice, lactation (vet.), milching (vet.), nervous disorder (vet.), nymphomania, painful diseases of rectum, paralysis, post-encephalitic parkinsonianism, purpurea, rheumatism (vet.), sciatica, skin diseases, stomach complaints, swelling of neck (vet.), tonsilitis, toothache, travel sickness, ulcers, and vicious indulgence (Beigh *et al.*, 2003; Chhetri, 2005; Chopra *et al.*, 1956; Dwivedi, 2003; Ganai & Nawachoo, 2003; Gogoi & Das, 2003; Jain, 1968, 1991; Jain *et al.*, 1994; Jha *et al.*, 1991; Manandhar, 1996; Singh, 1991; Srivastava *et al.*, 2003b; Subramani & Goraya, 2003; Upadhye *et al.*, 1994; Watt, 1972).

Debregeasia hypoleuca Wedd. (Pl. 18F)

Syn.: *D. salicifolia* (Don) Rendle; *D. bicolor* (Roxb.) Wedd.; *Boehmaria salicifolia* Don; *B. hypoleuca* Hochst.; *Urtica bicolor* Roxb.; *Missiessya hypopleuca* Wedd.; *Morocarpus salicifolius* Blume

Family: Urticaceae.

Vern. : Siaru, Syaru.

Regional Names:

P. : Pincho, Sansaru

U.P. : Sansaru, Siaru, Tusarro, Tushiari.

Distribution : N.W. Himalayas- 1,500-2,400 m.

Description : Sturdy shrubs having 3-nerved, long pointed leaves with lower surface white tomentose. Flowers yellow, unisexual. Drupes yellow, small, aggregated in heads.

Reproductive Cycle : May-August.

Habitat Ecology : Forests. Shrubberies.

Material Examined : HPU-EL-SKS-5592; Bhagpur (1,600 m), Chachian (1,300 m), Lolani (1,600 m), Rajeher (1,050 m), Sidhpur (1,100 m).

Parts Used : Stem. Leaves. Fruits.

Folk Uses : Fruits **edible**; also taken to dissolve **kidney stones**. Leaves given as **fodder** for cattle. Stem used as **fuel**.

Uses in Literature : Discribed in literature for edible purposes (*fruits*), cordage (*bark*), fishing lines, fodder, stomach pain, ulcers and killing maggots in wounds of cattle and goats (Agarwal, 2003; Ambasta, 1986; Arora & Pandey, 1996; Chandra, 1997; Kapur, 1996a; Kirtikar & Basu, 1984; Purohit & Samant, 1995; Rama Rao & Henry, 1996; Roy *et al.*, 1998; Singh & Kumar, 2000a; Sood & Kaushal, 2008; Watt, 1972).

***Deeringia amaranthoides* Merr.** (Pl. 19A)

Syn.: *D. celosiodes* R. Br.

Family: Amaranthaceae.

Vern. : Bhringya.

Distribution : N. India.

Description : Undershrubs. Leaves alternate, ovate-lanceolate. Flowers pale yellow-green, in panicles. Fruits red, globose.

Reproductive Cycle : July-October.

Habitat Ecology : Field edges.

Material Examined : HPU-EL- SKS-5662; Balah (900 m), Boda (1,000 m), Khaira (940 m), Jasun (1,000 m), Makol (900 m).

Part Used : Leaves.

Folk Use : Leaves cooked as a **vegetable**.

Uses in Literature : Used earlier for sores, and vegetable (leaves); and good as sternutatory (Agarwal, 2003; Ambasta, 1986; Arora & Pandey, 1996; Chopra *et al.*, 1956; Sharma & Rana, 2005; Sinha, 1996).

***Dendrocalamus hamiltonii* Nees & Arn. *ex* Munro** (Pl. 19B)

Syn.: *Bambusa monogyna* Griff.; *B. maxima* Ham.; *B. falconeri* Munro

Family: Poaceae.

Vern. : Maggar.

Regional Names:

Ass. : Kokua

Bo. : Pecha, Tama

H. : Kagzi-Bans.

Distribution : S.E. Asia. W. Himalayas.

Description : Bamboos having glabrous stem sheath and broadly-lanceolate, cuspidate leaves. Spikelets ovoid, depressed, purple. Anthers black, hairy, twisted.

Reproductive Cycle : Flowering sporadically.

Habitat Ecology : Forests. Shady areas.

Material Examined : HPU-EL-SKS-5534; Arla (1,100 m), Bhagpur (1,600 m), Chandpur (1,290 m), Malnu (1,000 m), Saparu (1,250 m).

Parts Used : Whole Plant. Leaves. Culms. Rootstock.

Folk Uses : Plant grown as **hedges**. Stem employed for **house building**, and also for making **winnowing pans** and **baskets**. Rootstock with basal part of culm used as **hockey stick**, and **fuel**. Young culms cooked as **vegetable**, and also **pickled**. Leaves given as **fodder** for milching cattle. '**Bans-slochan**' obtained from its culm taken to check **Ca deficiency**. One teaspoonful of watery extr. of its young shoots prescribed thrice a day for 4-5 days against **persistent fever**.

Uses in Literature : Recorded earlier as symbolic in religious ceremonies, vegetable (young culms); and for baskets, bow, building material, burning sensation after urination, clearing uterus after childbirth, cutting, fuel, headgear of bridegroom, house building, minor injury, swellings and water conduits (Ambasta, 1986; Arora & Pandey, 1996; Bennet *et al.*, 1991; Bhuyan, 2003; Borthakur 1996b; Chhetri, 2005; Chowdhery, 1996; Jain, 1991; Kumar *et al.*, 2004; Lalramnghinglova, 1996; Roy *et al.*, 1998; Saklani & Jain, 1994; Samant *et al.*, 2001; Sharma, 2003; Sharma & Rana, 2005; Srivastava *et al.*, 2003a-c; Watt, 1972).

***Didymocarpus pedicellata* R. Br.** (Pl. 19C,D)
Family: Gesneriaceae.

Vern. : Musakani.

Sanskrit & Regional Names:

Sans. : Shila, Pushpa

H. : Pathar phori.

Distribution : Foot-hills of Himalayas.

Description : Small herbs. Stem reduced to a point. Leaves ovate, crenate. Petiole long. Pods small.

Reproductive Cycle : July-March.

Habitat Ecology : Moist-shady rock-crevices.

Material Examined : HPU-EL-SKS-5573; Balah (900 m), Bhauda (1,050 m), Boda (1,000 m), Palampur (1,280 m), Tikkar (910 m).

Part Used : Leaves.

Folk Uses : Young curried leaves (5 g) in combination with leaf-juice (1/2 teaspoonful) of *Vitex nigundo*, a pinch of fruit powder of *Piper nigrum* and jaggery taken for 2 or 3 times a day as **diuretic** and **stomachic**.

Active Constituents : Pedicines and pedicellin isolated from *leaves*.

Uses in Literature : Reported so far as lithotriptic; and for yielding crystalline colouring matters including pedicin, toxic to fish, and perfumery (Ambasta, 1986; Jain, 2003a).

***Dioscorea bellophylla* Voigt** (Pl. 19E)

Syn.: *D. glabra* Roxb.; *D. cliffortiana* Lam.; *D. versicola* Herb.; *D. pulchella* Roxb.; *D. heterophylla* Roxb.; *D. decemangularis* Herb.; *D. thunga* Herb.; *Helmia bulbifera* Kunth

Family: Dioscoreaceae.

Vern. : Tardi.

Distribution : Sub-tropical Himalayas.

Description : A climber with opposite, ovate-lanceolate leaves. Flowers in spikes. Capsule membranous. Seeds winged.

Reproductive Cycle : September-November.

Habitat Ecology : Scrub jungles. Hedges.

Material Examined : HPU-EL-SKS-5539; Balah (900 m), Boda (1,000 m), Khaira (940 m), Jasun (1,000 m).

Part Used : Tuber.

Folk Uses : Tubers consumed as **vegetable**, **pickle**, and considered highly nutritious. Roasted ones prescribed to check **piles**.

Active Constituents : *Tuber* contains diosgenin.

Biological Activity : Antiinflammatory and antifertility activities found (+)ve.

Uses in Literature : Known in India as a tonic, anthelmintic, aphrodisiac, edible (tubers), stomachic; good for asthma, abdominal pains, flatulence, piles, dysentery, dyspepsia, headache, leucoderma, bronchitis, biliousness, tumours, sores (Ambasta, 1986; Arora & Pandey, 1996; Chandra, 1997; Jain, 1991; Kirtikar & Basu, 1984; Kumar, 2003; Lal *et al.*, 1996; Murthy *et al.*, 2003; Rawat & Chowdhury, 1998; Roy *et al.*, 1998; Said, 1997; Sharma & Rana, 2005; Singh, 1996b; Singh & Kumar, 2000a; Sinha, 1996; Sood & Prakash, 2007; Sood & Thakur, 2004; Watt, 1972).

***Dioscorea bulbifera* L.** (Pl. 19F)
Syn.: *D. pulchella* Roxb.; *D. pulchella* Roxb.;
D. pulchelia Hoken.; *D. tamnifolia* Salisb.; *Helmia bulbifera* Kunth
Family: Dioscoreaceae.

Vern. : Gangaradi.

English, Sanskrit & Regional Names:

Eng. : Black Bearing Yam
Sans. : Badar Kachna, Vishvaksenapriya
Ass. : Kathalu, Mati-alu
B. : Banalu, Bandorechalu
Bo. : Hadukaranda
G. : Salvinavelya, Suariya
H. : Bhirvolikanda, Karawakand
Kan. : Heggenasu, Onthalaigasu, Tung-Genasu
Mal. : Kattukachil
Mar. : Dukarakanda
Ori. : Pitalu
P. : Zaminkand
Tel. : Chedupaddudumpa, Malakakayapenda Lamu, Pannukelangu.

Distribution : Himalayas upto 2100 m.

Description : Climbers with ovate-cordate, alternate leaves. Flowers yellowish-green but fading later to dull red. Seeds winged at the base.

Reproductive Cycle : June-August.

Habitat Ecology : Scrub Jungles. Hedges. Waste places.

Material Examined : HPU-EL-SKS-5524; Bhagpur (1,600 m), Ladoh (980 m), Kathena (1,290 m), Malnu (1,000 m), Paror (940 m).

Parts Used : Bulbils. Tuber.

Folk Uses : Tubers and bulbils cooked as **vegetable**. Paste of tubers taken orally for 20-30 days to check **bleeding piles**.

Active Constituents : Three new furanoids nortiterpenes-diosbulbins A, B and diosbulbin C have been reported from *root tubers*.

Biological Activity : Anti-inflammatory and antifertility activities (+)ve.

Uses in Literature : Recorded in literature as contraceptives, edible (tubers), febrifuge, fish catcher, psychoactive plant, tonic; and for abdominal pain, asthma, boils, bone fractures, bulbils, burns, cancer, cough and cold, diarrhoea, dysentery, epistaxis, goiter, itching, gonorrhoea, jaundice, lactating disorders, piles, rheumatism and gout, snakebite, skin infections, sores, syphilis, and ulcers (Agarwal, 2003; Ambasta, 1986; Arora & Pandey, 1996; Bhattacharya *et al.*, 2005; Borthakur, 1996b; Brahmam *et al.*, 1996; Chowdhery, 1996;

D'Rozario *et al.*, 1999; Deshmukh & Rothe, 2003; Devi, 2003; Diwakar & Ansari, 2003; Dwivedi, 2003; Duke *et al.*, 2002; Girach *et al.*, 1986; Jain, 1991; Jain *et al.*, 1994; Katewa *et al.*, 2000; Kirtikar & Basu, 1984; Lal *et al.*, 1996; Lalramnghinglova, 1996, 2000; Mahato & Chaudhary, 2005; Mandal & Basu, 1996; Murthy *et al.*, 2003; Pandey & Joshi, 2001; Patole & Jain, 2002; Prajapati *et al.*, 2006; Rai & Upadhyay, 1996; Rama Rao & Henry, 1996; Rawat & Chowdhury, 1998; Roy *et al.*, 1998; Saini, 1996a; Sharma & Rana, 2005; Singh, 1999, 2000, 2003; Singh & Kumar, 2000a; Singh & Prakash, 1996; Singh & Zaheer, 1996; Sinha, 1996; Sinha & Sinha, 2001; Siwakoti & Varma, 1996; Sood & Prakash, 2007; Sood & Thakur, 2004; Upadhyay, 2003; Watt, 1972).

***Dioscorea oppositifolia* L.** (Pl. 20A)
Family: Dioscoreaceae.

Vern. : Daregal.

Regional Names:

B.	: Sauralu
H.	: Bhusa, Gazaria, Kantalu
Kan.	: Nuraigenasu
Mal.	: Nurunudkilaangu
Mar.	: Chatavali, Mandi, Ulasi
Ori.	: Pani-talu, Santal
Tam.	: Kallukkillangu
Tel.	: Adavidumpalige, Aretegalu, Arelige.

Distribution : Indian coast. Sri Lanka, Assam, Myanmar.

Description : Glabrous climbers with short rootstock, slender, branched, glabrous or pubescent stem, and coriaceous, opposite, simple leaves. Male flowers in dense, shortly pedunculate spikes. Female flowers solitary or fasciculate, in axillary spikes. Capsules suborbicular.

Reproductive Cycle : August-September.

Habitat Ecology : Climber on shrubs, trees in forests, wastelands.

Material Examined : HPU-EL-SKS-5531; Bhauda (1,050 m), Bhaura (940 m), Jasun (1,000 m), Thandol (950 m), Tikkar (910 m).

Part Used : Tubers.

Folk Uses : Roasted tubers **eaten** and considered highly **nutritious**.

Active Constituents : These are smilagenone, epismilagenin, deltonin and dioscin (*tubers*).

Uses in Literature : Known so far for reducing swelling; snakebite, scorpion-sting, and edible purposes (Ambasta, 1986; Arora & Pandey, 1996; Bennet *et al.*, 1991; Chopra *et al.*, 1956; D'Rozario *et al.*, 1999; Goud & Pullaiah, 1996a; Hosagoudar & Henry, 1996a, b; Katewa *et al.*, 2000; Kirtikar & Basu, 1984; Murthy *et al.*, 2003; Nadanakunjidam, 2003; Radhakrishnan *et al.*, 1996; Rao *et al.*, 1996; Reddy *et al.*, 1996; Saini, 1996a; Sivakumar *et al.*, 2003; Sharma & Rana, 2005; Sood & Prakash, 2007; Watt, 1972).

***Diospyros kaki* L.** (Pl. 20B)

Syn.: *D. chinensis* Blume; *D. schi-tse* Bunge; *D. roxburghii* Reveal; *D. costata* Blume; *Embryopteris kaki* G. Don

Family: Ebenaceae.

Vern. : Japani-phal.

English & Regional Names:

Eng. : Japanese Persimmon, Kaki Persimmon

Ass. : Dieng:Iong, Soh-Tang-Jong

H. : Halwa Tendu.

Distribution : Planted. W. & N. India.

Description : Small deciduous trees. Leaves alternate, simple, ovate glossy-green. Flowers cream coloured (female) and pink-tinged (male). Fruit spherical to squarish, orange red.

Reproductive Cycle : May-June.

Habitat Ecology : Edges of fields.

Material Examined : HPU-EL-SKS-5610; Banuri (1,150 m), Chimbalhar (1,150 m), Gopalpur (1,350 m), Sungal (1,050 m).

Part Used : Fruit.

Folk Uses : Fruits **astringent** to bowels and also **edible**.

Active Constituents : *Fruit* yields delphinidin, cyanidin, lycopene, catechin-3-gallate, gallocatechin and gallocatechin-3-gallate.

Uses in Literature : Recorded earlier as an astringent, edible; and for boxes, dyeing, ornamental works, tannin (raw fruits) (Ambasta, 1986; Jain, 1991; Watt, 1972).

***Diplazium esculentum* (Retz.) Sw.** (Pl. 20C)

Syn.: *Anisogonium esculentum* Presl.

Family: Athyriaceae.

Vern. : Lungadu.

Regional Names:

P. : Kuthjerkul

U.P. : Lengra, Kuthalla, Niwara.

Distribution : Throughout India upto 2,000 m.

Description : Ferns having erect, rhizome; long, bipinnate fronds; and long, erect, glabrous 30-60 cm stipes. Pinna more or less deeply lobed, subcoriaceous. Rachis often pubescent.

Reproductive Cycle : April-July.

Habitat Ecology : Damp-shady areas in forests, rocky slopes.

Material Examined : HPU-EL-SKS-5720; Bhagpur (1,600 m), Kathena (1,290 m), Lolani (1,600 m).

Part Used : Fronds.

Folk Uses : Circinately coiled fronds cooked as a **vegetable** and also **pickled**.

Active Constituents : *Aerial parts* contain esculentic acid, eriodictyol – 5 – O-methyl – 7 – O – β – D – xylosylgalactoside isolated.

Uses in Literature : Earlier used as pot-herb, salad; and for asthma, cough, diarrhoea, dysentery, easy delivery, dyspepsia, fever, haemoptysis, skin diseases, stomachache, toothache (carries) (Arora & Pandey, 1996; Banerjee & Ghora, 1996; Bennet *et al.*, 1991; Borthakur, 1996b; Chakraborty, 2003; Chakraborty *et al.*, 2003; Das, 1997, 2000; Das & Sharma, 2003; Lal *et al.*, 1996; Lalramnghinglova, 2002; Mao, 1993; Nwosu, 2001; Ranjan, 2003; Rawat & Chowdhury, 1998; Saklani & Jain, 1994; Singh & Kumar, 2000a; Singh & Viswanathan, 1996).

Diplocyclos palmatus (L.) C. Jeffrey (Pl. 20D)

Syn.: *Bryonia laciniosa* Sensu L.; *B. palmatus* L.; *Bryonopsis laciniosa* Sensu Naudin

Family: Cucurbitaceae.

Vern. : Shivlingi.

Sanskrit & Regional Names:

Sans. : Lingini
B. : Mala, Shivalingani
G. : Chivalingi, Shivalingi, Shivlingi
H. : Gargumaru, Ishwaralingi, Shivalingi
Kan. : Linga, Tonde balli, Malingana balli
Mal. : Neohmaka
Mar. : Kavadori, Kavalechedole, Vaduballi
Tel. : Lingadonda.

Distribution : Throughout The Old World Tropics from tropical Africa eastwards to Australia.

Description : Annual vines with glabrous stem, forked tendrils; and base cordate, black membranous, ovate-suborbicular leaves. Flowers small, greenish-yellow; female in fascicles, male solitary. Berries with eight vertical white streaks and few seeded.

Reproductive Cycle : September-October.

Habitat Ecology : Weed in maize fields. Shrubby areas.

Material Examined : HPU-EL-SKS-5644; Balah (900 m), Malahu (1,050 m), Nadli (1,000 m), Palampur (1,280 m), Rangru (910 m).

Part Used : Seeds.

Folk Uses : 2-4 seeds prescribed 7th day of menses to ensure conception for **male** child and also to check **miscarriage**.

Uses in Literature : Used for ensuring conception; and birth of male child, high fever, snakebite, wounds (Bhatt *et al.*, 1999, 2003b; Katewa *et al.*, 2001; Pawar & Patil, 2004; Singh & Pandey, 1998; Upadhyay & Chauhan, 2000).

Dodonaea viscosa Jacq. (Pl.20E)

Syn.: *D. angustifolia* L.; *D. dioica* Roxb.; *D. burmanniana* DC.; *D. pallida* Miq.; *D. micropoda* DC.; *D. wightiana* Blume; *D. petandra* Griff.; *D. heterophylla* Hortul.; *D. spathulata* Sm.; *D. arabica* Hochst. & Steud.; *Ptelea viscosa* L.

Family: Sapindaceae.

Vern. : Mehndu, Mehendadu.

Regional Names:

B.	: Bandari, Zakhumi
Bo.	: Jakhmi, Lutchmi
H.	: Alliar, Sonalta, Walaytinahndi
Kan.	: Bandara
Mal.	: Unnataruvi, Virali
Mar.	: Lutchmi
Ori.	: Mohra
P.	: Banmehndu, Mendhar, Sonatta
Tam.	: Velari, Virali
Tel.	: Bandaru, Pullenu.

Distribution : Afghanistan-H.P. upto 1,500 m.

Description : Shrubs. Leaves alternate, viscid with yellow resinous dots. Flowers greenish-yellow. Capsule yellowish-brown, compressed.

Reproductive Cycle : February-April.

Habitat Ecology : Open slopes. Scrub jungles.

Material Examined : HPU-EL-SKS-5560; Balah (900 m), Bhuana (1,400 m), Boda (1,000 m), Jasun (1,000 m), Khaira (940 m).

Parts Used : Leaves. Stem. Aerial Parts.

Folk Uses : Leaves chewed raw to check **diabetes** and its paste heals **cuts**, **burns** and hastens suppuration of **boils**. Poultice of leaves along with that of *Vitex negundo*, *Murraya koenigii*, and bark of *Litsea sebifera* along with a pinch of rock salt applied locally for **sprains** and **bone fractures**. Stem used as **fuel** and for making **walking sticks**. Aerial plant parts used for **thatching** purposes and **decorating** gates.

Biological Activity : Plant extr. exhibits cardio-inhibitory, and spasmolytic activity on smooth muscles and intestine. It also has sedative effect on virgin guinea pig uterus and hypotensive effect unaffected by atropine.

Uses in Literature : Used in India as aromatic, astringent, carminative, edible (leaves, seeds), febrifuge, firewood, fodder, manure, purgative, substitute for hops (*Humulus lupulus* L.) in the preparation of beer and yeast, sudofic in gout and rheumatism; and for bone fractures, broken legs of cows, burns, cholera, cold, cuts, engraving, menstrual flow, epilepsy, fermentations, intestinal worms, muscle pain, ringworm, sprains, swellings, ticks (vet.), tool-handles, turnery, walking sticks, worms (vet.), and wound (Agarwal, 2003; Ambasta, 1986; Bennet *et al.*, 1991; Chopra *et al.*, 1956; Geetha *et al.*, 1996b; Goud & Pullaiah, 1996b; Kirtikar & Basu, 1984; Jain, 1991; Kshirsagar, 2003; Kshirsagar *et al.*, 2003; Rama Rao & Henry, 1996; Saklani & Jain, 1994; Seetharam *et al.*, 1999; Sharma, 2003; Sharma & Sood, 1997; Singh & Kumar, 2000a; Subramaniam, 2000; Vishwanthan, 1997; Watt, 1972).

***Dolichos biflorus* L.** (Pl. 20F)

Family: Fabaceae.

Vern. : Kulath.

English, Sanskrit & Regional Names:

Eng. : Horse Gram, Koolhe, Madras Gram
Sans. : Kalarrinta, Kulathika, Kulitha, Tamabija
B. : Kurtikalai
Bo. : Hulga, Kulte, Kulti
G. : Kalathi, Kulit
H. : Gahat, Kulthe
Mal. : Muthera
Mar. : Kulith, Klthi
P. : Bothungt, Gagli, Guar, Kalat, Kulath, Rawan,
Tam. : Kollu
Tel. : Wulavatti, Uluva.

Distribution : Widely cultivated.

Description : Branched, twinning, glabrescent annuals having broadly lanceolate or oblong leaflets and yellow flowers. Pods compressed, recurved; seeds reniform, reddish brown.

Reproductive Cycle : June-October.

Habitat Ecology : Cultivated land.

Material Examined : HPU-EL-SKS-5766; Gopalpur (1,350 m), Malahu (1,050 m), Jasun (1,000 m), Sulah (950 m).

Parts Used : Seeds.

Folk Uses : Seeds consumed as **curry**, and prescribed thrice a week for 3 months for **kidney stone**.

Uses in Literature : Reported in literature as acrid, anthelmintic, antipyretic, appetiser, astringent, bitter, cattle feed, diuretic, emmanagogue, intestinal colic, sweetish; and for abdominal complaints, asthma, bronchitis, diseases of the brain, eye diseases, fattening, heart troubles, inflammations, kidney stone, leucoderma, liver troubles, piles, stranguary, tumours, and urinary discharges Ambasta, 1986; Kirtikar & Basu, 1984; Negi *et al.*, 1999; Shukla *et al.*, 2003).

Drimia indica (Roxb.) Jessop

Syn.: *Urginea indica* (Roxb.) Kunth; *U. wightiana* Hook.

Family: Liliaceae.

Vern. : Ban-Ganda.

English, Sanskrit & Regional Names:

- **Eng.** : Indian Squill, Tree Squill, Sea-onion
- **Sans.** : Vanapandam, Kolkanda
- **Bo.** : Kochinda
- **G.** : Jangli Piyas, Banpiyazi
- **Kan.** : Adairirulli
- **Mal.** : Kattulli
- **Mar.** : Ranacha Kanda, Rankanda
- **P.** : Kachwassal
- **Tam.** : Adavithagadda, Nakkavulligadda.

Distribution : W. Himalayas below 1,400 m.

Description : A glabrous bulbous herb having radical leaves and erect scapes. Flowers appearing before the leaves, drooping or spreading. Perianth bell-shaped, 6-parted; segments white with three green ribs in the centre. Capsules oblong, 3-valved. Seeds many, flat, black.

Reproductive Cycle : May-September.

Habitat Ecology : Shady slopes.

Material Examined : HPU-EL-SKS-5733; Balah (900 m), Boda (1,000 m), Khaira (940 m), Nadli (1,000 m), Rangru (910 m).

Part Used : Bulbs.

Folk Uses : Mixture of its crushed bulbs (1/4 part) alongwith 3-seeds of 'kali mirach' (*Piper nigrum*), handful leaves from 'tirmir, (*Zanthoxylum armatum*) and 'kali-basuti' (*Pogostemon benghalense*) mixed with wheat flour prescribed for **cough**, **fever**, **constipation** and as **diuretic** in livestock.

Active Constituents : *Bulb* contains 6-disacetoxyscillirasidin.

Uses in Literature : Reported earlier as alexiteric, anthelmintic, cardiotonic, cathartic, emetic, emmanagogue, expectorant, purgative, stimulant, stomachic, diuretic, hypoglycaemic, dropsy, rheumatism; and for bronchial catarrh, chronic bronchitis, headache, internal pains, leprosy, paralysis, scabies, sizing cotton cloth, skin diseases, warts and corns (Ambasta, 1986; Manandhar, 2003; Rastogi & Mehrotra, 1999).

***Duranta repens* L.** (Pl. 21A)

Syn.: *D. plumieri* Jacq.

Family: Verbenaceae.

Vern : Neel Kanta.

English, Sanskrit & Regional Names:

Eng. : Creeping Skin Flowers, Golden Dew Drop, Pigeon Berry

Sans. : Vriti Kanta

H. : Daranta.

Distribution : W. Himalayas: 200-1,600 m.

Description : A spiny shrub. Leaves opposite, broadly elliptic, entire. Spines stout. Flowers bright, lavender-blue with pale centers. Fruit golden-yellow, globular.

Reproductive Cycle : April-August.

Habitat Ecology : Very common as a hedge plant.

Material Examined : HPU-EL-SKS-5577; Dadh (1,100 m), Ghugar (1,250 m), Molichak (1,020 m), Panapari (1,100 m), Saparu (1,250 m).

Parts Used : Whole Plant. Stem.

Folk Uses : Plant grown as **hedges**. Stem used as **fuelwood**.

Uses in Literature : Described earlier as insecticidal; for wounds, sore throat, and turnery (Ambasta, 1986; Diwakar & Ansari, 2003; Rajendran & Sikarwar, 2003).

***Ehretia acuminata* R. Br.** (Pl. 21B)

Syn: *E. serrata* Roxb.; *E. pyrifolia* Don; *E. ovalifolia* Hassk.; *Cordia thyrsiflora* Sieb. & Zucc.

Family: Boraginaceae.

Vern. : Punna.

Regional Names:

- **Ass.** : Bual
- **B.** : Kula-aja
- **H.** : Arjun, Koda, Punjlawai, Punyam
- **P.** : Punna, Pursan, Sium
- **U.P.** : Arjun, Kur-kuria, Narra, Shaursi.

Distribution : Sub-Tropical Himalayas.

Description : Trees with alternate, ovate, sharply toothed leaves. Flowers white. Drupes red, globose with two stones.

Reproductive Cycle : May-July.

Habitat Ecology : Common in Wastelands.

Material Examined : HPU-EL-SKS-5587; Arla (1,100 m), Gopalpur (1,350 m), Lolani (1,600 m), Nanaun (900 m), Rajpur (1,100 m).

Parts Used : Leaves. Wood.

Folk Uses : Leaves used as **fodder** and wood for **furniture**, **agricultural implements**, and **building purposes**.

Uses in Literature : Known earlier for agricultural implements, building, gun-stocks, pickling, scabbards, sword hilts, tongue's sores and as edible and fodder (Ambasta, 1986; Arora & Pandey, 1996; Jain, 1991; Mudgal & Pal, 1980; Murthy *et al.*, 2003; Negi & Gaur, 1991; Purohit & Samant, 1995; Roy *et al.*, 1998; Samant *et al.*, 2001; Sood & Thakur, 2004; Watt, 1972).

Ehretia laevis Roxb. (Pl. 21C)

Syn.: *E. punctata* Roth; *E. affinis* Wall.; *E. dichotoma* Rottl.; *E. floribunda* Benth *ex* Royle; *Beurreria laevis* G. Don; *B. punctata* G. Don

Family: Boraginaceae.

Vern. : Bhankar.

Regional Names:

Ass.	: Hanbuok-arong
B.	: Tamboli
H.	: Chamror, Datrang, Darar, Koda
Kan.	: Adak, Begari, Kappura
M.P.	: Tamoiya, Chinor, Bhoiumbar
Mal.	: Chavandi
Mar.	: Datrang
Ori.	: Guachipo, Masonea
P.	: Sakar
Tam.	: Addula
Tel.	: Paldatam, Peddapulmera, Pogari.

Distribution : Throughout India; ascending to 1,000 m.

Description : Medium-sized trees. Leaves ovate, entire, and leathery. Flowers white, in one-sided slender spike. Drupes depressed, globose, distinctly 4-ribbed when dry.

Reproductive Cycle : January-June.

Habitat Ecology : Drier areas. Wastelands.

Material Examined : HPU-EL-SKS-5683; Boda (1,000 m), Bhuana (1,400 m), Jasun (1,000 m), Malahu (1,050 m).

Parts Used : Leaves. Bark . Wood.

Folk Uses : Leaves lopped for **fodder**. Bark slices cooked as '**pakodas**' and eaten to check '**badalbis**' (rashes on skin). Wood for **building works**, **agricultural implements** and as **fuel**.

Active Constituents : *Stem* yields bauerenol, bauerenol acetate and β-sitosterol.

Uses in Literature : Useful in India as cattle fodder; and for agricultural implements, brush-backs, building purposes, diarrhoea, diphtheria, edible purposes (fruit and inner bark eaten in times of scarcity), matchboxes, muscular and rheumatic pains, shoe lasts, splints, syphilis, and thatching (Ambasta, 1986; Jain, 1991; Negi & Gaur, 1991; Rama Rao & Henry, 1996; Saini, 1996b; Singh & Maheshwari, 1983; Watt, 1972).

Eleusine coracana Gaertn. (Pl. 21D)

Family: Poaceae.

Vern. : Mandal.

Sanskrit & Regional Names:

Sans. : Bahupatrica, Bhuchara, Guchha, Kanisha, Ragi, Raji
B. : Marua
Bo. : Nagli, Nangli
G. : Barto, Nagli
H. : Makra, Mandua, Marua, Rotka
Mal. : Muttari
Mar. : Nagli, Nachori
P. : Chalodra, Koda, Kodon, Kodra, Mandal
Tam. : Kelvaregu, Kayur
Tel. : Ragulu, Tamidelu.

Distribution : Cultivated in the tropics of Old World.

Description : Annuals upto 120cm tall having leaves with compressed loose sheaths and ligules of hairs. Spikes suberect, incurved. Spikelets congested, awnless; 3-6 flowered. Seeds globose, dark-brown.

Reproductive Cycle : June-September.

Habitat Ecology : Borders of agricultural fields.

Material Examined : HPU-EL-SKS-5480; Bhauda (1,050 m), Bhuana (1,400 m), Chhanchri (930 m), Panapari (1,100 m).

Part Used : Grains.

Folk Uses : Grain-flour relished as '**chapati**', and also considered good for **diabetes**, **heart** ailments, and **constipation**.

Uses in Literature : Considered acrid, alcoholic beverages, astringent, bitter, blood diseases cakes, cooling, edible, leprosy, liver tonic, diabetes, diarrhoea, pudding, sweet, tonic, and useful in biliousness (Caius, 2003; Drury, 1978; Kirtikar & Basu, 1984; Kumar & Rao, 2001; Mokat & Deokule, 2004; Rajan *et al.*, 2003; Rawat & Chowdhary, 1998; Srivastava *et al.*, 2003a).

***Emblica officinalis* Gaertn.** (Pl. 21E)

Syn.: *Phyllanthus emblica* L.; *Cicca emblica* Kurz.; *Dichelastina nodicaulis* Hance

Family: Euphorbiaceae.

Vern. : Ambla, Amla.

English, Sanskrit & Regional Names:

Eng.	: Emblic Myrobalan, Indian Goose Berry
Sans.	: Adiphala, Amalaka, Dhatri
B.	: Amla, Amlaki
G.	: Amla, Amlika, Aonla
H.	: Amla, Amlika, Aonla
Kan.	: Amalaka, Nelli Kayi
Mal.	: Nelli
Tam.	: Nelli
Tel.	: Amalakamu.

Distribution : Throughout India, often cultivated- 1,400 m.

Description : A deciduous tree with small, narrowly linear leaves resembling pinnate leaves and yellow axillary flowers. Fruits fleshy, globose and pale coloured.

Reproductive Cycle : April-September.

Habitat Ecology : Common along borders of agricultural fields, forests, open slopes.

Material Examined : HPU-EL-SKS-5527; Bhuana (1,400 m), Chhanchri (930 m), Khaira (940 m), Gharthun (900 m), Malahu (1,050 m).

Parts Used : Leaves. Bark. Fruit. Stem. Wood.

Folk Uses : Leaf-powder applied to heal **wounds**. Decoction of stem bark used for tanning **leather**. Fruits **eaten** raw and **pickled**; considered **diuretic**, **refrigerant**, **purgative**, **liver tonic**, **hair dye** and **conditioner** and also forms one of the ingredients of '**Trifla**' churan. Plant considered incarnation of Lord '**Vishnu**' hence its twig used for solemnizing marriage with *Ocimum sanctum* which is regarded as incarnation of '**Laxmi**'. Wood considered must for **cremation** purposes.

Active Constituents : *Fruits* rich in trigalloylglucose, terchebin, corilagin, ellagic acid, indole acetic acid and four other auxins-a1, a3, a4 & a5, two growth inhibitors R_1 & R_2, tannins and vitamin C.

Biological Activity : Water soluble fraction of methanol extr. of leaves exhibits anti-inflammatory activity in carrageenin and dextran-induced rat hind paw oedema.

Uses in Literature : Used so far as anticancerous, antiscorbutic, antispasmodic, aperient, appetizer (vet.), astringent, CNS stimulant, detergent, diuretic, edible (fruits eaten raw or cooked), expectorant, hair dye, hair shampoos, important constituent of 'chayawanprash' and 'triphala', inferior quality furniture, laxative, local drinks, magico-therapeutic to treat toothache, psychoactive plant, refrigerant, rich source of vitamin C, symbolic abode of God (sacred), tonic; and for agricultural implements, allaying thirst, anaemia, anthrax (vet.), asthma, bad colds, biliousness, bronchitis, burns, clearing muddy water, confusion of thoughts, conjunctiva, diabetes, diarrhoea, disturbed cerebral circulation attended by a burning sensation at the crown of head, dropsy (vet.), dyeing, dysentery (vet.), dyspepsia (vet.), enlarged liver and ascites, epistaxis (vet.), eye complaints, fever, fractured bones, gastropathies (vet.), genito-urinary discharges, gonorrhoea, haemorrhage, headache, heart complaints, horses in dysentery, impaction (vet.), indigestion (vet.), inflammation (vet.), insanity, insect bites, irritability of the bladder, jaundice, leucorrhoea, liver (vet.), loss of hair, madness, making preserve, menorrhagia, piles, pulmonary tuberculosis, pus formation in ear, stomach and intestine disorders, earache, reviving taste, rheumatism (vet.), rinderpest (vet.), scurvy, skin diseases, stomach complaints, stomach diseases (vet.), syphilis, tanning skin and garments, ulcer, vaginal complaints, vomiting, washing head, well rings, blackish-grey dye, and wound maggots (Agarwal, 2003; Ambasta, 1986; Aminuddin & Girach, 1991, 1993; Arora & Pandey, 1996; Banerjee & Pal, 1996; Beigh *et al.*, 2003; Bhatt *et al.*, 2003b; Borthakur, 1996a; Borthakur & Gogoi, 1994; Chakraborty & Bhattacharjee, 2003; Chandra, 1997; Chauhan, 1999; Chevallier, 1996; Chopra *et al.*, 1956; D'Rozario *et al.*, 1999; Das & Agarwal, 1991; Das & Sharma, 2003; Dey & Bahadur, 1973; Dwivedi, 2003; Ghosh *et al.*, 1996; Girach, 1992; Gogoi & Borthakur, 2001; Hebber *et al.*, 2003; Jain, 1968, 1991; Jain *et al.*, 1994; Jha & Kumar, 2003; Kalakoti, 2001; Khanna *et al.*, 1996c; Khare, 2004; Kothari & Moorthy, 1996; Kulkarni & Kumbhojkar, 1996; Kumar & Pullaiah, 2003; Kumar & Rao, 2003; Lalramnghinglova, 1996; Mandal & Basu, 1996; Mehrotra *et al.*, 1995; Mishra *et al.*, 1997; Murthy *et al.*, 2003; Negi *et al.*, 1993; Pandey & Verma, 2003; Panthi & Chaudhary, 2003; Pawar & Patil, 2004; Prajapati *et al.*, 2006; Rajendran *et al.*, 2003; Rama Rao & Henry, 1996; Rana *et al.*, 1996; Ranjan, 2003; Roy *et al.*, 1998; Sahni, 1998; Sahoo & Mudgal, 1995; Said, 1997; Sarin, 1996; Sen & Batra, 1997; Shanavashan *et al.*, 2003; Sharma, 2003; Sharma & Borthakur, 2003; Sharma & Rana, 2005; Sharma & Singh, 2001; Sharma & Sood, 1997; Sikarwar & Kaushik, 1992; Singh, 1996b, 1999, 2003; Singh & Kumar, 2000a; Singh & Pandey, 1998; Singh & Prakash, 1994; Singh *et al.*, 2003a; Sinha, 1996; Sinha & Sinha, 2001; Siwakoti & Siwakoti, 2000; Siwakoti & Varma, 1996; Sood & Thakur, 2004; Srivastava *et al.*, 2003b; Sudhakar & Vedavathy, 2000; Tirkey *et al.*, 2003a; Vihari, 1995; Viswanathan & Singh, 1996; Upadhyay & Singh, 2005; Thomas & Britto, 2003; Watt, 1972).

Engelhardia colebrookeana Lindl. (Pl. 21F)

Family: Junglandaceae.

Vern. : Samma.

Regional Names:

Ass. : Pasomasi, Deengparsomasi

P. : Timarrakh.

Distribution : Himalayas.

Description : Trees having alternate, odd-pinnate leaves. Flowers unisexual, both sexes on the same tree. Male flowers numerous, in long catkins; female ones few, in terminal clusters. Drupes fleshy, enclosing a woody 1-seeded nut.

Reproductive Cycle : April-October.

Habitat Ecology : Forests. Shady steep-slopes.

Material Examined : HPU-EL-SKS-5651; Balah (900 m), Boda (1,000 m), Bodhal (970 m), Jasun (1,000 m), Makol (900 m).

Parts Used : Leaves. Wood.

Folk Uses : Leaves used for **scouring** teeth, and wood for household **construction**.

Uses in Literature : Earlier known to be used as cattle fodder, and manure (ashes of leaves) **(Ambasta, 1986)**.

Equisetum debile Roxb. (Pl. 22A)

Family: Equisetaceae.

Vern. : Sau-greendi.

Regional Name:

P. : Irotak, Nari, Bandukei.

Distribution : Along shady streams all over India.

Description : Scrambling stems with long slender branches in whorls of 2-3. Internodes 3-10 cm long. Leaf-teeth long. Cone long, sessile, in funnel-shaped tip of the branch. Sporophylls peltate. Sporangia oblong, yellow.

Reproductive Cycle : July-September.

Habitat Ecology : Along Water Streams.

Material Examined : HPU-EL-SKS-5419; Bandla (1,280 m), Dharaman (1,100 m), Jiun (1,550 m), Lolani (1,600 m), Saliana (1,000 m).

Parts Used : Aerial Plant Parts.

Folk Use : Aerial plant parts fed to livestock to induce **rumination**.

Uses in Literature : Used as refrigerant, cooling medicine; and for gonorrhoea **(Ambasta, 1986).**

Eriobotrya japonica **Lindl.** (Pl. 22B)

Syn.: *Mesilpus japonica* Banks

Family: Rosaceae.

Vern. : Galaat, Logat, Lokaat.

English & Regional Names:

Eng. : Chinese Medlar, Japan Quince Loquat

H. : Logat

Kan. : Lakkote

Tam. : Ilakota.

Distribution : Usually cultivated in many parts of India.

Description : Trees with thick, 12-15 nerved coriaceous leaves and crowded, fragrant flowers. Fruits yellow, 2-5 cm long.

Reproductive Cycle : July-September.

Habitat Ecology : Borders of agricultural fields. Roadsides.

Material Examined : HPU-EL-SKS-5609; Bhaura (940 m), Dehan (1,130 m), Sidhpur (1,100 m), Saparu (1,250 m), Thandol (950 m).

Part Used : Fruit.

Folk Uses : Fruits **edible**, and also used in the preparation of **jams** and **jellies**.

Active Constituents : *Fruit pulp* contains new leucocyanin-loquatoside together with fructose, glucose and sucrose.

Biological Activity : Ursolic acid exhibits potent anticarcinogenic activity against sarcoma 180 cells and significantly suppresses c-myc oncogene expression.

Uses in Literature : Reported for rulers and drawing materials, jams, jellies, pies, preserves; and as desert, expectorant, and sedative (Ambasta, 1986; Watt 1972).

***Eriophorum comosum* Wall.** (Pl. 22C)
Family: Cyperaceae.

Vern. : Bagdella.

English Names:

Eng. : Cotton grass, Cotton sedge.

Distribution : Throughout India.

Description : Annual herbs with fibrous roots and three-sided, tufted, solid, terete culms which are leafy only near the base. Leaves narrow, sessile, sharp edged. Bracts longer than the umbel. Glumes many, acute. Nut 3-sided, pointed.

Reproductive Cycle : August-October.

Habitat Ecology : Moist-shady steep slopes in rock crevices.

Material Examined : HPU-EL-SKS-5467; Bandla (1,280 m), Chauki (900 m), Jiun (1,550 m), Lolani (1,600 m), Nanaun (900 m).

Parts Used : Aerial Plant Parts.

Folk Uses : Aerial plant parts employed for making **brooms**, and as **fodder**.

Uses in Literature : Reported earlier for cordages (Ambasta, 1986).

***Erythrina suberosa* Roxb.** (Pl. 22D)
Syn.: *E. glabrescens* Prain.; *E. alba* Roxb.; *E. nahasuta* Ham.; *E. reniformis* Ham.; *Micropteryx suberosa* Walp.
Family: Fabaceae.

Vern. : Paryara.

English, Sanskrit & Regional Names:

Eng. : Indian Coral Tree
Sans. : Bahupushpa, Kantaki, Parijata
Ass. : Mandal
B. : Palita Madar
Bo. : Pangara
G. : Jagriyo Khakharo
H. : Dauldhak, Mandara, Rungra
Kan. : Kaduparivala, Mulluharivana
Mal. : Kalyanamurrikku
Mar. : Pangra
Ori. : Bonopalpdhua, Piri
P. : Gulnastar, Pariara

Tam. : Mullumurukku, Munimoduga

Tel. : Badisa, Modugu.

Distribution : Often planted throughout India: 900-1,200 m.

Description : Trees with prickly branches, three leaflets, red flowers and glabrous, cylindrical stalked pods.

Reproductive Cycle : March-June.

Habitat Ecology : Forests. Ravines.

Material Examined : HPU-EL-SKS-5586; Bhauda (1,050 m), Bodhal (970 m), Tapa (945 m), Jasun (1,000 m), Saliana (1,000 m).

Parts Used : The Whole Plant. Twigs. Wood.

Folk Uses : Plant considered incarnation of **Lord Vishnu**. Juice extracted from toasted tender twig used as drops for **ear complaints** ('sailli'). Wood used for making drums and as **fuel** and **timber**.

Active Constituents : Ingredients are erythraline, erysodine, erysotrine, hypaphorine (*seeds*), erysotrine (*leaves*), delphinidin-3,5-diglucoside, cyanidin-3,5-digluco-side, erysodine, erysopine, erythraline, 11β-methoxyerythraline and its N-oxide, erythraline, 8-oxoerythrinine, erthratidine (*flowers*), myristic (0.13), palmitic (12.7), stearic (5.25), oleic (54.5), linoleic (20.44), arachidic (4.37) and behenic (2.62%) acids (*seed oil*), β-sitosterol, campesterol and stigmasterol (*seed coat*), and a new prenylated flavanone characterised as 5,7,3′,4′-tetrahydroxy-2′-(3″-methylbutyl-2″-enyl) flavonone (*roots*).

Biological Activity : Lecithin from seeds shows anti-H specificity.

Uses in Literature : Useful as antiseptic, bottle stoppers, cordage, crown cork liners, psychoactive plant, scabbards (wood); and for cordage, destroying maggots in ulcer, drums, dyeing, dysentery, insulation boards, jars storing ghee, laddles, lacquered boxes, matchboxes, packing cases, planking, sieve frames, splints, and water troughs (Ambasta 1986; Jain 1991; Jain *et al.*, 1994; Kshirsagar *et al.*, 2003; Watt, 1972).

***Eucalyptus tereticornis* Sm.** (Pl. 22E)
Syn.: *E. umbellata* Domin
Family: Myrtaceae.

Vern. : Safeda.

English & Regional Names:

Eng. : Blue Gum, Eucalyptus
H. : Safeda
Tam. : Karupuramaram.

Distribution : Introduced in India upto 1,200 m.

Description : Tall trees with smooth bark peeling in long thin layers; and erect, cylindrical, smooth stem. Leaves petiolated, alternate, lanceolate. Flowers small, pentamerous, pedicillate, umbellate. Fruits a capsule.

Reproductive Cycle : March-April.

Habitat Ecology : Roadsides. Water Canals.

Material Examined : HPU-EL-SKS-5562; Chornali (1,280 m), Darugnu (1,400 m), Gopalpur (1,350 m), Kathena (1,290 m), Panapari (1,100 m).

Parts Used : Leaves. Wood.

Folk Uses : Leaf-juice with mustard oil applied for **rheumatic** pains. Bath of water boiled with its leaves prescribed to check **post-delivery** and **rheumatic** pains. Wood used as **timber**.

Active Constituents : Active principles attributed to oleanoic and maslinic acids (*leaves*), borneol, camphene, cineole, cironellal, citronellyl acetate, p-coumarin, limonene, linalool, phellandrene, α- and β-pinenes, piperitone, γ-terpinene, terpen-1-en-4-ol and α-terpineol (*essential oil*), gallic acid, 7 β – D – glucosyloxy - 5-hydroxy – 2-methylchromone, β-sitosterol and (+) catechin (*wood*).

Uses in Literature : Useful in India for engineering construction, sleepers, ship building, and flooring (Ambasta, 1986).

***Eulaliopsis binata* (Ritz.) C. E. Hubb.** (Pl. 22F)

Syn.: *Ischaemum angustifolium* Hack.; *Pollinidium angustifolium* Hines; *P. binnatum* (Retz.) C. E. Hubb.

Family: Poaceae.

Vern. : Baggad.

Regional Names:

- **H.** : Bhabar, Baib, Babut, Sabai
- **Ori.** : Bagali, Babura.

Distribution : Throughtout India.

Description : A perennial tufted grass having wiry stem which is wooly at the base. Leaves long, sharply pointed, ligulate. Inflorescence simple or compound spike in cluster of 2-4. Base of outer glumes covered with golden or brown hairs.

Reproductive Cycle : July-November.

Habitat Ecology : Moist-Shady Steep Slopes. Rock Crevices.

Material Examined : HPU-EL-SKS-5659; Bhaura (940 m), Panapari (1,100 m), Purani-palam (1,100 m), Rangru (910 m), Tikkar (910 m).

Parts Used : Aerial Plant Parts.

Folk Uses : Aerial plant parts used for making **brooms**, **brushes** for white wash, **ropes**, and as **binder** in mud flooring; and also its ash blended with milk prescribed for **internal injury**.

Uses in Literature : Reported as raw material for paper, ropes, mats and strings (Ambasta, 1986).

***Euonymus hamiltonianus* Wall.** (Pl. 23A)

Syn.: *E. atropurpureus* Roxb.

Family: Celastraceae.

Vern. : Chamardei.

Regional Names:

- **Kash.** : Brahmani
- **Kum.** : Agnium
- **P.** : Chual, Naga, Rithu, Sidhera, Watal.

Distribution : Temperate Himalayas: Kashmir-Kumaon.

Description : Small-sized trees with terete branches and thin, membranous leaves. Flowers greenish-white. Capsule 4-lobed.

Flowering & Fruiting : August-September.

Habitat Ecology : Wastelands.

Material Examined : HPU-EL-SKS-5559; Balah (900 m), Khaira (940 m), Jasun (1,000 m), Tikkar (910 m).

Parts Used : Leaves. Wood.

Folk Uses : Leaves lopped for **fodder**. Wood used for **tool-handles** and as **fuel**.

Uses in Literature : Known in India as fodder and its wood for carving and turnery (Ambasta, 1986; Watt, 1972).

***Euphorbia royleana* Boiss.** (Pl. 23B)

Syn.: *E. pentagona* Royle

Family: Euphorbiaceae.

Vern. : Chhoo.

Regional Names:

H. : Shakarpitan, Suli, Thor

P. : Dandathor.

Distribution : N.W. Himalayas—1,800 m.

Description : Cactus-like shrubs having angled branches and sessile, spathulate leaves with thorny persistent stipules, yellow involucres and brown glands. Capsule glabrous.

Reproductive Cycle : June-July.

Habitat Ecology : Rocky dry slopes. Dry river valleys.

Material Examined : HPU-EL-SKS-5703; Bhuana (1,400 m), Chachian (1,300 m), Nadli (1,000 m), Nanaun (900 m), Panapari (1,100 m).

Parts Used : Latex. Stem.

Folk Uses : Stem pith washed in running water cooked as **vegetable** and 100 g of its decoction taken for one month for **indigestion** and **constipation**. It is also tied around distal end of tail to prevent further spread of infection locally called '**nanaain**' in cattle, and also fed to livestock early morning after conceiving to check **abortion**. Stem along with *Aloe vera* (whole plant) and *Cucurbita hispida* (fruits) hung on the rooftop of newly constructed house to ward off **evil spirits**. Crushed stem mixed with wheat-flour fed to stimulate **ovulation** in cattle. Curd of milk boiled with pieces of pith good against **body heat**. Latex applied locally to hasten **suppuration** of boils, and as **fish poison**.

Active Constituents : *Latex* yields triterpenoids, taraxerol, epitaraxerol, euphol, ellagic acid, euphol, taraxerol, sitosterol m-hydroxy benzoic acid, 7-hydroxy - 5, 4 - benzcoumarin, 7-methoxy - 3, 4 - benzcoumarin, 2′, 7 - dihydroxy - 3, 4 -benzcoumarin, α-amyrin, octa-cosanol, hexacosanol, tetracosanol, β-sitosterol, stigmasterol, and campesterol.

Biological Activity : Antiviral activity found to be (+) ve.

Uses in Literature : Described in literature as anthelmintic, cathartic (latex), fish poison, fuelwood, vegetable; and for boils, burns and cuts, diarrhoea and dysentery, indigestion, insect bite, tooth infection, ear complaints, leprosy, skin diseases, sprains, and wounds (Ambasta, 1986; Binoj & Balakrishnan, 1996; Jain, 1984, 1991; Jain *et al.*, 1991; Kapur & Nanda, 1996; Manandhar, 1987a; Panthi & Chaudhary, 2003; Shah & Joshi, 1971; Sharma, 2003; Singh, 1996b; Siwakoti & Varma, 1996; Sood & Thakur, 2004, Watt, 1972).

Fagopyrum cymosum Meisn. (Pl. 23C)

Syn.: *F. triangulare* Meisn.; *F. emarginatum* Meisn.; *Polygonum cymosum* Treviran.; *P. triangulare* Wall.; *P . emarginatum* Wall.; *P. dibotrys* Don; *P. volubile* Turca.; *P. rugosum* Ham.

Family: Polygonaceae.

Vern. : Fafru.

Regional Name:

P. : Banogal.

Distribution : Temperate Himalayas: 1,700-3,650 m; Khasi Hills: 1,350-2,700 m.

Description : Erect pubescent branching herbs. Leaves long-stalked, entire, broadly triangular. Flowers 2-sexual, small, white, in racemes. Nuts ovate, actually 3-cornered.

Reproductive Cycle : June-October.

Habitat Ecology : Wastelands. Damp places.

Material Examined : HPU-EL-SKS-5637; Darugnu (1,400 m), Kathena (1,290 m), Lolani (1,600 m), Purani-palam (1,100 m), Saparu (1,250 m).

Parts Used : Aerial Plant Parts.

Folk Use : Aerial plant parts consumed as **vegetable** ('saag').

Active Constituents : Quercitin, quercitrin and rutin present in herb.

Biological Activity : Quercetin found to inhibit the replication of herpes simplex virus in RK-13 cells, reduce leucocyte migration and heart rate in isolated rabbit heart.

Uses in Literature : Recommended so far as capillary fragility, colic, dye, emollient, pancakes, porridge, poultry feed, resolvent, vegetable; and for abdominal obstructions, diarrhoea and fluxes of all kinds (Agarwal, 2003; Ambasta, 1986; Arora & Pandey, 1996; Borthakur, 1996b; Chandra, 1997; Das & Agarwal, 1991; Jain *et al.*, 1991; Kapur & Nanda, 1996; Kapur & Singh, 1996; Kirtikar & Basu, 1984; Lal *et al.*, 1996; Manandhar, 1997; Roy *et al.*, 1998; Saini, 1996a; Samant *et al.*, 2001; Singh & Kumar, 2000a; Singh *et al.*, 1996a).

***Fagopyrum esculentum* (L.) Moench (Pl. 23D)**

Syn.: *Fagopyrum esculentum* Moench; *Polygonum fagopyrum* L.; *P. dioicum* Ham.; *P. emarginatum* Ham.; *P. emarginatum* Roth

Family: Polygonaceae

Vern. : Bhresh.

English, Sanskrit & Regional Names:

Eng. : Buchwheat

Sans. : Bras, Phaphra, Trabri, Ulgo, Ugal

Ass. : Doron

H. : Kotu, Kultu, Phapra

P. : Daran, Obal, Ogal, Phapher, Phulan.

Distributions : Throughout the Himalayas from W. Tibet-Sikkim. The Khasia Hills. The Nilghiri Hills.

Description : Small herbs. Flowers pink or white, in heads or compound cymes. Nuts pale.

Reproductive Cycle : August-October.

Habitat Ecology : Damp Slopes.

Material Examined : HPU-EL-SKS-5618; Darugnu (1,400 m), Dharaman (1,100 m), Jiun (1,550 m), Rajeher (1,050 m), Rajpur (1,100 m).

Parts Used : Aerial Parts.

Folk Use : Aerial parts cooked as '**saag**'.

Active Constituents : Rich in rutin, quercetin (*whole plant*), kaempferol-3-rhamnosyglucoside, p-cumaroyl-quinic and feruloylquinic (*leaves*).

Biological Activity : Increased biliary secretion and faecal excretion of bile acids have been confirmed.

Uses in Literature : Known in India as anthaemorrhagic, capillary fragility, dye, emollient, resolvent; and for lung diseases, rheumatic pains, constipation, stomach ailments, porridge, and poultry feed (Arora & Pandey, 1996; Ambasta, 1986; Chauhan, 1999; Chevallier, 1996; Duke *et al.*, 2002; Rajendran & Sikarwar, 2003; Roy *et al.*, 1998; Samant *et al.*, 2001; Sharma, 2003; Singh & Kumar, 2000a; Singh *et al.*, 1996b, 2001).

***Ficus benghalensis* L.** (Pl. 23 E,F)

Syn.: *F. indica* L.; *Urostigma benghalense* Gasp.

Family: Urticaceae.

Vern. : Badein, Bargad.

English, Sanskrit & Regional Names:

Eng. : Banyan

Sans. : Avaroha, Bahupada, Bhandira, Bhringi, Chira, Dhruva, Jatala, Jatiljati, Karmaja, Kshiri, Mahachhaya, Mana, Nandi, Nila, Nyagrodha, Padarohona, Raktaphala, Rohina, Skandhaja, Vaishravanodaya, Yakshaves, Yamapriya

Ass. : Bot

B. : Bar, Bot, But

G. : Vad, Vadlo, Vor

H. : Bar, Bargad, Bargat, Ber, Bor

Kan. : Ala, Alada, Mara, Vata

Mal. : Ala, Nyagrodham, Peral, Peralim, Peralu, Vatam

Mar. : Vada, Wad, War

Ori. : Boro, Boto

P. : Bargad, Bera, Bohar, Bohir, Bor

Tam. : Al, Ala, Alam, Eguvasam, Kadavam, Kagavaugam, Pali, Palu, Peral, Pudam, Pudavirujam, Vadavirukkam, Vekkurodam,

Tel. : Bhandiramu, Mari, Matti, Pedamatti, Rohi, Vrikshadanamu.

Distribution : Throughout India upto 1,400 m.

Description : A large-sized spreading tree with numerous aerial roots. Leaves leathery, ovate and pubescent. Figs red, sessile, axillary.

Reproductive Cycle : February-June.

Habitat Ecology : Commonly planted as a shade tree at roadsides.

Material Examined : HPU-EL-SKS-5415; Bhuana (1,400 m), Makol (900 m), Sungal (1,050 m), Tikkar (910 m).

Parts Used : Whole plant. Leaves. Fruits. Latex.

Folk Uses : Tree considered **sacred** and **worshipped**. Poultice of its toasted leaves with mustard oil, applied for underarm infection of **herpes**, and also over breast to check lactation in mothers. Latex applied locally for **toothache**, **cuts**, and **cracked** heels; and also used as **chewing gum**. Fruits **edible** and stimulate **digestion**.

Active Constituents : Active properties of *stem-bark* due to methyl ethers of leucoanthocyanins-delphinidin-3-0-α-L-rhamnoside (I), pelargonidin-3-0-α-L-rhamnoside (II) and leucocyanadin-3-O-β-D-galactosylcello-bioside (III) along with a methyl ether of leucoanthocyanidin, 20-tetra triacontene-2-one, pentatriacontan-5-one and 6-heptatriaconten - 10 - one, β-sitosterol - α - D -glucoside and mesoinositol.

Biological Activity : Bark exhibits hypoglycaemic activity in both normal and alloxan diabetic rats.

Uses in Literature : Reported earlier as an antidote, aphrodisiac, astringent, diaphoretic, fodder, edible (fruits, leaves, shoots, bark powder), tonic, tooth-brush; and for atrophy, abortion, bangy-poles, bird lime, bleeding and swelling of gum, boils, bone fractures, brass and bronze ornaments, burning sensation of burnt part of body, cart-shafts, cart-yokes, cholera, coarse ropes, conception, cough, cracked or inflamed heel and toes, cuts, diabetes, diarrhoea, dysentery, eye complaints, fever, fibre, furniture, gastric (vet.), gonorrhoea, haemoptysis, increasing the viscosity of seminal fluid, insanity, knee pain, leucorrhoea, long and luxuriant growth of hair, lumbago, menorrhagia, neck injuries (vet.), neuritis, obstinate vomiting, oxidation of copper, pains and bruises, paper-pulp, piles, pregnancy, rheumatism, ringworm, scabies, sexual diseases, skin diseases, snakebite, sores in mouth, spermatorrhoea, syphilis, vomiting, toothache, urinary complaints, well-curbs, wood in tent-poles, yokes, and in religious ceremonies. (Agarwal, 2003; Ambasta, 1986; Arora & Pandey, 1996; Bennet *et al.*, 1991; Bhatt *et al.*, 2001; 2003a; Bondya & Sharma, 2004; Borthakur *et al* 2004; Chopra *et al.*, 1956; D'Rozario *et al.*, 1999; Datt, 1996; Dey & Bahadur, 1973; Dwivedi, 2003; Girach, 1992; Hembrom & Goel, 2005; Huidrom, 1996; Jain, 1991; Jain & Sharma, 1996; Katewa *et al.*, 2000, 2001; Khanna & Kumar, 2000; Khare, 2004; Kirtikar & Basu, 1984; Kulkarni & Kumbhojkar, 1996; Kumar, 2003; Mahato & Chaudhary, 2005; Manandhar, 1996; Masih, 2003; Mitaliya *et al.*, 2003; Mohanty & Padhy, 1996; Mukherjee & Namhata, 1990; Pal *et al.*, 2000; Panthi & Chaudhary, 2003; Parrotta, 2001; Prajapati *et al.*, 2006; Rajendran, 2003; Rajendran *et al.*, 2003; Ranjan, 2000; 2003; Rama Rao & Henry, 1996; Roy *et al.*, 1998; Sahni, 1998; Sahoo & Mudgal, 1995; Saini, 1996b; Saklani & Jain, 1994; Samant *et al.*, 2001; Sen & Batra, 1997; Sen & Behera, 2003; Seetharam *et al.*, 1999; Singh, 2000, 2003; Singh & Kumar, 2000a; Singh & Pandey, 1996, 1998; Singh & Prakash, 1994; Singh & Rao, 2003; Sinha 1996; Sinha & Sinha, 2001; Sivakumar *et al.*, 2003; Siwakoti & Varma, 1996; Subramaniam, 2000; Thomas & Britto, 2003; Tigga *et al.*, 2003; Tirkey *et al.*, 2003a).

Ficus carica L. (Pl. 24A)

Family: Urticaceae.

Vern. : Anjeer.

English, Sanskrit & Regional Names:

Eng. : Anjira, Manjula, Kaodumbar

Sans. : Avaroha, Bahupada, Bhandira, Bhringi, Chira, Dhruva, Jatala, Jatiljati, Karmaja, Kshiri, Mahachhaya, Mana, Nandi, Nila, Nyagrodha, Padarohona, Raktaphala, Rohina, Skandhaja, Vaishravanodaya, Yakshaves, Yamapriya

B. : Anjir

Bo. : Anjira

G. : Anjir

H. : Anjir

Mal. : Simayatti

Mar. : Anjir

P. : Fagari, Fagu, Faguri, Kimri

Tel. : Simayatti

U.P. : Anjura, Manjimedi, Medi, Simayatti, Teneyatti.

Distribution : North-West and the Daccan.

Description : A shrub or small deciduous tree with branchlets and underside of leaves pubescent or tomentose. Leaves cordate, upper side rough. Fruit a fig.

Reproductive Cycle : February-June.

Habitat Ecology : Field borders. Wasteland. Forests.

Material Examined : HPU-EL-SKS-5672; Bhawarna (1,100 m), Khaira (940 m), Jasun (1,000 m), Pahra (950 m).

Part Used : Fruits.

Folk Uses : Ripe fruits **edible**, and also considered **purgative** in nature.

Active Constituents : Bergapten and psoralen, taraxasterol, β-sitosterol, lupeol, β-amyrin isolated from *leaves*.

Uses in Literature : Useful as alexeteric, antipyretic, aphrodisiac, demulcent, diuretic, emollient, expectorant, laxative, nutritive, purgative, sweet, tonic; and for boils, carbuncles, cooling, cough, diseases of head and blood, inflammations, leprosy, leucorrhoea, liver and spleen, nose bleeding, pain in the chest, paralysis, piles, ringworm, thirst, and weakness (Ambasta, 1986; Beigh *et al.*, 2003; Chopra *et al.*, 1956; Dey & Bahadur, 1973; Duke *et al.*, 2002; Harborne & Baxter, 2001; Kapur & Singh, 1996; Khanna & Kumar, 2000; Khare, 2004; Kirtikar & Basu, 1984; Prajapati *et al.*, 2006; Roy *et al.*, 1998; Said, 1997; Saini, 1996a; Sen & Batra, 1997; Singh, 2003; Tigga *et al.*, 2003; Watt, 1972).

Ficus clavata Wall.

Syn.: *F. trachycarpa* Miq.; *F. candata* Wall.; *F. chincha* Roxb.; *F. subincisa* Buch.-Ham. *ex* Sm.

Family: Urticaceae.

Vern. : Ban Feguda.

Distribution : Lower hills of Himalayas—ascending to 1,500 m.

Description : Erect shrubs having narrow long pointed leaves. Figs axillary, shortly-stalked, pale yellow on ripening.

Reproductive Cycle : June-September.

Habitat Ecology : Shady Ravines. Forests.

Material Examined : HPU-EL-SKS-5708; Balah (900 m), Bhaura (940 m), Bhauda (1,050 m), Khaira (940 m).

Parts Used : Fruits. Leaves.

Folk Uses : Ripe fruits **edible**, and leaves used as **fodder**.

Uses in Literature : Known to be used as fodder and vegetable (Ambasta, 1986; Jain, 1991).

Ficus foveolata Wall. (Pl. 24B)

Syn.: *F. sarmentosa* Ham.; *F. erecta* Miq.; *F. nipponica* Franch. & Savat.; *F. thunbergii* Maxim.; *F. impressa* Benth.; *F. wrightii* Benth.; *F. luducca* Roxb.; *F. pubigera* Wall.; *F. ludens* Wall.; *Pogonotrophe pubigera* Miq.; *P. verrucosa* Miq.; *P. reticulata* Miq.

Family: Urticaceae.

Vern. : Phaegdi.

Regional Name:

P. : Garelu, Karmbal, Phogri.

Distribution : Outer Himalayas: 700-2,400 m.

Description : A shrub creeping over rocks and forming mats or climbing over trees. Leaves pubescent or glabrous. Figs axillary, solitary. Achenes oblong-reniform.

Reproductive Cycle : March-July.

Habitat Ecology : Steep banks.

Material Examined : HPU-EL-SKS-5607; Chhanchri (930 m), Jasun (1,000 m), Tapa (945 m), Tikkar (910 m).

Parts Used : Leaves. Fruits.

Folk Uses : Known to be used as **fodder** (leaves) and **edible** (fruits).

Uses in Literature : Reported in India for edible purposes, fodder, and cold (Ambasta, 1986; Jain, 1991; Roy *et al.*, 1998; Sood & Thakur, 2004; Watt, 1972).

Ficus glomerata Roxb. (Pl. 24C)

Syn.: *F. chitagonga* Miq.; *F. racemosa* Wall.; *F. mollis* Miq.; *F. goolereca* Roxb.; *Covellia glomerata* Miq.; *C. mollis* Miq.

Family: Urticaceae.

Vern. : Umbrein.

English, Sanskrit & Regional Names:

Eng. : Cluster Fig
Sans. : Jantuphala, Phalasambendha, Suprastish-thida
Ass. : Jangedumuru, Yagyadimru
B. : Jagyadumar
Bo. : Umbar
G. : Umbro
H. : Gular, Umarai
Kan. : Athi
Kash. : Rumbal
Mal. : Rumadi, Umbaram
Ori. : Dimri, Odumboro
P. : Batbar, Kakallam, Kathagular
Tam. : Adavu, Anjiram, Malaiyinmmunivam, Tumbaram
Tel. : Atti, Medi.

Distribution : Sub-Himalayan tract; outer hill ranges upto 1,000 m.

Description : Trees having tapering leaves which are pubescent above and glabrous beneath. Receptacles reddish showing depressed umbilicous.

Reproductive Cycle : April-July.

Habitat Ecology : Forests. Borders of agricultural fields.

Material Examined : HPU-EL-SKS-5627; Bhuana (1,400 m), Khaira (940 m), Pahra (950 m), Tapa (945 m).

Parts Used : Whole Plant. Leaves. Wood. Latex.

Folk Uses : Plant considered **sacred** and **worshipped**. Wood used as **fuel.** Leaves lopped for **fodder**, forms one of the ingredients of '**havan**', and also employed as such during **religious** and **death ceremonies**. Latex applied as a lubricant over thread used for spinning '**charkha**'(**spinning wheel**). Also, one teaspoonful of latex taken daily for **digestive problems** and improving **health**.

Active Constituents : Active properties due to lupeol, friedelin, β-sitosterol and stigmasterol, ceryl behenate, lupeol, its acetate, α-amyrin acetate (*trunk bark*), a tetracyclic triterpene, glucanol acetate (*leaves*) and leucopelarogonidin-3-0-α-rhamnopyrano-side, β-sitosterol and lupeol (*root-bark*).

Biological Activity : A glycoside rich fraction from leaves reported to have hypotensive, vasodilator and cardiac depressive effects. Also, found to be antiprotozoal, hypoglycaemic and anticancerous.

Uses in Literature : Useful in India as astringent, breakfast food, carminative, edible (fruits), febrifuge, fodder, galactagogue, powerful tonic, sacred, spermator, stomachic, toothbrush, vegetable, vulnerary, wood for shoring wells; and for asthma, bilious affections, blisters, blood dysentery, blood dysentery (vet.), blood tissues, boils, bronchitis, cancer, curing piles, dermatitis, diabetes, diarrhoea, diseases of kidney and spleen, dislocation of joints, door, dry cough, dyeing, fuse-box, glandular enlargements, gonorrhoea, gum, habitual abortion, haemoptysis, healthy foetus (vet.), hydrocele, lactating mothers to increase the secretion of milk, laryngitis, leprosy, leucorrhoea, matchboxes, menorrhagia or excessive bleeding during menstruation, mouth sores, mumps, muscular pain, nose bleeding, pain in throat, piles, plague (vet.), planks and shutters, pneumonia, promoting fertility, pulmonary disease in children, rinderpest (vet.), scabies, scorpion sting, skin crack on heels and lips, smallpox, tone, sperm deficiency, vitality of the body, urinary and urinogenital and venereal complaints, washing wounds, well frames, worms; and also for making effigy if corpse is lost or not available (Ambasta, 1986; Bhandary & Chandrashekar, 2002; Bhatt *et al.*, 2001; Bhattacharya *et al.*, 2005; Chatterjee & Pakrashi, 1991; Chopra *et al.*, 1956; Dwivedi, 2003; Hosagoudar & Henry, 1996a; Islam & Jha, 2003; Jain, 1991; Jain *et al.*, 1991; Jha *et al.*, 1996; Jha & Kumar, 2003; Katewa *et al.*, 2000; Khanna *et al.*, 1996b; Kirtikar & Basu, 1984; Kshirsagar & Singh, 2000; Kumar, 1996; Kurian, 1995; Malhotra & Moorthy, 1973; Mishra *et al.*, 1996a; Mitaliya *et al.*, 2003; Mokat & Deokule, 2004; Nath & Maiti, 2003; Painuli & Maheshwari, 1996; Panthi & Chaudhary, 2003; Purohit & Samant, 1995; Radhakrishnan *et al.*, 1996; Rajendran *et al.*, 1996; Rama Rao & Henry, 1996; Ranjan, 2000; Roy *et al.*, 1998; Saini, 1996a; Samant *et al.*, 2001; Seetharam *et al.*, 1999; Sharma, 1996; Sharma *et al.*, 2000; Singh, 2000; Singh & Pandey, 1998; Sinha, 1996; Sinha & Sinha, 2001; Siwakoti & Siwakoti, 2000; Watt, 1972).

***Ficus hispida* L.** (Pl. 24D)

Syn.: *F. oppositifolia* Willd.; *F. daemonum* Koem.; *F. prominens* Wall.; *F. mollis* Willd.; *F. scabra* Jacq.; *Covellia daemonum* Miq.; *C. setulosa* Miq.; *C. courtallensis* Miq.; *C. wightiana* Miq.; *C. assamica* Miq.; *C. dasycarpa* Miq.; *Sycomorpha roxburghii* Miq.

Family: Urticaceae.

Vern. : Dagrein.

Sanskrit & Regional Names:

Sans.	: Bhadrodumbarika, Kushthaghni, Phalguni
Ass.	: Khoskadumar
B.	: Dumoor, Kakdumar
Bo.	: Dhedu, Mira
G.	: Dhedaumuro, Janglianjir
H.	: Gobla, Katugularia, Totmila
Kan.	: Adavi - atti, Kadatti
Mal.	: Erumanakku, Peyalti
Mar.	: Bhokada, Dherumbar, Kharoti
Ori.	: Bhaidimiri, Korotosamo
P.	: Daduri, Rumbal
Tam.	: Ottanalam, Sonatti
Tel.	: Boddamatti, Kukkabodda
U.P.	: Ban-gular.

Distribution : H.P. – Bhutan: 450-1,100 m.

Description : Small trees with broadly ovate opposite leaves. Figs globose, pale yellow and shortly-stalked.

Reproductive Cycle : June-August.

Habitat Ecology : Rocky Slopes. Occasionally Along Borders of Agricultural Fields.

Material Examined : HPU-EL-SKS-5437; Balah (900 m), Chhanchri (930 m), Jasun (1,000 m), Rangru (910 m).

Parts Used : Latex. Leaves. Fruits.

Folk Uses : Ripe fruits **edible** and unripe ones cooked as **vegetable**. Crushed fruits mixed with wheat flour fed to check **abortion** in livestock. Leaves lopped for **fodder**. Latex applied over boils to hasten **suppuration**, and also given to infants along with mother's milk for **sound sleep**.

Uses in Literature : Known to be used in India as an antiperiodic, astringent, cattle fodder, edible (fruits), emetic, galactagogue (vet.), fibre, jam, purgative, tonic, vegetable (leaves sometimes cooked with pork); and for blood dysentery, bubo (vet.), curries, cuts, delivery (vet.), diarrhoea (vet.), earache, easy shedding of placenta (vet.), female diseases, giddiness, gonorrhoea, headache, milching cattles to dry up their milk, preserving foetus in the womb, removal of placenta (vet.), stomach ulcers, ulcers in mouth, warts, and wounds (Agarwal, 2003; Ambasta, 1986; Arora & Pandey, 1996; Bennet *et al.*, 1991; Das, 2000; Das & Agarwal, 1991; Dey & Bahadur, 1973; Devi, 2003; Geetha *et al.*, 1996b; Girach *et al.*, 1996, 1997; Jain, 1991; Khare, 2004; Lalramnghinglova, 2002; Mahato & Chaudhary, 2005; Mohanty & Padhy, 1996; Nath & Maiti, 2003; Rama Rao & Henry, 1996; Ranjan, 2003; Rao, 1990; Roy *et al.*, 1998; Saini, 1996a, b; Sharma *et al.*, 2000; Singh *et al.*, 1996b; Singh & Kumar, 2000a; Tigga *et al.*, 2003; Sinha, 1986, 1987; Siwakoti & Varma, 1996; Sood & Thakur, 2004; Subramaniam, 2000; Watt, 1972).

Ficus infectoria Roxb.

Syn.: *F. lucescens* Blume; *F. lacor* Buch.- Ham.

Family: Moraceae

Vern. : Pallakharein.

Sanskrit & Regional Names:

- **Sans.** : Palaksha
- **B.** : Pakar, Pakur
- **G.** : Pepri
- **H.** : Kahimaql, Kaim, Keol
- **Kan.** : Basari, Juwi, Kari-basari
- **Mal.** : Bakri, Chakkiala, Chela
- **Mar.** : Bassari, Dhedumbara, Gandhaumbara
- **Tam.** : Jovi, Kallal, Kasurigatti, Suvi
- **Tel.** : Badijuwi, Jati.

Distribution : N. W. Himalayas.

Description : A large spreading tree. Leaves dark-green, shining, entire, oblong-lanceolate. Fruit a fig.

Reproductive Cycle : February-June.

Habitat Ecology : Rocky slopes. Wasteland.

Material Examined : HPU-EL-SKS-5584; Chhanchri (930 m), Nadli (1,000 m), Pahra (950 m), Sulah (950 m), Tikkar (910 m).

Parts Used : Leaves. Wood.

Folk Uses : Leaves for performing various **religious** and death **ceremonies**, and as **fodder**. Also, it constitutes one of the ingredients of '**havan**'. Wood considered **sacred**.

Uses in Literature : Described earlier as edible (fruits), fodder (succulent and relished by animals), paper-pulp; and for gargle, vulnary and wound (Ambasta, 1986).

Ficus palmata Forsk. (Pl. 24E)

Syn.: *F. caricoides* Roxb.; *F. pseudo-sycamorus* Decne.; *F. virgata* Roxb.

Family: Urticaceae.

Vern. : Dhuda, Khasra, Phegura.

Regional Names:

G.	: Pepri
H.	: Anjiri, Khabara
P.	: Daholia, Jamir, Phaguro
Tel.	: Manjimedi
U.P.	: Ban-Gular.

Distribution : Himalayas: upto 2,300 m.

Description : Small trees. Leaves broadly ovate, toothed towards the top. Figs axillary, stalked, solitary, pear-shaped and purple colured.

Reproductive Cycle : April-September.

Habitat Ecology : Rocky Slopes. Along Agricultural Fields.

Material Examined : HPU-EL-SKS-5429; Bandla (1,280 m), Dadh (1,100 m), Jasun (1,000 m), Malahu (1,050 m), Panapari (1,100 m).

Parts Used : Aerial Plant Parts. Bark. Fruits. Latex.

Folk Uses : Ripe fruits **edible** and considered good for checking **diabetes**. Tender aerial plant parts consumed as **vegetable** and for checking **cough** and **constipation**. 10 g of bark ash mixed with 2 teaspoonful of 'suhaaga' (borax) and 100 g of honey taken 2-3 times daily for 7-10 days as an **expectorant**. Latex applied locally for drawing out **thorne**.

Active Constituents : *Leaves* yield β-sitosterol and bergapten.

Uses in Literature : Known so far as cattle fodder, demulcent, edible (cooked fruits and leaves), laxative; and for bladder, building purposes, diseases of lungs, drawing out thorns, dysentery, high blood pressure, indigestion, jam, pimples, squash (Ambasta, 1986; Ansari, 1991; Arora & Pandey, 1996; Aswal,

1996; Bennet *et al.*, 1991; Chopra *et al.*, 1956; Das & Agarwal, 1991; Jain, 1991; Kapur & Nanda, 1996; Kirtikar & Basu, 1984; Lal *et al.*, 1996; Murthy *et al.*, 2003; Negi *et al.*, 1993; Parmar & Kaushal, 1982; Rana *et al.*, 1996; Roy *et al.*, 1998; Saini, 1996a; Samant *et al.*, 2001; Singh, 1996b; Singh & Kumar, 2000a; Singh & Pandey, 1998; Sinha, 1996; Watt, 1972).

***Ficus religiosa* L.** (Pl. 24F)

Syn.: *F. affinior* Griff.; *Urostigma religiosum* Gasp.; *U. affine* Miq.

Family: Urticaceae.

Vern. : Bad, Peepal.

English, Sanskrit & Regional Names:

Eng. : Peepul Tree, Pipul Tree
Sans. : Chaityavriksha, Pavitraka, Vriksharaja, Yajnika
Ass. : Ahot
B. : Ashathwa
Bo. : Arli, Jari
G. : Pipar, Pipro
H. : Pipal, Pipli
Kan. : Arali, Ashwuattha
Kash. : Bad
Mal. : Arayal, Chalodalam
Mar. : Ashvatha, Pimpla
Ori. : Oshwottho, Pippolo
P. : Bhor, Pipal
Tam. : Achuvattam, Podoroganam, Vanagardi
Tel. : Ashvaddhamu, Bogaha.

Distribution : Widely cultivated in the Himalayas upto 1,400 m.

Description : A large glabrous tree with long-petioled, coriaceous leaves which are shining above and dry beneath. Receptacle dark purple. Figs red.

Reproductive Cycle : March-October.

Habitat Ecology : Near Water Sources. Hilltops.

Material Examined : HPU-EL-SKS-5578; Chauki (900 m), Dharaman (1,100 m), Makol (900 m), Rajeher (1,050 m), Sidhpur (1,100 m).

Parts Used : Whole Plant.

Folk Uses : Plant considered **sacred** and **worshipped** as incarnation of Lord **Brahma**. Its used in '**havan**' for performing **religious** ceremonies. One teaspoonful paste of its ash and that of *Ficus palmata* taken with honey every morning and at bedtime to relieve **cough**.

Active Constituents : *Bark* contains tannins, β-sitosterol and D-glucoside.

Biological Activity : Stem-bark exhibits antiprotozoal and anthelmintic activity.

Uses in Literature : Reported in literature as abortifacient, absorbent in inflammatory swelling, alterative, antidote to snakebite, astringent, colouration material and fixative, contraceptive, edible (fruits and tender leaf buds eaten in times of scarcity), fodder, laxative, psychoactive plant, sacred tree (worshipped—cutting tree more sinful than killing a saint), tonic, tree of fertility; and for after-pain (postnatal), aphthae, asthma, bird-lime, blisters, boils, bowels, bronchitis (vet.), burns, carbuncle, cholera, chronic liver disorders, constipation (vet.), cough (vet.), dyeing, dysentery, eczema, enhancing memory, fever, fibre, fistula, foot and mouth disease (vet.), gonorrhoea, gum, heamaturia, leucorrhoea, matchboxes, obstinate hiccups, paintings (Mithila), over intoxication by opium, packing cases, paralysis, rheumatism, scabies, skin affections, smallpox, sores in mouth, spermatorrhoea, tanning, throat infection, toothache, ulcer, urinary complaint, vermifuge (vet.), whooping cough, and wounds (vet.) (Ambasta, 1986; Arora & Pandey, 1996; Beg *et al.*, 2005; Bennet *et al.*, 1991; Bhatt *et al.*, 2001, 2003b; Bhattacharya *et al.*, 2005; Borthakur, 1996a; Chatterjee & Pakrashi, 1991; Chopra *et al.*, 1956; D'Rozario *et al.*, 1999; Das & Agarwal, 1991; Diwakar & Ansari, 2003; Dwivedi, 2003; Gogoi & Borthakur, 1991; Gogoi *et al.*, 2003b; Hosagoudar & Henry, 1996a; Jain, 1991; Jain *et al.*, 1991, 1994; Jha & Basak, 1994; Kalpesh *et al.*, 2004; Khanna, 2002; Khanna *et al.*, 1996a, c; Khare, 2004; Kirtikar & Basu, 1984; Kumar *et al.*, 2003; Masih, 2003; Mishra *et al.*, 1996a; Mitaliya & Bhatt, 2003; Molla & Roy, 1985; Nayak *et al.*, 2003; Panthi & Chaudhary, 2003; Pawar & Patil, 2004; Prajapati *et al.*, 2006; Rama Rao & Henry, 1996; Roy *et al.*, 1998; Sahni, 1998; Saini, 1996a, b; Samant *et al.*, 2001; Satapathy & Brahmam, 1996; Shah, 2005; Sharma, 1996; Sikarwar, 1996; Sikarwar & Kaushik, 1992; Singh, 2000; Singh & Kumar, 2000a; Singh & Prakash, 1996; Singh & Rao, 2003; Singh & Zaheer, 1996; Sinha, 1996; Sinha & Sinha, 2001; Sivakumar *et al.*, 2003; Siwakoti & Varma, 1996; Thomas & Britto, 2003; Virjee *et al.*, 1984; Watt, 1972).

***Ficus roxburghii* Wall. *ex* Miq.** (Pl. 25A)
Syn.: *F. auriculata* Lour.; *F. macrophylla* Griff.; *F. regia* Miq.; *F. scleroptera* Griff.; *Covellia macrophylla* Miq.
Family: Urticaceae.

Vern. : Traimbal.

Regional Names:

B. : Demur, Doomoor

H. : Timla, Tirmal

P. : Dadari, Tirnal, Trimal, Trimbal, Urmul.

Distribution : Outer Himalayas—ascending to 1,600 m.

Description : Spreading trees. Leaves stalked, broadly ovate. Figs stalked, clustered on short thick, leafless branches, red or dull purple when ripe.

Reproductive Cycle : April-August.

Habitat Ecology : Wastelands. Rocky Slopes. Forests.

Material Examined : HPU-EL-SKS-5432; Bhauda (1,050 m), Nadli (1,000 m), Nanaun (900 m), Panapari (1,100 m), Rajpur (1,100 m).

Parts Used : Leaves. Bark. Fruit.

Folk Uses : Leaves used for making **meal-plates** and as **fodder** for milching **livestock**. Ripe fruits **edible**, and unripe ones cooked as **vegetable**. Bark paste applied on **boils**.

Active Constituents : Constituents isolated are β-sitosterol, fridelin and epifriedelanol (*plant*).

Uses in Literature : Used in India as edible (fruits), fodder, galactagogue (vet.), vegetable; and for breast ulcers, curries, delivery (vet.), jams, leucoderma, meal plates for feasts in villages, ringworm and ropes (bark) (Arora & Pandey, 1996; Bennet *et al.*, 1991; Jain, 1991; Parmar & Kaushal, 1982; Rama Rao & Henry, 1996; Roy *et al.*, 1998; Singh & Kumar, 2000a; Watt, 1972).

***Ficus rumphii* Blume** (Pl. 25B)

Syn.: *F. cordifolia* Roxb.; *Urostigma cordifolium* Miq.; *U. rumphii* Miq.

Family: Urticaceae.

Vern. : Badarein.

Regional Names:

	Ass.	: Jakri, Pakri
	B.	: Gelasvattna
	H.	: Gagjaira, Pikan
	Kan.	: Betaarali
	Mar.	: Ashta, Payal
	P.	: Badha, Pulakh, Rumbal.

Distribution : Throughout India—ascending to 1,650 m.

Description : Trees with long-petioled, broadly-ovate leaves and axillary black figs.

Reproductive Cycle : April-July.

Habitat Ecology : Shrubberies. Open Slopes.

Material Examined : HPU-EL-SKS-5570; Bodhal (970 m), Bhuana (1,400 m), Malnu (1,000 m), Paror (940 m), Tapa (945 m).

Part Used : Leaves.

Folk Use : Leaves lopped as **fodder**.

Uses in Literature : Known to be used as anthelmintic (fruit juice), charcoal, edible (fruits), fibre, fodder (leaves and twigs), fuel, gum, sacred; and for asthma, lice (vet.), rearing lac insect, snakebite, vomiting, worms, wounds (vet.) (Ambasta, 1986; Chatterjee & Pakrashi, 1991; Chopra *et al.*, 1956; Jain, 1991; Kirtikar & Basu, 1984; Murthy *et al.*, 2003; Purohit & Samant, 1995; Roy *et al.*, 1998; Samant *et al.*, 2001; Tigga *et al.*, 2003; Watt 1972).

***Ficus semicordata* Buch.-Ham. *ex* Sm.** (Pl. 25C)

Syn.: *F. cunia* Buch.-Ham. *ex* Roxb.; *F. conglomerata* Roxb.

Family: Urticaceae.

Vern. : Khanurein, Khasarkhurein.

Distribution : India. S. China. Burma. S.E. Asia: 2,000 m.

Description : A small-spreading tree. Leaves rough, unequally lobed at the base, the lower lobe more rounded and much longer and with reddish brown fruit.

Reproductive Cycle : May-September.

Habitat Ecology : Forests.

Material Examined : HPU-EL-SKS-5702; Bhuana (1,400 m), Boda (1,000 m), Rangru (910 m), Saparu (1,250 m), Thandol (950 m).

Parts Used : Leaves. Fruits.

Folk Uses : Leaves used as **sandpaper** and also lopped for **fodder**. Ripe fruits **edible**, unripe ones cooked as **vegetable**.

Uses in Literature : Described in India as edible (fruits), fodder, jams, sandpaper (leaves), vegetable; and for bladder complaints, choking in throat (vet.), dysentery, leprosy, and viscereal obstructions (Arora & Pandey, 1996; Chandra, 1997; Jain, 1991; Jain & Saklani, 1992; Kumar, 2003; Lalramnghinglova, 1996; Negi & Gaur, 1991; Sahu, 1996; Singh & Kumar, 2000a; Siwakoti & Varma, 1996; Tigga *et al.*, 2003; Watt, 1972).

***Flacourtia indica* (Burm. f.) Merr.** (Pl. 25D)

Syn.: *F. ramontchi* L. Herit; *F. sapida* Roxb.; *F. sepiara* Roxb.; *F. latifolia* Cooke; *F. occidentalis* Blatter; *Gmelina indica* Burm. f.

Family: Flacourtiaceae.

Vern. : Kaangu.

English, Sanskrit and Regional Names:

Eng. : Governor's Plum, Madagascar Plum
Sans. : Wadukantaka
B. : Binja, Tambat
Bo. : Kakad, Swadu
G. : Kankod
H. : Bhanber, Kakein, Katar
Kan. : Hattarimullu, Humunki
M.P. : Kankar
Mar. : Kaker, Paker
Ori. : Bonisha, Potnabonisho
P. : Kakao, Kukai
Tam. : Mullikkarai, Sottaikala
Tel. : Bontakandregu, Nakkaneredu
U.P. : Katai.

Distribution : Throughout India upto 1,000 m.

Description : A small tree armed with axillary thorns. Leaves variable, elliptic-oblong. Flowers greenish-yellow. Fruits dark brown to red, oval.

Reproductive Cycle : March-April.

Habitat Ecology : Edges of Agricultural Fields. Wasteplaces. Drier Slopes.

Material Examined : HPU-EL-SKS-5435; Jasun (1,000 m), Rangru (910 m), Sulah (950 m), Thandol (950 m), Tikkar (910 m).

Parts Used : Leaves. Fruits. Bark. Seeds. Wood.

Folk Uses : Ripe fruits **edible**. Leaves given as **fodder** for goats. Wood used for making **coal**, **agricultural implements**, and as **fuel**. One teaspoonful of decoction from its bark and seeds of *Trachyspermum ammi* prescribed twice a day for **digestive disorders**.

Active Constituents : These are tannin and flacourtin, a phenolic glucoside ester (*bark*), new butyrolactone lignan disaccharide-romontoside diphyllin-4-O-[β-D-glucosyl (1→4)]-β-2,3-di-O-methyl-D-methyl-D-xyloside (*heartwood*).

Biological Activity : Hypotensive activity and effect on CNS found positive.

Uses in Literature : Recorded in literature as an antidote to snakebite, appetiser, astringent, cattle fodder, diuretic, edible fruits, tarts, tonic after delivery; and for agricultural implements, backache in ladies, bite of mad dog, burning sensation in chest, carbuncle, chest pain, cholera, dysentery, eczema, enlarged spleen, excess body heat, facilitating childbirth, fodder, foot and mouth disease (vet.), gout, intermittent fever, leucorrhoea, liver disorders, pickling, reshaping abdomen after delivery, rheumatic pains after delivery in women, skin diseases, smallpox, sores, syphilis, tanning, tonsillitis, turnery, and wounds (Arora & Pandey, 1996; Ambasta, 1986; Aminuddin & Girach, 1996; Bennet *et al.*, 1991; Bhatt *et al.*, 2003a; Brahmam & Saxena, 1990; Das & Agarwal, 1991; Datt, 1996; Jain, 1991; Jain *et al.*, 1991; Katewa & Arora, 1997; Katewa *et al.*, 2000; Kirtikar & Basu, 1984; Maheshwari *et al.*, 1986; Murthy *et al.*, 2003; Nadanakunjidam, 2003; Nayak *et al.*, 2003; Nazarudeen *et al.*, 1996; Negi & Gaur, 1991; Painuli & Maheshwari, 1996; Parrotta, 2001; Parmar & Kaushal, 1982; Rajendran *et al.*, 1996; Rama Rao & Henry, 1996; Ramachandran & Nair, 1981; Roy *et al.*, 1998; Saini, 1996a; Shah & Singh, 1990; Sharma & Singh, 2001; Sharma *et al.*, 2000; Sikarwar & Kaushik, 1992; Singh, 1996b; Singh & Kumar, 2000a; Singh & Pandey, 1998; Sood & Thakur, 2004; Sudhakar & Vedavathy, 2000).

Foeniculum vulgare Gaertn. (Pl. 25E)

Syn.: *F. capillaceum* Gillb.; *F. officinale* All.; *F. dulce* Bauh; *F. panmorium* DC.; *Anethum foeniculum* L.; *A. panmorium* Roxb.; *Ozodia foeniculacea* Wight & Arn.

Family: Apiaceae.

Vern. : Meethi-saunf.

English, Sanskrit & Regional Names:

Eng. : Fennel
Sans. : Madhurika
B. : Mauri, Pan-muhori
Bo. : Bari-Shopha
H. : Bari-Saunf, Saunf
Kan. : Badi-Sopu
Mar. : Badishep
Tam. : Sohikire
Tel. : Pedda-jila-kurra.

Distribution : Cultivated throughout India; rarely as an escape along fields.

Description : Glabrous herbs upto 90 cm tall. Leaves 3-4 pinnate. Umbels having 15 or more rays. Vittae conspicuous.

Flowering & Fruiting : August-September.

Habitat Ecology : Cultivated. Often an escape in Wasteplaces and Roadsides.

Material Examined : HPU-EL-SKS-5622; Dharaman (1,100 m), Gopalpur (1,350 m), Kathena (1,290 m), Saliana (1,000 m).

Parts Used : Leaves. Fruits.

Folk Uses : Leaves cooked as **vegetable**. Fruits used for flavouring **sweet-dishes**, **jams** and **jellies**; and its powder taken orally as **astringent** and against **digestive** ailments.

Active Constituents : Isolated constituents are anethole, fenchone, anisalehyde and feniculin (*plant*), columbianetin, 7-hydroxy- and 6,7-dihydroxy-coumarins, marmesin, osthenol, psoralen, scoparone, seselin, xanthotoxin, α- and β-amyrins, oleanolic acid, Δ^5-avenasterol, campesterol, Δ^7-stigmasterol, stigmastadienol, linoleic, oleic, palmitic and petroselinic acids (*fruits / seeds*).

Biological Activity : Seeds and fruits antimicrobial.

Uses in Literature : Considered useful as an aromatic, carminative, colic, diuretic, salad, stimulant, sudorific, vegetable, vermicide; and for burning sensation in body, chest diseases, cough, dysentery, fever, flatulence, headache, menstruation and spleen complaints, soups, sauces, spleen and kidney ailments, thirst, toothache and wounds (Ambasta, 1986; Bhalla *et al.*, 1982; Dar *et al.*, 1984; Pal & Paul, 2003; Sharma, 2003; Uniyal, 1968).

***Fragaria indica* Andrz.** (Pl. 25F)

Syn.: *Duchesnea indica* Focke

Family: Rosaceae.

Vern. : Bhoo-Akha.

English & Regional Names:

- **Eng.** : Indian or mock strawberry
- **H.** : Pahari rashbhari
- **P.** : Banphal, Ingrach, Kanzar, Palzor, Tawai

Distribution : Throughout the Himalayas.

Description : Soft silky herbs having perennial rootstock; nearly erect stems and long stalked, digitately compound leaves. Leaflets 2, sessile, ovate, toothed. Flowers yellow, axillary or few in terminal cymes. Fruits red globose. Achenes glabrous.

Reproductive Cycle : June-August.

Habitat Ecology : Shady and Moist places.

Material Examined : HPU-EL-SKS-5704; Bhawarna (1,100 m), Gharthun (900 m), Holta (1,200 m), Lahla (930 m), Purani-palam (1,100 m).

Part Used : Fruits.

Folk Use : Ripe fruits **relished** by children.

Uses in Literature : Reported to be used as edible (fruits) and low ground cover (Ambasta, 1986; Arora & Pandey, 1996; Bennet *et al.*, 1991; Parmar & Kaushal, 1982; Roy *et al.*, 1998; Singh & Kumar, 2000a; Sood *et al.*, 2001).

***Fumaria indica* (Hausskn.) Pugsley** (Pl. 26A)

Syn.: *F. vaillantii* Loisel. var. *indica* Hausskn.; *F. vaillantii* Loisel.

Family: Fumariaceae.

Vern. : Papda.

English, Sanskrit & Regional Names:

Eng. : Fine-leaved Fumitory
Sans. : Araka, Charaka, Katupatra, Panshu
B. : Bansulpha
Bo. : Pitpatra
G. : Khasudlio, Pitpapda
H. : Pitpapda
Mar. : Pitpapda
Tam. : Tura
Tel. : Chata-rashi.

Distribution : Spread throughout India.

Description : A diffuse herb with much divided leaves and flowers in lax racemes. Fruits globose, indehiscent.

Flowering & Fruiting : October-March.

Habitat Ecology : Moist slopes. Cultivated fields.

Material Examined : HPU-EL-SKS-5552; Banuri (1,150 m), Chachian (1,300 m), Dehan (1,130 m), Holta (1,200 m), Sidhpur (1,100 m).

Parts Used : Whole Plant. Aerial Plant Parts.

Folk Uses : 1 teaspoonful of plant decoction prescribed thrice a day for 4-5 days to check **fever**. Aerial parts employed as **fodder**.

Active Constituents : Protopine, nonacosanol, sitosterol, n-alkanes and sterols (*leaves, stem*); sanguinarine, protopine, cryptopine, fumaridine, fumaramine, perfumidine, perfumine, d and dl-bicuculline, dl-tetrahydrocoptisine, fumariline, fumarilicine, narceimine, dehydrocheilanthifoline, copticine and narlumidine have been isolated.

Biological Activity : Herb extr. antiarrythmic in rabbits.

Uses in Literature : Useful for bodyache, diarrhoea, fever, flue, indigestion, liver complaints, skin diseases, melasma, malaria, cough, mouth ulcers and skin diseases; and as an anthelmintic, aperient, blood purifier, cool drink, diaphoretic, diuretic, laxative, and vegetable (Chandra, 1997; Chopra *et al.*, 1956; Dhiman, 1976; Drury, 1978; Dwivedi, 2003; Hembrom & Goel, 2005; Shah, 2005; Sharma, 2003; Shukla *et al.*, 2003; Srivastava *et al.*, 2003b; Uphof, 1968; Watt, 1972).

Galium aparine L. (Pl. 26B)

Family: Rubiaceae.

Vern. : Jhanjheedi.

English Names:

Eng. : Cleavers, Goosegrass.

Distribution : Kashmir - Sikkim (Temperate Himalayas).

Description : Trailing herbs. Peduncles axillary, 3-flowered. Fruits covered with hooked bristles.

Flowering & Fruiting : July-August.

Habitat Ecology : Grassy slopes. Wastelands. Screes.

Material Examined : HPU-EL-SKS-5560; Bhauda (1,050 m), Chornali (1,280 m), Malnu (1,000 m), Paror (940 m), Sungal (1,050 m).

Parts Used : Aerial Parts.

Folk Uses : Aerial parts consumed as **vegetable** and **fodder**.

Active Constituents : *Aerial parts* contain iridoid glycoside, asperuloside.

Biological Activity : Extr. of the plant lowers arterial pressure up to 50%.

Uses in Literature : Known in India as antiscorbutic, diuretic, edible (shoots), substitute for tea, coffee; and for cancer, dysuria, eczema, kidney stones, psoriasis, refrigerant, swollen lymph nodes, and urinary problems (Agarwal, 2003; Ambasta, 1986; Aswal & Mehrotra, 1994; Chevallier , 1996; Chopra *et al.*, 1956; Harborne & Baxter, 2001; Kapur & Nanda, 1996; Karnick, 1994; Prajapati *et al.*, 2006; Roy *et al.*, 1998; Sood & Thakur, 2004).

Geranium nepalense Sweet (Pl. 26C)

Syn.: *G. radicans* DC.; *G. pallidium* Royle; *G. patens* Royle; *G. affine* Wight & Arn.; *G. arnottianum* Steud.

Family: Geraniaceae.

Vern. : Laljari.

English & Regional Names:

Eng. : Nepal Geranium, Nepalese Cranesbill

Kash. : Roel

H. : Bhanda

P. : Bhanda.

Distribution : Temperate Himalaya.

Description : Branched hairy annuals with small, yellowish-white flowers and 5-gonal leaves. Capsule 5-lobed.

Flowering & Fruiting : March-May.

Habitat Ecology : Rock-crevices. Drier areas. Wastelands.

Material Examined : HPU-EL-SKS-5619; Dehan (1,130 m), Sungal (1,050 m), Tapa (945 m), Sulah (950 m), Thakurdwara (1,150 m).

Parts Used : Aerial Parts.

Folk Use : Decoction (1 teaspoonful) of aerial parts taken regularly for one month to improve **immune system**.

Active Constituents : Active constituents are geranin, kaempferol 7-rhamnoside and kaempferitrin.

Biological Activity : EtOH (50%) extr. of the plant hypotensive.

Uses in Literature : Used so far as an astringent, diuretic, dye, kidney ailments; and for cuts, jaundice, stomach complaints, toothache, ulcer, wounds (Aswal & Mehrotra, 1994; Jain & Goel, 2005; Kant & Dutt, 2005; Krishna & Singh, 1987; Sharma, 2003; Uniyal & Chauhan, 1973b; Uphof, 1968; Watt, 1972).

Glycine max Merr. (Pl. 26D)

Syn.: *G. soja* Sieb. & Zucc.; *G. hispida* Maxim.; *Soja max* Piper

Family: Fabaceae.

Vern. : Soyabean.

English & Regional Names:

- **Eng.** : Soyabean
- **Ass.** : Patanijokra
- **B.** : Garjkalai
- **H.** : Bhat, Bhatwar, Bhetmas, Ramkurthi

Distribution : Foothills of Himalayas.

Description : Herbaceous annuals having stem coated with brown hairs, trifoliate leaves; and white flowers with conspicuous bracts. Pods constricted, hairy.

Reproductive Cycle : July-September.

Habitat Ecology : Cultivated lands. Also as an escape.

Material Examined : HPU-EL-SKS-5479; Chornali (1,280 m), Nanaun (900 m), Sidhpur (1,100 m), Tanda (1,120 m).

Part Used : Seeds.

Folk Uses : Seeds used as **pulse** and its **flour** considered highly **nutritious** and **antidiabetic**.

Uses in Literature : Earlier reported as adhesive, breakfast foods, canned, chocolate bars, cooking oil, cosmetics, detergents, enamal industries, fertilizer, forage plant, ice-creams, paints, pharmaceuticals, salads, stabilising agents in foods, varnish, vegetable; and for leather goods, leather finishes, insecticidal sprays, plywood glue, soaps, and textiles sizes (Ambasta, 1986).

***Gnaphalium affine* L.** (Pl. 26E)

Family: Asteraceae.

Vern. : Gadua.

Distribution : N. W. Himalayas.

Description : Erect herbaceous annuals having simple or branched tomentose stem. Leaves sessile, tomentose, lanceolate. Flowers yellow in cymose heads.

Reproductive Cycle : April-September.

Habitat Ecology : Grass-fields.

Material Examined : HPU-EL-SKS-5590; Banuri (1,150 m), Chauki (900 m), Lahla (930 m), Malnu (1,000 m), Panchrukhi (990 m).

Part Used : Flowers.

Folk Uses : Flowers **offered** to appease Goddess **Parvati**.

***Gonatanthus pumilus* (D. Don) Engl. & Krause** (Pl. 26F)

Syn.: *G. sermentosus* Klotzsch; *Caladium pumilum* D. Don

Family: Araceae.

Vern. Name : Aala.

Distribution : Temperate Himalayas.

Description : Herbaceous annuals. Rootstock tuberous, branches slender. Bulbils small with fibres. Spathe yellow. Flowers fragrant. Berries clustered on spadix.

Reproductive Cycle : June-August.

Habitat Ecology : Marshy lands. Water Streams.

Material Examined : HPU-EL-SKS-5715; Chauki (900 m), Darugnu (1,400 m), Gopalpur (1,350 m), Sungal (1,050 m), Thandol (950 m).

Parts Used : Leaves. Tubers.

Folk Uses : Leaves consumed as **vegetable** and '**patrodu**'(fried leaf rolls). Tubers also as **vegetable** and **pickled**.

Uses in Literature : Known for its edible tubers (Arora & Pandey, 1996; Sharma & Rana, 2005).

Gossypium arboreum L. (Pl. 27A)

Syn.: *G. indicum* Tod.; *G. neglectum* Tod.; *G. sanguineum* Hassk.; *G. intermedium* Tod.; *G. obustifolium* Roxb.; *G. nanking* Meyen; *G. arboreum* L.

Family: Malvaceae.

Vern. : Kapah.

English, Sanskrit & Regional Names:

Eng. : Tree Cotton
Sans. : Karpasamu
Bo. : Deokapas
H. : Deokapas, Nurma
Kan. : Hathi
Mal. : Chemparutti
Mar. : Devakapusa
Ori. : Karposa, Kopa
P. : Kapas
Tam. : Sembarutti
Tel. : Patti, Karpasamu.

Distribution : Plains of India upto 1,800 m.

Description : Shrubs. Leaves palmately 5-7 lobed. Flowers purple or yellow with a crimson centre. Capsule opening by 3-5 valves. Seeds with curved embryo.

Reproductive Cycle : July-September.

Habitat Ecology : House courtyards.

Material Examined : HPU-EL-SKS-5402; Bhatu (1,040 m), Bhaura (940 m), Jasun (1,000 m), Paror (940 m), Rangru (910 m).

Part Used : Seeds.

Folk Uses : Cotton fibres used for making **sacred thread** ('janau'), **stuffing** pillows and quilts, and also for making spun **cloth**. Overnight soaked seeds considered highly **nutritious** for cattle.

Active Constituents : Active properties due to (-) gossypol (*seeds*).

Biological Activity : Gossypium-8-O-rhamnoside along with quercitin-3-7- and 3′-O-glucoside of quercitin exhibit antibacterial activity against *Pseudomonas maltophila* and *Enterbacter cloacae*.

Uses in Literature : Known earlier as soothing agent, symbolic for making sacred thread; and for conjunctivitis of infants, earache, fever, fibre, gonorrhoea, insanity, labour complaints, skin eruptions, strangulary, and patellar dislocation (Khanna *et al.*, 1996b; Sharma & Borthakur, 2003; Watt, 1972).

Grevillea robusta A. Cunn. *ex* R. Br. (Pl. 27B)

Family: Proteaceae

Vern. : Silver Oak.

English & Regional Name:

Eng. : Silky Oak, Silver Oak

Tam. : Savukkumaram.

Distribution : Planted as a shade tree in gardens and avenues.

Description : Evergreen trees with pinnate leaves. Leaflets deeply pinnate. Flowers orange in clusters. Fruits obliquely-oblong, compressed, brown.

Reproductive Cycle : March-July.

Habitat Ecology : Often planted along roadsides in Himalayas.

Material Examined : HPU-EL-SKS-5436; Bindraban (1,270 m), Chornali (1,280 m), Gopalpur (1,350 m), Saparu (1,250 m), Thakurdwara (1,150 m).

Parts Used : Whole Plant. Wood.

Folk Uses : Plant grown for its **aesthetic** value; also considered an excellent source of **nectar** for bee keeping. Wood extensively employed for **ornamental works**.

Active Constituents : *Wood* yields revile, sitosterol, epilyeniresinol, 2-O-glucoside, rhamnocitrin-3-O-rutinoside and kaempferol-3-O-rutinoside.

Uses in Literature : So far described as edible, green manure, gum; and for floors, furniture, ornamental panelling, parquet, plywood, tannin, toys, and veneering (Ambasta, 1986; Sinha, 1996).

Grewia disperma Rottb. in Spreng. (Pl. 27C)

Syn.: *G. glabra* Blume Bijdr; *G. laevigata* (Non Vahl) Mast.

Family: Tiliaceae.

Vern. : Byulee.

Distribution : Sub-Himalayan tracks. Eastern & S. India.

Description : Small trees. Leaves oblong-lanceolate. Flowers white. Drupes, black when ripe.

Reproductive Cycle : September-November.

Habitat Ecology : Along the fields.

Material Examined : HPU-EL-SKS-5648; Balah (900 m), Khaira (940 m), Jasun (1,000 m), Makol (900 m), Tikkar (910 m).

Parts Used : Leaves. Stem. Bark. Fruits.

Folk Uses : Leaves lopped for **fodder** and increase **fat** content of milk in milching cattle. Bark after retting good for making **ropes**. Mucous after soaking bark in water overnight, used for washing **hair** and **smoothening** mud-flours. Stem employed as good **fuel**. Ripe fruits **eaten** and considered **diuretic** in nature.

Uses in Literature : Reported for edible fruits (Ambasta, 1986; Saklani & Jain, 1994).

Grewia elastica Royle (Pl. 27D)

Syn.: *G. vestita* Wall.; *G. subinaequalis* DC.; *G. obtecta* Wall.; *G. asiatica* L. var. *vestita* Mast.

Family: Tiliaceae.

Vern. : Dhaman.

Sanskrit & Regional Names:

Sans. : Dhanvana, Dharmana
Ass. : Man-bijal
B. : Dhamani
H. : Bimla, Dhaman, Dhamun, Pharsia
Ori. : Mirgi chara
P. : Dhaman, Phalwa.

Distribution : N.W. and Tropical Himalayas.

Description : Small trees. Leaves ovate, rough. Flowers greenish-yellow, in densely crowded cymes. Drupes round, stalked.

Reproductive Cycle : April-July.

Habitat Ecology : Common along agricultural fields.

Material Examined : HPU-EL-SKS-5621; Balah (900 m), Chhanchri (930 m), Jasun (1,000 m), Rangru (910 m), Tikkar (910 m).

Parts Used : Leaves. Fruits. Wood. Bark.

Folk Uses : Leaves lopped for **fodder**. Ripe fruits **edible**. Bark employed for **cordage** Wood employed for making **bows**, and as **fuelwood**.

Biological Activity : Stem-bark CNS depressant.

Uses in Literature : Known as astringent, coolant, edible (fruits), stomachic; timber used for tool-handles, shafts, and bark for ropes (Ambasta, 1986; Arora & Pandey, 1996; Bennet *et al.*, 1991; Das & Agarwal, 1991; Murthy *et al.*, 2003; Purohit & Samant, 1995; Roy *et al.*, 1998).

Grewia optiva Drumm. *ex* Burret (Pl. 27E)

Syn.: *G. oppositifolia* Roxb.

Family: Tiliaceae.

Vern. : Beul.

Regional Names:

- **H.** : Biul, Biung, Bhimal
- **Kan.** : Thidsal
- **P.** : Behal, Dhaman, Pharwa.

Distribution : N. W. Himalayas upto 1,800 m.

Description : A medium-sized tree having 3-nerved, toothed, elliptic leaves. Petals yellowish-white. Drupes olive-green, black when ripe.

Reproductive Cycle : April-June.

Habitat Ecology : Edges of Agricultural Fields. Wasteplaces.

Material Examined : HPU-EL-SKS-5434; Balah (900 m), Jasun (1,000 m), Panapari (1,100 m), Tapa (945 m), Thandol (950 m).

Parts Used : Leaves. Branches. Bark. Fruits. Wood.

Folk Uses : Leaves lopped for **fodder** for milching cattle to increase **fat** content of milk. Fibers from branches after retting used for making **ropes**. Wood used mostly in **religious ceremonies**. Bark used for inducing **rumination** in livestock; its extr. used for **washing** hair. Ripe fruits **eaten** and have **diuretic** properties.

Uses in Literature : Recorded in literature as edible (fruit), fuel, galactagogue (vet.), lubricant during childbirth, nutritive fodder, shampoo, symbolic torch in religious festivals; and for agricultural implements, axe-handles, banghy-poles, basketry, boils, blisters, bone fractures (vet.), bows, clothing, cordage, cot frames, cough, dyeing, eruptions, fever, gastric problem (vet.), increasing milk-giving powers, indigestion (vet.), material requirement, oar-shafts, paddles, ropes, tool and axe handles (Agarwal, 2003; Ambasta, 1986; Arora & Pandey, 1996; Chandra, 1997; Hosagoudar & Henry, 1996a; Jain, 1991; Kapur, 1996a; Lata, 1996, 1997; Manandhar, 1995; Murthy *et al.*, 2003; Rana *et al.*, 1996; Sharma, 2003; Singh, 1996b; Singh & Kumar, 2000a; Sood & Thakur, 2004, Watt, 1972).

Hamiltonia suaveolens Roxb. (Pl. 27F)

Syn.: *H. scabra* Don; *H. mysorensis* Wight & Arn.; *H. propinqua* Decne.; *Lasianthus tubiflorus* Blume; *Spermadictyon suaveolens* Roxb.; *S. azureum* Wall.

Family: Rubiaceae.

Vern. : Padaar.

Regional Names:

- **Bo.** : Gidasawa, Gidesa
- **Kum.** : Padera
- **M.P.** : Mahabal
- **Ori.** : Janamirgiri, Pitondi
- **P.** : Kanera, Kantalu, Muskei, Tulenni
- **Tel.** : Kondamuritidi.

Distribution : Tropical and Sub-Tropical Himalayas.

Description : Erect shrubs upto 3 m tall with blue flowers in terminal trichotomous panicles. Seeds 3-quetrous.

Flowering & Fruiting : September-November.

Habitat Ecology : Moist slopes. Screes.

Material Examined : HPU-EL-SKS-5422; Bhuana (1,400 m), Boda (1,000 m), Chhanchri (930 m), Gharthun (900 m), Ladoh (980 m).

Part Used : Stem.

Folk Use : Stem used as **fuel**.

Uses in Literature : Used earlier for diarrhoea, fuel, ulcers and wounds (Ambasta, 1986; Pal & Srivastava, 1976; Watt, 1972).

Helicteres isora L.

Family: Sterculiaceae.

Vern. : Ban-beuli.

English, Sanskrit & Regional Names:

Eng. : East Indian Screw Tree
Sans. : Avartani, Mriga-singa
B. : Atmora, Antamora
G. : Murdasingh
H. : Murarphali, Jonkaphal, Bhendu
Kan. : Yadamuri, Kavargi
Mal. : Kaivum, Isvarmuri
Mar. : Kewan, Kevani, Varkati
Ori. : Murmuria, Murimuri, Kaval
P. : Mororphali
Tam. : Valampiri, Kaiva
Tel. : Nuliti, Kavanchi, Syamuli, Gubadarra
Urdu : Mororphali.

Distribution : Punjab to Sri Lanka.

Description : Shrubby trees having young shoots clothed with stellate hairs, bigamous leaves. Erect, branched stem and bilabiate flowers. Bracts small, hairy. Petals red, unequal. Follicles-5, beaked. Seeds numerous wrinkled.

Reproductive Cycle : September-November.

Habitat Ecology : Scrub forests.

Material Examined : HPU-EL-SKS-5543; Balah (900 m), Bhuana (1,400 m), Chhanchri (930 m), Khaira (940 m), Makol (900 m).

Parts Used : Leaves. Stem.

Folk Uses : Leaves as excellent **fodder** for milching cattle as it enhances **fat content** of milk and also **milk yield**. Stem as **fuelwood**.

Uses in Literature : Reported so far as antigalactagogue, antidote to snake-venom, appetizer, astringent, colic, cordage, demulcent, expectorant, fibre for canvas, fodder, gunpowder charcoal, intestinal complaints, vermicide; and for abdominal complaints, amoebiasm, diabetes, diarrhoea, dysentery, flatulence, sores of ear, stomach affections, scabies, and wounds (Ambasta, 1986; Bhatt *et al.*, 2003b; Deshmukh & Rothe, 2003; Kirtikar & Basu, 1984; Jain *et al.*, 2003; Kumar & Lal, 2003; Nadanakunjidam, 2003; Nayak *et al.*, 2003).

Heteropogon contortus (L.) Beaub. *ex* Roem. & Schult.

Syn.: *Andropogon contortus* L.

Family: Poaceae.

Vern. : Lambia ghaa.

English, Sanskrit & Regional Names:

Eng. : Spear grass, Bellary grass
B. : Kher
G. : Dabhjuliyum
H. : Kher, Kumeria, Parwa, Sura
Kan. : Kari Vanugada Hullu, Sunkari Hullu
Mar. : Gantegawta
Ori. : Dauria, Sinkola
P. : Suryata, Sarwala
Tam. : Oosi Pullu, Pani Pullu, Karunsi Pullu
Tel. : Eddi gaddi.

Distribution : Mediterranian region.

Description : Perennial grasses. Stems 30-150 cm long. Internodes densely tufted, slender. Leaves linear, acuminate, flat, suberect. Sheaths compressed. Racemes long. Spikelets closely imbricating.

Reproductive Cycle : July-October.

Habitat Ecology : Open Shady Slopes.

Material Examined : HPU-EL-SKS-5650; Andretta (980 m), Bhatu (1,040 m), Bindraban (1,270 m), Malahu (1,050 m), Purani-palam (1,100 m).

Parts Used : Aerial Parts. Leaves. Root.

Folk Uses : Aerial parts used as **fodder**. Leaves chewed for **toothache**. 2-3 teaspoonful of root decoction taken thrice a day for a week as **diuretic**.

Uses in Literature : Reported as stimulant, diuretic, hay and silage (Ambasta, 1986; Kirtikar & Basu, 1984).

Hibiscus rosa sinensis L. (Pl. 28A)

Family: Malvaceae.

Vern. : Dunaj-phul.

English, Sanskrit & Regional Names:

Eng. : Chinese Hibiscus, Shoe Flower
Sans. : Hariballabha, Odhrapushpa
Ass. : Joba
B. : Jiwa, Oru
Bo. : Jasavanda
G. : Jasuva
H. : Gulhar, Jasum, Jasut
Kan. : Dasavala
Mal. : Ayamparutti, Jampa
Mar. : Dasindachapula, Jasavanda
Ori. : Mondaro, Ongribollika
P. : Jasum.

Distribution : Cultivated throughout India.

Description : Shrubs. Leaves coarsely toothed at apex and entire at base. Corolla red. Capsule—many seeded.

Reproductive Cycle : February-October.

Habitat Ecology : Cultivated in house and temple courtyards.

Material Examined : HPU-EL-SKS-5440; Bhagpur (1,600 m), Bhawarna (1,100 m), Jiun (1,550 m), Ladoh (980 m), Sungal (1,050 m).

Parts Used : Leaves. Flo ›rs.

Folk Uses : 3-4 ml of leaf-juice taken orally for 15-days to check **leucorrhoea**; its paste with curd applied to check **baldness** and for hastening **suppuration of boils**. Flowers **offered** to deities.

Active Constituents : Active properties due to quercetin - 3 -diglucoside 3,7 - diglucoside, cyaniding - 3, 5 -diglucoside and cyanidin-3-sophoroside - 5 – glucoside (*yellow flowers*), and kaempferol – 3 - xylosylglucoside (*ivory white flowers*).

Biological Activity : Hypotensive and antispasmodic activities found to be (+) ve.

Uses in Literature : Described earlier as abortifacient, anodyne, demulcent, dye, emollient, laxative, refrigerant drink in fevers, symbolic (offering in chronic diarrhoea), vegetable (flowers eaten raw or pickled); and for antifertility, appendicitis, asthma, bleeding piles, boils, bronchitis, burns spots, cancerous swellings, childbirth, colouring syrups, consumption, cough, developing foetus (vet.), diarrhoea, flu, gonorrhoea, hair dye, headache, impotency, inflammation, irritable conditions of genito-urinary tract, kidney diseases, marriage ceremonies (flowers), menstrual complaints, mumps, offerings and adornments, postnatal haemorrhage, bronchitis, whooping cough, tuberculosis, scabies (vet.), skin diseases, soothing internal and external wounds, sores, and syphilis (Ambasta, 1986; Banerjee, 1996; Banerjee & Pal, 1996; Chowdhury *et al.*, 1996; Datt, 1996; Gill *et al.*, 1993; Girach, 1992; Girach *et al.*, 1996; Huidrom, 1996; Jain, 1991; Jha & Basak, 1994; Jha *et al.*, 1996; Kumar *et al.*, 2003; Kurian, 1995; Maliya, 2004; Nath & Maiti, 2003; Pandey, 2003; Pawar & Patil, 2004; Prasad *et al.*, 1996; Ranjan, 2003; Rawat & Chowdhury, 1998; Saikia & Nath, 2003; Sen & Behera, 2003; Sharma, 2003; Singh & Pandey, 1998; Thomas & Britto, 2003; Vihari, 1995; Watt, 1972).

***Holarrhena antidysentrica* Wall.** (Pl. 28B)

Syn.: *H. pubescens* Wall.; *H. codaga* G. Don; *H. malaccensis* Wight; *Echites antidysentrica* Roxb.; *E. pubescens* Buch.-Ham.; *Wrightia antidysentrica* Graham

Family: Apocynaceae.

Vern. : Keor, Inderjaun.

English, Sanskrit & Regional Names:

Eng. : Antidysentric Holarrhena, Conessi Bark, Kurchi
Sans. : Girimallika,Mallikapushpa, Shakraparyaya
Ass. : Dudcory, Dhutkhuri
B. : Kurchi, Titaindaraju
Bo. : Dolakura, Pandhrakura
G. : Dhowda, Indrajwannujad
H. : Dhudi, Karvaidrajau
Kan. : Beppak, Koodsaloo, Korchei
Kash. : Keor, Kewar
M.P. : Kudo, Kurato
Mal. : Kaippakkotakappala, Venpala
Mar. : Bedaki, Pandharakuda
Ori. : Kherwa, Patrukurwan, Potrakaya
P. : Kewar, Koeva, Kogur
Tam. : Erukkalaipalai, Vattagam
Tel. : Istrarakupala, Peddankuda, Vistarakuphala.

Distribution : Throughout India; ascending to 1,200 m.

Description : A small deciduous tree. Leaves with rounded base, arched. Flowers white, inodorous. Seeds glabrous, crowned with a tuft of long silky hairs.

Reproductive Cycle : June-February.

Habitat Ecology : Open Grassy Slopes.

Material Examined : HPU-EL-SKS-5447; Dadh (1,100 m), Gharthun (900 m), Panapari (1,100 m), Rangru (910 m), Tikkar (910 m).

Parts Used : Leaves. Fruits. Seeds. Bark.

Folk Uses : Leaves used for performing various **religious ceremonies**. Seeds employed for performing '**havan**'; and also 250 ml of its extract taken for one month to relieve **body heat** and **bleeding piles**. Crushed fruits (100 gm) cooked as '**halva**' and prescribed early morning for seven days to check irregular **menstrual** cycle. Bark (1 kg) given orally to buffaloes for 4-5 days to cure **prolapsed uterus**.

Active Constituents : Alkaloids isolated from *bark* are conessine, kurchicine, kurchine, conamine, conkurchine, holarrhenine, holarrhinine, holarrhine and lettocine.

Biological Activity : Antiprotozoal, anticancerous, hypoglycaemic and spasmolytic activities found +ve.

Uses in Literature : Known in India as an anthelmintic, antidysenteric, antiperiodic, antipyretic, antitetanic, aphrodisiac, appetiser (vet.), astringent, bitter, carminative, conceptive, diuretic, expectorant, febrifugal, fodder, lithnotropic, pen holders, psychoactive plant, spermator, tonic, vegetable (flowers, seeds); and for abnormal pain, anaemia, anthrax (vet.), asthma, biliousness, blisters (vet.), boils, bowels, bronchitis, bronchopneumonia, carved boxes, chest complaints, cold, combs, constipation, cough, curdling milk, cuts, dandruff, diarrhoea, diarrhoea (vet.), dropsy, dyeing, dyspepsia, easing delivery, eczema, epilepsy, fatigue, flatulence, furniture, gastric disorder, gout, gums, haematuria, haemorrhages after childbirth, hallucinations, headache, increases secretion of milk, inflammation of uterus, influenza, killing lice, lessening inflammation to strengthen and giving tone to the stomach, leucoderma, liver ailments, lumbago or pain in the lumbar region, mathematical instruments, menorrhagia, neck sore (vet.), ornamental beads worn around the neck as charm, paralysis, picture frames, piles, plague, reviving taste, rheumatic pain, skin affections, spleen complaint, spoons, stomach ulcers, scorpion and snakebite, stomachache, swellings, toys, malarial fever, nose bleeding, ulcers, vomiting and nausea, worms, and wounds (vet.) (Agarwal, 2003; Ambasta, 1986; Aminuddin & Girach, 1996; Baruah & Sarma, 1984; Bennet *et al.*, 1991; Borthakur & Sarma, 1996; Bose, 1984; Brahmam & Saxena, 1990; Chauhan, 1999; Chopra *et al.*, 1956; Datt, 1996; Dey & Bahadur, 1973; Dwivedi, 2003; Girach, 1992; Girach & Aminuddin, 1995; Gopal & Chandra, 2003; Gupta,

1981; Islam & Jha, 2003; Jadeja, 1999; Jain, 1968; Jain *et al.*, 1991, 1994; Karnick, 1994; Khare, 2004; Kothari & Moorthy, 1996; Kumar, 1996; Kurian, 1995; Lalramnghinglova, 1996; Masih, 2003; Panthi & Chaudhary, 2003; Prakash, 2001; Rama Rao & Henry, 1996; Ranjan, 2003; Sanyal, 1994; Sarin, 1996; Sarma *et al.*, 2001 ; Shah *et al.*, 1983; Sharma, 2003; Sharma & Sood, 1997; 1999, 2003; Singh & Kumar, 2000a; Singh & Rao, 2003; Siwakoti & Varma, 1996; Srivastava *et al.*, 2003c; Subramani & Goraya, 2003; Thomas & Britto, 2003; Tirkey, 2004; Tirkey *et al.*, 2003a; Upadhyay & Chauhan, 2000; Upadhye *et al.*, 1994; Varma, 1997; Watt, 1972).

***Hypericum oblongifolium* Choisy** (Pl. 28C)

Syn.: *H. cernuum* Roxb. *ex.* D.Don ; *H. speciosum* Wall.; *H. govanianum* Wall.

Family: Hypericaceae.

Vern. : Peelan.

Distribution : W. Himalayas: 800-2,100 m.

Description : Glabrous shrubs. Leaves ovate, each pair of leaves placed at right angles to the next pair, above or below it. Flowers yellow. Capsule conical.

Reproductive Cycle : April-October.

Habitat Ecology : Shady Slopes.

Material Examined : HPU-EL-SKS-5646; Boda (1,000 m), Gopalpur (1,350 m), Panchrukhi (990 m), Sulah (950 m), Thakurd-wara (1,150 m).

Part Used : Leaves.

Folk Use : Leaves as **fodder** for goat.

Uses in Literature : Recorded earlier as an aromatic, stimulant; and for eczema, old sores, and toothache (Kapur, 1996a).

Impatiens balsamina L. (Pl. 28D)

Family: Geraniaceae.

Vern. : Teur.

English & Regional Names:

Eng. : The Garden Balsam
B. : Dupati
Bo. : Terada
H. : Gulmendi
Ori. : Mujethi
P. : Halu, Tatura, Tilphar.

Distribution : Throughout India.

Description : Annual pubescent herbs, 30-90 cm tall with alternate leaves and axillary flowers. Spur long. Seeds reticulate.

Reproductive Cycle : July-September.

Habitat Ecology : Abundant in fields. Wastelands.

Material Examined : HPU-EL-SKS-5500; Bhawarna (1,100 m), Dharaman (1,100 m), Pahra (950 m), Rajeher (1,050 m), Tikkar (910 m).

Parts Used : Leaves. Flower.

Folk Uses : Flower paste applied externally to cure **burns** and **sore toes**. Leaves used as a substitute for '**henna**'.

Active Constituents : Properties due to cyanidin, leucocyanidin, myricetin, quercetin, kaempferol, 2-methoxy-1,4-naphthaquinone (*flowers*); caffeic, p-coumaric, ferulic, gentisic, p-hydroxybenzoic, protocatechuic, salicylic, sinapic, syringic and vanillic acids; 4-(2-carboxyphenyl)-4-oxobutyric acid, 2-hydroxy-1,4-naphthaquinone and naphtholglycosidase (*plant*).

Biological Activity : 50% EtOH extr. of the flowers exhibits marked antibiotic activity against *Sclerotinia fructicola* and *Colletotrichum lindemuthianum.*

Uses in Literature : Reported to be used earlier as an ammenorrhoea, cathartic, demulcent, diuretic, edible (*leaves, seeds & cooking oil*), emetic, joint pain, laxative, neuralgic, tonic, ulcers; and for burns, colour extraction, dysentery, normal delivery, and snakebite (Agarwal, 2003; Ambasta, 1986; Bennet *et al.*, 1991; Chopra *et al.*, 1956; Deshmukh & Rothe, 2003; Gogoi & Das, 2003; Jha & Basak, 1994; Kapur & Nanda, 1996; Kirtikar & Basu, 1984; Mahato & Chaudhary, 2005; Parrotta, 2001; Ranjan, 2003; Roy *et al.*, 1998; Singh & Singh, 2003; Siwakoti & Verma, 1996; Sood & Thakur, 2004).

***Indigofera cassioides* Rottl. *ex* DC.** (Pl. 28E).
Syn.: *I. pulchella* auct. (non Roxb.) Baker; *I. purpurescens* Roxb.; *I. ellittica* Roxb.; *I. violacea* Roxb.; *I. arborea* Roxb.
Family: Fabaceae.

Vern. : Kanthi.

Regional Names:

H. : Hakna, Sakena
Kan. : Gogge
Kum. : Sakina
Mar. : Baroli
Ori. : Girili
Tam. : Narinji
Tel. : Siralli, Vuyye.

Distribution : Commonly throughout the hills of India.

Description : An erect shrub with 11-12 leaflets and bright pink flowers arranged in racemes. Pod glabrous.

Reproductive Cycle : March-May.

Habitat Ecology : Drier Areas. Open Slopes.

Material Examined : HPU-EL-SKS-5643; Bhagpur (1,600 m), Bhuana (1,400 m), Khaira (940 m), Nadli (1,000 m), Bhaura (940 m).

Parts Used : Leaves. Flowers.

Folk Uses : Leaves lopped as **goat-fodder**. Flower relished as '**pakodas**' and considered good against **digestive ailments**.

Uses in Literature : Used earlier for conception, cough, epilepsy, menorrhagia, mucous in urine, swelling of face in labour and as vegetable (Bennet *et al.*, 1991; Jain, 1991; Watt, 1972).

***Ipomoea batatas* (L.) Lam.** (Pl. 28F)
Family: Convolvulaceae.

Vern. : Shakarkanni.

English, Sanskrit & Regional Names:

Eng. : Sweet Potato, Pichhila, Pindula
Sans. : Kandagranthi
Ass. : Bogaalu, Goriaalu, Rangaalu
B. : Latalu, Rangalu

Bo. : Ratalu, Shakarkandi
G. : Kanangi, Sakaria
H. : Mitaalu, Shakarkand
Kan. : Chu
Mal. : Kapakalenga
P. : Shakarkand
Tam. : Sakkareivelleikilangu, Vattiki
Tel. : Chelagada.

Distribution : A native of America. Cultivated in India.

Description : Prostrate annual glabrous herbs with red tubers and ovate-cordate, acute, angular leaves. Peduncles long, many-flowered. Flowers purple, corolla funnel-shaped. Seeds glabrous.

Reproductive Cycle : June-March.

Habitat Ecology : Agricultural Fields. Met as an Escape in Damp Places.

Material Examined : HPU-EL-SKS-5530; Bandla (1,280 m), Bodhal (970 m), Dadh (1,100 m), Saliana (1,000 m), Tapa (945 m).

Part Used : Root tubers.

Folk Use : Root tubers **relished** after roasting.

Active Constituents : Furanoterpenoid-dehydroipomeamarone is olated from *roots.*

Uses in Literature : Considered earlier as adhesive, aphrodisiac, cooling, laxative, sweet; and for burning sensation, chest and lungs infection, diarrhoea, fevers, sizing paper, strangury, textiles, thirst, urinary discharges, and vegetable (Agarwal, 2003; Ambasta, 1986; Chopra *et al.*, 1956; Duke *et al.*, 2002; Kirtikar & Basu, 1984; Prajapati *et al.*, 2006; Roy *et al.*, 1998; Saklani & Jain, 1994; Siwakoti & Varma, 1996; Sood & Prakash, 2007; Tigga *et al.*, 2003; Watt, 1972).

Ipomoea turbinata Lag. (Pl. 29A)

Syn.: *Calonyction muricatumi* (L.) G. Don; *Convolvulus muricatus* (L.); *Calonyction muricatus* (L.) Jacq.; *Ipomoea muricata* (L.) Jacq.

Family: Convolvulaceae.

Vern. : Ghondali.

Regional Names:

B. : Michai
Bo. : Gariya
G. : Garayo
H. : Michai

Mal. : Mukkattikay
Mar. : Bhonwari
Tam. : Kattutali.

Distribution : Pan-tropical in distribution.

Description : Large annual glabrous vines. Leaves ovate, glabrous, apex acute to acuminate. Flowers pedicellate. Corolla funnel-shaped, rose-purple. Capsule with ovoid seeds.

Reproductive Cycle : July-September.

Habitat Ecology : Maize Fields. Damp Places.

Material Examined : HPU-EL-SKS-5463; Chhanchri (930 m), Molichak (1,020 m), Paror (940 m), Sulah (950 m).

Part Used : Pedicel.

Folk Use : Pedicels cooked as **vegetable**.

Uses in Literature : Recorded earlier as cathartic, edible (pedicles; roots and fruits unripe as well as ripe), febrifuge, insecticide and purgative (Ambasta, 1986; Chandra, 1997; Parrotta, 2001; Reddy & Raju, 2000; Sharma & Rana, 2005; Sood & Prakash, 2007; Sood & Thakur, 2004).

Jasminum dispermum **Wall.** (Pl. 29B)

Family: Oleaceae.

Vern. : Chameli.

Distribution : Kashmir - S.W. China: 900-2,500 m.

Description : Climbing woody shrubs with imparipinnate leaves. Leaflets 3-5. Flowers white, numerous, fragrant. Fruits dark purple, elliptic.

Reproductive Cycle : April-November.

Habitat Ecology : Shrubberies. Ravines.

Material Examined : HPU-EL-SKS-5595; Chandpur (1,290 m), Chimbalhar (1,150 m), Darugnu (1,400 m), Holta (1,200 m), Sungal (1,050 m).

Parts Used : Leaves. Flowers.

Folk Uses : Leaf-paste good against **wounds**. Flowers **offered** to appease **Lord Shiva**.

Biological Activity : EtOH (50%) extr. of aerial parts anticancerous and diuretic.

Uses in Literature : Used in India for its leaves (vegetable) and fruit (Mao, 1993; Negi & Gaur, 1991).

Jasminum grandiflorum L. (Pl. 29C)

Syn.: *J. aureum* Don; *J. officinale* L.; *J. officinale* L. var. *grandiflorum* (L.) Stokes; *J. officinale* var. *grandiflorum* L.

Family: Oleaceae.

Vern. : Malti.

English, Sanskrit & Regional Names:

Eng. : Common White Garden Jasmine, Spanish Jasmine, White Jasmine, Catalonian Jasmine

Sans. : Bahugandha, Gajavhaya, Punyagandha, Shankhayuthika

B. : Chameli, Jati

Bo. : Chambeli

G. : Chambeli

H. : Chaluba

Kan. : Ajjige, Jaji, Jati, Mallige

Kash. : Chirichog, Kiri

Mar. : Pichakam, Pichakamulla

P. : Bansu, Naugri, Somum

Tam. : Manbadanam, Mullai, Pedarmalligai, Pichi

Tel. : Jagi, Malati

U.P. : Chameli.

Distribution : Afghanistan-S.W. China; S.E. Asia: 900-2,500 m; cultivated.

Description : A climbing shrub with angular, ribbed branches and imparipinnate, opposite leaves. Leaflets 3-7, terminal longer than others. Flowers white, fragrant; in terminal panicles. Berries with crimson juice.

Reproductive Cycle : April-September.

Habitat Ecology : Shrubberies. Ravines.

Material Examined : HPU-EL-SKS-5538; Dehan (1,130 m), Panchrukhi (990 m), Sidhpur (1,100 m), Thakurdwara (1,150 m), Thandol (950 m).

Parts Used : Twigs. Leaves. Flowers.

Folk Uses : Handful paste of its tender twigs with leaves of *Carissa congesta*, *Pogostemon benghalense, Capsicum annuum*, *Piper nigrum* and bulbs of *Allium cepa* prescribed for **fever** in livestock. Flowers **offered** to appease Goddess 'Parvati'. Leaves chewed for **toothache**.

Active Constituents : Active properties of *flowers* due to cis-3-hexenol, 2-vinylpyridine, indole, myrcene, linalool, geranyl linalool, α-terpineol, geraniol, linalyl acetate, nerolidol, phytol isophytol, farnesol, engenol, benzyl alcohol, p-cresol, methyl benzoate benzyl cyanide, benzyl acetate, methyl dihydrojasmonate methyl anthranitate, lis-jasmone, methyl N-methyl anthranitate, vanillin, cis-3-hexenyl benzoate, benzyl benzoate, methyl palmalte and methyl linolate.

Uses in Literature : Reported so far as an anthelmintic, aphrodisiac, audient, cooling, diuretic, emmenagogue, hair oils, offering, psychoactive plant, social taboo (girls place flames in their hair when they become pregnant); and for apthae, attars, corns between the toes, dermatitis, headache, jasmine D1, otorrhoea, stoping breast milk production, mouth cleansing, perfumed waters, religious offerings, sexual impotency, skin eruptions, toothache, treatment of ear discharge, ulceration of mouth (Ambasta, 1986; Jain *et al.*, 1994; Kapur & Singh, 1996; Rana *et al.*, 1996; Shome *et al.*, 1996; Singh & Pandey, 1998; Srivastava *et al.*, 2003c; Thakur, 2005; Thomas & Britto, 2003; Vihari, 1995; Watt, 1972).

Jasminum humile **L.** (Pl. 29D)

Syn.: *J. chrysanthemum* Roxb.; *J. revolutum* Sims; *J. bignoniaceum* Wall.; *J. wallichainum* Lindl.; *J. inodorum* Jacq.; *J. pubigerum* D. Don var. *glabrum* DC.

Family: Oleaceae.

Vern. : Chameli.

English, Sanskrit & Regional Names:

Eng. : Yellow Jasmine, Italian Jasmine, Nepal Jasmine, Golden Jasmines

Sans. : Svarnajuthica, Hemapushpika

B. : Svarnajui

Bo. : Hemapushpika

H. : Peelajui, Peeli Chameli, Malto

Kan. : Hasarumallige

Mal. : Ponmallika

Tam. : Semmalligai

Tel. : Pachche-Adavimalle.

Distribution : Sub-Tropical Himalayas: 1,500-3,000 m.

Description : Erect stout shrubs. Leaves alternate. Leaflets 3-7, oblong or ovate. Flowers yellow, in dense cymes. Fruits a juicy black berry.

Reproductive Cycle : May-September.

Habitat Ecology : Shrubberies. Forests.

Material Examined : HPU-EL-SKS-5608; Bhagpur (1,600 m), Darugnu (1,400 m), Saparu (1,250 m), Tanda (1,120 m), Tikkar (910 m).

Parts Used : Whole Plant. Flowers.

Folk Uses : Plant considered **sacred** and grown in house and temple courtyards. Flowers offered to **appease** Lord **Ganesha** and its offering considered equivalent to that of gold.

Uses in Literature : Described in literature as antidote in poison cases, aromatic essential oil, astringent, tonic to heart; and for chronic sinuses, dyeing, skin diseases, wounds, ulcers, stomatitis, toothache, earache, eye diseases, fistulas, headache, ringworm affections, skin diseases, stomatitis, toothache, ulcerations, and wounds (Ambasta, 1986; Chauhan, 1999; Chopra *et al.*, 1965; Jain, 1991; Sharma, 2003; Watt, 1972).

***Jasminum multiflorum* (Burm. f.) Andrews** (Pl. 29E)

Syn.: *J. pubescens* (Retz.) Willd; *J. hirsutum* Willd.; *J. multiflorum* Roth; *J. congestum* Wall.; *J. bracteatum* Wight.; *Mogorium pubescens* Lam.; *Nyctanthes multiflora* Burm. f.; *N. pubescens* Retz.

Family: Oleaceae.

Vern. : Chameli.

English, Sanskrit & Regional Names:

- **Eng.** : Common Jasmine, Downy Jasmine, Musk Jasmine
- **Sans.** : Atimukha, Attahasa, Attapuspaka, Damana, Kunda, Madhya, Mahamoda, Makaranda, Manodna, Manohara, Manorama, Mughya, Muktapushpa, Tarapushpa, Vanahasa
- **H.** : Chameli, Kunda, Kundo, Kundphul
- **Mal.** : Gujari, Kattuchirakamulla, Kundam, Kurukuttimula, Kuruna
- **Mar.** : Mogra
- **Ori.** : Kontabelo
- **Tam.** : Magarandam, Malligai
- **Tel.** : Adavimalle, Gujari, Kundamu, Molla.

Distribution : Throughout India.

Description : Scandent, densely villous shrubs having calyx teeth with yellow hairs, and white flowers in dense cymes. Fruits a berry.

Reproductive Cycle : November-April.

Habitat Ecology : Shrubberies. Forests.

Material Examined : HPU-EL-SKS-5563; Balu (1,030 m), Malahu (1,050 m), Nanaun (900 m), Panapari (1,100 m), Sidhpur (1,100 m).

Parts Used : Leaves. Flowers.

Folk Uses : Flowers offered to various **deities**. Leaf-paste applied for **suppuration of boils**.

Active Constituents : These are friedelin, lupeol, betulin, α-amyrin and betulinic, ursolic and oleanolic acids (*leaves*). Jasmolactones A, B, C and D, multifloroside, multiroside and 10-hydroxyoleoside-11-methyl ester, jasmultiside and multiflorin (*aerial parts*).

Uses in Literature : Known so far as an antidote to snakebite; and for burns, chest pain, cough and cold, hounding dogs (vet.), offerings and adornments, pin-worms, religious and ceremonial functions, scabies, stomachache, ulcers, and urine blockage (Ambasta, 1986; Dagar, 2003; Datt, 1996; Gogoi & Borthakur, 1991; Jain, 1991; Kapur & Nanda, 1996; Sharma, 2003; Watt, 1972).

***Jatropha curcas* L.** (Pl. 29F)

Syn.: *J. moluccana* Herb.

Family: Euphorbiaceae.

Vern. : Jamalghota, Rattanjot.

English, Sanskrit & Regional Names:

Eng. : Physic Nut, Purging Nut
Sans. : Kananaerand, Nepala, Paravatayeranda
B. : Erandagachh, Safed ind
Bo. : Maraharalu
G. : Jangliarundi, Ratanjota
Kan. : Maranarulli
Mal. : Kaakavenako
Mar. : Mogalieranda
Ori. : Baigab
P. : Jamalgota, Japhrota, Rattanjot
Tam. : Kaatamunck
Tel. : Pepalam.

Distribution : Throughout tropics. Cultivated and naturalized in India and Sri Lanka.

Description : Small deciduous trees. Leaves alternate, 3- or 5-lobed. Flowers yellowish-green, in terminal cymes. Fruits black.

Reproductive Cycle : February-April.

Habitat Ecology : Waste places. Roadsides. Field borders.

Material Examined : HPU-EL-SKS-5428; Bhauda (1,050 m), Bhuana (1,400 m), Boda (1,000 m), Chauki (900 m), Khaira (940 m).

Parts Used : Leaves. Seeds. Latex.

Folk Uses : Leaf-juice applied locally for **piles**. Latex applied to heal **wounds**. Seed-paste with pure ghee good for softening **cracked heels**. Seeds taken in small quantity as **purgative**.

Active Constituents : Important constituents are vitexin, isovitexin, campesterol, β-sitosterol-3β,7β-diol and l-triacontanol (*leaves*), arachidic, linoleic, myristic, oleic, palmitic and stearic acids; arabinose, galactose, glucose, rhamnose, xylose, galacturonic acid and β-sitosterol (*seeds*).

Biological Activity : 50% EtOH extr. of aerial parts CNS depressant and diuretic.

Uses in Literature : So far described as an abortifacient, anasarca, antihaemorrhagic, antiseptic, haemostatic, insecticidal, itching, lactagogue, rubefacient, stimulant, stomachic, symbolic, styptic, supperative; and for anaemia, animal-bites, astringent, biliousness, burns, cancer, carbuncle, congestion of chest, cramps, diseases of heart, dropsy, dye, eczema, hair growth, herpes, fish poison, inflammation, leprosy, piles, pleurisy, pneumonia, poisonous, purgative, rheumatism, ringworm, scabies, sores, syphilis, tooth troubles, toothbrush, tumours, ulcers, whitlow and wounds (Bhuyan, 2003; Chopra *et al.*, 1956; Dey, 1973; Dwivedi, 2003; Jain, 1991; Joshi & Joshi, 2005; Kumar & Rao, 2003; Kshirsagar, 2003; Kshirsagar *et al.*, 2003; Maliya, 2005; Mokat & Deokule, 2004; Nath & Maiti, 2003; Rajore & Batra, 2003; Sharma, 2003; Subramani & Goraya, 2003).

***Juglans regia* L.** (Pl. 30 A,B)

Syn.: *J. regia* L. var. *kumaonica* Cas. DC.; *J. arguta* Wall.

Family: Juglandaceae.

Vern. : Akhrot, Khod.

English, Sanskrit & Regional Names:

Eng. : Common Walnut Tree, European Walnut, Himalayan Walnut, Walnut Tree

Sans. : Akhoda, Akschota

Ass. : Kabsing

B. : Akhrut

G. : Akhrot

H. : Akhor, Akhrot, Khor

	Kash.	: Akhor, Krot
	Mar.	: Akroda
	P.	: Akhori, Charmagz, Kabotang
	Tam.	: Akrottu
	Tel.	: Akrotu.

Distribution : Kashmir-S. E. Tibet: 1,500-3,000 m.

Description : Trees. Leaves imparipinnate, aromatic. Male and female flowers borne on the same tree. Drupes ovoid with green, thick, fleshy rind. Nut woody, wrinkled, two-valved.

Reproductive Cycle : April-October.

Habitat Ecology : Forests. Widely Cultivated in W. Himalayas.

Material Examined : HPU-EL-SKS-5433; Bhagpur (1,600 m), Jiun (1,550 m), Kathena (1,290 m), Lolani (1,600 m), Molichak (1,020 m).

Parts Used : Leaves. Fruit. Bark. Wood.

Folk Uses : Kernels **edible**. Bark, leaves and fruit rind used for **scouring teeth** and **coloring lips**. Wood good for making **furniture**.

Active Constituents : Properties due to 5-hydroxy-1,4-napthoquinone, berberine (0.08%) (*stem-bark*).

Biological Activity : Juglone inhibits growth of liver ascites carcinoma and S-180 solid tumour in mice at 0.1 mg/ml *in vitro.*

Uses in Literature : Used earlier as an alexipharmic, alterative in scrofula, anthelmintic, aphrodisiac, astringent, cattle feed, cholagogue, detergent, edible (kernels), insect repellent for storage of grains, laxative, nerve tonic, psychoactive plant, toothbrush, tooth powder; and for bed wetting, carving, chewing to colour the lips, confectionery (pastries), decolourising charcoal and activated carbon, diarrhoea, dispelling flies, drawing instruments, dyeing, eczema, foot and mouth disease (vet.), frames, frost bite, furniture, gun-stocks, handlooms, herpes, hoof (vet.), ice-creams, lacquer work, leucoderma, menorrhagia, musical instruments, oil for culinary purposes and illuminant, oil paints, ornamental works, paints, ploughs, printing ink, rheumatism, rickets, scrofula, shafts, sore throat, sores of toes, stupefying fish, syphilis, and tanning (Agarwal, 2003; Ambasta, 1986; Aswal, 1996; Bennet *et al.*, 1991; Beigh *et al.*, 2003; Chakraborty, 2003; Chauhan, 1999; Chhetri, 2005; Chopra *et al.*, 1956; Das & Agarwal, 1991; Dey & Bahadur, 1973; Jain, 1991; Jain *et al.*, 1994; Kant & Dutt, 2005; Kapur & Singh, 1996; Kirtikar & Basu, 1984; Lalramnghinglova, 1996; Negi & Pant, 1990; Negi *et al.*, 1993; Pandey & Joshi, 2001; Prajapati *et al.*, 2006; Purohit & Samant, 1995; Rana *et al.*, 1996; Saini, 1996a, b; Sharma *et al.*, 2000, 2003a, b; Singh 1991, 1999; Singh & Kumar, 2000a; Viswanathan & Singh, 1996; Watt, 1972).

Justicia adhatoda L. (Pl. 30C)

Syn : *Adhatoda vasica* Nees; *A. zeylanica* Medic.

Family: Acanthaceae.

Vern. : Basuti, Basunti.

English, Sanskrit & Regional Names:

Eng. : Malabarnut leaf
Sans. : Bhisangmata, Raktappittaghni
Ass. : Herbuksha
B. : Bakas, Baasaka patra
Bo. : Adalsa, Vasuka
G. : Adsoge, Arduosa, Paandru
H. : Adalsa, Vasaka, Banasaa
Kan. : Adusoge
Mal. : Attalotakam
Mar. : Baksha
Ori. : Rottoomuli
P. : Bhekar
Tam. : Adadoli, Kattumurunghai
Tel. : Addasaramu
U.P. : Adusa, Arusa, Rusa, Roos.

Distribution : Distributed in India from Punjab-Assam: 500-1,600 m.

Description : A strong smelling gregarious shrub with shortly stalked, ovate-lanceolate leaves and white streaked flowers dotted with pink. Capsule club-shaped.

Reproductive Cycle : April-November.

Habitat Ecology : Wastelands, shrubberies; common.

Material Examined : HPU-EL-SKS-5506; Bhuana (1,400 m), Boda (1,000 m), Balah (900 m), Khaira (940 m), Rangru (910 m).

Parts Used : Leaves. Flowers.

Folk Uses : Petal ash kept in earthen pot along with 'maghan' (*Piper longum*) for 15 days taken with honey for **asthma**. 1-2 teaspoonful of leaf-juice along with ginger-juice and honey prescribed as an **expectorant.** Also leaf-juice used as nasal drops to check **headache** due to cold. Powder from roasted tender leaves mixed with 'ajwain' (*Trachispermum ammi*) taken to relieve **cough**. Leaf-paste applied to **heal** injuries. Leaves warmed in sunlight wrapped around banana for ripening; as **bedding** for cattle in rainy season; and dumped with dung to make **manure**. 1 teaspoonful of decoction prepared by boiling

1.5 g of its roots with 2 litres of water taken every morning and night to cure **paralysis** and **bodyache**. Flower decoction taken as an **expectorant**; also these form an important component of local wine, '**sur**'.

Active Constituents : Chief principles are vascicine, vasinone, vascicone, vascine, β-sitosterol, tritriacontane, vasicinine, vascicinol, 4 new quinazoline alkaloids-vascicoline, adhatodine, vasciolinone and anistone (*leaves*) and vasicinone (*inflorescence*).

Biological Activity : Bronchodilatory, uterotonic, hypotensive and abortifacient activities found to be (+)ve.

Uses in Literature : Known so far in India as anthelmintic, antipyretic, antiseptic, antispasmodic, aromatic, beads for rosaries, beverage, blood purifier, bronchodilator, electuary, expectorant (vet.), febrifuge, fuel for brick-burning, green manure, gunpowder charcoal, ingredient of 'chyawanprash', insect repellent for stored grains, psychoactive plant, refrigerant in fever, uterine tonic (vet.), vegetable, yellow dye; and for abortion, arthritis, asthma, bleeding from nose, boils, childbirth, chronic bronchitis, chronic cold, consumption, cough, curing eczema, diarrhoea (vet.), diseases of mouth, dislocation of joints, dropsical affections attended with anaemia, dyeing, dysentery (vet.), dysuria or painful and difficult urination, easing delivery, eye pain, flatulence, glandular tumours, gonorrhoea, heart troubles, horses for dysentery, infusion, injuries, intestinal worms, jaundice with chronic liver disorders, joint pains, killing domestic flies, leucoderma, loss of memory, malaria, mastitis (vet.), neuralgia, opthalmia, packaging, phthisis, pulmonary affections, relieving body pain after delivery, removal of foetus (vet.), respiratory troubles, rheumatism (vet.), scabies (vet.), skin affections, sores of eye, stomach (vet.), stomachache, swelling, tuberculosis, tumours, venereal complaints, vomiting, whooping cough, wrapping paddy in preparation of rice beer 'yeast' and also to stop post partum haemorrhage (Ambasta, 1986; Aswal, 1996; Beg *et al.*, 2005; Bhatt *et al.*, 2003a; Bhuyan, 2003; Billore *et al.*, 1998; Bondya & Sharma, 2004; Borthakur, 1993; Borthakur & Sarma, 1996; Bose, 1984; Chandra, 1997; Chevallier, 1996; Chowdhury *et al.*, 1996; Dey & Bahadur, 1973; Diwakar & Ansari, 2003; Dixit & Pal, 1996; Dixit & Kumar, 2003; Dwivedi, 2003; Girach, 1992, 2000; Girach *et al.*, 1996; Gogoi *et al.*, 2003a; Gopal & Chandra, 2003; Harborne & Baxter, 2001; Huidrom, 1996; Jain, 1968, 1991; Jamir, 1997; Kalakoti, 2001; Kapur & Singh, 1996; Katewa & Arora, 1997; Khare, 2004; Kothary & Moorthy, 1996; Kumar & Rao, 2001; Kumar *et al.*, 2003; Kurian, 1995; Lal *et al.*, 1996; Lalramnghinglova, 1996; Mahato & Chaudhary, 2005; Mahato & Mahato, 1996; Mao, 1993; Mehrotra *et al.*, 1995; Mishra *et al.*, 1996a, 1997; Negi *et al.*, 1993; Padhy & Dash, 2003; Pandey, 2003; Panthi & Chaudhary, 2003; Prajapati *et al.*, 2006; Prakash, 2001; Prasad *et al.*, 1996; Raghupathy & Mahadevan, 1991; Rajendran *et al.*, 2003; Rama Rao & Henry, 1996; Rana *et al.*, 1996; Ranjan, 1996, 2000, 2003; Roy *et al.*, 1998; Saini, 1996a, b; Saklani & Jain, 1994; Samant *et al.*, 2001; Sen & Batra, 1997; Sharma, 2003; Sharma & Borthakur, 2003; Sikarwar, 1996; Singh & Pandey, 1998; Singh & Singh, 2003; Singh *et al.*, 1999; Siwakoti & Varma, 1996; Sood & Thakur, 2004; Srivastava *et al.*, 2003a, c; Subramani & Goraya, 2003; Swami & Gupta, 1996; Thomas & Britto, 2003; Tirkey, 2004; Tirkey *et al.*, 2003b; Varma, 1997; Verma & Chauhan, 2006; Watt, 1972).

***Kalanchoe pinnata* (Lam.) Pers.**

Syn.: *Bryophyllum calycinum* Salisb.; *B. pinnatum* (Lam.) Oken

Family: Crassulaceae.

Vern. : Gand-patar.

Regional Names:

- **B.** : Koppata
- **G.** : Ghayamaroi
- **H.** : Zakham-haiyat
- **Tel.** : Sima-jamudes.

Distribution : Throughout India.

Description : Perennial glabrous herbs having erect, hollow stem. Leaves simple, rarely compound with 3 leaflets, opposite, fleshy. Flowers pendulous, red; in panicle. Follicles 4, many-seeded.

Reproductive Cycle : December-March.

Habitat Ecology : Moist-shady places.

Material Examined : HPU-EL-SKS-5541; Darugnu (1,400 m), Holta (1,200 m), Jasun (1,000 m), Rajeher (1,050 m), Saparu (1,250 m).

Parts Used : Whole Plant. Leaves.

Folk Uses : Plant grown for **ornamental** purposes. Leaves applied as such to hasten **suppuration** of **boils**. Fresh leaves also chewed empty stomach in morning to lower **blood sugar**, and dissolving **kidney stone**.

Uses in Literature : Reported to be used an antidote, bites of venomous insects, blood dysentery, boils, bruises, headache, hypertension, gall bladder and kidney stones, influenza, swelling, ulcers and wounds (Ambasta, 1986; Jamir, 1997; Nath & Maiti, 2003; Ranjan, 2003).

***Lannea coromandelica* (Houtt.) Merr.** (Pl. 30D)
Syn.: *L. grandis* Engl.; *Odina woodier* Roxb.
Family: Anacardiaceae.

Vern. : Kahemble.

Regional Names:

- **B.** : Jiyal, Jeol, Bhadi
- **G.** : Mavedi
- **H.** : Jhingan, Kaimil, Mohin
- **Kan.** : Ajasringi, Kuratige, Gojal
- **Mar.** : Moi, Shimti
- **Tam.** : Wodier, Kalasan
- **Tel.** : Appriyada.

Distribution : Throughout India and valley below Shimla, ascending to 1,700 m.

Description : A tree having imparipinnate leaves, unisexual flowers. Disc annular, 4-5 lobed. Stamens 8-10. Styles 3-4. Drupes small, reniform crowned by the styles.

Reproductive Cycle : March-September

Habitat Ecology : Forests. Open Slopes. Borders of Agricultural Fields.

Material Examined : HPU-EL-SKS-5457; Balah (900 m), Bhauda (1,050 m), Boda (1,000 m), Jasun (1,000 m), Khaira (940 m),

Parts Used : Leaves. Wood. Bark.

Folk Uses : Leaves employed as **fodder**. Wood used as **fuel**. Bark-paste good for treating **cuts** and **wounds**.

Active Constituents : Main ingredients are quercetin, isoquercetin, morin and ellagic acid (*flowers*), phlobatannins, hydroxyanthraquinones (*stem-bark*); leucocyanidin, leucodelphinidins (*leaves*); dl-epicatechin, leucocyanidin (*bark*), quercetin and rutin (*leaves*).

Uses in Literature : Useful in India as an adhesive, agricultural implements, antidote to irritation caused by *Mucuna pruriens,* astringent, cuts, edible, expectorant, fever, fodder, furniture, gum troubles, haemostatic, healing wounds, house-building, packing cases, railway sleepers, skin eruption, timber, toothache and wounds (vet.) (Ambasta, 1986; Augustine & Sivadasan, 2005; Bhatt *et al.*, 2003c; Reddy *et al.*, 2003a; Sharma, 2003; Singh & Pandey, 1998; Sood & Kaushal, 2008; Subramani & Goraya, 2003; Upadhyay *et al.*, 1994).

Lantana camara L. (Pl. 30E)

Syn.: *L. aculeata* L.; *L. myxa* L.; *L. camara* L. var. *aculeata* (L.) Moldenke

Family: Verbenaceae.

Vern. : Phulanu, Zahar.

English & Regional Names:

Eng. : Lantana, Wild Sage
G. : Ghanidalia
H. : Desi Lantana
Kan. : Hesike, Kakke, Nata Hu Gida
M.P. : Raimuniya
Mal. : Arippu
Mar. : Chadrung, Ghaneri
Ori. : Naga Airi
P. : Desi Lantana
Tam. : Arippu
Tel. : Pulikampa.

Distribution : Himalayas: ascending to 1,500 m.

Description : Rambling shrubs having numerous recurved prickles on branches. Leaves opposite, toothed. Flowers orange-yellow in long-stalked heads. Drupes black shining.

Reproductive Cycle : April-June.

Habitat Ecology : Semi-Naturalised and Cultivated Areas.

Material Examined : HPU-EL-SKS-5522; Andretta (980 m), Gharthun (900 m), Holta (1,200 m), Kathena (1,290 m), Nanaun (900 m).

Parts Used : Whole Plant. Stem. Leaves.

Folk Uses : Plant grown as **hedges**. Leaf-paste applied over affected parts to get relief from **sprain**. Leaves put in granaries for repelling **insects**. Stem used for **cleaning teeth**.

Active Constituents : Isolated constituents are lantadene A, lantadene B, lantalonic acid, lantic acid, ursonic acid, leucine, valine, tyrosine, glycine (*leaves*), geraniol, linalool, cedrene, caryophylene, farnerol, cis-nerolidol and trans-nerolidol (*essential oil*).

Biological Activity : Acid extr. of shoot exhibits antibacterial activity. Antitumour activity of verbascoside *in vitro* might partly be due to inhibition of protein kinase.

Uses in Literature : Known to be used earlier as antiseptic, antispasmodic, carminative, diaphoretic, edible (fruit), fish bait, mosquito repellent, poison (vet.), stimulant, toothbrush; and for anaemic patients, basketry, boils, burns, chest diseases, chronic inflammation of skin, cold, cough, cuts, diabetes, diarrhoea, dysentery, dyspepsia (vet.), fevers, fistulae, fits, gum care, green manure, infection of respiratory tract, itching, jaundice (vet.), mouth ulcers, mumps, pulmonary problems, pustules, rabies, retaining humus in the soil, rheumatism, ringworm, sprains, stored grains, swellings, tetanus, tuberculosis, tumours, ulcers, vulnerary, whooping cough, and wounds (Ambasta, 1986; Aminuddin & Girach, 1991; Arora & Pandey, 1996; Augustine & Sivadasan, 2005; Bhattacharya, 1996; Chopra *et al.*, 1956; Duke *et al.*, 2002; Dwarakan & Ansari, 1996; Dwivedi, 2003; Geetha *et al.*, 1996a; Girach & Aminuddin, 1995; Gogoi & Das, 2003; Goud & Pullaiah, 1996a; Henry *et al.*, 1996; Hosagoudar & Henry, 1996b, c; Jain, 1991, 1996; Kirtikar & Basu, 1984; Kshirsagar, 2003; Kshirsagar *et al.*, 2003; Kumar & Goel, 1998; Kumar *et al.*, 2003; Kurian, 1995; Lal *et al.*, 1996; Lalramnghinglova, 1996; Masih, 2003; Mishra *et al.*, 1996b; Nayak *et al.*, 2003; Roy *et al.*, 1998; Saini, 1996b; Saklani & Jain, 1994; Samant *et al.*, 2001; Seetharam *et al.*, 1999; Singh & Kumar, 2000a; Singh & Singh, 2003; Sinha & Sinha, 2001; Sood & Thakur, 2004; Srivastava *et al.*, 2003c; Subramaniam, 2000; Upadhyay & Singh, 2005; Watt, 1972).

***Lathyrus aphaca* L.** (Pl. 30F)

Family: Fabaceae.

Vern. : Mater-phali, Khinnu, Sudu.

English & Regional Names:

- **Eng.** : Yellow-flowered pea
- **B.** : Janglimatar, Masur-chuna
- **H.** : Jangli Matar
- **P.** : Gagla, Rawan, Rawari.

Distribution : Temperate Regions.

Description : Annuals. Rachis ending in a tendril. Leaflets abortive. Flowers at the end of long axillary stalks, yellow. Pods wingless.

Reproductive Cycle : April-May.

Habitat Ecology : Moist Places.

Material Examined : HPU-EL-SKS-5620; Chornali (1,280 m), Lahla (930 m), Malnu (1,000 m), Saparu (1,250 m), Thakurdwara (1,150 m).

Parts Used : Aerial Parts. Seeds.

Folk Uses : Aerial parts given as **fodder**, and tender ones as **vegetable**. Seeds used as **pulse**, and its flour used in **culinary** preparations.

Uses in Literature : Described earlier as an astringent, famine food, fodder, psychoactive, narcotic; and resolvent, wild relative of legumes (Ambasta, 1986; Dwivedi, 2003; Kirtikar & Basu, 1984; Manandhar, 1997; Jain *et al.*, 1994; Saini, 1996a; Samant *et al.*, 2001; Sharma & Rana, 2005; Sood & Thakur, 2004).

Lathyrus sativus **L.** (Pl. 31A)
Family: Fabaceae.

Vern. : Kalan.

English, Sanskrit & Regional Names:

Eng. : Chicking Vetch
Sans. : Lanka, Sandika, Tripula
B. : Kassur, Khesri, Teora
G. : Lang, Langue
H. : Kesari, Kassar, Kassur, Khesari, Kasari
P. : Chural, Karas, Karil, Kisari, Mattar.

Distribution : From S. Caucasus - N. India.

Description : Herbs with winged stem, two, linear leaflets and solitary, reddish purple flowers. Pods oblong with brown seeds marbled with red.

Reproductive Cycle : May-August.

Habitat Ecology : Wheat Fields.

Material Examined : HPU-EL-SKS-5604; Chornali (1,280 m), Gopalpur (1,350 m), Gopalpur (1,350 m), Ladoh (980 m), Pahra (950 m), Paror (940 m).

Part Used : Seeds.

Folk Uses : Seeds cooked as a '**dal**' (curry) and its flour used for making '**pakodas**'.

Active Constituents : *Seeds* afford isovenasterol, clerosterol, 24 – epislerosterol, homoarginine and γ-aminobutyric acid

Uses in Literature : Described in literature as an astringent, cathartic, flour, pulse, tonic; and for biliousness (Ambasta, 1986; Ballabh & Chaurasia, 2006; Das & Agarwal, 1991; Dey & Bahadur, 1973; Jha *et al.*, 1996; Kirtikar & Basu, 1984; Saini, 1996b; Samant *et al.*, 2001; Sharma & Rana, 2005).

***Lawsonia inermis* L.** (Pl. 31B)
Syn.: *Lawsonia alba* Lam.; *L. purpurea* Lam.; *L. spinosa* L.
Family: Lythraceae.

Vern. : Maindi.

Sanskrit & Regional Names:

Sans. : Kuravaka, Mendhi, Mendiki, Sakachara
B. : Mendi, Menedi, Shudi
G. : Medi, Mendi
H. : Hena, Mehndi, Mhindi
Kan. : Gorante, Gorantlu
Mar. : Mendi, Mendhi
Ori. : Manghati, Muljuyati
P. : Menndi, Nakrize, Ninna, Panwar
Tam. : Aivanam, Marithondi, Marutonri
Tel. : Goranta, Pachchapeddagoranta, Iveni.

Distribution : Throughout India (commonly cultivated).

Description : Branched shrubs having four-angled branches and opposite, acute, sharp pointed leaves. Flowers small, fragrant, white or pinkish.

Reproductive Cycle : July-October.

Habitat Ecology : Common as a Hedgeplant in Gardens and Lawns.

Material Examined : HPU-EL-SKS-5713; Bodhal (970 m), Gopalpur (1,350 m), Nanaun (900 m), Rangru (910 m), Sidhpur (1,100 m).

Part Used : Leaves.

Folk Uses : Leaf-paste applied against **burns** and also over head and soles to relieve **body heat** and **headache**. Leaf-decoction (2 teaspoonful twice a day for one month) prescribed for **kidney problems**, and also its gargle good against **sore throat**.

Active Constituents : Isolated constituents from *leaves* are lawsone, laxathone II, luteolin, γ-sitosterol and its 3-O-glucoside.

Biological Activity : Anti-inflamatory activity confirmed.

Uses in Literature : Recorded to be used earlier as antifertility, antipyretic, astringent, brain tonic, diuretic, hair dye, hair oil; and for boils, bronchitis, burning sensation, calculous affections cooling effect, dental care, diarrhoea, diseases of spleen, eczema, emetic, enlargement of spleen, expectorant, foot and mouth diseases, fungal infection of feet, hair growth, headache, hemicrania, hiccups, inflammation of wounds, insanity, jaundice, kidney problem, leucoderma, lumbago, opthalmia, perfumes, scabies, skin diseases, sores, spermatorrhohea, stomatitis, tool-handles and ulcers (Ambasta, 1986; Bhandary & Chandrashekar, 2002; Bhatt *et al*., 2001; Dwivedi, 2003; Ganai & Nawachoo, 2003; Girach *et al*., 1997; Jha & Kumar, 2003; Katewa & Arora, 1997; Kirtikar & Basu, 1984; Kshirsagar *et al*., 2003; Kumar *et al*., 2003; Maya *et al*., 2003; Mohanty & Padhy, 1996; Pawar & Patil, 2004; Rajendran *et al*., 2003; Ranjan, 2003; Saikia & Nath, 2003; Sen & Behera, 2003; Singh & Pandey, 1998; Thomas & Britto, 2003; Vihari, 1995).

***Lepidagathis cuspidata* Nees** (Pl. 31C)
Family: Acanthaceae.

Vern. : Billi.

Distribution : Tropical India.

Description : Small shrubs upto 70 cm having white flowers dotted with purple spots, which are borne in dense spikes. Leaves attentuate at both ends. Calyx 5-partite. Capsule with minute seeds.

Reproductive Cycle : December-June.

Habitat Ecology : Rock-Crevices. Wastelands. Dry Slopes.

Material Examined : HPU-EL-SKS-5680; Bodhal (970 m), Gopalpur (1,350 m), Nanaun (900 m).

Part Used : Rhizome.

Folk Uses : Extr. (50 ml) of rhizome prescribed for one week against **body heat**, and its paste massaged over gums to check **toothache**.

Uses in Literature : Used as a tonic in fever; and for skin affections and sores (Chopra *et al*., 1956; Sood & Thakur, 2004; Srivastava *et al*., 2003c).

***Lepidium sativum* L.** (Pl. 31D)
Family: Brassicaceae.

Vern. : Halein.

English, Sanskrit & Regional Names:

Eng. : Garden Cress

Sans. : Ashalika, Bhadra, Chandrika, Darakrishna, Kalamesha, Nandini, Suvasura, Vasupreshpa

B. : Aleverie, Halim

Bo. : Ahaliva

	G.	: Asahio, Halim
	H.	: Chaunsar, Halim, Hurf
	Mar.	: Ahliva
	P.	: Halim, Tezak
	Tam.	: Aliverai
	Tel.	: Adalavitulu
	Urdu	: Halim.

Distribution : Cultivated throughout India.

Description : Glabrous erect annuals upto 45 cm with sessile cauline leaves and radical ones long-petioled. Flowers white, in long racemes. Pods notched at the apex.

Reproductive Cycle : March-May.

Habitat Ecology : Moist Shady Slopes. Often Cultivated.

Material Examined : HPU-EL-SKS-5634; Chimbalhar (1,150 m), Gopalpur (1,350 m), Panchrukhi (990 m), Sidhpur (1,100 m), Tapa (945 m).

Parts Used : Leaves. Seeds.

Folk Uses : Leaves consumed as **vegetable**. Seeds (7-8) put in eyes to check **opthalmic injuries** and to exclude **dust** (eye). 150 ml decoction of seeds prescribed for 7-10 days to treat **pneumonia**; also its poultice applied locally on chest for the same.

Active Constituents : These are glucotropoeolin, 4-methoxyglucobrassicin, sinapine, sinapic acid, β-sitosterol, benzylcyanide, calmodulin, sinapoyglucose, esters of caffeic, p-coumaric, ferulic and quinic acids (*plant*); and lepidine, N,N′-dibenzyl urea, N-N′-dibenzylthiourea, sinapic acid and its choline ester; arachidic, behenic and lignoceric acids (*seeds*).

Biological Activity : Crushed leaves antibacterial against *Bacillus subtilis* and *Micrococcus pyogenes* var. *aureus*. The volatile oil has pronounced oestrogen activity.

Uses in Literature : Described so far as an antiscorbutic, colic, diabetes, stomachic, vegetable; and for asthma, bronchitis, cough, diarrhoea, dysentery, edible oil, flatulence, mucus in urine, pain, piles, pneumonia, and syphilis (Bhatt *et al.*, 2003a; Chatterjee & Pakrashi, 1991; Chopra *et al.*, 1956; Dey, 1973; Jain & De, 1966; Jain & Tarafder, 1970).

Leucaena glauca Benth. (Pl. 31E)

Syn.: *L. leucocephala* Lam. de Wet.; *Mimosa lecocephala* Lam.; *Acacia frondosa* Willd.; *A. caringa* Ham.; *A. biceps* DC.; *A. leucocephala* DC.

Family: Fabaceae.

Vern. : Alseenia, Raseenia.

English & Regional Names:

- **Eng.** : White Popinae, Lead Tree
- **G.** : Lasobaval, Pardesi Baval, Vilaytibaval
- **Mal.** : Takaranniram
- **Ori.** : Rajokasundiri
- **Tam.** : Tagarai-Virai.

Distribution : Throughout India.

Description : Small trees with 10-15 pairs of leaflets, whitish flowers and linear flat, shining pods.

Reproductive Cycle : September-November.

Habitat Ecology : Hedges. Along Boundaries of Fields.

Material Examined : HPU-EL-SKS-5546; Jasun (1,000 m), Kathena (1,290 m), Makol (900 m), Pahra (950 m), Rangru (910 m).

Parts Used : Leaves. Wood.

Folk Uses : Leaves form an excellent **fodder** for milching cattle. Wood employed as **fuel**.

Active Constituents : Constituents isolated from *leaves* are kaempferol-3-arabinoside, kaempferol-3-xyloside, quercetin-3-arabinoside quercetin-3-rhamnoside and quercetin-3-galactoside.

Uses in Literature : Used earlier as source of fibre, fodder, ornaments, protein, stimulant, tonic, vegetable; and for baskets, flutes, whistles, carotene, fancy bags, and poultry rations (Ambasta, 1986; Gill & Nyawuame, 1994; Watt, 1972).

***Leucas cephalotes* Spreng.** (Pl. 31F)
Syn.: *Leucas capitata* Desf.; *Phlomis cephalotes* Roth
Family: Lamiaceae.

Vern. : Bhedu.

Sanskrit & Regional Names:

Sans.	: Dranapushpi
B.	: Barahalkusa
G.	: Kubi, Kubo
H.	: Dhurpisag, Delotona, Gonna
Mar.	: Deokumba, Shetvad, Tumba
P.	: Chatra, Guldoda, Maldoda, Phuman, Sisalius
Tel.	: Peddatummi, Tumni.

Distribution : Deccan Peninsula, 650-1,000 m.

Description : Annual pubescent hairy herbs with quadrangular stem and ovate membranous leaves. Flowers sessile, in globose dense terminal whorl. Nutlets long, obovoid-oblong, rounded at the apex.

Reproductive Cycle : July-October.

Habitat Ecology : Wastelands.

Material Examined : HPU-EL-SKS-5561; Boda (1,000 m), Jasun (1,000 m), Sungal (1,050 m), Tikkar (910 m).

Part Used : Head (inflorescence).

Folk Uses : Crushed inflorescence alongwith seeds of 'Brahamjeeri' (*Centratherum anthelminticum*), 'lal-mirach' (*Capsicum annuum*), and onion-tuber (*Allium cepa*) recommended for **constipation**, **stomach disorders**, **cough**, and for inducing **rumination** in livestock.

Active Constituents : Active properties due to β-sitosterol glucoside is (*plant*), labellenic acid (octadeca-5,6-dienoic acid) (*seed oil*).

Uses in Literature : Described so far as anthelmintic, antiseptic, appetiser, diaphoretic, dyspepsia, insecticidal, laxative, stimulant; and for asthma, bronchitis, burning sensation, burns, cold, cough, fevers, gummosis, heating, indigestible, inducing sound sleep, inflammations, jaundice, liver disorders, mouth ulcers, pain in half skull, paralysis, redness of eyes, scorpion sting, stomachache, swelling, typhoid, urinary disorders, and worm infestation of eye (Ambasta, 1986; Caius, 2003; Chauhan, 1999; Dwivedi, 2003; Khanna, 2002; Kirtikar & Basu, 1984; Kumar & Goel, 1998; Pandey & Verma, 2002; Panthi & Chaudhary, 2003; Rajendran *et al.*, 2003; Sharma, 2003; Shukla *et al.*, 2003; Singh & Pandey, 1998; Srivastava *et al.*, 2003a; Subramani & Goraya, 2003).

***Linum usitatissimum* L.** (Pl. 32 A,B)
Family: Linaceae.

Vern. : Allain.

Sanskrit & Regional Names:

Sans. : Atasi
B. : Masina
G. : Alsi
H. : Alsi, Tisi
Kan. : Agasi
Mar. : Javas
Tam. : Alivirai
Tel. : Arisi.

Distribution : Throughout India.

Description : Herbaceous annuals with fibrous stem and simple, lanceolate, entire leaves. Flowers delicate, bisexual, blue or mauve. Fruits a capsule. Seeds flat.

Reproductive Cycle : November-April.

Habitat Ecology : Cultivated Land. An Escape in Open.

Material Examined : HPU-EL-SKS-5625; Chauki (900 m), Gopalpur (1,350 m), Purani-palam (1,100 m), Saparu (1,250 m).

Part Used : Seeds.

Folk Uses : Seeds roasted and considered highly **nutritious**; its oil used for **cooking** purposes, and oil-cakes given as **feed** for milching cattle. Seeds kept in eyes to **clean** it.

Active Constituents : *Seeds* afford cholesterol campesterol (26), stigmastirol (7), sitosterol (41), 5-dehydro-avenasterol (13), cycloaretenol (9) and 24-methylenecycloartanol (2%).

Uses in Literature : Known so far as aphrodisiac, astringent, blood purifier, brain and heart tonic, demulcent, diuretic, edible, emmenagogue, emollient, galactagogue, hard to digest, hot, oily, rheumatic, sedative, slightly bitter, sweet, tonic; and for backache, bad for eyesight, biliousness, causes loss of appetite, cough and cold, gonorrhoea, gouty swellings, inflammation, kidney ailments, lead to impotency, leprosy, ulcers, urinogenital disorders, and urinary discharges (Ambasta, 1986; Kirtikar & Basu, 1984; Sen & Batra, 1997; Sharma, 2003; Upadhyay & Chauhan, 2003).

Litchi chinensis Sonn. (Pl. 32C)

Syn.: *Nephelium litchi* Camb.; *Dimocarpus* Lour.

Family: Sapindaceae.

Vern. : Litchi.

English & Regional Names:

- **Eng.** : Litchi, Lychee
- **Bo.** : Lichi
- **H.** : Litchi
- **Ori.** : Lishi
- **Tam.** : Ilichi.

Distribution : Widely Cultivated in Orchards of Bihar, Bengal & U.P.

Description : Small trees. Leaves alternate with 4-12 leaflets. Flowers minute. Fruits red with fleshy aril. Nuts brittle.

Reproductive Cycle : April-June.

Habitat Ecology : Cultivated.

Material Examined : HPU-EL-SKS-5624; Banuri (1,150 m), Bhatu (1,040 m), Nanaun (900 m), Sulah (950 m).

Part Used : Aril.

Folk Uses : Aril **edible**, and its juice has **astringent** properties.

Active Constituents : *Fruits* possess cyanidin-3-rutinoside, cyanidin-3-glucoside and malvidin-3-acetylglucoside.

Uses in Literature : Known earlier in India as anodyne in neuralgic disorders and orchitis, edible (fruit), and for offerings (Ambasta, 1986; Gogoi & Borthakur, 1991).

Litsea monopetala (Roxb.) Pers. (Pl. 32D)

Syn.: *L. polyantha* Juss.; *Tetranthera monopetala* Roxb.; *T. macrophylla* Wall.; *T. alnoides* Miq.; *T. fruticosa* Herb.; *T. verticillata* Herb.; *T. semecarpifolia* Wall.; *T. hexantha* Sieb.; *Tomex pubescens* Herb.

Family: Lauraceae.

Vern. : Guaan.

English, Sanskrit and Regional Names:

- **Eng.** : Common Gray Mango Laurel
- **Sans.** : Gajapippali
- **Ass.** : Sualu, Muga
- **B.** : Barakurchita, Barakukar

H.	: Gwa, Kakuri, Kari, Karkawa, Katmarra, Katmedh, Katmonia, Kerauli, Marda, Meda, Papria, Patoia, Randkari, Sangran, Singraf, Singran
Mar.	: Ranamba
Ori.	: Kulya, Kurukochhodiya, Mahachhodiya
P.	: Gwa, Harien, Medalakri, Rian
Tam.	: Maida Lagadil, Muchaippeyetti, Pisinbattau
Tel.	: Meda, Nara, Naramamidi.

Distribution : Foothills of Himalayas, ascending to 1,000 m.

Description : A medium-sized tree with pubescent branches, alternate leaves and white flowers in umbels. Drupes ovoid, black when ripe.

Reproductive Cycle : March-July.

Habitat Ecology : Wastelands. Rocky Wet Slopes.

Material Examined : HPU-EL-SKS-5566; Bhauda (1,050 m), Boda (1,000 m), Chhanchri (930 m), Khaira (940 m), Panapari (1,100 m).

Parts Used : Leaves. Fruits. Wood.

Folk Uses : Leaves lopped for **fodder**. Ripe fruits **edible**. Wood used for making **tool-handles**, **agricultural implements**, **building works** and **cheap furniture**.

Active Constituents : Polysaccharide isolated from *leaves* contains xylose and arabinose in molar ratio of 1-4 and an arabinoxylan consisting of xylose and arabinose in a molar ratio of 1-2.

Uses in Literature : So far described as an astringent, edible (fruits), feed for silkworms, vegetable; and for agricultural implements, boils, bone fractures (vet.), candle manufacturing, chronic diarrhoea, chronic dysentery, cuts, fodder, furniture, giddiness, house building, plywood, relieving bodyache, rheumatism, sore (vet.), ulcer (vet.), and whitlow (Ambasta, 1986; Chatterjee & Pakrashi, 1991; Chopra *et al.*, 1956; Chowdhury *et al.*, 1996; Das & Agarwal, 1991; Jain, 1991; Jha & Varma, 1996; Kirtikar & Basu, 1984; Murthy *et al.*, 2003; Purohit & Samant, 1995; Rama Rao & Henry, 1996; Singh *et al.*, 1996b; Sood & Kaushal, 2008; Watt, 1972).

***Litsea sebifera* Pers.** (Pl. 32E)

Syn.: *L. glutinosa* (Lour.) C. B. Rob.; *L. chinensis* Lam.; *L. tetranthera* Pers.; *L. multiflora* Blume; *Neolitsea chinensis* (Lam.) Chun; *Tetranthera lauriflora* Blume; *T. mutiflora* Blume; *T. racemosa-umbellata* Blume; *T. roxburghii* Blume; *T. captitata* Herb.; *T. apetala* Roxb.; *Tomex tetranthera* Willd.; *T. sebifera* Willd.; *Sebifera glutinosa* Lour.; *Laurus involucrata* Koenig.

Family: Lauraceae.

Vern. : Rhaanh.

English, Sanskrit & Regional Names:

- **Eng.** : Common Tallow Laurel
- **Sans.** : Adhavara, Bahurandhrika, Bhavya, Dhira, Jivani, Madhura, Manichhidra, Medarsa, Medodbhava, Purusadantika, Shalyada Snehavati, Svalpaparni, Vibhavari
- **Ass.** : Bagnara, Heluka
- **B.** : Garur, Kukurchita, Maidalakri, Ratun
- **Bo.** : Chickana, Maidalakadi, Maidalakri
- **H.** : Garbijaur, Maidalakri, Medh, Menda, Singrauf
- **M.P.** : Menda
- **Mar.** : Maidalakadi, Miri
- **Ori.** : Jaisanda
- **P.** : Chamana, Gwa, Maidasak, Medachob, Medalakri, Medasak, Rian
- **Tam.** : Elumpurukki, Mushaippeyetti, Uralli
- **Tel.** : Naramamidi, Narra alagi.

Distribution : Throughout warmer regions of India; lower Himalayas.

Description : A medium-sized tree with subterminal, coriaceous, simple leaves which are pale when dry. Fruits pea-sized.

Reproductive Cycle : May-September.

Habitat Ecology : Steep Slopes.

Material Examined : HPU-EL-SKS-5697; Balah (900 m), Boda (1,000 m), Bhuana (1,400 m), Khaira (940 m).

Parts Used : Leaves. Bark. Wood.

Folk Uses : Warm oil smeared leaves tied as plaster to treat **sprain** and **fracture**. Bark paste applied for **rheumatic pains** and **bone fractures**. Leaf extr. used as **eye-drops** for healing **wounded** eye. Wood for **furniture**.

Active Constituents : Chief ingredients are naringerin, naringen, kaempferol-3-and 7-glucosides, quercetin and its 3-thamnoside pelargonidin-3 and 5-glycosides, cystine, glycine, L-alanine, β-alanine, valine, tyrosine proline phenylalanine and leucine (*leaves*), two new alkaloids-sebiferine and litseferine (*trunk-bark*).

Uses in Literature : Useful in India as anodyne, antidote to venomous animals, aphrodisiac, demulcent, edible, emollient, feebly balasamic, mild astringent, styptic; and for agricultural implements, boils, bone fractures (vet.), bruises, candles, chest pain, cold, cooling, cough, diarrhoea, dysentery, dyspepsia, fibres, floor boards, furniture, house building, incense industry, indigestion (vet.), inflammation, leucorrhoea, oars, packing cases, pain in umbilicus region, rheumatic gouty joints, skin diseases, sprain, and wounds (Ambasta, 1986; Chowdhury *et al*., 1996; Jain, 1991; Rama Rao & Henry, 1996; Shah & Singh, 1990; Watt, 1972).

Litsea semicarpifolia Wall. (Pl. 32E)

Family: Lauraceae.

Vern. : Pahadi-sannanh.

Distribution : Foothills of Himalayas

Description : Medium-sized trees. Leaves simple, entire, large. Fruits small, globose.

Reproductive Cycle : April-September.

Habitat Ecology : Lower Shady Part of Sharp Hills.

Material Examined : HPU-EL-SKS-5567; Balah (900 m), Boda (1,000 m), Bhuana (1,400 m), Khaira (940 m), Makol (900 m).

Parts Used : Leaves. Wood.

Folk Uses : Leaves employed as an excellent **fodder**. Wood used for **household construction**.

Loranthus falcatus L. (Pl. 33A)

Syn.: *L. longiflorus* Desr.

Family: Loranthaceae.

Vern. : Parand.

Sanskrit & Regional Names:

Sans. : Vanda, Vrikshabhaska, Vrikshadani, Vriksharuha

B. : Buramanda

G. : Vando

H. : Banda

Mal. : Itti, Ittukani, Pulanni, Utia

Mar. : Vanda

P. : Amut, Banda, Pand, Panda, Panand
Tam. : Kamaricham, Pulluri, Pulluruvi, Uchi, Uruvi
Tel. : Badanika, Bajinika, Jiddu, Yelinga.

Distribution : Throughout India, Myanmar, Sri Lanka-Australia.

Description : A bushy branched parasite having grey, smooth bark and thick, coriaceous, opposite leaves. Flowers in unilateral racemes. Berries ovoid-oblong, crowned by cup-shaped calyx.

Reproductive Cycle : July-December.

Habitat Ecology : As parasite on *Vitex nigundo*, *Mangifera indica* and *Bauhinia variegata.*

Material Examined : HPU-EL-SKS-5403; Balah (900 m), Boda (1,000 m), Nadli (1,000 m), Rangru (910 m), Tapa (945 m).

Part Used : Leaves.

Folk Uses : Leaves used as **fodder** for goat. Two and half leaves fed to cure tail infection, '**nanai**' in cattle.

Uses in Literature : Considered a narcotic, substitute for betelnut; and for asthma, constipation, mania, menstrual trouble, and wounds (Chatterjee & Pakrashi, 1991; Kirtikar & Basu, 1984).

***Lotus corniculatus* L.** (Pl. 33B)

Family: Fabaceae.

Vern. : Chidi.

English Names:

Eng. : Birdsfoot, Trefoil.

Distribution : Sub-tropical Regions.

Description : Perennial glabrous herbs with slender, decumbent or ascending stem. Leaflets-5, obovate, sessile. Flowers yellow. Calyx bell-shaped. Pods cylindrical, straight.

Reproductive Cycle : April-September.

Habitat Ecology : Open Places. Grasslands.

Material Examined : HPU-EL-SKS-5591; Chachian (1,300 m), Dharaman (1,100 m), Molichak (1,020 m), Saparu (1,250 m).

Part Used : Flowers.

Folk Use : Flowers offered to **appease** Goddess Parvati.

Uses in Literature : Reported in literature as good soil-binder free from pest and diseases, fodder, dye, hay, good source of carotene, silage, skin troubles, and bee plant (Ambasta, 1986; Kapur, 1996a; Singh & Kumar 2000).

Luffa acutangula (L.) Roxb. (Pl. 33C)

Syn.: *Cucumis acutangulus* L.; *L. acutangula* var. *amara* (Roxb.) Cl.; *L. kleinii* Wight and Arn.; *L. umbellata* (Klein *ex* Willd.) Roem.

Family: Cucurbitaceae.

Vern. : Jangli-ghangher.

English, Sanskrit & Regional Names:

Eng.	: Ridged or Ribbed Goure
Sans.	: Jhongaka, Koshataki, Torai, Jalini, Sukosha
B.	: Jhinga, Sataputi, Kalitori, Turi
Bo.	: Gonsali, Jinga, Sirola, Turai
G.	: Ghisoda
H.	: Kalitori, Jhingatori
Kan.	: Hirekaya
Mal.	: Pichenga, Djinii
Mar.	: Shirola, Dodaki, Sataputi
Tam.	: Pirkankai, Peyppichukku, Pikunkai
Tel.	: Birakaya, Burkai
Urdu	: Torai.

Distribution : Throughout the greater part of India.

Description : Monoecious climbers having 5-angled, glabrous stem; tendrils 3-fid and palmately 5-7 angled leaves. Male flowers in racemes; female ones solitary. Fruits 15-30 cm long, longitudinally ribbed with 10 sharp angles. Seeds ovoid-oblong, black, compressed.

Reproductive Cycle : May-September.

Habitat Ecology : Agricultural fields. Naturalized in outskirts of forests.

Material Examined : HPU-EL-SKS-5690; Chachian (1,300 m), Pahra (950 m), Paror (940 m), Rajeher (1,050 m), Sungal (1,050 m).

Part Used : Fruits.

Folk Uses : Fruits cooked as a **vegetable**, and dried ones used as 'loofah'**(bath sponge)**.

Active Constituents : Cucurbitacins B and E, and oleanolic acid isolated from *fruits*.

Biological Activity : Aq. extr. of seeds anticancerous.

Uses in Literature : Useful as amenorrhoea, anthelmintic, antibilious, antidote to snakebite, antipyretic, decorative piece, edible, emetic, purgative, scrubber, stomachic, sweet; and for asthma, biliousness, bronchitis, conjunctivitis, cooling, diabetes, flatulence, hemorrhoids, leprosy, oleaginous, ringworm, splenitis, and uraemia (Agarwal, 2003; Ambasta, 1986; Bhandary & Chandrashekar, 2002; Bhatt *et al.*, 2003a; Chopra *et al.*, 1956; Das & Agarwal, 1991; Dey & Bahadur,

1973; Girach *et al.*, 1997; Karnick, 1994; Katewa *et al.*, 2001; Kirtikar & Basu, 1984; Kulkarni & Kumbhojkar, 1996; Padhy & Dash, 2003; Parrotta, 2001; Prajapati *et al.*, 2006; Rastogi & Mehrotra, 1993; Roy *et al.*, 1998; Sahu, 2005; Said, 1997; Sarangi & Sarin, 1996; Sharma & Rana, 2005; Singh & Pandey, 1998; Varma, 1997).

***Luffa cylindrica* (L.) Roem.**

Syn.: *L. aegyptica* Mill.; *L. pentandra* Roxb.; *L. racemosa* Roxb.; *L. clavata* Roxb.; *L. petola* Ser.; *L. cattu-picinna* Ser.; *L. parvala* Wall.; *L. gosa* Wall.; *L. hederacea* Wall.; *L. satpatia* Wall.; *Bryonia cheirophylla* Wall.; *Momordica cylindrica* L.; *M. luffa* L.

Family: Cucurbitaceae.

Vern. : Ghi-ghangher.

English, Sanskrit & Regional Names:

Eng. : Gourd Towel, Spinge Gourd, Vegetable Sponge, Washrag Sponge

Sans. : Aibhi, Brihatkoshtaki, Dhamargava, Dhirghapatolika, Ghoshaka, Hastiparna, Hastighosha, Hartikoshataki, Mahakoshataki, Mahaphala, Rajakoshataki, Sapitaka

Ass. : Bhatkarel, Bhol

B. : Dhindul, Hastighasha

Bo. : Ghosali, Gonsali, Parula, Turi

G. : Galaka, Martigonsali, Turia

H. : Ghiaturai, Nenua, Purula

Kan. : Tuppahirekai

Mal. : Kattupechal

Mar. : Ghadaghosali, Ghosali, Parosi

P. : Ghiatori, Ghiaturia, Ghigandoli

Tam. : Pichukku, Pikku

Tel. : Guttibira, Netibira, Nunebira.

Distribution : Throughout India.

Description : Climbers with weak stem. Leaves 5-lobed, reticulated, long-petioled, dotted. Flowers unisexual, yellow. Fruits cylindrical green. Seeds flat.

Reproductive Cycle : May-September.

Habitat Ecology : As an escape near habitations.

Material Examined : HPU-EL-SKS-5636; Dehan (1,130 m), Gharthun (900 m), Gopalpur (1,350 m), Rangru (910 m), Tapa (945 m).

Part Used : Fruits.

Folk Uses : Fruits cooked as **vegetable**, dried ones used as **bath sponge**.

Active Constituents : Major saponins–lucyosides A, E, F, I, J, K, L and M isolated from *fruits*.

Uses in Literature : Recorded as an anthelmintic (vet.), carminative, cathartic, diuretic, edible, emetic, galactagogue, fibre, purgative, vegetable; and for abnormal stoppage of the menses, anaemia, cold (vet.), cough (vet.), diarrhoea, dyspepsia (vet.), expulsion of worms, fibre, indigestion (vet.), iron deficiency, killing white ants, liver diseases, piles, prickly heat, wounds, infected teeth, swelling of lungs and wound of legs, snakebite, tonsilitis, varanus, mad dog-bite, and HIV patients (Ambasta, 1986; Bhandary & Chandrashekar, 2002; Bondya & Sharma, 2004; Jain, 1991; Jha *et al.*, 1996; Katewa *et al.*, 2001; Khan, 2005; Kulkarni & Kumbhojkar, 1993; Kurian, 1995; Ramdas *et al.*, 2000; Satapathy & Brahmam, 1996; Singh & Maheshwari, 1992; Singh & Pandey, 1998; Watt, 1972).

Mallotus philippinensis (Lam.) Muell.-Arg. (Pl. 33D)

Syn.: *Rottlera tintorii* Roxb.; *R. aurantiaca* Hook. & Arn.; *R. affinis* Hassk.; *R. montana* Wall.; *R. mollis* Wall.; *Croton philippensis* Lam.; *C. punctatus* Retz.; *C. coccineus* Vahl; *C. montanus* Willd.; *C. distans* Wall.

Family: Euphorbiaceae.

Vern. : Kaamal.

English, Sanskrit & Regional Names:

Eng. : Monkey Face Tree, Kamala Tree
Sans. : Bahupushpa, Pikaksha, Raktachurnaka
Ass. : Gangai, Jagguru
B. : Kamlagundi, Tung
Bo. : Kamala, Ruhin
G. : Kapilo
H. : Kambhal, Rohni, Rula
Kan. : Kunkumadara
Kash. : Kambil
Mal. : Chenkolli, Piponnakam, Poonagam
Mar. : Shendri
Ori. : Bosontogundi, Punango
P. : Kamal, Rulya
Tam. : Kabilappodi, Kurangumanjanatti
Tel. : Adavigubbtuda, Nagarakta.

Distribution : S. E. Asia: upto 1,800 m.

Description : Small evergreen trees with male and female flower on separate plants. Leaves ovate-lanceolate and their lower surface densely dotted with minute red glands. Capsule 3-cornered covered with crimson powder.

Reproductive Cycle : April-October.

Habitat Ecology : Moist Forests.

Material Examined : HPU-EL-SKS-5692; Balah (900 m), Chhanchri (930 m), Khaira (940 m), Makol (900 m), Nadli (1,000 m).

Parts Used : Leaves. Seeds. Bark. Stem. Wood.

Folk Uses : Leaves used as **forage** for cattle. Wood used as **fuel**. Glandular powder on seeds blended with mustard oil considered good against **cutaneous diseases**; and one teaspoonful of it blended with buttermilk prescribed for 4-5 days to check **constipation**. Also, along with jaggery taken at bedtime for 7 days to check **digestive disorders**. Bark tied around neck, wrist and feet to check **rashes** on skin locally called 'badalbis'. Stem coloured and worshipped on the day of '**Holi**' festival.

Active Constituents : Active properties due to rottlerin, isorottlerin, corotoxigenin, coroglaucigen (*seeds*), betulin-3-acetate, lupeol, lupeol acetate, sitosterol, bergenin (*heartwood*), rottlerin, isoallorottlerin and red compound (*leaves*).

Biological Activity : Aq. extr. of red resinous powder (0.03 g) exhibits lithotropic effect in adult albino rats.

Uses in Literature : Described in Indian liiterature as an alternative to sendur, anthelmintic, anthelmintic (vet.), antioxidant for ghee, aphrodisiac, blood purifier, cathartic, contraceptive, dye, fodder, fuel, hepatic, manure, purgative, styptic, tonic for pregnant women, red dye from it adorned by women on forehead as a sign of being married, vermifuge; and for anorexia, boils, lactation in cow, controlling allergic infections, downer cow syndrome (vet.), dysentery, fever, fits, giddiness, hemiplegia, herpes, house decoration, hypocaemia (vet.), jaundice, leprous eruptions, matchboxes, parasitic skin diseases, rapid drying of varnish (seeds used), rheumatism, snakebites, sores, stomachache, tannins, thoracic pains, ulcers, and wound (vet.) (Ambasta, 1986; Banerjee & Pal, 1996; Chhetri, 1994; Hosagoudar & Henry, 1996c; Jain, 1968, 1991; Kapur & Singh, 1996; Kumar & Rao, 2003; Maliya, 2005; Mukherjee & Namhata, 1990; Nayak *et al.*, 2003; Purohit & Samant, 1995; Rama Rao & Henry, 1996; Rana *et al.*, 1996; Shanavashan *et al.*, 2003; Sharma, 2003; Shome *et al.*, 1996; Singh, 1996a; Singh & Kumar, 2000b; Singh & Prakash, 1994; Tirkey, 2004; Tirkey *et al.*, 2003b; Tiwari & Tiwari, 1996; Vihari, 1995; Watt 1972).

***Mangifera indica* L.** (Pl. 33 E,F)
Syn.: *M. domestica* Gaertn.
Family: Anacardiaceae.

Vern. : Aam, Amb.

English, Sanskrit & Regional Names:

Eng. : Cuckoo's Joy, Cupid's Favourite, Mango Tree, Spring Tree
Sans. : Adisairrabha, Manmathalaya, Parapushtamahotsava
Ass. : Ghariam
B. : Ambra
Bo. : Am, Amba
G. : Ambo, Amri
H. : Am, Amb
Kan. : Mavu
Mal. : Amram, Gomanne, Nattumavu
Mar. : Amba
Ori. : Ambo, Boula, Chuto
P. : Am, Amb, Mawashi
Tam. : Adishelarayam, Kilimukkuma
Tel. : Gujjumamidi, Makandamu.

Distribution : Near habitations. Forests.

Description : A large tree with dark-green, shining, alternate, and oblong-lanceolate leaves. Flowers very small, yellow, odorous, in panicles. Drupes one seeded, yellowish in colour.

Reproductive Cycle : March-July.

Habitat Ecology : Planted Along Roadsides. Field Borders. Open Slopes.

Material Examined : HPU-EL-SKS-5684; Bhauda (1,050 m), Jasun (1,000 m), Rangru (910 m), Tapa (945 m).

Parts Used : Whole Plant. Leaves. Fruits. Seeds. Latex. Wood.

Folk Uses : Tree considered **sacred**, and wood used for performing '**havan**' in various religious ceremonies. Wood also used for **house construction**. Leaves good for **scouring** teeth and **toothache**; also its garland hung on door in all **religious ceremonies**. Latex from petiole applied to treat **sty**. Ripe fruits **eaten** raw, cooked as **vegetable**, and its pulp used for making '**aampapad**' (traditional candies). Unripe ones **pickled**, and used for making '**aamchoor**' after drying in the sun. Seed powder (1 tabespoonful) with water prescribed early morning regularly to check **diabetes**; and its paste locally applied to check **hairfall**.

Active Constituents : Constituents isolated are homomangiferin and characterised as 2-β-D-glucopyranosyl-3-methoxy-1, 6, 7-trihydroxyxanthone, leucine, tyrosine and valine (*leaves*), protocatechuic acid, catechin and mangiferin (*bark*), alanine, glycine, γ-aminobutyric acid, kinic and shikimic acids (*leaves, bark*), β-sitosterol, (+) catechin, (±)catechin, (-) epicatechin and leucocyanidin (seed coat), stearic acid isolated (seeds), exudate methyl, ethyl, propyl, butyl, amyl and iso-butryl alcohols (*leaf oil*), isomyl alcohol, α-pinene, β-pinene, myrcene, limonene and fenchone (*pericarp oil*), and methyl propyl, isopropyl, butyl and isomyl alcohols (*seed oil*).

Biological Activity : Gallic acid (1 mg/ml) and quercetin (4 mg/ml) showed marked anti-influenza virus activity.

Uses in Literature : Known so far as abortifacient, anthelmintic, antiscorbutic, diabetes, antisyphilitic, appetiser, astringent, colic (vet.), cremation wood, diuretic, edible (fruits, vegetable, pickle, 'chutney', 'amchur', jams, jellies, squash, juice, 'am papad'), febrifuge, laxative, magico-therapeutic to vanish sty, porridge, restorative tonic, sacred (Lord Buddha reposed under this tree), stomachic, symbolic (abode of god); and for abdominal pain, agricultural implements, anasarca, anthrax (vet.), antifertility, arthritis, asthma, blood cancer, cholera, constipation (vet.), cough, cuts, decoration in marriages and festivals, diabetes, diarrhoea, diarrhoea (vet.), dropsy, drums and other musical instruments, dyeing, dysentery containing slight mucous, expelling placenta, extraneous affections, eye complaint, furniture, gingivitis, gum, healthy and long hair, house construction, liver disorder, loo fever, malaria, menorrhagia, piles, plague, pneumonia (vet.), pregnant woman to beget fair complexioned babies, preparing lakes, rabies (vet.), rheumatism, scabies, skin diseases, style, sunstroke, swelling of throat (vet.), tanning, throat diseases, toothache, ulcers, ulcerated tongue, uterine haemorrhage, wounds, writing inks, yellow dye (Agarwal, 2003; Ambasta, 1986; Arora & Pandey, 1996; Bennet *et al.*, 1991; Bhatt *et al.*, 2003c; Bhattacharya *et al.*, 2005; Borthakur & Gogoi, 1994; Borthakur *et al.*, 2004; Chowdhury *et al.*, 1996; D'Rozario *et al.*, 1999; Dey & Bahadur, 1973; Diwakar & Ansari, 2003; Ganesan & Kesavan, 2003; Ghosh *et al.*, 1996; Gill *et al.*, 1993, 1997; Gogoi *et al.*, 2003a; Goud & Pullaiah, 1996a; Harborne & Baxter, 2001; Henry *et al.*, 1996; Hosagoudar & Henry, 1996 a, c; Jain, 1991; Jain *et al.*, 1997; Jha *et al.*, 1996; Kapur *et al.*, 1996; Khare, 2004; Kirtikar & Basu, 1984; Kumar & Goel, 1998; Kumar & Rao, 2001, 2003; Kumar *et al.*, 2003; Lalramnghinglova, 1996; Manandhar, 1996; Masih, 2003; Mishra *et al.*, 1996b; Mohanty & Padhy, 1996; Nayak *et al.*, 2003; Padhy & Dash, 2003; Pandey & Varma, 2003; Pruthi, 1976; Rajan *et al.*, 2003; Rajendran *et al.*, 2003; Roy *et al.*, 1998; Sahni, 1998; Sahu, 1996; Samant *et al.*, 2001; Saikia & Nath, 2003; Saini, 1996a; Sen & Behera, 2003; Shah

& Singh, 1990; Sharma, 2003; Sharma & Rana, 2005; Sharma *et al.*, 2000; Sikarwar, 1996; Singh & Prakash, 1994; Srivastava *et al.*, 2003b; Thomas & Britto, 2003; Tirkey *et al.*, 2003b; Verma & Chauhan, 2006; Watt, 1972).

Medicago polymorpha Roxb. (Pl. 34A)
Syn.: *M. denticulata* Willd.; *M. canescens* Graham
Family: Fabaceae.

Vern. : Khukhani.

Regional Name:

P. : Maina.

Distribution : Tropical Regions.

Description : Robust herbs. Leaflets toothed. Stipules laciniated. Flowers yellow, peduncles short. Pods spirally-twisted, covered with hooked spines, indehiscent.

Reproductive Cycle : February-April.

Habitat Ecology : Barren Fields. Edges of Cultivated Areas. Moist Places.

Material Examined : HPU-EL-SKS-5685; Arla (1,100 m), Chachian (1,300 m), Chornali (1,280 m), Saparu (1,250 m), Thakurdwara (1,150 m).

Parts Used : Aerial plant parts.

Folk Uses : Aerial plant parts used as **pot-herb** and **fodder**.

Uses in Literature : Known so far as a pot herb, nutritive, cattle-fodder and vegetable (Ambasta, 1986; Arora & Pandey, 1996; Kayal *et al.*, 2003; Roy *et al.*, 1998; Singh & Singh, 1981; Watt, 1972).

***Melia azedarach* L.** (Pl. 34B)
Syn.: *M. sempervirens* Sw.; *M. bukayan* Royle
Family: Meliaceae.

Vern. : Drek.

English, Sanskrit & Regional Names:

Eng. : Barbados Lilac, Bead Tree, Indian Lilac, Persian Lilac

Sans. : Brihan Nimba, Mahadroksha, Vishamushtika

Ass. : Thamage

B. : Ghornim, Mahanim

Bo. : Bakayan, Vilayatinim

	G.	: Bakan limbodo
	H.	: Bakayan, Deikna, Mahanimb
	Kan.	: Arevebu, Hutchu Bevu
	Mal.	: Malaveppu
	Mar.	: Bakananimb, Limbara
	P.	: Bakain, Chein, Kacheu
	Tam.	: Malaivembu, Sigarinimbam
	Tel.	: Turukavepa, Vet- tivapa.

Distribution : Throughout India: 600-2,700 m.

Description : A large-sized tree having bi- or tripinnate axillary leaves. Flowers lilac, honey scented, in white panicles. Drupes yellow, wrinkled, subglobose, 1-seeded.

Reproductive Cycle : May-October.

Habitat Ecology : Grassy slopes. Wastelands.

Material Examined : HPU-EL-SKS-5404; Balah (900 m), Bhauda (1,050 m), Chhanchri (930 m), Jasun (1,000 m), Ladoh (980 m).

Parts Used : Leaves. Flowers. Seeds. Wood.

Folk Uses : Poultice of flower good for **headache**. Paste of roasted seeds in mustard oil applied against **baldness**. Leaves used as bedding to check **jaundice**. Wood good for **agricultural implements**, **tool handles**, **arms of bed**, and **building works**.

Active Constituents : Kaemferol-3-L-rhamno-D-glucoside, and rutin isolated from *leaves*.

Biological Activity : Azedarachin A and 12-O-acetylazedarachins A and B exhibit antifeedant activity against larvae of *Spodoptera exigua*.

Uses in Literature : Used earlier as an alexipharmic, anthelmintic (vet.), antilithic, antiseptic, astringent, blood purifier, brain tonic, cathartic, deobstruent, diuretic, diuretic (vet.), edible, emmenagogue, febrifuge, fodder, fuelwood, laxative, psychoactive plant, resolvent, stomachic, symbolic abode of god, tonic, vermifuge, vermifuge (vet.); and for bile influences in cattle, cold, curry, dilatation of the pupil, dyeing, enlarged lymph nodes and glands, fever (vet.), furniture, gout, hair fall, headache, hypertension, hysteria, leprosy, making idols, malarial fever, nausea, necklaces, nervous headache, paintings (Mithila), preserving dried fruit and clothing from insects, promoting conception, rheumatism, rosaries, scrofula, skin diseases, small pox (vet.), spasm, spleen enlargement, sprain (vet.), swelling (vet.), tuberculosis, tumours, ulcer (vet.), wood, and wounds (Ambasta, 1986; Bhatt *et al.*, 2003a; Borthakur & Sarma, 1996; Chhetri, 1994; Drury, 1978; Dwivedi, 2003; Goud & Pullaiah, 1996b; Jain, 1991; Jain *et al.*, 1994; Jha & Basak, 1994; Jha *et al.*, 1991; Joshi &

Joshi, 2005; Kapur & Singh, 1996; Kurian, 1995; Lalramnghinglova, 1996; Nadanakunjidam & Abirai, 2003; Pearson & Brown, 1932; Rana *et al.*, 1996; Rawat & Chowdhury, 1998; Shah, 2005; Sharma, 1996, 2003; Siwakoti & Varma, 1996; Srivastava *et al.*, 2003a; Watt, 1972).

***Mentha longifolia* (L.) Huds.** (Pl. 34C)

Syn.: *M. spicata* L. var. *longifolia* L.; *M. sylvestris* L.; *M. royleana* Wall. *ex* Benth.; *M. longifolia* var. *royleana* Rech. f.; *M. longifolia* var. *royleana* (Benth.) Raiz. & Sax.

Family: Lamiaceae.

Vern. : Jungli pudina.

English, Sanskrit & Regional Names:

Eng. : Horse mint

Sans. : Ajirnahara, Pudina, Rochani, Ruchishya,Shakashobana, Sugandhipatra, Vantihara, Vyanjana

Bo. : Pudina, Vartalau

H. : Podina

P. : Koshu, Pudnakushna, Vien, Yura

Urdu : Pudina.

Distribution : Temperate W. Himalayas.

Description : Hoary-tomentose, aromatic perennial herbs. Stem robust. Whorls in terminal spikes. Bracts lanceolate. Nutlets reticulate.

Reproductive Cycle : November-February.

Habitat Ecology : Water sources. Rivulets. Waste places.

Material Examined : HPU-EL-SKS-5507; Darugnu (1,400 m), Gopalpur (1,350 m), Jiun (1,550 m), Purani-palam (1,100 m), Thandol (950 m).

Part Used : Leaves.

Folk Uses : Leaves used against **digestive** disorders; also employed for making '**chutney**'. Its juice with lemon allaying **vomiting**.

Active Constituents : *Essential oil* possesses diospherol, piperitenone oxide, diosphenolene, piperitone, piperitenone, limonene and cineol.

Uses in Literature : Recorded earlier as an anthelmintic, antiseptic, astringent, carminative, cooling medicine, digestive, stimulant; and for bronchitis, diarrhoea, dysentery, fever, flavouring, headaches, heart diseases, heat apoplexy, killing maggots, liver and womb pain rheumatism, stomach disorders, and vomiting (Agarwal, 2003; Ambasta, 1986; Chauhan, 1999; Chopra *et al.*, 1956; D'Rozario *et al.*, 1999; Dwivedi, 2003; Gaur *et al.*, 1983; Jain, 1968; Kapur & Singh, 1996; Kirtikar & Basu, 1984; Koelz, 1979; Samant *et al.*, 2001; Shah, 2005; Sharma & Rana, 2005; Singh & Kumar, 2000b; Singh *et al.*, 1996a; Sood & Thakur, 2004; Uphof, 1968; Watt, 1972).

***Mentha piperita* L.** (Pl. 34 D)

Family: Lamiaceae.

Vern. : Bhura-pudina, Pipermint.

English & Regional Names:

Eng. : Brandy Mint, Peppermint

H. : Gamathi Phudina, Paparaminta

P. : Vilayeti Podina.

Distribution : Cultivated or as an escape throughout India.

Description : An erect fragrant perennial herb. Leaves petioled, coarsely serrate. Flower whorls in terminal spikes. Bracts minute. Nutlets dry.

Reproductive Cycle : July-September.

Habitat Ecology : Commonly near water channels, moist places

Material Examined : HPU-EL-SKS-5509; Bhagpur (1,600 m), Chauki (900 m), Gharthun (900 m), Kathena (1,290 m), Sungal (1,050 m).

Part Used : Leaves.

Folk Uses : Leaves relished as '**chutney**' and considered **stomachic**.

Active Constituents : Various constituents of *oil* are methyl acetate 20%, menthyl acetate 55.8%, menthol 9.1%, menthone and hydrocarbons 15%.

Uses in Literature : So far reported as a carminative, diaphoretic, source of pepper-mint oil, spasmolytic, stimulant, stomachic; and for breast cancer, burns, diarrhoea, digestive disorders, diphtheria, dysentery, eczema, flatulence, local pains, nausea, neuralgia, ringworm, sore throat, ulcer and vomiting (Agarwal, 2003; Ambasta, 1986; Chopra *et al.*, 1956; D'Rozario *et al.*, 1999; Duke *et al.*, 2002; Harborne & Baxter, 2001; Jain, 1991; Karnick, 1994; Kirtikar & Basu, 1984; Prajapati *et al.*, 2006; Sharma, 2003; Sharma & Rana, 2005; Singh *et al.*, 1996b; Sood & Thakur, 2004; Uphof, 1968; Watt, 1972).

Michelia champaca L. (Pl. 34E)

Syn.: *M. rufinervis* DC.; *M. doldsopa* Ham.; *M. aurantiaca* Wall.; *M. rheedi* Wight

Family: Magnoliaceae.

Vern. : Chamba, Champa.

English, Sanskrit & Regional Names:

Eng. : Champak, Golden Champa, Yellow Champa

Sans. : Anjana, Atigandhka, Bhramavatithi, Bhringmohi, Chambunala, Champeya, Deepapushpa, Gandhaphali, Hemapushpa, Hemavhaya, Kamabana, Katu, Kumara, Kusuma, Kusumadhivata, Nagapushpa, Patichampaka, Ugragandha, Vanadapika, Vavlabdha

Ass. : Phulchopa, Phulsopa, Titasoppa

B. : Champa, Champaka

Bo. : Champa, Chapha

G. : Champo, Pitochampo, Raechampo, Sonchampo

H. : Champ, Champa, Champac, Champaca

Kan. : Sampige, Kola-Sampige

Mal. : Champakam

Mar. : Kudchampa, Pivalachampa, Sonachampa

Ori. : Chompa, Chompoko

P. : Chamba, Chamoti, Chamuti

Tam. : Amariyam, Sambagam, Shampangi, Vandumarmalar

Tel. : Champakmu, Champeyamu, Gangaravi, Heman-gamu, Kanjanamu, Sampamgi.

Distribution : Planted in outer Himalayas upto 2000 m.

Description : Evergreen trees having alternate, simple leaves which are tapering at ends. Flowers creamish-yellow, scented. Etaerio of follicles attached on cone-like receptacle. Seeds large.

Reproductive Cycle : April-October.

Habitat Ecology : Planted around houses in villages and temples.

Material Examined : HPU-EL-SKS-5609; Chimbalhar (1,150 m), Gopalpur (1,350 m), Lahla (930 m), Purani-palam (1,100 m), Purani-palam (1,100 m).

Parts Used : Whole Plant. Fruits. Seeds.

Folk Uses : Plant grown for its **aesthetic** value. Poultice of fruits and seeds applied for healing **cracked skin**. Powdered pods form one of the constituents of 'batna' (**face pack**). Seed oil considered good against **rheumatism** and **headache**.

Biological Activity : Magnosprengerine and salicifoline possess muscle relaxant and hypotensive activity.

Uses in Literature : Known earlier as an abortifacient, antiperiodic, antispasmodic, astringent, carminative, colic, diuretic, emmenagogue, expectorant, febrifuge, edible (fruit), oil seeds, purgative, stimulant, stomachic, symbolic (sacred in offering), tonic, yellow dye, vermifuge; and for asthma, boils, bronchitis, canoes, carbuncle, carriage work, cephalalgia, cholera, cold, colouring garments, cough, delirium, door panels, dysentery, dyspepsia, furniture, gonorrhoea, gout, headache, house building, inflammation, infusion, mania, menorrhagia, mucus in urine, native drums, nausea, opthalmia, perfumery and attar preparations, planking, pthisis, removing lice, renal diseases, sculptures, sores, spermatorrhoea, timber, ulcer (Ambasta, 1986; Banerjee, 1996; Chatterjee & Pakrashi, 1991; Chhetri, 2005; Jain, 1991; Khanna *et al.*, 1996b; Kumar & Rao, 2003; Lalramnghinglova, 1996; Nath & Maiti, 2003; Nayak *et al.*, 2003; Padhy & Dash, 2003; Ranjan, 2003; Rao, 1996; Rawat & Chowdhary, 1998; Saini, 1996a; Sharma, 2003; Vihari, 1995; Watt, 1972).

Micromeria biflora (Buch.- Ham.) Benth. (Pl. 34F)

Syn.: *M. ovata* Beck; *Thymus biflorus* Buch. - Ham.

Family: Lamiaceae.

Vern. : Janglee ajwain.

English Name: : Indian Wild Thyme.

Distribution : More or less throughout Temperate and Tropical Himalayas.

Description : Herbs with short stems and small, sessile leaves. Whorls axillary, flowers unisexual, small, pink. Calyx 13-nerved. Nutlets smooth.

Reproductive Cycle : January-July.

Habitat Ecology : In crevices of rocks.

Material Examined : HPU-EL-SKS-5613; Bhauda (1,050 m), Dadh (1,100 m), Gopalpur (1,350 m), Sungal (1,050 m), Tikkar (910 m).

Part Used : Leaves.

Folk Uses : Leaves chewed as **stomachic**, and also considered good for **cough** and **cold**.

Uses in Literature : Regarded to be used earlier as flavouring, stomachic, carminative; urinary disorders and for cough, cold, fever, infested wounds of cattle and postnatal care (Abraham, 1981; Ambasta, 1986; Lal *et al.*, 1996; Nautiyal *et al.*, 2003; Sharma & Rana, 2005; Singh & Kumar, 2000a; Singh *et al.*, 1980, 2001; Sood & Thakur, 2004).

Milletia extensa Benth. *ex* Baker (Pl. 35A)

Syn.: *M. auriculata* Baker *ex* Brandis; *Robinia macrophylla* Roxb.; *Pongamia macrophylla* Graham; *Otostema macrophylla* Benth.; *Amerimnum pellidum* Ham.

Family: Fabaceae.

Vern. : Salod.

Regional Names:

- **H.** : Ganj, Gauj
- **Mal.** : Vallimurutali
- **Ori.** : Arkawla, Kissi
- **Tel.** : Kondatangedutige.

Distribution : The Himalayas.

Description : A woody climber. Leaflets glabrescent above and pale beneath. Corolla dull-white. Pods straight, very hard, hairy. Seeds flat.

Reproductive Cycle : July-September.

Habitat Ecology : Open Slopes. Scrub jungles. Ravines.

Material Examined : HPU-EL-SKS-5711; Bhaura (940 m), Khaira (940 m), Malahu (1,050 m), Panapari (1,100 m), Tikkar (910 m).

Parts Used : Leaves. Stem.

Folk Uses : Leaves lopped for **fodder**. Rope from its stem used as stand ('**tarede**') for earthen pots.

Uses in Literature : Reported to be used as a fermenting agent, fibre yielding, fish poison, fodder (vet.), insect repellent, piscicide, symbolic (magic cure for fever); and for cattle's sores, collage (vet.), killing bed bugs (vet.), spleen diseases, toothache, and washing wounds (Ambasta, 1986; Aminuddin & Girach, 1991; Diwakar & Ansari, 2003; Jain, 1991; Joshi & Joshi, 2005; Purohit & Samant, 1995; Rao, 1996; Sikarwar, 1996; Siwakoti & Varma 1996).

Mirabilis jalapa L. (Pl. 35B)
Family: Nyctaginaceae.

Vern. : Gulbansi.

English, Sanskrit & Regional Names:

Eng. : Four O'Clock Flower, Marvel of Peru
Sans. : Krishnakeli, Sandhyakali
B. : Gulabas, Krishnakeli
Bo. : Gubhaji
H. : Gulabbas
Mal. : Antimalari
P. : Abasi, Gulabbas
Tam. : Antinarulu, Pattarachi
Tel. : Batharachi, Chandrakanta
Urdu : Guleabbas.

Distribution : Spread throughout India, often planted.

Description : Leafy perennials upto 75 cm high. Stem jointed at nodes. Leaves cordate. Deep-green-flowers tubular, pentamerous, fragrant, opening in evening. Achenes elliptical, leathery, black achene.

Reproductive Cycle : June-September.

Habitat Ecology : Moist localities. Waste places near habitations.

Material Examined : HPU-EL-SKS-5476; Bhagpur (1,600 m), Jiun (1,550 m), Nadli (1,000 m), Rajeher (1,050 m), Tapa (945 m).

Parts Used : Leaves. Flowers. Root tubers.

Folk Uses : Flowers as an **offering** to various deities. Leaf-paste applied against **boils**. Root tubers taken as **purgative** in the form of pickle.

Active Constituents : Active constituents isolated from *flowers* are miraxanthins I, II, III and IV, indicaxanthin and vulgaxanthin.

Biological Activity : Whole plant spasmolytic.

Uses in Literature : Recorded earlier as an aphrodisiac, antipyretic, diuretic, edible (tubers), purgative, sacred, spasmolytic, tonic, ornamental and vegetable; and for inflammation, boils, blisters, cold at childbirth, earache, enlarged lymph nodes, piles, phlegmons, scorpion bite, skin freckles, stomachache, sunstroke, and urticaria (Ambasta, 1986; Chatterjee & Pakrashi, 1991; Dastur, 1970; Dwarakan & Ansari, 1996; Gogoi & Borthakur, 1991; Henry *et al.*, 1996; Hosagoudar & Henry, 1996a; Islam, 1996; Jain, 1991; Kirtikar & Basu, 1984; Kurian, 1995; Mahato & Chaudhary, 2005; Molla & Roy, 1996;

Nadanakunjidam, 2003; Saklani & Jain, 1994; Samant *et al.*, 2001; Sanyal, 1994; Singh & Kumar, 2000a, b; Sood & Prakash, 2007; Sood & Thakur, 2004; Upadhye *et al.*, 1994).

***Momordica charantia* L. (Pl. 35C)**

Syn.: *M. humilis* Wall.; *M. muricata* DC.; *M. senegalensis* Lam.

Family: Cucurbitaceae.

Vern. : Karela.

English, Sanskrit & Regional Names:

Eng.	: Bitter Gourd
Sans.	: Brihadavalli, Karavellaka
Ass.	: Kakiral
B.	: Baramasiya, Potikakar
Bo.	: Karla
G.	: Karela, Karola
H.	: Karela, Karola
Kan.	: Hagal
Mal.	: Kaipavalli, Pavakacheti
Mar.	: Karale
Ori.	: Karena
P.	: Karela, Karila
Tam.	: Pavakkachedi
Tel.	: Mattakakata, Tellakakara.

Distribution : Throughout India.

Description : Climbers having orbicular, petioled, alternate leaves which are deeply-lobed, glabrous. Flowers yellow, solitary, unisexual. Berries spindle-shaped with warty swellings. Seeds compressed.

Reproductive Cycle : July-October.

Habitat Ecology : Cultivated.

Material Examined : HPU-EL-SKS-5508; Bodhal (970 m), Dehan (1,130 m), Gopalpur (1,350 m), Rajpur (1,100 m), Thandol (950 m).

Parts Used : Fruit. Leaves. Seeds.

Folk Uses : Fruit powder taken to check persistent **fever**. Unripe fruits consumed as **vegetable** and also **pickled**. Leaf and fruit juice considered **antidiabetic**, **cardio-tonic**; and also applied externally over anus for **piles** and **pin-worms** in kids. Seeds eaten to check **pulmonary** ailments.

Active Constituents : *Fruits* rich in momoridocosides G, F1, F2, I, K and L.

Biological Activity : Positive hypoglycaemic activity of fruits due to charautin-β-sitosterol-glucoside and stigmast-5,25-diene-3β-0-glucoside.

Uses in Literature : Recorded in India as an abortifacient, anthelmintic, astringent, blood purifier, colic, carminative, emetic, febrifuge, laxative, stomachic, tonic, vegetable; and for bilious affection, bleeding, blood cancer, burning of the soles of the feet, cooling, cutaneous diseases, diabetes, diseases of spleen and liver, dissipating melancholy and gross humours, dysentery, eczema, gout, headache, haemorrhoids, hypertension, jaundice, leprosy, malaria, night-blindness, oedema, peptic ulcer (vet.), poisonous bite (vet.), poultry diseases, psoriasis, rheumatic pains, scabies, scorpion sting, swollen breasts, snakebite, sores, water worms, wound maggots (Ambasta, 1986; Arinathan *et al.*, 2003; Bajpayee & Dixit, 1996; Ghosh *et al.*, 1996; Gill *et al.*, 1993; Jain, 1991; Jha *et al.*, 1996; Kothari & Moorthy, 1996; Kulkarni & Kumbhojkar,1993, 1996; Lalramnghinglova, 1996; Mahato & Mahato, 1996; Nadanakunjidam & Abirai, 2003; Negi & Gaur, 1991; Rama Rao & Henry, 1996; Sharma, 2003; Shome *et al.*, 1996; Singh & Pandey, 1998; Siwakoti & Varma, 1996; Srivastava *et al.*, 2003b; Subramani & Goraya, 2003; Thomas & Britto, 2003; Watt, 1972).

Momordica dioica Roxb.

Family: Cucurbitaceae.

Vern. : Meetha Karela.

English, Sanskrit & Regional Names:

Sans. : Avandhya, Deri, Diva, Ishwari
Ass. : Batkarila
Bo. : Karatola, Kurtoli
G. : Kantoli, Kuntola
H. : Kaksa, Gol Kandra
Kan. : Karlekai
Mal. : Erimapasel
Mar. : Kartoli
P. : Dharkarela, Kirara
Tam. : Paluppakkay
Tel. : Agakara, Potukandulu.

Distribution : Throughout India to Sri Lanka, ascending to 1,700 m.

Description : Perennials with tuberous roots, glabrous stem, simple, striate tedrils, and membranous leaves. Flowers yellow, with oblong-lanceolate petals. Fruit long, echinate with soft spines.

Reproductive Cycle : June-August.

Habitat Ecology : Edges of cultivated fields. Outskirts of forests.

Material Examined : HPU-EL-SKS-5712; Darugnu (1,400 m), Paror (940 m), Saliana (1,000 m), Sulah (950 m), Tikkar (910 m).

Part Used : Fruits.

Folk Use : Fruits cooked as **vegetable**.

Uses in Literature : Reported so far as anthelmintic, edible, laxative, stomachic; and for asthma, bowel complaints, bronchitis, diabetes, diptheria, fever, jaundice, skin diseases, snakebite, tumours, urinary complaints, vegetable, and piles. (Agarwal, 2003; Ambasta, 1986; Arora & Pandey, 1996; Bhatt *et al.*, 2003a, b; Bennet *et al.*, 1991; Borthakur, 1996b; Deshmukh & Rothe, 2003; Dwivedi, 2003; Jadeja, 1999; Kirtikar & Basu, 1984; Mishra *et al.*, 1996a; Mitaliya *et al.*, 2003; Murthy *et al.*, 2003; Nayak *et al.*, 2003; Painuli & Maheshwari, 1996; Radhakrishnan *et al.*, 1996; Rajendran & Henry, 1994; Sahu, 1996; Said, 1997; Sharma & Rana, 2005; Singh & Pandey, 1998; Srivastava *et al.*, 2003a).

Moringa oleifera Lam. (Pl. 35D)

Syn.: *M. polygana* DC.; *M. pterygosperma* Gaertn.; *M. zeylanica* Pers.; *Hyperanthera moringa* Vahl; *H. decandra* Willd.; *Guilandina moringa* L.

Family: Moringaceae.

Vern. : Sunanha.

English, Sanskrit & Regional Names:

Eng. : Drum Stick Tree, Indian Horse Raddish
Sans. : Damasamalu, Tikshnandhaka
Ass. : Saijna, Sohjna
B. : Sajina
Bo. : Mangai, Sanga
G. : Midhosaragaro, Seyla
H. : Mungna, Saonjna
Kan. : Nugge
Mal. : Muringa, Muringa Elai, Murinna
Mar. : Murinna, Tishagandha
Ori. : Achajhada, Managachajhada
P. : Saonjna
Tam. : Karunjanam, Tavuselam
Tel. : Mochakamu, Sajana.

Distribution : Throughout foothills of W. Himalayas upto 1,100 m.

Description : A medium-sized tree usually with 3-pinnate leaves. Leaflets opposite, pale beneath. Flowers pale-white, honey-scented. Pods pedunculous. Seeds winged at the angles.

Reproductive Cycle : February-April.

Habitat Ecology : Hilly slopes. Forests.

Material Examined : HPU-EL-SKS-5675; Balah (900 m), Bhauda (1,050 m), Boda (1,000 m), Jasun (1,000 m), Tikkar (910 m).

Parts Used : Inflorescence. Leaves. Gum. Pods. Wood.

Folk Uses : Flower buds and pods consumed as **vegetable** and '**chutney**', and considered **blood purifier**. Gum oozing out of stem taken in '**halwa**' (pudding) for **backache**, **post-delivery pains**, and **dysentery**. Leaves given as **fodder** for goats and sheep. Wood used as **fuel**.

Active Constituents : Properties attributed to aldotriuronic acid, arabinose, galactose, glumronic acid, rhamnose, mannose, xylose (*gum*), aspartic acid, glutamic acid, glycine, threonine, alanine, valine, leucine, isoleucine, histidine, lysine, arginine, phenyl alanine, tryptophan, cysteine, methimine (*leaves*), alanine, arginine, glycine, serine, threonine, valine, glutamic and aspartic acids (*flowers, fruits*), 4-hydroxymellein, vanillin, octacosanoic acid, β-sitosterol and β-sitosterone (*stem*), benzyl isothiocyanate (*seeds*), and four auxin-like factors isolated (*fast growing fruits*).

Biological Activity : 4(α-L-rhamnosyloxy) benzyl isothiocyanate exhibits anti-bacterial activity against *Mycobacterium phlei* (40 μM/1) and *Bacillus subtilis* (56 μM/I).

Uses in Literature : Described in India as an abortifacient, anthelmintic, antilithic, antipyretic, antirheumatic, antiscorbutic, aphrodisiac, appetiser, cardiac stimulant, carminative, cholagogue, cooking oil, diuretic, edible (fruit and pods), emetic, flatulent colic, illuminant, lubricants, psychoactive plant, rubefacient, stimulant, stomachic, symbolic (in folk proverbs), tonic, vesicant; and for acute traumic pain, anthrax (vet.), articular pains, asthma, baldness, blood pressure, boils (vet.), calculus affections, calico printing, catarrhal affections, cholera, circulating stimulants, cold, cosmetics, cough, cuts, debility of nerves, delivery, dental caries, dermatitis, dispersing swellings, dog bite, dropsy, dyeing, dysentery, dyspepsia (vet.), ear complaints, eczema, enlarged spleen or liver, epilepsy, expelling placenta, eye (vet.), eye aliments, fits (vet.), fodder, foot sores, glandular swellings, gout, gum boils, headache, hoarseness, hydrocele, hysteria, increasing flow of bile, injuries, intermittent fevers, internal and deep seated inflammation, leprosy, madness, mats, menstrual complaints, otalgia, paper or cordage, paralysis, patches, perfumeries, pneumonia, postnatal, pustules, reducing labour pains in pregnant women, rheumatism (vet.), scabies, scrofula, scurvy, smallpox, soreness of mouth and throat, swelling (vet.), syphilis, tetanus, toothache, typhoid, ulcer (vet.), urinary complaints, venomous bites, and wounds (Ambasta, 1986; Banerjee, 1996; Bhatt *et al.*, 2003b; Chatterjee & Pakrashi, 1991; Chauhan, 1999; Geetha *et al.*, 1996b; Hembrom, 1996; Henry *et al.*, 1996; Hosagoudar & Henry, 1996b; Jain, 1991; Jain *et al.*, 1994; Jha & Kumar, 2003; Jha *et al.*, 1996; Molla & Roy, 1996; Mukherjee & Namhata, 1990; Nadanakunjidam & Abirai, 2003; Negi & Gaur, 1991; Padhy & Dash, 2003; Prasad *et al.*, 1996; Rajendran *et al.*, 2003; Rama Rao & Henry, 1996; Shah, 2005; Saini, 1996b; Sharma, 2003; Shome *et al.*, 1996; Singh, 1996b; Siwakoti & Varma, 1996; Sood & Kaushal, 2008; Srivastava *et al.*, 2003a; Thomas & Britto, 2003; Upadhye & Kulkarni, 2004; Vishwanathan, 1997; Viswanathan & Singh, 1996; Watt, 1972).

***Morus alba* L.** (Pl. 35E)

Syn.: *M. serrata* Wall.; *M. tatarica* L.

Family: Urticaceae.

Vern. : Toot.

English, Sanskrit & Regional Names:

Eng. : Mulberry or White Mulberry
Sans. : Tula
B. : Tut
Bo. : Chinni
G. : Shetur
H. : Chun
Kan. : Hipnerle
Kash. : Tut
Mar. : Ambat, Tut
Ori. : Tuticoli
P. : Tut, Tutri
Tam. : Pattuppuchi
Tel. : Pippalipand Chettu, Reshmechattu
U.P. : Shahtoot.

Distribution : Drier areas of W. Himalayas upto 3,300 m (cultivated).

Description : Trees with lanceolate, toothed leaves; flowers inconspicuous greyish-white in male and female catkins. Fruit sorosis, black when ripe.

Reproductive Cycle : March-May.

Habitat Ecology : Common along agricultural fields.

Material Examined : HPU-EL-SKS-5612; Dharaman (1,100 m), Gharthun (900 m), Holta (1,200 m), Rajeher (1,050 m), Tikkar (910 m).

Parts Used : Leaves. Fruits. Wood. Inflorescence.

Folk Uses : Fruits **edible** and considered **refrigerant**. Tender leaves cooked as '**pakodas**'. Leaves as an excellent **fodder** for rearing silk worms and livestock. Wood employed for making handles of **agricultural implements**.

Active Constituents : Active ingredients are heptanoic (0.047), caprylic (0.21), pelargonic (0.047), capric (0.07), myristic (0.105), palmitic (11.85), palmitoleic (0.188), methyl heptadecanoic (0.14), stearic (4,46), oleic (12.67), linoleic (68.3), nonodecanoic (0.641), linilenic (0.98%), acids (*fruits*), two phytoalexins-moracin C and moracin D (*shoot*), and astragalin (*leaves*).

Biological Activity : Mulberrifuran A exhibits antibacterial activity against *Staphylococcus aureus* and *Fusarium roseum* whereas Moran A exhibits hypoglycaemic effect in normal and alloxan induced hyperglycaemic mice.

Uses in Literature : Known earlier as an anthelmintic, diaphoretic, edible (fruits), expectorant, fever, fodder, purgative, refrigerant, sore throat, vermifuge; and for agricultural implements, boats, buildings, dyspepsia, feeding silk worms, fever, furniture, headache, inflammed eyes, melancholia, paper, silk worm and sore throat (Agarwal, 2003; Ambasta, 1986; Arora & Pandey, 1996; Bennet *et al.*, 1991; Chevallier, 1996; Das & Agarwal, 1991; Dey & Bahadur, 1973; Harborne & Baxter, 2001; Jain, 1991; Kirtikar & Basu, 1984; Kshirsagar & Singh, 2000; Negi & Gaur, 1991; Parrotta, 2001; Prajapati *et al.*, 2006; Roy *et al.*, 1998; Saini, 1996b; Samant *et al.*, 2001; Sharma & Rana, 2005; Singh & Kumar, 2000a; Sood & Thakur, 2004; Watt, 1972).

Mucuna pruriens (L.) DC. (Pl. 35F)

Syn.: *M. prurita* Hook.; *M. utilis* Wall.; *Caryopogon pruriens* Roxb.; *Dolichos pruriens* L.

Family: Fabaceae.

Vern. : Gazal bel.

English, Sanskrit & Regional Names:

Eng. : Cowhage, Cowitch

Sans. : Durabhigraha, Galrabhange

B. : Akolshi, Bichchoti

Bo. : Kuhili
G. : Kanchan, Kivanch
H. : Goncha, Kivachh, Kaunch
Kan. : Nasukunni, Hasagunigida
Mal. : Naykkurma
Mar. : Kuhili
Ori. : Alokushi
P. : Gunchagaji, Kanaucha
Tam. : Attumabuttar, Punaiyavarai
Tel. : Pravrishayani, Telladuradagondi.

Distribution : Throughout India.

Description : Climbers with membranous leaflets. Corolla purplish. Pods hairy, pale brown and turgid.

Reproductive Cycle : July-September.

Habitat Ecology : On trees. Ravines. Slopes.

Material Examined : HPU-EL-SKS-5414; Bhuana (1,400 m), Nadli (1,000 m), Rangru (910 m), Thandol (950 m), Tapa (945 m).

Parts Used : Seeds. Fruits.

Folk Uses : Roasted seeds (10-15) taken fortnightly to check **impotency** in males. Its flour used for making '**halva**' (pudding) and prescribed to check **post-delivery stress**. Stinging fruits used for **repelling rats**.

Active Constituents : *Seeds* contain mucadine, mucunadinine, mucuadininene, prurienidine mucunin, mucunadine, nicotine, L-dopa, glutarthiane lecithin, gallic acid, a glucoside and aminoacids.

Biological Activity : Plant extr. exhibits marked hypotensive and hypocholestrolaemic effects in normal rats.

Uses in Literature : Known earlier as an anthelmintic, aphrodisiac, appetizer, asthma, astringent, diuretic (vet.), edible (pods, leaves, roasted seeds), emmenagogue, lactagogue (vet.), local stimulant, mild vesicant, nervine tonic, psychoactive plant, purgative, stimulant, vegetable, vermifuge (vet.); and for anasarca, blisters (vet.), boils (vet.), bone fractures, cancer, cough, digestion, dropsy, dysentery, elephantiasis, eye diseases, facial paralysis, fertility, fever, general debility, hemiplegia or paralysis on one side of the body and hallucinations accompanied with high fever, impotency, intestinal (vet.), leucorrhoea, madness, pain, nervous disorders (vet.), night dreams, regulating menstruation, renal affections, ringworm, scabies (vet.), scorpion and honey bee bites, seminal weakness, snakebite, sores, spermatorrhoea, syphilis, ulcer (vet.), urinary complaints, worm (vet.), and wounds (vet.) (Ambasta, 1986; Aminuddin & Girach, 1991; Bhandary & Chandrashekar, 2002; Borthakur &

Sarma, 1996; Chandra, 1997; Das & Agarwal, 1991; Dey & Bahadur, 1973; Diwakar & Ansari, 2003; Dwivedi, 2003; Gill & Nyawuame, 1994; Girach *et al.*, 1997; Gopal & Chandra, 2003; Jain, 1991; Jain & Patole, 2001; Jha & Varma, 1996; Jha *et al.*, 1991; Khare, 2004; Kurian, 1995; Manandhar, 2003; Murthy *et al.*, 2003; Nadanakunjidam & Abirai, 2003; Nayak *et al.*, 2003; Pal *et al.*, 2000; Parrotta, 2001; Prasad *et al.*, 2003; Radhakrishnan *et al.*, 1996; Rama Rao & Henry, 1996; Rana *et al.*, 1996; Sahu, 1996; Saklani & Jain, 1994; Samant *et al.*, 2001; Sanyal, 1994; Sarangi & Sahu, 2005; Satapathy & Brahmam, 1996; Sharma, 2003; Singh, 1991, 1999, 2000, 2001, 2003; Singh & Pandey, 1998; Singh & Zaheer, 1996; Sinha, 1996; Sinha & Sinha, 2001; Siwakoti & Siwakoti, 2000; Siwakoti & Varma, 1996; Srivastava *et al.*, 2003a, b; Sudhakar & Vedavathy, 2000; Watt, 1972).

Murraya koenigii **Spreng.** (Pl. 36A)
Syn.: *Bergera koenigii* L.
Family: Rutaceae.

Vern. : Ghandhla, Kadi-Patta.

English, Sanskrit & Regional Names:

Eng. : Curry Leaf Tree
Sans. : Alakavhaya, Krishnanimba, Surabhinimba
Ass. : Bishahari, Narasingha
B. : Barsunga, Kariaphyli
Bo. : Goranimb, Karrinim
G. : Kadhilimbdo
H. : Barsunga, Gandhela, Katnim
Kan. : Karibevu
Mal. : Kariveppilei
Mar. : Jhirang, Karepacha, Kudianim
Ori. : Basango, Sanomorosinghi
P. : Bowala, Gandanim
Tam. : Karuvembu, Karuvepallai
Tel. : Karepeka, Karivepaku.

Distribution : Throughout India, ascending to 1,650 m.

Description : Shrubs with gland dotted leaves which are petiolate, imparipinnate crowded towards the end of branches. Leaflets 5-7. Flowers white, pedicellate, fragrant, pentamerous, in cymes. Berries ovoid, black. Testa hairy.

Reproductive Cycle : March-June.

Habitat Ecology : Often planted.

Material Examined : HPU-EL-SKS-5628; Boda (1,000 m), Khaira (940 m), Makol (900 m), Nanaun (900 m), Tikkar (910 m).

Parts Used : Leaves. Stem. Fruit.

Folk Uses : Ripe fruits **edible**, considered **blood purifier**. Stem used for **scouring teeth** and as **fuelwood**. Tender leaves and inflorescence along with condiments relished as '**chutney**'. Also leaves used as **condiment** for flavouring curries, and poultice of its paste good against **sprain**. Water boiled with its leaves prescribed for bath against **post-delivery** and **rheumatic** pains.

Active Constituents : These are mahanimbine, koenimbine, koenigicine (*fruits, leaves*), girinimbine, muconicine, cyclo mahanimbine and mahanimibidine (*leaves*).

Biological Activity : Koenoline exhibits cytotoxic activity against ICB cell culture. Roots antiamoebic, and hypotensive whereas aerial parts spasmolytic.

Uses in Literature : Used in India as an anthelmintic, carminative, edible (vegetable), febrifuge, flavouring, laxative, stimulant for eruptions, stomachic, tonic; and for acute abdominal pain, agricultural implements, bites of poisonous animals, bodyache, bleeding, bone fractures, burning pains, chronic liver disorders associated with alcoholism and dyspepsia, culinary purposes, cuts, diarrhoea, dropsy, dysentery, fertility (vet.), flavouring curries (leaves), grooms as ornamental, herpes, hysyteria, joint pains, malarial fever, postpartum, purifying blood, renal pains, rinder pest (vet.), scurvy, skin eruptions, kidney problems, snakebite, swellings in various tract of the body, toning stomach, tuberculosis, vomiting, and worms (Ambasta, 1986; Aminuddin & Girach, 1996; Arora & Pandey, 1996; Borthakur, 1996b; Chandra, 1997; Chauhan, 1999; Chevallier, 1996; Chopra *et al.*, 1956; Das & Choudhary, 2003; Gill *et al.*, 1997; Gogoi & Borthakur, 1991; Gogoi *et al.*, 2003a, b; Jain, 1991; Kirtikar & Basu, 1984; Kurian, 1995; Lal *et al.*, 1996; Mahato & Chaudhary, 2005; Manandhar, 1995; Mehtani & Ingle, 1999; Mohanty & Padhy, 1996; Nayak *et al.*, 2003; Parrotta, 2001; Parmar & Kaushal, 1982; Rama Rao & Henry, 1996; Ranjan, 2000, 2003; Roy *et al.*, 1998; Saklani & Jain, 1994; Samant *et al.*, 2001; Sharma, 2003; Singh, 1996b; Singh & Kumar, 2000a; Siwakoti & Varma, 1996; Sood & Thakur, 2004; Srivastava *et al.*, 2003a; Verma & Chauhan, 2006; Watt, 1972).

***Murraya paniculata* (L.) Jack.** (Pl. 36B)
Syn.: *Chalcas paniculata* L.;
C. sumatrana Roem.; *Murraya exotica* L.; *M. sumatrana* Roxb.
Family: Rutaceae.

Vern. : Gandhala.

Sanskrit & Regional Names:

Sans. : Surabhininiba
Ass. : Kamini
B. : Kamini
Bo. : Chulajuli, Manchulajuti
H. : Kamini, Marchula
Kan. : Pandry
Mar. : Kunti, Pandari
Ori. : Barsan, Bhursunga, Basango, Pandari
Tam. : Keriveppilai, Konji, Panjandale
Tel. : Nagagolunga, Simaikkonji.

Distribution : Almost throughout India.

Description : A small evergreen tree with smooth, yellowish-white bark. Leaves alternate, imparipinnate. Leaflets 3-7, entire, glabrous. Flowers fragrant, campanulate, solitary or in terminal and axillary corymbs; berries oblong or ovoid, red on ripening.

Reproductive Cycle : March-June.

Habitat Ecology : Wastelands. Barren areas.

Material Examined : HPU-EL-SKS-5699; Bhatu (1,040 m), Bodhal (970 m), Bhuana (1,400 m), Chhanchri (930 m), Molichak (1,020 m).

Parts Used : Leaves. Stem. Fruits.

Folk Uses : Ripe fruits **edible**. Stem used for **cleaning** teeth. Leaves consumed as '**chutney**', and for **flavouring** curries. These are also employed as **bedding** for cattle during rainy season.

Active Constituents : Mexoticin (*stem bark*); coumarin, coumarrayin and 3, 3′, 4′, 5, 5,′ 6, 7-heptamethoxyflavone, semi-β-carotenone, β-carotenone (*fruits*).

Biological Activity : Keonoline exhibited cytotoxic activity.

Uses in Literature : So far used in India as astringent, edible, refrigerant, stomachic, tonic, stimulant; and for agricultural implements, bodyache, bone fractures, cough, cuts, dropsy, dysentery, diarrhoea, hysteria, flavouring, joint pains, perfume, rheumatism, and snakebite (Ambasta, 1986; Aminuddin &

Girach, 1996; Beg *et al.*, 2005; Bennet *et al.*, 1991; Brahmam & Saxena, 1990; Brahmam *et al.*, 1996; Chopra *et al.*, 1956; Kirtikar & Basu, 1984; Lalramnghinglova, 1996; Manandhar, 1996; Parrotta, 2001; Ranjan, 2003; Saklani & Jain, 1994; Samant *et al.*, 2001; Sharma, 2003; Singh & Kumar, 2000a; Verma *et al.*, 1995).

***Musa sapientum* L.** (Pl. 36C)

Syn.: *M. paradisiaca* L.

Family: Musaceae.

Vern. : Kela.

English, Sanskrit and Regional Names:

Eng. : The Banana, Plantain

Sans. : Kadli, Rambha

B. : Kach, Kala, Kula

Bo. : Kel, Kela

G. : Kela

H. : Kach-Kula , Kela, Maoz-kulo

Kan. : Bale, Balenaru

Mal. : Kadali, Kel, Vazha

Mar. : Kadali, Kel, Vazhai

P. : Kela, Khela, Muz

Tam. : Valei, Vazhaip-pazham, Vazhai

Tel. : Amti, Anali, Anati, Arati, Ariti, Bonta-ariti Chakrakeli-ariti, Nalla-arii.

Distribution : Bihar, E. Himalayas, Sri Lanka (cultivated throughout India and the Tropics).

Description : Pseudostem upto 4.5 m. Leaves oblong, long; spike decurved, drooping. Lower bracts with female flowers, above these are male flowers. Fruits oblong, 3-gonous. Seeds numerous, black or brownish black.

Reproductive Cycle : Throughout the year.

Habitat Ecology : Humus rich soil. Shady places.

Material Examined : HPU-EL-SKS-5452; Gharthun (900 m), Nanaun (900 m), Panchrukhi (990 m), Saparu (1,250 m), Thandol (950 m).

Parts Used : Whole Plant. Bracts. Leaves. Fruits. Stem.

Folk Uses : Plant considered **sacred** and considered equivalent to four '**Vedas**', also worshipped. Bract juice good against **mouth blisters**. Fried chips ('kattalu') of unripe fruit relished as **vegetable**, and prescribed for **measals** and also its pericarp employed for making '**chutney**'. Leaves also used for making **mats**. Leaves and stem given as **fodder** for milching cattle and considered **refrigerant**.

Active Constituents : Constituents isolated are dopamine, dopa-noradrenalin, serotonina and caffeic, cinnamic, p-coumaric, ferulic, gallic and protocatechuic acids, campesterol, β-sitosterol (*flowers*) 3 – methyl pentanoate, acetate, butanoate, 3 – methyl butanoate and pentanoate esters isolated (*fruits*).

Uses in Literature : Useful earlier as abortifacient, anthelmintic, antiscorbutic, astringent, cattle feed, coolant, demulcent, diuretic, purgative, tonic, vegetable; and for burns, biliousness, blood diseases, burning sensation, bronchitis, diabetes, diarrhoea, dysentery, dyspepsia, epilepsy, fever, gonorrhoea, hypertension, hysteria, kidney troubles, leprosy, leucorrhoea, menstrual disorders, scabies, sore throat, and for vaginal discharges (Agarwal, 2003; Ambasta, 1986; Bhatt *et al.*, 2001; Bose, 1984; Chakrabarty & Balakrishnan, 2003; Chopra *et al.*, 1956; Chowdhury *et al.*, 1996; Dagar & Dagar, 1996; Gill *et al.*, 1993, 1997; Hembrom, 1996; Jha & Basak, 1994; Khanna, 2002; Kirtikar & Basu, 1984; Kurian, 1995; Lalramnghinglova, 1996; Mohanty & Padhy, 1996; Nadanakunjidam & Abirai, 2003; Padhy & Dash, 2003; Pandey & Joshi, 2001; Prajapati *et al.*, 2006; Rajendran *et al.*, 2003; Rawat & Chowdhury, 1998; Saikia & Nath, 2003; Saini, 1996a; Saklani & Jain, 1994; Sen & Behera, 2003; Sen *et al.*, 2002; Sharma *et al.*, 2000; Singh, 2000; Singh & Pandey, 1998; Singh & Singh, 2003; Siwakoti & Siwakoti, 2000; Subramani & Goraya, 2003; Thomas & Britto, 2003; Upadhye & Kulkarni, 2004; Verma *et al.*, 1995; Watt, 1972).

Neolitsea umbrosa (Nees) Gamble (Pl. 36D)

Syn.: *Litsea umbrosa* Nees var. *consimilis* (Nees) Hook. f.; *L. consimilis* Nees; *Litsea umbrosa* (Nees) Nees; *Tetradenia pallens* Don; *T. umbrosa* Nees; *T. pallens* Don

Family: Lauraceae.

Vern. : Gulab-jamun.

Regional Names:

Kash. : Chirindi

P. : Chindi, Chirudi.

Distribution : Temperate & Sub-tropical Himalayas: 1,500-2,700 m.

Description : Glabrous trees having lower surface of leaves covered with white bloom. Flowers pale-yellow. Fruits rounded with a thin pulpy pericarp, black when fully dried.

Reproductive Cycle : March-May.

Habitat Ecology : Forests.

Material Examined : HPU-EL-SKS-5605; Bindraban (1,270 m), Chornali (1,280 m), Gopalpur (1,350 m), Molichak (1,020 m), Thakurdwara (1,150 m).

Parts Used : Fruits. Wood.

Folk Uses : Fruits **edible**. Wood employed for making cheap **furniture**.

Uses in Literature : Used earlier as edible, illuminant, and for house construction, itching and skin affections (Ambasta, 1986; Jain, 1991; Kapur, 1996a; Watt, 1972).

***Nerium indicum* Mill.** (Pl. 36E)

Syn.: *N. odorum* Sol.; *N. odoratum* Lam.; *N. latifolium* Mill.

Family: Apocynaceae.

Vern. : Ghaneera.

English, Sanskrit & Regional Names:

Eng. : Indian Oleander, Sweet Scented Oleander
Sans. : Asvamaraka, Pratihasa, Vishavrykshanka
Ass. : Karavi, Karbira, Karvir
B. : Karabi
Bo. : Ganira, Kanhera
G. : Kanera
H. : Kaner, Karber, Kuruvira
Kan. : Kanagalu
Kash. : Gandeela, Gandula
M.P. : Arali, Kanave Eram, Kattalari
Mal. : Chuvannarali, Satraprasam
Mar. : Kaneri
Ori. : Konera, Koniyoro, Krobyo
P. : Ganhira, Kanhira
Tam. : Asuvabari, Irattaichegappayalari, Vellalari
Tel. : Ettaganneru, Kasturipatha.

Distribution : N. W. Himalayas: 600-1,500 m.

Description : Perennial shrubs with long, pointed, leathery leaves. Flowers pink, fragrant, in terminal panicled cymes. Follicles paired, cylindrical. Seeds linear, hairy.

Reproductive Cycle : April-November.

Habitat Ecology : Foothills of W. Himalayas.

Material Examined : HPU-EL-SKS-5549; Bhawarna (1,100 m), Dadh (1,100 m), Ghugar (1,250 m), Saparu (1,250 m), Thakurdwara (1,150 m).

Parts Used : Whole Plant. Stem-bark.

Folk Uses : Plant grown as **ornamental** tree. Cow ghee boiled with its crushed bark, considered good for **skin diseases**.

Active Constituents : These are neriodorin, nerioderin, karabin, scopoletin, and scopoline (*bark*).

Biological Activity : Alc. extr. exhibits antibacterial activity.

Uses in Literature : Known earlier as an abortifacient, anthelmintic, attenuant, cardiac poison, carminative, diaphoretic, febrifuge, fodder, gunpowder, insect repellent, insecticide, opthalmia, ornamental garlands, powerful heart poison, rat poison, resolvent, symbolic (sacred), vermic (vet.); and for asthma, blisters, boils, bronchitis, copious lachrymation, dysentery, gangrene, epilepsy, gum troubles, hooka tubes, leprosy, menorrhagia, cardiac poison, pthisis, ringworm, skin eruptions like herpes, snakebite, swellings, ulcers, worms, and wounds (Ambasta, 1986; Chauhan, 1999; Dastur, 1970; Girach *et al.*, 1996; Jain, 1991; Kapur & Singh, 1996; Khanna *et al.*, 1996a; Kurian, 1995; Ranjan, 2003; Saini, 1996a; Sharma, 2003; Shukla *et al.*, 2003; Thomas & Britto, 2003; Watt, 1972).

***Ocimum basilicum* L.** (Pl. 36F)

Syn.: *O. pilosum* Willd.; *O. album* L.; *O. minimum* Burm.; *O. hispidum* Lam.; *O. menthaefolium* Benth.; *O. caryophyllatum* Roxb.;*Plectranthus barrelieri* Spreng.

Family: Lamiaceae.

Vern. : Bhabhri.

English, Sanskrit & Regional Names:

Eng. : Basil, Common Basil, Sweet Basil
Sans. : Ajagendhika, Asurasa, Barbara, Karahi
B. : Babuitulsi, Debunsha
G. : Damaro, Nasabo
H. : Bahari, Kalitulsi
Kan. : Kamkasturi, Sajjebiya
Mal. : Pachcha
Mar. : Marva, Sabja
Ori. : Dhala Zulasi
P. : Baburi, Rehan
Tam. : Tirnutpatchie, Tirunitru
Tel. : Rudrajada, Vipudipatri.

Distribution : Spread throughout India.

Description : Softly hairy, strongly-scented, erect annuals. Flowers small, pinkish-white, in densely racemose whorls. Nutlets ellipsoid, pitted.

Reproductive Cycle : June-September.

Habitat Ecology : Commonly cultivated; rarely as an escape in moist slopes.

Material Examined : HPU-EL-SKS-5406; Darugnu (1,400 m), Tanda (1,120 m), Malnu (1,000 m), Pahra (950 m), Purani-palam (1,100 m), Tanda (1,120 m).

Parts Used : Leaves. Inflorescence.

Folk Uses : Decoction of inflorescence (1 teaspoon, thrice daily) taken for **flatulence**, **dysentery** and other **digestive ailments**. Leaves used for **flavouring** salt; and its paste also good for massaging gums to check **toothache**.

Active Constituents : Composition of *oil* from fresh flowering herbs shows 65.3% linalool and small amounts of cineole, eugenol, sesquiterpenes and d-terpene.

Biological Activity : Alc. extr. of leaves and flowers exhibits antibacterial activity against *Micrococcus pyrogenes* var. *aureus*.

Uses in Literature : Known so far as an anthelmintic, antipyretic, aphrodisiac, appetizer, carminative, condiment, demulcent, diaphoretic, edible (seeds, leaves, stem), expectorant, nasal douche, psychoactive, stimulant and stomachic; and for alc. intoxication, blindness, bowl complaints, cholera, cough, convulsion, cramp, diarrhoea, dropsy, dysentery, earache, epilepsy, fever, fits, headache, labour complaints, piles, respiratory troubles, ring worm, sinuses, sores, snake-bite and wounds (Agarwal, 2003; Ambasta, 1986; Bhatt *et al.*, 1999, 2003a; Billore *et al.*, 1998; Chevallier, 1996; Chopra *et al.*, 1956; Chun-Lin & Jieru, 1995; Das & Agarwal, 1991; Duke *et al.*, 2002; Girach, 1992; Jain, 1991; Jain *et al.*, 1994; Jamir, 1997; Katewa *et al.*, 2000; Khanna *et al.*, 1996a; Khare, 2004; Kirtikar & Basu, 1984; Kumar *et al.*, 2003; Kurian, 1995; Lal *et al.*, 1996; Mandal & Basu, 1996; Molla & Roy, 1996; Nayak *et al.*, 2003; Parrotta, 2001; Prajapati *et al.*, 2006; Ranjan, 2003; Rajendran *et al.*, 2003; Rawat & Chowdhury, 1998; Roy *et al.*, 1998; Saini, 1996b; Samant *et al.*, 2001; Seetharam *et al.*, 1999; Sharma, 2003; Singh, 2000; Singh & Kumar, 2000a; Singh *et al.*, 1999; Saklani & Jain, 1994; Sood & Thakur, 2004).

Ocimum gratissimum L. (Pl. 37A)

Family: Lamiaceae.

Vern. : Ban-ajwain.

English, Sanskrit & Regional Names:

Eng. : Large basil, Lemon basil, Shrubby basil

Sans. : Ajaka, oshkleshi, Nidralu, Ramatulsi

Ass. : Sugandhi, Suprassanka, Suvarka, Vanabarbarika, Vishghna, Vriddhatulasi

B. : Ramulshi, Ramtulsi

Bo. : Ramalutasa, Ramtulasi

G. : Avachibavachi, Ramtulsi

H. : Bantulsi, Malalulu, Ramatulsi

P. : Banjere

Tam. : Elumichantulasi, Perunlitlasi

Tel. : Nimmatulasi, Ramtulasi.

Distribution : Throughout India.

Description : Shrubs with subquadrangular branches; acute, crenate-serrate, pubescent leaves and greenish-yellow flowers in racemes. Nutlets subglobose, brown.

Reproductive Cycle : June-August.

Habitat Ecology : Shady places along water steams.

Material Examined : HPU-EL-SKS-5718; Chornali (1,280 m), Jiun (1,550 m), Lolani (1,600 m), Purani-palam (1,100 m), Tanda (1,120 m).

Parts Used : Leaves. Inflorescence.

Folk Use : Decoction of leaves and inflorescence (three teaspoons daily) prescribed for **digestive ailments** both in human beings and livestock.

Active Constituents : Eugenol, β – caryophyllene, germacrene D, α – copaene, humulene, β – bourbonene, γ – muurolene and ð – cadinene isolated from *leaf oil.*

Biological Activity : Extr. of leaves raised arterial blood pressure in rats.

Uses in Literature : Known so far as antiemetic, antiseptic, aphrodisiac, cardiac stimulant, carminative, diaphoretic, demulcent, diuretic, edible (leaves), expectorant, purgative, psychoactive, tonic; and for burns, cough, digestive problem, dysentery, eczema, fever, headache, neuralgia, paralysis, rheumatism, seminal weakness, tonic, swellings, brain, heart and liver diseases, leucorrhoea, malaria, scorpion sting and scurvy (Agarwal, 2003; Arora & Pandey, 1996; Augustine & Sivadasan, 2005; Chopra *et al.*, 1956; Das & Agarwal, 1991; Gill *et al.*, 1997; Girach *et al.*, 1996; Kirtikar & Basu, 1984; Jain *et al.*, 1994; Jha &

Kumar, 2003; Parrotta, 2001; Prajapati *et al.*, 2006; Said, 1997; Sharma, 2003; Sharma & Rana, 2005; Singh *et al.*, 1997).

Ocimum kilimandscharicum Guerke (Pl. 37B)

Family: Lamiaceae.

Vern. : Tulsi.

English & Regional Names:

Eng. : Camphor Basil

B. : Karpur Tulsi

H. : Kapur Tulsi.

Distribution : Throughout India.

Description : Pubescent undershrubs with ovate leaves, and flowers in whorls on long villose racemes. Nutlets ovoid, black.

Reproductive Cycle : August-September

Habitat Ecology : In courtyards of houses and temples.

Material Examined : HPU-EL-SKS-5536; Andretta (980 m), Bhagpur (1,600 m), Bhuana (1,400 m), Chimbalhar (1,150 m), Rangru (910 m).

Parts Used : Whole Plant. Leaves.

Folk Uses : Plant considered sacred and considered incarnation of goddess **Laxmi**; also grown to keep away **evil spirits**. Leaves used for **flavouring** tea, and its decoction (four to five teaspoons daily) helpful in checking **pulmonary** ailments.

Active Constituents : *Leaves* yield an essential oil 6.13% of which 50-75% is camphor.

Uses in Literature : Reported earlier as a mosquito repellent, and for metallic luster on ceramic bodies (Ambasta, 1986).

***Ocimum sanctum* L.** (Pl. 37C)

Syn.: *O. inodorum* Burm.; *O. monachorum* L.; *O. tenuiflorum* L.

Family: Lamiaceae.

Vern. : Sailli.

English, Sanskrit and Regional Names:

Eng. : Holy Basil, The Sacred Basil

Sans. : Ajaka, Bahupatri, Krishna Tulasi, Manjarika, Parnasa, Subhaga, Surasa, Tulashi, Tulasi, Vrinda

B. : Kala-Tulsi, Kural, Tulsi, Varanda

Bo. : Tulas, Tulasa

G. : Talasi

H. : Baranda, Kala-Tulsi, Tulsi, Varanda

Kan. : Tulashi-Gida, Vishnu Tulasi

Mal. : Krishnatulasi, Trittavu, Tulasi

Mar. : Kalitulasi, Tulasa, Tulas-icha-jhada

P. : Bantulsi, Tulsi

Tam. : Tulashi, Tulasi

Tel. : Graggera - chettu, Krushna- Tulasi, Tulasi.

Distribution : Throughout India, ascending Himalayas to 2,000 m.

Description : Erect softly hairy, strong scented herbs. Leaves oblong-obtuse or acute, entire or subserrate. Racemes very slender. Nutlets broadly oblong, slightly compressed, pale red-brown.

Reproductive Cycle : August-September.

Habitat Ecology : Courtyards of temples and houses.

Material Examined : HPU-EL-SKS-5501; Bhatu (1,040 m), Bindraban (1,270 m), Chachian (1,300 m), Chauki (900 m), Saparu (1,250 m).

Parts Used : Whole Plant. Leaves. Seeds. Inflorescence.

Folk Uses : Plant considered an incarnation of **Laxmi** (the goddess of wealth). Seeds employed as **mosquito repellent**. Decoction of leaf and inflorescence (four teaspoons daily) taken for **cough, indigestion**, and as **liver tonic**.

Active Constituents : These are cardiene, 1, 8−eineole, β−carotene, ursolic acid, eugenol, limoneme (*leaf*); cardinene, β -caryophyllene, eugenol and humulene (*flower oil*).

Uses in Literature : Reported so far as antiperiodic, anthelmintic, alexiteric, antipyretic, cholagogoue, demulcent, diaphoretic; and for asthma, blood diseases, bronchitis, catarrh, cutaneous diseases, dislodging maggots, earache, foul smells, heart diseases, leucoderma, lumbago, malarial fever, urinogenital disorders, strangury and vomiting (Ambasta, 1986; Arinathan *et al.*, 2003; Borthakur *et al.*, 2004; Caius, 2003; Dwivedi, 2003; Kirtikar & Basu, 1984; Kumar *et al.*, 2003; Maya *et al.*, 2003; Rajendran *et al.*, 2003; Ranjan, 2003; Saikia & Nath, 2003; Singh & Singh, 2003).

***Opuntia dillenii* Haw.** (Pl. 37D)

Syn.: *Cactus indicus* Roxb.

Family: Cactaceae.

Vern. : Daandi-chhoo.

English, Sanskrit & Regional Names:

Eng. : Prickly Pear, Slipper Thorn
Sans. : Bahudagdhika, Kandashaka, Nagaphana
B. : Nagphana, Phenimana
G. : Chorhathola
H. : Haththathoira, Nagphani
Kan. : Papaskalli
Mal. : Nagmullu, Palakkalli
Mar. : Chapal, Nagphana, Samar
Ori. : Nagophenia, Poturiyabiju
P. : Chhittarthohar
Tam. : Manjarnagadali, Pattanadugalli
Tel. : Nagajemudu, Nagamullu.

Distribution : Planted and naturalized.

Description : Shrubs with thick branches which are flat and dotted with hairy tufts of short spinescent bristles. Flowers orange-red, large. Berries reddish purple. Seeds many.

Reproductive Cycle : April-June.

Habitat Ecology : Edges of grass fields. Ravines. Drier areas.

Material Examined : HPU-EL-SKS-5698; Balah (900 m), Banuri (1,150 m), Bhauda (1,050 m), Bhuana (1,400 m), Dadh (1,100 m), Tikkar (910 m).

Parts Used : Whole Plant. Fruits.

Folk Uses : Plant grown as a **hedge**. Ripe fruits **edible**, **astringent**, and also taken to treat **leucorrhoea** and **constipation**.

Active Constituents : *Flowers* yield isoquercitin, β-sitosterol, opuntiol, opuntiol acetate and a glucoside narcissin.

Biological Activity : Ether extr. of the stem exhibits some antibiotic properties.

Uses in Literature : Used earlier as an antipyretic, carminative, demulcent, edible (fruits), expectorant, hedge, laxative, opthalmia, paper material, purgative, refrigerant, stomachic, symbolic in religious ceremonies of 'Santhals' (spines); and for allaying heat, asthma, biliousness, boils, bronchitis, chest pains, controlling spasmodic cough, dyeing, enlargement of spleen, expelling worms, gonorrhoea, hastening formation of pains in the body, inflammation, leucoderma, loss of consciousness, pains, piles, pimples, promoting flow of urine, scorbutic ulcers, skin diseases, sores, swellings, syphilis, tumours, ulcers, urinary complaints, and whooping cough (Ambasta, 1986; Arora & Pandey, 1996; Billore *et al.*, 1998; Chatterjee & Pakrashi, 1991; Chopra *et al.*, 1956; Das & Agarwal, 1991; Dey & Bahadur, 1973; Jain, 1991; Karnick, 1994; Kirtikar & Basu, 1984; Kurian, 1995; Parmar & Kaushal, 1982; Sood & Thakur, 2004; Sudhakar & Vedavathy, 2000; Watt, 1972).

***Origanum vulgare* L.** (Pl. 37E)

Family: Lamiaceae

Vern. : Ban-Ajwain, Kah-Juainhe.

English & Regional Names:

- **Eng.** : Common or Wild Marjoram
- **H.** : Mirzanjosh, Sathara
- **Kash.** : Ven
- **P.** : Mirzanjosh
- **Tel.** : Mrzanjosha.

Distribution : Temperate Himalayas, (Kashmir-Sikkim).

Description : Aromatic hairy perennial herbs upto 60 cm. Leaves ovate-entire, petiolated. Flowers pale-purple on corymbose cymes, polygamous with upper lip notched, bracteate. Fruits of 4 nutlets.

Reproductive Cycle : July-September.

Habitat Ecology : Open slopes. Grassy meadows. Drier areas.

Material Examined : HPU-EL-SKS-5472; Bhauda (1,050 m), Boda (1,000 m), Jasun (1,000 m), Rangru (910 m), Tikkar (910 m).

Parts Used : Aerial parts.

Folk Uses : Aerial parts including flowers used as a **spice**, and also mixed with wheat flour and prescribed for **stomach disorders** in cattle. Its decoction taken as **stomachic**.

Active Constituents : Leptosidin, naringin, luteolin, apigenin, peonidin, catechol, α – pinene, p – cymene, thymol, carvacrol, β

– caryophyllene, β – bisabolene, γ – terpinene have been isolated (*whole plant*).

Uses in Literature : Known earlier as an antiseptic, diuretic, diaphoretic, emmenagogue, condiment, stimulant, pot herb, hair tonic, febrifuge, vegetable; and for bronchitis, cold, cough, cuts, diarrhoea, flatulence, hysteria, influenza, menstrual complaints, skin boils, toothache, tonsillitis, local drinks (Ambasta, 1986; Arora & Pandey, 1996; Aswal, 1996; Chandra, 1997; Chauhan, 1999; Chevallier, 1996; Chopra *et al.*, 1956; Gaur *et al.*, 1983; Gupta, 1981; Harborne & Baxter, 2001; Kaul, 1997; Kapur & Singh, 1996; Prajapati *et al.*, 2006; Rana *et al.*, 1996; Rawat & Pangtey, 1987; Said, 1997; Samant *et al.*, 2001; Sharma, 2003; Singh & Kumar, 2000a; Sood *et al.*, 2001).

Oroxylum indicum **Vent.** (Pl. 37F)

Syn.: *Spathodea indica* Pers.; *Calosanthes indica* Blume; *Bignonia petandra* Lour.; *B. indica* L.

Family: Bignoniaceae.

Vern. : Tat-Palanga.

English, Sanskrit & Regional Names:

Eng. : Indian Trumpet Flower
Sans. : Adhvantashatrava, Svarnavalkana
Ass. : Bhatgila, Dingari, Kering, Toguna
B. : Nasona
Bo. : Phalphura, Tetu
G. : Podval
H. : Assarsauna, Sauna
Kan. : Bunepale, Sonepatta, Tigdu
Mal. : Palakapayani
Mar. : Taitu
Ori. : Phompania
P. : Miringa, Tatmorang
Tam. : Paluiyudaichi, Vangam
Tel. : Dundilamu, Mundakaparnamu.

Distribution : Throughout India, ascending to 1,350 m.

Description : Small trees. Flowers infundibular, yellowish-cream. Capsule thick, sword-like, margins incurved. Seeds large, winged.

Reproductive Cycle : May-August.

Habitat Ecology : Forests. Scrub jungles.

Material Examined : HPU-EL-SKS-5517; Arla (1,100 m), Chandpur (1,290 m), Darugnu (1,400 m), Jasun (1,000 m), Sungal (1,050 m).

Parts Used : Stem. Bark. Seeds. Pods.

Folk Uses : Unripe fruits consumed as **vegetable**. Mustard oil boiled with its powdered seeds applied for **rheumatic** pains. Decoction of seeds (daily for seven to nine days) also good against **constipation**. Crushed bark applied as poultice for healing **fractured bones** in animals as well as human beings; also fed to cattle mixed with wheat flour for **internal injuries** and **fractures**. Stem used as **fuelwood**.

Active Constituents : Active principles attributed to baicelin its 6-glycuronide and 7-glucoronide, scutellarein, its 7-glucuronide, oroxylin A, chrysin and scutellarein-7-rutinoside (stem-bark), prunetin and β-sitosterol (*heartwood*), glucuronide-oroxidin, synthesis of biacatein-6-glucuronide and 7-glucuronide (*seeds*), fixed oil (25.0) and crude protein (7.9%).

Uses in Literature : Known to be used earlier as anodyne, antidote to snakebite, appetizer, antiseptic, astringent, black dye, carminative, colic, diaphoretic, ingredient of 'chywanprash', pickle, psychoactive plant, purgative (seeds), stomachic, symbolic (garlands of idols), tonic, vegetable (corolla, young shoots, unripe fruits), vermicide, and for affections of the brain, blood dysentery, body pain, bone fracture (vet.), cholera, chronic dysentery, crack in nipples, diarrhoea, dropsy, dysentery (vet.), dyspepsia, dysuria, enlarged spleen, epilepsy, headache, inducing conception, inflammatory diseases of chest, leprosy patients, liver ailments, madness, miscarriage, oedema, otorrhoea, pains, piles, pimples, polyuria, post-delivery pain, rheumatism, rutting (vet.), scabies, scorpion sting, smallpox, snakebite, sore backs of herds and bullocks, sores of nails and soles (vet.), tanning, tuberculosis, ulcers, urticaria, vitality, and wounds (vet.) (Agarwal, 2003; Ambasta, 1986; Arora & Pandey, 1996; Borthakur, 1996b; Borthakur & Sarma, 1996; Brahmam & Saxena, 1990; Brahmam *et al.*, 1996; Chakrabarty & Balakrishnan, 2003; Chhetri, 2005; Chopra *et al.*, 1956; Chun-Lin & Jieru, 1995; Chowdhery, 1996; Dey & Bahadur, 1973; Diwakar & Ansari, 2003; Girach, 1992, 2000, Girach & Aminuddin, 1995; Jain, 1991, 2003; Jain *et al.*, 1994; Jha & Kumar, 2003; Kemp, 2003; Kothari & Moorthy, 1996; Kurian, 1995; Lalramnghinglova, 2000, 2002; Mehrotra *et al.*, 1995; Mitaliya & Bhatt, 2003; Mohanty & Padhy, 1996; Molla & Roy, 1996; Mukherjee & Namhata, 1990; Pal *et al.*, 2000; Pandey *et al.*, 1996; Prajapati *et al.*, 2006; Rajendran *et al.*, 1996; Rama Rao & Henry 1996; Reddy *et al.*, 1996; Roy *et al.*, 1998; Samant *et al.*, 2001; Sarin, 1996; Sen & Behera, 2003; Sharma, 2003; Singh, 1996b, 1999, 2001; Siwakoti & Varma, 1996; Srivastava *et al.*, 2003b; Upadhye *et al.*, 1994; Viswanathan & Singh, 1996; Watt, 1972).

Ougeinia dalbergioides Benth. (Pl. 38A)

Syn.: *O. oojeinensis* (Roxb.) Hochr.

Family: Fabaceae.

Vern. : Sannanh.

English, Sanskrit & Regional Names:

Eng. : Sandan
Sans. : Tinsha
B. : Tinish
Bihar : Pandan
G. : Tanach
H. : Panan, Panjam, Sandan, Tinsha
Kan. : Kurimutal
Mal. : Malavinna
Mar. : Kalaphulas, Tiwas
Ori. : Bandhona, Banjan
Tam. : Narivengai
Tel. : Tellamotuku.

Distribution : Outer Himalayas and sub-Himalayan tract upto 1,500 m.

Description : Medium-sized deciduous trees upto 12 m tall and 1.35 m in girth; leaves pinnately trifoliate. Leaflets coriaceous. Flowers white or pink, in dense clusters. Pods, linear-oblong, flat, light brown slightly contracted between the seeds.

Reproductive Cycle : March-May.

Habitat Ecology : Shady steep slopes.

Material Examined : HPU-EL-SKS-5617; Balah (900 m), Boda (1,000 m), Khaira (940 m), Malahu (1,050 m), Rangru (910 m).

Parts Used : Leaves. Flowers. Wood.

Folk Uses : Leaves as an excellent **fodder**. Flowers used for making '**pakodas**'. Wood employed for **furniture**, and **household construction**.

Uses in Literature : Known earlier as acrid, hot, anthelmintic, astringent to bowels, febrifuge, fishpoision, vegetable (flowers after boiling); and for anaemia, biliousness, burning sensation, cart-making, diarrhoea, dysentery, fodder, leucoderma, timber, urinary discharges, ulcers, blood and skin diseases (Bennet *et al.*, 1991; Drury, 1978; Kirtikar & Basu, 1984; Purohit & Samant, 1995; Singh & Pandey, 1998; Sood & Kaushal, 2008; Watt, 1972).

Oxalis corniculata L. (Pl. 38B)

Syn.: *O. repens* Thunb.; *O. pusilla* Salisb.; *O. villosa* Bieb.; *O. foliosa* Blatt.

Family: Geraniaceae.

Vern. : Malori.

English, Sanskrit & Regional Names:

Eng. : Indian Sorrel, Yellow Oxalis
Sans. : Ambashta, Amlika, Amlotaja, Changeri, Shuklika
Ass. : Chengeritenga
B. : Amrul, Chalmori
Bo. : Ambuti, Butisarpati
H. : Amrul, Anboti, Seh
Kum. : Amrul, Chalmori
Mal. : Poliyarala
Mar. : Ambuti, Bhin-sarpali
P. : Amlika, Amrul
Tam. : Paliakiri
Tel. : Ambotikura.

Distribution : Common throughout India in the warmer regions.

Description : Hairy annuals with trailing stems and trifoliolate leaves. Flowers yellow. Capsule 5-angled, beaked. Seeds ribbed.

Reproductive Cycle : April-November.

Habitat Ecology : Shady moist areas.

Material Examined : HPU-EL-SKS-5653; Balu (1,030 m), Bhaura (940 m), Chauki (900 m), Darugnu (1,400 m), Lolani (1,600 m).

Parts Used : Aerial plant parts. Leaves.

Folk Uses : Aerial plant parts cooked as **pot-herb**. Also, extensively employed to **clean silver and copper** utensils; eaten raw as a rich source of **Vit. C**. Leaf paste applied to **cholesterol plaque**, and also its juice used as eye-drop to check **opthalmic diseases**.

Active Constituents : Constituents isolated are malic acid (*stem*); tartaric and citric acids (*stem, leaves*).

Biological Activity : Plant hypoglycaemic.

Uses in Literature : Known earlier as an antidote to *Datura* poisoning, antipyretic, antiscorbutic, appetizing, astringent, edible (raw leaves, fruits, seeds and vegetable), stomachache, refrigerant, sacred and stomachic; and for anaemia, boils, blisters, swelling of belly, corn, cough, cuts, diarrhoea, digestion, dysentery, dyspepsia, epilepsy, eye complaints, fever, insect-bite, jaundice, malaria, muscular pain, piles, physical weakness, rickets, scurvy, skin diseases, snakebite, swelling, toothache, tympanitis, wart and wounds

(Ambasta, 1986; Bajpayee & Dixit, 1996; Balasubramaniam & Prasad, 1996; Banerjee & Ghora, 1996; Bennet *et al.*, 1991; Bondya *et al.*, 2006; Borthakur *et al.*, 2004; Bose, 1984; Chopra *et al.*, 1956; Das, 2000; Dey & Bahadur, 1973; Drury, 1978; Dwivedi, 2003; Gogoi & Borthakur, 1991; Gogoi & Das, 2003; Gogoi *et al.*, 2003a; Hosagoudar & Henry, 1996a; Jain, 1991; Kapur & Singh, 1996; Kapur & Srivastava, 1996; Khanna, 2002; Khare, 2004; Kumar *et al.*, 2003; Mahato & Chaudhary, 2005; Maliya, 2004; Maya *et al.*, 2003; Mohanty & Padhy, 1996; Murthy *et al.*, 2003; Nadanakunjidam, 2003; Negi *et al.*, 1993; Panthi & Chaudhary, 2003; Patole & Jain, 2002; Prakash, 1999, 2001; Prajapati *et al.*, 2006; Rana *et al.*, 1996; Ranjan, 2000; Rawat & Chowdhury, 1998; Roy *et al.*, 1998; Saklani & Jain, 1994; Samant *et al.*, 2001; Sarma *et al.*, 2001; Sharma, 2003; Singh, 2000; Singh & Kumar, 2000a; Singh & Maheshwari, 1992; Singh & Singh, 2003; Singh *et al.*, 1997, 1999; Sinha, 1996; Sood & Thakur, 2004; Srivastava *et al.*, 2003a; Swami & Gupta, 1996; Thomas & Britto, 2003; Upadhyay & Singh, 2005; Varma, 1997; Watt, 1972).

***Panicum antidontale* Retz.** (Pl. 38C)

Family: Poaceae.

Vern. : Barua.

English & Regional Names:

- **Eng.** : Blue Panic
- **Bo.** : Git, Male, Girni
- **G.** : Dhusdo, Dhusghas, Dun, Dusto
- **H.** : Ghemor, Gunara, Jamur
- **P.** : Baru, Ghamur, Gharam, Ghirri, Mangrur
- **Tam.** : Nassiam pillu, Pinsu pillu.

Distribution : Common in India.

Description : Perennial tall grasses having smooth woody stem and linear, glabrous, membranous leaves. Inflorescence (panicles) drooping.

Reproductive Cycle : July-October.

Habitat Ecology : Borders of agricultural fields. Wastelands.

Material Examined : HPU-EL-SKS-5490; Bhatu (1,040 m), Bhuana (1,400 m), Chhanchri (930 m), Dharaman (1,100 m), Gharthun (900 m).

Parts Used : Aerial plant parts. Root.

Folk Uses : Roots form one of the ingredients ('**dahli**') for making local **wine**. Root-juice employed as ear-drops to check **ear-infection** ('**saili**'). Aerial parts given as **fodder**.

Uses in Literature : Recorded to be used for burns, erosion control, fumigating wounds, hydrophobia, throat affections; and as fodder, edible, disinfectant in smallpox, windbreak (Ambasta, 1986; Caius, 2003; Kirtikar & Basu, 1984; Parrotta, 2001; Singh & Pandey, 1998; Srivastava *et al.*, 2003c).

Pennisetum americanum (L.) Leeke (Pl. 38D)

Syn.: *P. alopecroides* Spreng.; *P. cereale* Trin.; *P. cylindrium* Sw.; *P. giganteum* Ten.; *P. typhoideum* Rich.; *P. violaceum* Hort.; *P. violaceum* Rich.; *C. spicatus* Cav.

Family: Poaceae.

Vern. : Bajara.

English, Sanskrit & Hindi Names:

Eng. : The Bulrush, The Cumboo, The Spiked Millet
Sans. : Agradhanya, Nali, Sajaka, Vajarika
B. : Bajra, Lahra
G. : Bajara
H. : Bajra, Bajri, Bajera, Lahra, Kara-jonar
Kan. : Sajjai, Sajji
Mal. : Kampam, Mattari
Mar. : Bajri
P. : Bajra, Bajza
Tam. : Kambu
Tel. : Gantelu, Ganti, Peddaganti, Sajja.

Distribution : Largely cultivated in India.

Description : Erect annuals possessing simple stem and lanceolate, glabrous leaves. Inflorescence spike-like, cylindrical, greenish-yellow. Peduncle more or less woolly below the spike. Anthers linear. Grains oblong, obovoid or pyriform.

Reproductive Cycle : June-July.

Habitat Ecology : Cultivated land.

Material Examined : HPU-EL-SKS-5464; Banuri (1,150 m), Bindraban (1,270 m), Dadh (1,100 m), Jasun (1,000 m), Saliana (1,000 m).

Parts Used : Aerial parts. Grains.

Folk Uses : Aerial parts employed as **fodder**. Powdered grains relished as '**roti**' and considered to have **warming** effect as well as good against **heart** diseases and **piles**.

Active Constituents : These are myristic, palmitic, stearic, oleic and linoleic acids (*seeds*).

Uses in Literature : Considered aphrodisiac (women), appetizer, edible, fuel, fodder, tonic; and useful for bloating abdomen and diseases of heart. (Ambasta, 1986; Caius, 2003; Kirtikar & Basu, 1984., Kumar *et al.*, 2004; Parrotta, 2001; Sen & Batra, 1997).

Pentanema indicum L. (Pl. 38E)

Syn.: *Vicoa auriculata* Cass.; *V. indica* (L.) DC.

Family: Asteraceae.

Vern. : Jugaali-di-dua.

Regional Names:

- **G.** : Sona Sali
- **H.** : Hannavarikasoppu, Mugulisoppu
- **Kan.** : Muguti-soppu, Hannavarikasoppu
- **Tam.** : Jimikipoo, Mookuti, Poondu.

Distribution : Deciduous forest zones throughout India upto 1,800 m.

Description : Erect, pubescent herbs. Leaves ovate-lanceolate, acute. Flower heads yellow, solitary. Achenes angled.

Reproductive Cycle : Throughout year.

Habitat Ecology : Common in abandoned fields and drier areas.

Material Examined : HPU-EL-SKS-5496; Bhauda (1,050 m), Boda (1,000 m), Khaira (940 m), Rangru (910 m), Tikkar (910 m).

Parts Used : Whole Plant. Aerial plant parts.

Folk Uses : Plant paste applied for **bone fractures** and also taken to treat **stomach ailments**. Aerial plant parts along with wheat dough given orally to induce **rumination** in cattle.

Active Constituents : *Plant* contains germacranolide-vicolide D, and 5,6,3′-trihydroxy-7,4′-dimethoxyflavone.

Uses in Literature : Reported so far for dysentery expeling worms, indigestion, fracture, inducing sterility; and as fodder (Ambasta, 1986; Dwivedi, 2003; Rastogi & Mehrotra, 1999).

Perilla frutescens **(L.) Britt.** (Pl. 38F)
Family: Lamiaceae.

Vern. : Ban-Juainhe.

Distribution : Tropical and Temperate Himalayas.

Description : Erect fragrant herbs possessing grooved, angular stem and ovate-lanceolate leaves. Flowers in axillary and terminal racemes, white. Nutlets dry.

Reproductive Cycle : May-August.

Habitat Ecology : Along water streams.

Material Examined : HPU-EL-SKS-5696; Bhagpur (1,600 m), Chachian (1,300 m), Palampur (1,280 m), Saliana (1,000 m), Sulah (950 m).

Parts Used : Aerial Plant Parts.

Folk Use : Decoction of aerial plant parts (2 teaspoons daily) good for **stomach disorders**.

Active Constituents : Isolated constituents are isoegomaketone, cis – α – farnesene, oleic, linoleic, linolenic, sitosterol and farnesol.

Biological Activity : Ether extr. exhibits antifungal activity.

Uses in Literature : Known so far in India as edible, flavorant; and for diarrhoea, fever, headache, influenza, cough, lungs trouble and threatened abortion (Arora & Pandey, 1996; Bennet *et al.*, 1991; Harborne & Baxter, 2001; Pandey & Joshi, 2001; Prajapati *et al.*, 2006; Rawat & Chowdhury, 1998; Roy *et al.*, 1998; Samant *et al.*, 2001; Singh & Kumar, 2000a; Sharma & Rana, 2005).

Persea odoratissima **(Nees) Kosterm.** (Pl. 39A)
Syn.: *Machilus odoratissima* (Wall. *ex* Nees) Nees
Family: Lauraceae.

Vern. : Bhadrol.

Regional Names:

H. : Kawala

P. : Badroor, Dalchini, Kalban, Kaula, Mithpatta, Taura.

Distribution : Himalaya: 1,500-2,100 m.

Description : Evergreen trees. Leave long-pointed, elliptic-lanceolate. Flowers pale-yellow, solitary. Fruits ellipsoid, dark-brown.

Reproductive Cycle : March-April.

Habitat Ecology : Forests.

Material Examined : HPU-EL-SKS-5597; Dharaman (1,100 m), Rajpur (1,100 m), Molichak (1,020 m), Purani-palam (1,100 m), Saparu (1,250 m).

Parts Used : Leaves. Wood.

Folk Uses : Leaves lopped for **fodder**. Wood employed for **furniture** and **house construction**.

Uses in Literature : Reported to be good for house construction, rearing silk worms, and tea-boxes (Ambasta, 1986; Watt, 1972).

Phoenix acaulis **Buch.- Ham.** *ex* **Roxb.** (Pl. 39B)

Family: Arecaceae.

Vern. : Khajoori.

English, Sanskrit & Regional Names:

Eng. : Dwarf date palm
Sans. : Bhukharjuri
H. : Jangli khajur, Khajur, Pind Akhajur, Aaathakal
Mar. : Boichind, Shevra, Shilind
Ori. : Kojiri
P. : Khajuri
Tel. : Yita.

Distribution : Found in the outer Himalayas, the Khasi and Naga hills, central India and peninsular India.

Description : A short stemless palm having pinnate leaves and flowers borne in compressed, branched spadices. Berries bright red-blue, black when ripe. Seeds oblong, grooved.

Reproductive Cycle : April-June.

Habitat Ecology : Wastelands. Dry slopes.

Material Examined : HPU-EL-SKS-5528; Bhaura (940 m), Bhuana (1,400 m), Jasun (1,000 m), Khaira (940 m), Malahu (1,050 m).

Parts Used : Fruits. Stem. Leaves.

Folk Uses : Fruits and stem-pith **edible** with **cooling** effect. Leaves given as **fodder** for milching cattle for enhancing **lactation**; and also used for making **brooms**, **mats** and **fencing**.

Uses in Literature : Known earlier as edible (flour of the root, fruits, stem pith, tender peduncles and terminal leaf buds); fodder; and for mats, brooms and ropes (Agarwal, 2003; Ambasta, 1986; Arora & Pandey, 1996; Bennet *et al.*, 1991; Kumar, 1996, 2003; Parrotta, 2001; Patole & Jain, 2002; Roy *et al.*, 1998; Samant *et al.*, 2001; Singh, 1996b).

***Phoenix sylvestris* Roxb.** (Pl. 39C)

Syn.: *Elate sylvestris* L.; *Katou indel* Ham.

Family: Arecaceae.

Vern. : Khajoor.

English, Sanskrit & Regional Names:

Eng. : Date Sugar Palm, Indian Wine Palm, Wild Date Palm
Sans. : Bhumikharjurika, Kakakarkati, Skandhappala
B. : Kajar, Kejur
Bo. : Khajuri, Khjura
G. : Kajuri, Kharak
H. : Khaji, Salma, Thakil
Kan. : Ichalu
Mal. : Intappana, Kattinta
Mar. : Boichand, Sindikajuri
Ori. : Khorjuri, Khorjuro
P. : Khaji, Khajur
Tam. : Ichamanai, Periyayinju
Tel. : Peddayita.

Distribution : H.P.- W. Nepal. India. Myanmar: 150-1,500 m.

Description : Trees with columnar, rough, unbranched stem and pinnate leaves having spiny stalk. Flowers small, unisexual in spadices. Fruits yellowish-reddish brown. Seeds hard.

Reproductive Cycle : May-June.

Habitat Ecology : Open dry hills. Rocky slopes. Scrub forests.

Material Examined : HPU-EL-SKS-5529; Balah (900 m), Bhuana (1,400 m), Chhanchri (930 m), Panapari (1,100 m), Tikkar (910 m).

Parts Used : Fruits. Leaves. Pith.

Folk Uses : Ripe fruits **edible** and have **cooling effect**. Leaflets considered an excellent **fodder** for **milching cattle** to increase **fat contents** of milk; also used for making **brooms, mats** and for **thatching purposes**. Pith relished for its **sweet-taste** and **cooling** effect.

Uses in Literature : Useful so far as cardiotonic, constipative, cooling beverage, edible (buds, stem, fruits), restorative, nervine tonic, psychoactive plant; and for abdominal complaints, ague, bakery, beverage, brooms, cough, diabetes, diarrhoea, dysentery, fattening, fevers, fodder, gonorrhoea, gum trouble, haematuria, heart complaints, respiratory diseases, roofing, intoxicating alcoholic drink, loss of consciousness, basketry, mats and bags, menstrual complaint, nervous debility, paper making, pickling, relief in pain during pregnancy, ropes, thatching, toddy tapped from apical region of the trunk, toothache, venereal disease, and vomiting (Ambasta, 1986; Arora & Pandey, 1996; Bennet *et al.*, 1991; Bhatt *et al.*, 2003a; Das & Agarwal, 1991; Goud & Pullaiah, 1996a; Harborne & Baxter, 2001; Hebber *et al.*, 2003; Jain, 1991; Jain *et al.*, 1994; Katewa *et al.*, 2000; Khare, 2004; Kirtikar & Basu, 1984; Kumar & Rao, 2001; Mitaliya & Bhatt, 2003; Parrotta, 2001; Patole & Jain, 2002; Ragupathy & Mahadevan, 1991; Rama Rao & Henry, 1996; Ravishankar & Henry, 1992; Roy *et al.*, 1998; Sahu, 1996; Singh, 1996b; Singh & Kumar, 2000a; Singh & Pandey, 1998; Singh *et al.*, 1999; Sood & Kaushal, 2008; Upadhyay & Chauhan, 2000; Viswanathan & Singh, 1996; Watt, 1972).

***Phyllanthus urinaria* L. (Pl. 39D)**

Syn.: *P. leprocarpus* Wight; *P. alatus* Blume; *P. cantoniensis* Hornem.; *P. mucronatus* Herb.; *P. racemosus* Herb.; *P. muricatus* Herb.; *P. polyphyllus* Herb.; *P. echinatus* Herb.

Family: Euphorbiaceae.

Vern. : Amli.

Sanskrit & Regional Names:

- **Sans.** : Adhyanda, Ajata, Buhuphala, Jhatamala
- **B.** : Hazarmani
- **G.** : Kharsadabonyaanmali
- **H.** : Hazarmani, Lalbhuinanvalah
- **Mal.** : Chirukizhukanelli, Chukannakizhanelli
- **Mar.** : Lalmundajanvali
- **Tam.** : Shivappunelli
- **Tel.** : Ettausirika.

Distribution : Throughout India.

Description : Tall herbaceous annuals upto 60 cm with distichously imbricate membranous leaves, angled branches, peltate stipules and yellowish flowers. Capsule depressed.

Reproductive Cycle : August-September.

Habitat Ecology : Moist places. Weed in cultivated land.

Material Examined : HPU-EL-SKS-5417; Bhawarna (1,100 m), Dharaman (1,100 m), Makol (900 m), Panchrukhi (990 m), Sungal (1,050 m).

Parts Used : Fruits.

Folk Uses : Powdered fruits (1 teaspoon daily) considered **diuretic** and also good against **jaundice**.

Active Constituents : Known to possess astragalin, ellagic tannin, quercetin, quercitrin, isoquercitrin and rutin (*plant*).

Uses in Literature : Recorded earlier as an appetiser, astringent, diuretic, febrifuge and tonic, and for gonorrhoea, sleeplessness and urinogenital diseases in children (Chopra *et al.*, 1956; Jain, 1991; Uphof, 1968; Watt, 1972).

Pinus roxburghii **Sarg.** (Pl. 39E)

Syn.: *P. longifolia* Roxb. *ex* Lamb.

Family: Pinaceae.

Vern. : Chir, Cheel.

English, Sanskrit & Regional Names:

Eng. : Himalyan Long Leaved Pine, Three Leaved Pine
Sans. : Bhadradaru, Putikashtha
Ass. : Telia
B. : Saralakashtha
G. : Sarala Deodara
H. : Chir, Dhupasarala
Kash. : Salla
Mal. : Charalam, Putikashtam
Mar. : Saraladeodara
Ori. : Sorolokatha
P. : Anandher, Chir, Drabchir
Tam. : Suruldevadari
Tel. : Devadaru, Saralu.

Distribution : Afghanistan-Bhutan: 1,000-2,000 m.

Description : Large-sized trees having 22-30 cm long acicular leaves in cluster of three. Cones ovoid-conical, solitary. Seeds winged.

Reproductive Cycle : March-May.

Habitat Ecology : Forming extensive forests.

Material Examined : HPU-EL-SKS-5565; Bhuana (1,400 m), Chandpur (1,290 m), Gopalpur (1,350 m), Lolani (1,600 m), Panapari (1,100 m).

Parts Used : Leaves. Twigs. Cones. Seeds. Resin. Wood.

Folk Uses : Seeds roasted and **eaten**. Twigs employed for **decorating** gates. Wood employed for making **furniture, house building** and as **fuel**. Female cones burnt to serve as **torch** and **fuel**. Resin from stem good for healing **cracked** feet and soles. Pine needles used for making **brooms**; also mixed with mud to serve as a binder for **mud walls**.

Active Constituents : Important constituents are pine oil (*needles*), α-pinene, β-pinene, abietic acid and isopimaric acid (*turpentine from xylem resin*).

Biological Activity : Oil from 'chir' pine needles exhibits antibacterial activity against a number of organisms including *Staphylococcus aureus*.

Uses in Literature : Reported to be used in India as an antiseptic, carminative in flatulent colic, colouring matter, denaturants, diaphoretic, diuretic, edible (seeds), expectorant, fuelwood, insecticide, psychoactive plant, rubefacient in various thematic affections such as lumbago, arthritis and neuralgia, stimulant; and for abscess, bangles, baskets, boils, bone fractures, buboes, burning of body, charcoal, chronic bronchitis, constipation, cough, cracks in sole of feet, fevers, furniture, gangrene of lungs, general care of hovers, gonorrhoea, house building, inflammation of abdomen, itching, leprosy, mange (vet.), match industry, minor haemorrhages on tooth-sockets, musical instruments, nose, painting, perfumery, phthisis, pine oil in various pharmaceuticals, preparation of plasters-ointments and fumigants, rafters, resin, rheumatism, worm infestations, shingles, skin diseases, snake bite, sport-goods, sprain, sprain (vet.), swelling, synthetic pine oils, tanning, tea chests, torches, turpentine, tympanitis, typhoid, and ulcers (Agarwal, 2003; Ambasta, 1986; Chauhan, 1999; Chopra *et al.*, 1956; Jain, 1968, 1991; Jain *et al.*, 1994; Khare, 2004; Kirtikar & Basu, 1984; Prajapati *et al.*, 2006; Roy *et al.*, 1998; Saklani & Jain, 1994; Sensarma, 1991; Sharma, 2003; Singh, 2001; Singh & Kumar, 2000a; Sood & Thakur, 2004; Srivastava *et al.*, 2003c; Watt, 1972).

***Pistacia khinjuk* Stocks** (Pl. 39F)

Syn.: *P. chinensis* Bunge; *P. chinensis* Bunge var. *integrimma* (Stewart) Zohary; *Pistacia integrimma* Stewart; *Rhus integrimma* Wall.; *R. kakra-singee* Royle

Family: Anacardiaceae.

Vern. : Kakad Singhi.

English, Sanskrit & Regional Names:

Eng. : Crab's Claw

Sans. : Chandraspada, Kasavinashini

Ass. : Kakaisrnghi
B. : Kakra
G. : Kakra
H. : Kakra
Kan. : Kakadsasingi, Karkatasrini
Kash. : Gurgu, Kakkar
Mal. : Karkatasringhi
Mar. : Kakra
Ori. : Kakadashringi, Kakadashrungi
P. : Kakkar, Kakkeram.
Tam. : Karkata Singhi
Tel. : Kakarsingi, Karkatashrungi.

Distribution : W. Himalayas: 800-2,500 m.

Description : A medium-sized tree. Leaves odd or even pinnate. Flowers red, without petals in panicles. Drupes glabrous, wrinkled, grey when ripe.

Reproductive Cycle : March-May.

Habitat Ecology : Riversides. Cultivated areas.

Material Examined : HPU-EL-SKS-5585; Bhaura (940 m), Chhanchri (930 m), Jasun (1,000 m), Kathena (1,290 m), Rangru (910 m), Thandol (950 m).

Parts Used : Leaves. Galls. Wood.

Folk Uses : Leaves lopped for **fodder**. Ash from crooked galls on leaves blended with honey and black pepper (*Piper nigrum*) given with honey to cure **cough**, **asthma**, and as an **expectorant**. Wood employed for performing '**havan**' during religious ceremonies.

Biological Activity : Gall oil antibacterial, antiprotozoal, antispasmodic and CNS depressant.

Uses in Literature : Known earlier as an aphrodisiac, astringent, edible (kernels), antidote to scorpion sting, anti-inflammatory, dye, expectorant, fodder, snake venom, skin diseases and tonic; and for asthma, cough, dysentery, dyspepsia, fever, leprosy, phthisis, tanning leather, ulcer, and skin diseases. wood for construction and furniture works (Agarwal, 2003; Ambasta, 1986; Chandra, 1997; Chauhan, 1999; Chopra *et al.*, 1956; Dey & Bahadur, 1973; Gupta, 1981; Jain, 2003a; Kant & Dutt, 2005; Kapur & Nanda, 1996; Kirtikar & Basu, 1984; Negi *et al.*, 1985; Prajapati *et al.*, 2006; Sarin, 1996; Sharma & Rana, 2005; Singh, 2001; Sinha & Sinha, 2001; Sood & Thakur, 2004).

***Plantago lanceolata* L.** (Pl. 40A)

Syn.: *P. attenuata* Wall.

Family: Plantaginaceae.

Vern. : Safed Isbagol.

English & Regional Names:

Eng. : Fighting Cocks, Ribgrass, Ripple grass

B. : Baltung, Bartung

H. : Baltanga.

Distribution : The Himalayas: Kashmir - H.P.

Description : Perennial stemless herbs having ribbed, radical leaves and deeply furrowed scapes. Flowers small, numerous, green in spikes. Sepals ciliate. Bracts acuminate. Seeds black.

Reproductive Cycle : January-April.

Habitat Ecology : Wastelands. Borders of agricultural fields.

Material Examined : HPU-EL-SKS-5407; Bhawarna (1,100 m), Chachian (1,300 m), Chornali (1,280 m), Jiun (1,550 m), Lolani (1,600 m).

Parts Used : Leaves. Inflorescence.

Folk Uses : Leaves and inflorescence chewed for healing **mouth ulcers** and as **blood purifier**. Also leaf-paste good against **boils**.

Active Constituents : *Seed* contains tannin and fatty oil.

Biological Activity : Extr. of young leaves antibacterial.

Uses in Literature : Known so far as an antibacterial, astringent, blood purifier, diuretic, haemostatic and purgative (seeds with sugar), and for asthma, boils, cough, diabetes, earache, gonorrhoea, pulmonary diseases, rheumatism, wounds, sores and swellings (Agarwal, 2003; Ambasta, 1986; Beigh *et al.*, 2003; Chopra *et al.*, 1956; Dar *et al.*, 1984; Das & Agarwal, 1991; Duke *et al.*, 2002; Jha & Kumar, 2003; Kapur & Singh, 1996; Kaul, 1997; Lal *et al.*, 1996; Samant *et al.*, 2001; Sharma, 2003; Singh & Kumar, 2000a; Sood & Thakur, 2004).

Plumbago zeylanica L. (Pl. 40B)
Syn.: *P. auriculata* Blume; *Thela alba* Lour.
Family: Plumbaginaceae.

Vern. : Chitra.

English, Sanskrit & Regional Names:

Eng. : Ceylon Lead Wort, White Flowered Leadwort
Sans. : Analanania, Chitraka, Jataveda, Vaishavan-ara
B. : Chita, Chitruk
Bo. : Chitrack
G. : Chitaro, Gholochitro
H. : Chitarak, Chitra, Chiti
Kan. : Chitramula, Vahni
Mal. : Tumpokotuveli
Mar. : Chitraka, Chitramula
Ori. : Chitramulo, Krishanu
P. : Chitrak
Tam. : Adigarradi, Vellaikkodiveli
Tel. : Chitramulamu, Tellachitramulamu.

Distribution : Throughout India, ascending to 1,650 m.

Description : Rambling undershrubs with glandular leaves and white corolla. Capsule enclosed in the persistent calyx, opening transversely near the base.

Reproductive Cycle : April-September.

Habitat Ecology : Moist slopes. Edges of forests.

Material Examined : HPU-EL-SKS-5686; Bhauda (1,050 m), Boda (1,000 m), Dadh (1,100 m), Jasun (1,000 m), Rangru (910 m).

Parts Used : Stem. Roots.

Folk Uses : Stem useful for **scouring** teeth and roots for **toothache**. Root paste applied to hasten suppuration of **boils**.

Active Constituents : Active ingredients are plumbagin, 3-chlorplumbagin and a new substance-3, 3′-biplumbagin, a new binaphthaquinone–chitranone, zeylinone, isozeylinone, elleptinone, droserine, zeylanone, isozeylanone, plumbagic acid (*root*) and isoschinandone, sitosterol, plumbazeylone, a new dihydronaptho-quinone-dihydroserone-tyrosine (4.5), threonine (4.1), alanine (3.5), histidine (3.4), glycine (1.7), methionine (0.5) and hydroxyproline (0.2%) (*aerial parts*).

Biological Activity : Plumbagin (0.75 mg/kg) active against P388 lymphocytic leukaemia and at 4 mg/kg exhibits antibacterial and antifungal activity against a wide variety of bacteria and fungi. Besides aq. and alc. extr. has been observed to be a powerful aphicide.

Uses in Literature : Descibed earlier as an abortifacient, acrid poison, antiperiodic, aphicide, appetiser, diuretic, expectorant, expellent of phlegmatic tumours, psychoactive plant, stimulant, symbolic (worn on arm in body pain), uterine tonic (vet.); and for abdominal pain, abscess, anasarca, arthritis, black-water fever, boils, bone fracture (vet.), chest pain, contraction of uterus after childbirth, diarrhoea, diarrhoea (vet.), dropsy, dysentery, dyspepsia, ecbolic action in delayed labour, eczema, elephantiasis, enlargement of spleen, hydrocele, infection in horses, headache, heart-burn, influenza, inhibiting/preventing conception, kala-azar, leprosy, malaria, muscular pain, night blindness, pains after birth, paralysis, piles, rheumatic pains, severe headache, skin diseases (vet.), skin disorder, snakebite, spleen complaint, sprains, stomach disorders, stomatitis (vet.), vesicant properties, and wounds (Ambasta, 1986; Balasubramaniam & Prasad, 1996; Bhandary & Chandrasekar, 2002; Bhandary *et al.*, 1996; Borthakur, 1993; Brahmam & Saxena, 1990; Brahmam *et al.*, 1996; Chauhan, 1999; Chevallier, 1996; Chopra *et al.*, 1956; Choudhary, 1996; Dey & Bahadur, 1973; Duke *et al.*, 2002; Gill *et al.*, 1993; Girach *et al.*, 1996; Gogoi & Das, 2003; Goud & Pullaiah, 1996b; Hembrom, 1996; Hembrom & Goel, 2005; Jain, 1991, 2003; Jain & Sharma, 1996; Jain *et al.*, 1994; Jha & Varma, 1996; Khare, 2004; Kothari & Moorthy, 1996; Mahato & Chaudhary, 2005; Manandhar, 1987b; Nadanakunjidam & Abirai, 2003; Nayak *et al.*, 2003; Padhy & Dash, 2003; Prakash, 1999, 2001; Prajapati *et al.*, 2006; Pullaiah & Kumar, 1996; Rama Rao & Henry, 1996; Ranjan, 1996; Rao *et al.*, 1996; Rawat & Chowdhury, 1998; Reddy *et al.*, 1996; Said, 1997; Saklani & Jain, 1994; Sarin, 1996; Sensarma, 1991; Sharma & Borthakur, 2003; Sikarwar, 1996; Singh, 1991, 1999; Singh & Pandey, 1998; Singh *et al.*, 1996b; Siwakoti & Varma, 1996; Swami & Gupta, 1996; Srivastava *et al.*, 2003b, c; Subramaniam, 2000; Tirkey, 2004; Varma, 1997; Watt, 1972).

***Pogostemon benghalense* (Burm. f.) Kuntze** (Pl. 40C)

Syn.: *P. plectranthoides* Desf.; *Origanum benghalensis* Burm. f.; *O. indicum* Roth; *Mentha secunda* Roxb.; *M. fruticulosa* Roxb.

Family: Lamiaceae.

Vern. : Kali Basunti.

Regional Names:

B. : Bakoha, Jin, Jui-Lata

Bo. : Pangla

Ori. : Dumobadottoka

Tel. : Kasurijang.

Distribution : H. P.- E. Nepal, upto 1,300m.

Description : Strongly-scented, pubescent shrubs. Leaves opposite, oblanceolate. Flowers pinkish.

Reproductive Cycle : December-April.

Habitat Ecology : Wastelands. Scrub forests.

Material Examined : HPU-EL-SKS-5664; Balah (900 m), Banuri (1,150 m), Chhanchri (930 m), Gharthun (900 m), Rajpur (1,100 m).

Parts Used : Leaves. Roots.

Folk Uses : Leaves as such or with black pepper, *Allium cepa* and *Trachyspermum ammi* given for inducing **rumination** and checking **fever** (vet.). Root ash mixed with honey good against **cough**.

Uses in Literature : Known so far in literature as antihaemorrhagic, drink, edible (eaten in times of scarcity), febrifuge, haemostatic, green manure, stimulant, stomachic, styptic; and for bleeding, body pain, boils, cuts, detergent, diarrhoea, diarrhoea (vet.), and wounds (Ambasta, 1986; Brahmam & Saxena, 1990; Datt, 1996; Jain, 1991; Kothari & Moorthy, 1996; Mahato & Chaudhary, 2005; Mokat & Deokule, 2004; Rama Rao & Henry, 1996; Siwakoti & Varma, 1996).

***Polygala chinensis* L.** (Pl. 40D)
Syn.: *P. tlephioides* Willd.; *P. brachystachya* DC.
Family: Polygonaceae.

Vern. : Tone-baedu.

Regional Names:

B. : Maradu
G. : Pilibhonyasna
H. : Meradu, Miragu
Mar. : Negri, Phuntani.

Distribution : Throughtout India, ascending to 1,700 m.

Description : Pubescent annuals having erect or diffuse, 5-15 cm long stem, sessile leaves and yellow flowers crowded in axillary racemes. Calyx persistent. Capsule fringed, narrowly winged.

Reproductive Cycle : July-September.

Habitat Ecology : Shady places in hills.

Material Examined : HPU-EL-SKS-5468; Balah (900 m), Khaira (940 m), Makol (900 m), Malahu (1,050 m), Rangru (910 m).

Parts Used : Aerial Plant Parts. Seeds.

Folk Uses : Aerial plant parts given as **fodder**. Kids love to **play** with its seeds.

Active Constituents : Isolated constituents are suchilactone, chisulactone and helioxanthin.

Uses in Literature : Reported so far for to asthma, chronic bronchitis, catarrhal affections, fevers, dizziness; and as edible, tonic, expectorant, and emetic (Ambasta, 1986; Kirtikar & Basu, 1984; Rastogi & Mehrotra, 1999).

***Polygonum barbatum* L.** (Pl. 40E)
Syn.: *P. horemanni* Meisn.; *P. rivulare* Koen.; *P. marmoramae* Herb.;
P. fluviatile Herb.; *P. stagninum* Buch. - Ham. ex Meisn.;
Persicaria barbatum (L.) Hara
Family: Polygonaceae.

Vern. : Ghaneeri.

Regional Names:

B. : Bekhunjubaz
Mal. : Vellutamodelamukku
Mar. : Dhaktasheral

	P.	: Narri
	Tam.	: Atalari
	Tel.	: Kondamalle, Niruganneru.

Distribution : Throughout the warmer regions of India.

Description : Tall erect herbs with stout branches and alternate, lanceolate leaves. Flowers small, pink, bisexual and in spikes. Bracts ciliate. Nutlets trigonous, winged. Seeds flat.

Reproductive Cycle : April-July.

Habitat Ecology : Common along water sources. Wet places.

Material Examined : HPU-EL-SKS-5412; Balu (1,030 m), Bodhal (970 m), Darugnu (1,400 m), Gopalpur (1,350 m), Sungal (1,050 m).

Parts Used : Aerial Plant Parts.

Folk Uses : Aerial plant parts used as **fodder**; its paste applied locally for **skin ailments**.

Uses in Literature : Recorded earlier as an astringent, cicatrizant, emetic, fish poison, fodder, pot herb, purgative, tonic, and for snakebite. (Ambasta, 1986; Goel *et al.*, 1984; Megoneitso & Rao, 1983; Pal, 1984; Pal & Srivastava, 1976).

***Populus ciliata* Wall.** (Pl. 40F)

Syn.: *P. pyriformis* Royle; *P. rotudifolia* Griff.

Family: Salicaceae.

Vern. : Poplar.

English & Regional Names:

	Eng.	: Himalayan Poplar
	H.	: Pahari Peepal, Tilaunja
	P.	: Dudfrass, Falish, Kramali.

Distribution : Pakistan - S.W. China: 2,100-3,600 m.

Description : Trees. Leaves possessing ciliate teeth. Flowers in drooping racemes. Capsules ovoid. Seeds many with tuft of silky hairs.

Reproductive Cycle : April-July.

Habitat Ecology : Stream sides. Drier areas in coniferous forests.

Material Examined : HPU-EL-SKS-5588; Bandla (1,280 m), Bodhal (970 m), Chornali (1,280 m), Sulah (950 m), Sungal (1,050 m).

Parts Used : Whole Plant. Leaves. Wood.

Folk Uses : Plant grown as **avenue** tree. Leaves used as **fodder**. Wood employed for **furniture** and **building works**.

Uses in Literature : Reported earlier to be used as blood purifier, fodder, fuel, stimulant, tonic; and for bone fractures, constructions, cricket bats, fruit crater, furniture, limbs, matchboxes, packing, splints, turnery, veneers, and water troughs (Ambasta, 1986; Jain, 1991; Kapur & Nanda, 1996; Purohit & Samant, 1995; Singh *et al.*, 1996a; Watt, 1972).

***Populus nigra* L.**

Syn.: *P. nigra* var. *italica* (Munchh) Koehne.

Family: Salicaceae.

Vern. : Poplar.

English & Regional Names:

Eng. : Black Poplar, Lombardy Poplar

Kash. : Frasht

P. : Buins, Do, Farsh, Frast, Kramadi, Makkal, Pakshubut.

Distribution : N.W. Himalayas.

Description : Large-sized trees forming a narrow cylindric crown. Leaves cordate, crenate, petioled. Flowers small, unisexual, naked in pendulous catkins. Capsule oval. Seeds hairy.

Reproductive Cycle : April-July.

Habitat Ecology : Planted along roadsides.

Material Examined : HPU-EL-SKS-5693; Arla (1,100 m), Bindraban (1,270 m), Chachian (1,300 m), Ghugar (1,250 m), Saparu (1,250 m).

Parts Used : Whole Plant. Leaves. Wood.

Folk Uses : Plant grown as **shade tree**. Leaves lopped as **fodder**. Wood used in **building works** and as **fuel**.

Uses in Literature : So far known as coacher, depurative, fodder; and for bone fractures, carving, colds, cross-beams, furniture, haemorrhoids, hair growth, matches, poles, round boxes for grapes, small boxes, timber, truck, and barrio trays (Ambasta, 1986; Jain, 1991; Singh *et al.*, 1996a; Watt, 1972).

***Porana paniculata* Roxb.** (Pl. 41A)
Family: Convolvulaceae.

Vern. : Chhube-beil.

Regional Name:

U.P. : Belkamu, Safed bel.

Distribution : Throughout India, ascending to 1,000 m.

Description : Softly tomentose twining shrubs with very long, drooping lateral branches and stalked, long pointed leaves. Flowers numerous, small, white in terminal panicles. Corolla funnel-like. Capsule hairy. Seeds one, glabrous.

Reproductive Cycle : October-February.

Habitat Ecology : Wastelands. Rocky slopes.

Material Examined : HPU-EL-SKS-5665; Bhuana (1,400 m), Chhanchri (930 m), Dehan (1,130 m), Khaira (940 m), Malahu (1,050 m).

Parts Used : Leaves. Stem.

Folk Uses : Leaves given as **fodder**. Stem used for making **rope**.

Uses in Literature : Used to make baskets, and flowers for bridal bouquets (Ambasta, 1986).

***Portulaca grandiflora* Hook.** (Pl. 41B)
Syn.: *P. oleracea* L.; *P. suffruticora* Thwaites
Family: Portulacaceae.

Vern. : Kulfa.

English, Sanskrit & Regional Names:

Eng. : Common Purslane, Garden Purselane, The Sun Plant
Sans. : Brhalloni, Ghotika, Loni, Lonika, Lunica
B. : Baralooniya
G. : Ghol, Loni, Mahotiluni
H. : Baralaniya, Khursa, Kulfa
Kan. : Duddagooni, Soppu, Duda-gorai
Mal. : Koricchira
Mar. : Bhuigholi, Kurfah, Mhotighol, Pasalai
P. : Lonak, Dhami
Tam. : Karikeerai, Parpukare, Parukire, Pullikirai
Tel. : Ganga-pavilikura, Pappukura, Peddapavila.
Urdu : Khurfah.

Distribution : Common weed throughout India.

Description : Glabrous annual herbs having prostrate, succulent stem and rounded-truncate leaves with very short petiole. Flowers sessile, yellow, bisexual in terminal clusters. Capsule with reniform seeds.

Reproductive Cycle : July-August.

Habitat Ecology : Vacant plots. Fields. Weed in cultivated areas.

Material Examined : HPU-EL-SKS-5495; Bandla (1,280 m), Bodhal (970 m), Darugnu (1,400 m), Gopalpur (1,350 m), Tanda (1,120 m).

Parts Used : Aerial Parts.

Folk Uses : Aerial parts consumed as a **vegetable** ('saag') and considered good for **heart**, **kidney** and **liver** ailments as well as **diuretic** in nature.

Active Constituents : Noradrenaline, 4-(2-aminoethyl) pyrocatechol, 3-(3,4-dihydroxyphenyl) alanine, carbohydrates, lipids, glycosides, alkaloids, sterols, triterpenes and flavanoids have been isolated (*leaves, stem*).

Uses in Literature : Considered so far as bug destroyer, edible, laxative, muscle relaxant, pot-herb, refrigerant, scurvy, stomachic, tonic, urinary discharges, vermifuge; and for boils, bronchitis, burning sensation of hands and feet, chest pain, diarrhoea, diseases of liver, spleen and kidney, dysentery, epilepsy, gonorrhoea, headache, inflammations, leprosy, malaria, musculo-skeletal disorders, pneumonia, sore nipples, stranguary, tetanus, toothache, tumours and ulcers (Agarwal, 2003; Ambasta, 1986; Arora & Pandey, 1996; Bajpayee & Dixit, 1996; Bennet *et al.*, 1991; Bondya *et al.*, 2006; Borthakur, 1996b; Caius, 2003; Chatterjee & Pakrashi, 1991; Chopra *et al.*, 1956; Devi, 2003; Diwakar & Ansari, 2003; Duke *et al.*, 2002; Gogoi & Das, 2003; Gogoi *et al.*, 2003a; Harborne & Baxter, 2001; Jha *et al.*, 1996; Katewa *et al.*, 2000; Kaul, 1997; Khare, 2004; Kirtikar & Basu, 1984; Kurian, 1995; Maya *et al.*, 2003; Mitaliya & Bhatt, 2003; Murthy *et al.*, 2003; Nayak *et al.*, 2003; Parrotta, 2001; Pawar & Patil, 2004; Prajapati *et al.*, 2006; Prasad *et al.*, 1999; Punjani, 2002; Ranjan, 2000, 2003; Rawat & Chowdhury, 1998; Roy *et al.*, 1998; Said, 1997; Saini, 1996b; Saklani & Jain, 1994; Samant *et al.*, 2001; Sarkar *et al.*, 2003; Sharma, 2003; Singh, 1996b; Singh & Kumar, 2000a; Singh & Pandey, 1998; Sinha, 1996).

Prinsepia utilis Royle (Pl. 41C)
Family: Rosaceae.

Vern. : Bhekhal.

Regional Names:

H. : Bekhra, Cherara, Dhatila, Jhatela

P. : Bekhul, Gurinda, Kharnglara, Kharnguru, Phulwara, Tatua.

Distribution : Temperate Himalaya: 1,200-2,700m.

Description : Glabrous thorny shrubs. Leaves simple, lanceolate, leathery. Flowers white in axillary racemes. Drupes purple, one-seeded.

Reproductive Cycle : June-October.

Habitat Ecology : Forests. Shrubberies. Hedges.

Material Examined : HPU-EL-SKS-5655; Chornali (1,280 m), Palampur (1,280 m), Jiun (1,550 m), Kathena (1,290 m), Lolani (1,600 m),

Parts Used : Seeds. Stem. Roots.

Folk Uses : Branches hung on main door to ward off **evil spirits**. Root paste along with *Piper nigrum* and cow's urine applied to hasten **suppuration** of **boils**. Seed oil applied over locally to relieve **rheumatic pain**.

Active Constituents : Active principles due to new lignan-prinsepiol-along with epicatelhin and β-sitosterol-β-glucoside (*stems*), 1.8% myristic, 15.2% palmitic, 4.5% stearic, 0.9% lignoceric, 32.6% oleic and 45.6% linoleic acids (*seed oil*).

Uses in Literature : Used earlier as cooking edible (seed oil), illuminant, insect repellent, fuelwood, rubefacient, tonic; and for bodyache, burns, cuts, hydrogenation, joint pain, pain due to fatigue, rheumatism, stomach disorders, soap making, walking sticks, and wounds (Ambasta, 1986; Chauhan, 1999; Chopra *et al.*, 1956; Jain, 1991; Kapur & Srivastava, 1996; Kirtikar & Basu, 1984; Nautiyal *et al.*, 2003; Negi *et al.*, 1999; Panda, 1996; Purohit & Samant, 1995; Rana *et al.*, 1996; Roy *et al.*, 1998; Samant *et al.*, 2001; Sharma, 2003; Sood & Thakur, 2004; Verma & Chauhan, 2006; Watt, 1972).

***Prunus cerasoides* D. Don** (Pl. 41D)

Syn.: *P. puddum* (Wall.) Roxb.; *Cerasus puddum* Ser.

Family: Rosaceae.

Vern. : Paaja, Pajja.

English, Sanskrit & Regional Names:

Eng. : Himalayan Cherry
Sans. : Padmagandhi, Padmavriksha
Ass. : Dieng-Soh-Iong-Krem
B. : Padmak
Bo. : Padma-Kasta
G. : Padmak, Padmakathi
H. : Padam, Phaja
Mar. : Padmakastha
P. : Amalguchh, Pajja.

Distribution : H.P. - S.W. China: 1,200-2,500 m.

Description : Small-sized trees having glossy, long-pointed, toothed leaves. Flowers long-stalked, pink. Drupes ovoid, long, yellow or red.

Reproductive Cycle : November-May.

Habitat Ecology : Forests; often planted by wayside resting places.

Material Examined : HPU-EL-SKS-5670; Bhagpur (1,600 m), Chandpur (1,290 m), Dadh (1,100 m), Panchrukhi (990 m), Thandol (950 m).

Parts Used : Leaves. Stem. Fruit. Bark.

Folk Uses : Leaves lopped as **fodder** for livestock, and fomentation of its warm decoction good against **sprain** and **internal injuries**. Stem used for making **walking sticks, dip pens** and tooth brush for **scouring**. Bark paste used as plaster to heal **fractures**. Ripe fruits **edible**.

Active Constituents : Chief ingredients are chyrsophanol, emodin and their 8β-D-glucosides, orientalone, physcion, puddumin A, padmakastein and its derivatives, β-sitosterol behenate, tectochrysin, genistein, leucocyanidin and 4`-glucoside of genkwanin (*stem bark*), naringenin, apigenin, β-sitosterol, sakuranetin, prunetin and genkawanin (*stems*).

Biological Activity : Spasmolytic activity found positive.

Uses in Literature : Reported in literature as an analgesic, carminative, conceptive, edible (fruits), expectorant, febrifuge, psychoactive plant, tonic; and for asthma, bone dislocation and fractures, building, burning sensation of the body, asthma, cold, cough, cuts, diarrhoea, fodder, furniture, hair oil, hemicrania, hoof (vet.), injury (vet.), internal injuries, liquor preparation, marriages, muscular pain, pregnancy, religious ceremonies, rheumatic pains, rosaries, seminal weakness, sprains, stomach problem, tanning, venereal disease, walking sticks, and wounds (Agarwal, 2003; Ambasta, 1986; Ansari, 1991; Arora & Pandey, 1996; Bennet *et al.*, 1991; Bhuyan, 2003; Chauhan, 1999; Chhetri, 2005; Chopra *et al.*, 1956; Das & Agarwal, 1991; Dash *et al.*, 2003; Dey, 1980; Jain, 1991; Jain *et al.*, 1991, 1994; Karnick, 1994; Khare, 2004; Kirtikar & Basu, 1984; Lal *et al.*, 1996; Lalramnghinglova, 1996; Negi *et al.*, 1999; Prajapati *et al.*, 2006; Purohit & Samant, 1995; Roy *et al.*, 1998; Sharma, 2003; Singh & Kumar, 2000a; Sinha, 1996; Sood & Kaushal, 2008; Uniyal, 1968; Watt, 1972).

***Prunus domestica* L.** (Pl. 41E)

Syn.: *P. insititia* L.; *P. communis* Huds. var. *insititia* (L.) Hook. f.

Family: Rosaceae.

Vern. : Aloo Bukhara, Palam.

English & Regional Names:

Eng. : Plum Tree, Prum Tree, Common Plum
B. : Alubukhara
G. : Alubukhara
H. : Alubukhara.

Distribution : W. Himalaya.

Description : Trees branched and without spines. Receptacles hollow. Flowers stalked, pink. Drupes ovoid.

Reproductive Cycle : April-August.

Habitat Ecology : Planted.

Material Examined : HPU-EL-SKS-5694; Bindraban (1,270 m), Chachian (1,300 m), Chornali (1,280 m), Rangru (910 m), Sidhpur (1,100 m).

Parts Used : Fruits. Stem.

Folk Uses : Ripe fruits **edible** as such**, pickled** and used for making **jams** and **candies**. Stem employed as **fuelwood**.

Active Constituents : These are galacturmic acid, arabinose, galactose, xylose and rhamnose (*fruits*).

Biological Activity : Naringenin exhibits significant antiulcer activity in experimentally induced ulcers in pylorusligated rats (132 mg/kg) rats.

Uses in Literature : Known in India as aperient, astringent, cathartic, demulcent, digestive, edible (fruit), laxative, nutritious, purgative, refreshing, refrigerant; and for bilious states of the system, cabinet work, cold, confectionery, desserts, fever, illuminating purposes, inlay and turnery, jams, skeleton of paper matchboxes (Agarwal, 2003; Ambasta, 1986; Chopra *et al.*, 1956; Kirtikar & Basu, 1984; Singh & Kumar 2000a; Watt, 1972).

Prunus persica **Benth. & Hook. f.** (Pl. 41F)

Syn.: *Amygdalus persica* L.; *A. colinus* Wall.; *Persica vulgaris* Mill.; *P. saligna* Royle

Family: Rosaceae.

Vern. : Aadu, Aarnei, Aru.

English & Regional Names:

- **Eng.** : Peach Tree, Nectarine
- **H.** : Aru
- **Ori.** : Pishu
- **P.** : Aru, Bembi, Chirmaru, Katherti.

Distribution : N.W. Himalayas, cultivated.

Description : Glabrous trees having simple, toothed leaves and pink, pentamerous, bisexual flowers which are either solitary or in cymose. Drupes downy with irregularly furrowed stones.

Reproductive Cycle : February-May.

Habitat Ecology : Commonly planted near habitations.

Material Examined : HPU-EL-SKS-5503; Chauki (900 m), Gopalpur (1,350 m), Nanaun (900 m), Rajeher (1,050 m), Thakurdwara (1,150 m).

Parts Used : Whole Plant. Fruit. Seed.

Folk Uses : Plant used as a stock for **grafting** scion of almond (*Prunus amygdalus*). Fruits **edible**, and also **pickled**. Seed oil used for **body massage** for rheumatic pains.

Biological Activity : Alc. extr. of embryo and immature seeds exhibits analgesic and anti-inflammatory activities in rats.

Uses in Literature : Reported in India as an abortifacient, anthelmintic, antiemetic, antiscorbutic, cattle feed, demulcent, dessert, diuretic, dye, eczema, edible (fruits, kernels), emmenagogue, fertilizer, insecticidal, illuminant (oil), peach nectar, peach brandy, psychoactive plant, purgative, table fruit, tonic, vegetable; and for beverages, biliousness, blood diseases, building purposes, canning, carbonizing, charcoal, cough, deftness, earache, eczema, foot, eye and mouth disease, headache, hoof rot of cattle, itching and scabies, lessoning thirst, leucoderma, pickling, piles, removing bad smell of mouth and sputum, rheumatism, scabies, stomach troubles, and whooping cough (Agarwal, 2003; Ambasta, 1986; Arora & Pandey, 1996; Bennet *et al.*, 1991; Chopra *et al.*, 1956; Harborne & Baxter, 2001; Jain, 1991; Jain *et al.*, 1994; Kapur & Singh, 1996; Lal *et al.*, 1996; Nautiyal *et al.*, 2003; Parmar & Kaushal, 1982; Purohit & Samant, 1995; Rawat & Chowdhury, 1998; Roy *et al.*, 1998; Said, 1997; Sharma, 1996; Sharma & Rana, 2005; Singh & Kumar, 2000a; Sood & Thakur, 2004; Watt, 1972).

***Psidium guajava* L.** (Pl. 42A)

Family: Myrtaceae.

Vern. : Amrood.

English, Sanskrit & Regional Names:

Eng. : Guava tree
Sans. : Aprithaktvacha, Ovidhavija
Ass. : Madhuriam
B. : Goaachhi, Peyara
Bo. : Jam, Perala
G. : Jamrud, Peru
H. : Amrud, Safed Safari
Kan. : Pera, Koyya
Mal. : Koyya, Malakkapera
Mar. : Jamba, Tupkel
Ori. : Bodojambo, Julabojamo
P. : Anjirzarad, Amrud
Tam. : Segappugnoyya, Vellaikoyya
Tel. : Ettajama, Tellajama.

Distribution : Widely cultivated throughout India.

Description : Small trees having opposite, ovate-oblong leaves. Petals white. Fruit globose or pear-shaped. Pulp yellow with white or deep pink.

Reproductive Cycle : July-September.

Habitat Ecology : Cultivated.

Material Examined : HPU-EL-SKS-5719; Chimbalhar (1,150 m), Darugnu (1,400 m), Darugnu (1,400 m), Panapari (1,100 m), Sulah (950 m), Thandol (950 m).

Parts Used : Leaves. Fruit.

Folk Uses : Fruits **edible** and considered good against **conspitation**, and **diabetes**. Leaves chewed for curing **mouth ulcers**, **gum ailments** and **scouring teeth**.

Active Constituents : Active properties due to sesquiguavaene, ellagic acid, ellagitannin-guavin β and guavin A, C and D, galacturonic acid, galactose, arabinose with traces of xylose, rhamnose (*fruits*), and quercitin-3-α-arabinopyranoside (*leaves*).

Biological Activity : Spasmolytic activity of leaf mainly due to presence of quercetin. Gallocatechin exhibits antimutagenic activity against UV-induced mutation in *Escherichia coli*.

Uses in Literature : Reported earlier in India as an antibiotic, antiemetic, antinarcotic to overcome the influence of liquor, aperient, astringent, colic, colouration material in 'Mithila' paintings, edible, laxative, tongue cleaner, tonic; and for carbuncle, cholera, chronic fever, chronic infantile diarrhoea, chutneys, cooling, cough and cold, deep cuts, dyeing, dyspepsia, engraving, gonorrhoea, headache, inflammation, intestinal haemorrhage, jam, jaundice, ketchup, lac-turnery, malaria, marmalades, menstrual disorders, mouth wash for swollen gums, piles, pulmonary disorders, pustules in mouth, scurvy, sores, spear handles and instruments of various kinds, sprains, stomachache, stones, tanning, ulcer and wounds (vet.) (Ambasta, 1986; Augustine & Sivadasan, 2005; Bhandary *et al.*, 1996; Bhuyan, 2003; Borthakur, 1993; Chowdhury *et al.*, 1996; Gill *et al.*, 1993; Girach & Aminuddin, 1995; Hosagoudar & Henry, 1996 a, b; Jain, 1991, 2003; Jain *et al.*, 1997; Jha & Basak, 1994; Kumar & Rao, 2003; Lalramnghinglova, 1996; Panda, 1996; Pandey, 2003; Rajan *et al.*, 2003; Ranjan, 2003; Rawat & Chowdhary, 1998; Shome *et al.*, 1996; Singh & Maheshwari, 1992; Siwakoti & Varma, 1996; Thomas & Britto, 2003; Tirkey *et al.*, 2003a; Upadhyay & Singh, 2005; Verma *et al.*, 1995; Viswanathan & Singh, 1996; Watt, 1972).

***Punica granatum* L.** (Pl. 42B)

Syn.: *P. nana* L.

Family: Punicaceae.

Vern. : Daadu.

English, Sanskrit & Regional Names:

Eng. : Pomengrate Tree
Sans. : Dantabijaka, Pindapushpa, Vrittaphala
Ass. : Dalim
B. : Dalimgachh, Dalimphal
Bo. : Anara, Dalimba
G. : Dadam, Daramphal
H. : Anar, Dharimb
Kan. : Daalimbe

Mal.	: Pumatalam, Uruyampalam
Mar.	: Dalimba
Ori.	: Dalimbo
P.	: Anar, Danu, Daruna
Tam.	: Kalumalmandalai, Tusagam
Tel.	: Dadimanu, Dalimma, Puvvudanimma
U.P.	: Anar, Rumman.

Distribution : W. Himalayas: 700-2,700 m.

Description : A small tree, armed with axillary thorns. Leaves simple, opposite, whorled. Petals red. Fruits woody, globular, brownish-orange.

Reproductive Cycle : May-September.

Habitat Ecology : Gregarious on dry lime soils.

Material Examined : HPU-EL-SKS-5445; Bhagpur (1,600 m), Chandpur (1,290 m), Jiun (1,550 m), Lolani (1,600 m), Molichak (1,020 m).

Parts Used : Whole Plant. Rind. Seeds.

Folk Uses : Plant considered **sacred** and **worshipped** on religious ceremonies. Seeds **edible** and considered rich source of **vit. C**; its paste mixed with mint leaves, salt and pepper relished as '**chutney**'. Powdered rind taken one teaspoonful every morning with honey for seven days against **piles** and **cough**.

Active Constituents : Active properties due to cyanidin-3-gluoside delphinidin-3-glucoside and 3,5-diglucoside (*seed coat*), punicalin, punicalagin, punicafolin, granatin A, granatin B, conlagin, strictinin, 1,2,4,6-tetra-0-galloyl, β-D-glucose, a new hydrolysable tannin 2-0-galloyl-4,6(S,S)-galloyl-D-glucose (I) (*leaves*), 28% gallotannic acid together with a yellow colouring matter (*rind*), glucose, fructose, maltose, starch, carotene, thiamine, riboflavin, nicotinic acid, vitamin, pectin and amino acid (*seeds*).

Biological Activity : Fruit extr. exhibits high antibacterial and antiviral activity.

Uses in Literature : Known so far as an anthelmintic, anti-vomiting agent, astringent, cardiac, colouration material in 'Mithila' paintings, condiment, edible (seeds, 'anardana') expectorant, good source of vit. C, laxative, souring agent, stomachic, styptic and cicatrizing, symbolic (offering), tonic, yellowish dye, and for bleeding after childbirth and miscarriage, brain affections, bronchitis, chest troubles, cholera, congestion of the liver, conjunctivitis, controlling cholesterol level, cough, diarrhoea and dysentery, dyeing, dyspepsia, enlarged spleen, epistaxis, eye complaint, fever, gastric, general debility, gums, tone to the stomach, heart disorders, in dysentery, increasing sexual vigour, inflammation of the urinary tract and haemorrhage, iron, jaundice, kidney stones, leucorrhoea, malaria, menstrual disorders, nose bleeding, pimples, refreshing drink, sore throat, tannin, uterine diseases, and wine (Ambasta, 1986; Anonymous, 1990; Arinathan *et al.*, 2003; Arora & Pandey, 1996; Beigh *et al.*, 2003; Chevallier, 1996; Chopra *et al.*, 1956; Dagar & Dagar, 1996; Das & Agarwal, 1991; Dey, 1980; Dey & Bahadur, 1973; Ganai & Nawachoo, 2003; Ghosh *et al.*, 1996; Hosagoudar & Henry, 1996 a, c; Jain, 1991; Jain *et al.*, 1997; Jha & Basak, 1994; Kapur & Singh, 1996; Katewa *et al.*, 2001; Khanna *et al.*, 1996b; Kirtikar & Basu, 1984; Kulkarni & Kumbhojkar, 1993; Kumar & Rao, 2001, 2003; Kurian, 1995; Lalramnghinglova, 1996; Mahato & Mahato, 1996; Mohanty & Padhy, 1996; Pandey & Joshi, 2001; Parmar & Kaushal, 1982; Prajapati *et al.*, 2006; Pruthi, 1976; Purohit & Samant, 1995; Rana *et al.*, 1996; Ranjan, 2003; Roy *et al.*, 1998; Said, 1997; Saklani & Jain, 1994; Sanyal, 1994; Seetharam *et al.*, 1999; Sharma & Borthakur, 2003; Sharma & Rana, 2005; Shome *et al.*, 1996; Shukla *et al.*, 2003; Singh, 1996b, 1999; Singh & Kumar, 2000a; Singh & Pandey, 1998; Sood & Thakur, 2004; Verma & Chauhan, 2006; Tirkey *et al.*, 2003b; Watt, 1972).

***Pyrus communis* L.** (Pl. 42C)

Syn.: *P. pyrifolia* Nakai var. *culta* Nakai; *P. sinensis* Hort.

Family: Rosaceae.

Vern. : Nashpati.

English & Regional Names:

- **Eng.** : Pear, Wild Pear
- **H.** : Nashpati
- **Kash.** : Batang, Kistabahira
- **P.** : Batank, Nashpati, Tangi
- **Tam.** : Perrikkay
- **Tel.** : Berikaya
- **U.P.** : Bagugosha.

Distribution : Cultivated in N. W. Himalayas.

Description : Trees. Leaves upto 6.5 cm long, simple, deciduous. Flowers white, scented. Pome pyriform, yellow.

Reproductive Cycle : June-August.

Habitat Ecology : Cultivated.

Material Examined : HPU-EL-SKS-5722; Jiun (1,550 m), Kathena (1,290 m), Lahla (930 m), Lolani (1,600 m), Purani-palam (1,100 m).

Part Used : Fruit.

Folk Uses : Fruits **eaten** raw and considered **astringent** and **laxative** in nature, also employed for making **pickles** and **jams**.

Biological Activity : Lupeol, β-sitosterol, betulin, betulinic acid, hydroguinone and arbutin exhibited bacteriostatic activity against *Escherichia coli, Salmonella typhi, Shigella flexneri* and *Staphylococcus aureus.*

Uses in Literature : Used in India as an edible, psychoactive plant; and for acute fever, beverages, carving, engraving purposes, malaria, mathematical and drawing instruments, turning, and wines (Ambasta, 1986; Gill *et al.*, 1993; Jain *et al.*, 1994; Watt, 1972).

Pyrus pashia Ham. (Pl. 42D)

Syn.: *P. variolosa* Wall.; *P. verruculosa* Bertol.; *P. nepalensis* Herb. Ham. *ex* Hort.

Family: Rosaceae.

Vern. : Kainth.

Regional Names:

- **Ass.** : Chalthei, Soh-Jhur, Soh-Shur
- **H.** : Mehal, Mol
- **P.** : Batangi, Kaenth, Shehgal, Tang.

Distribution : Afghanistan-S.W. China, Myanmar: 750-2,700 m.

Description : Small-sized trees with crenate, long-pointed; leaves. Flowers white, tinged with pink. Fruits dark-brown, globular.

Reproductive Cycle : May-August.

Habitat Ecology : Shrubberies, cultivated areas.

Material Examined : HPU-EL-SKS-5451; Andretta (980 m), Bhagpur (1,600 m), Bhuana (1,400 m), Chornali (1,280 m), Jasun (1,000 m).

Parts Used : Whole Plant. Fruits. Stem. Leaves.

Folk Uses : Plant used as **stock** for grafting scion of almond (*Prunus amygdalus*). Extr. of overnight soaked fruits used as hair shampoo to check **dandruff**. Ripe fruits **edible** and unripe ones chewed for healing **mouth ulcers**. Leaves considered good **fodder** for goats. Stem good for making **walking sticks** and **handles** of agricultural tools.

Active Constituents : These are friedelin, β-sitosterol (*bark*), n-triacontane, myricyl alcohol as chloride and sitosterol (*leaves*).

Biological Activity : Aerial parts anticancerous and CNS depressant.

Uses in Literature : Recorded in India for bobbins, catarrh (vet), combs, edible purposes, eye complaint, injury (vet), mouth infection, textile mills, tobacco pipes, walking sticks; and as fodder and fuel (Ambasta, 1986; Arora & Pandey, 1996; Bennet *et al.*, 1991; Jain, 1991; Lal *et al.*, 1996; Purohit & Samant, 1995; Rana *et al.*, 1996; Rao, 1990; Roy *et al.*, 1998; Sahni, 1998; Sharma & Rana, 2005; Samant *et al.*, 2001; Singh, 1996b; Singh & Kumar, 2000b; Watt, 1972).

Pyrus serotina Rehder (Pl. 42E)

Family: Rosaceae.

Vern. : Bada-kainth.

Distribution : Throughout mid hill regions.

Description : Trees with rough, light brown coloured bark. Leaves cordate, stipulated, serrated, acuminate, glossy. Flowers white. Fruits globular, brown. Seeds dark brown, pyriform.

Reproductive Cycle : March-May.

Habitat Ecology : Along field borders.

Material Examined : HPU-EL-SKS-5401; Chandpur (1,290 m), Darugnu (1,400 m), Gopalpur (1,350 m), Holta (1,200 m), Sungal (1,050 m).

Part Used : Fruits.

Folk Uses : Ripe fruits **edible** and unripe ones used for making **pickle** and **jams**.

Uses in Literature : Known for its edible fruits (Parmar & Kaushal, 1982; Sharma & Rana, 2005).

Quercus leucotrichophora A. Camus (Pl. 42 F)
Syn.: *Q. incana* Roxb.; *Q. dealbata* Wall.; *Q. lanata* Sm.; *Q. oblongata* Don
Family: Fagaceae.

Vern. : Baan.

English & Regional Names:

Eng. : Grey Oak, Himalayan White Oak, Kumaon Oak, Ban Oak
H. : Ban, Banji
Kash. : Iri, Shiddar, Silaspuri
P. : Ban, Daghunban, Kharpatarserai.

Distribution : Pakistan - C. Nepal. Sri Lanka. Myanmar: 1,200-2,400 m.

Description : A large-sized tree. Leaves stalked, ovate-lanceolate, spinous towards the tip. Nut ovoid.

Reproductive Cycle : April-October.

Habitat Ecology : Gregarious and common in forests.

Material Examined : HPU-EL-SKS-5601; Chandpur (1,290 m), Chornali (1,280 m), Paror (940 m), Saliana (1,000 m), Thakurdwara (1,150 m).

Parts Used : Leaves. Wood.

Folk Uses : Leaves considered an excellent **fodder** for enhancing **lactation** in milching cattle. Wood good for **house construction**, **furniture** and **agricultural implements**.

Active Constituents : *Leaves* afford quercetin and its 3-0-dissacharide.

Biological Activity : Hypotensive activity and effect of CNS found positive.

Uses in Literature : Useful in India as an astringent, diuretic, edible (acorns), fodder; and for agricultural implements such as ploughs, handles and beams, asthma, building, charcoal, confectionery, diarrhoea, gonorrhoea, hardboards, indigestion, scabies, skin diseases, tanning, urinary complaints (Ambasta, 1986; Jain, 1991; Jain *et al.*, 1991; Kapur & Nanda, 1996; Negi & Gaur, 1991; Purohit & Samant, 1995; Sharma, 2003; Uniyal & Chauhan, 1973b; Watt, 1972).

***Randia dumetorum* (Retz.) Lam.** (Pl. 43A)

Syn.: *R. nutans* DC. & Wight & Arn.; *R. brandisii* Gamble; *R. stipulosa* Miq.; *R. spinosa* Blume; *R. malabarica* Wall.; *Gardenia nutans* Roxb.; *G. spinosa* L.;*G. longispina* Roxb.; *G. floribunda* Roxb.; *G. dumetorum* Retz.; *G. stipularis* Rottl. & Willd.; *G. glabra* Brandis; *G. propinqua* Brandis; *Posoqueria dumetorum* Roxb.; *P. nutans* Roxb.; *P. longispina* Roxb.; *P. floribunda* Roxb.; *Ceriscus malabaricus* Gaertn.; *Catuneregum nutans* (DC.) Tiruv.; *Xeromphis spinosa* (Thunb.) Keay

Family: Rubiaceae.

Vern. : Rada.

English, Sanskrit & Regional Names:

Eng. : Common Emetic Nut
Sans. : Bastishodhana, Ramachhardanaka
Ass. : Bahmona, Gural, Mon
B. : Madan
Bo. : Gehala, Gelaphal
G. : Medhola
H. : Arar, Madanphal
Kan. : Banegara, Kara, Mangri
Kash. : Kirkla, Kokoa
Mal. : Karalikkaya, Kattunaranna
Mar. : Gelaphela, Monigeli
Ori. : Potova
P. : Mandkolla, Mindhal
Tam. : Karadam, Murkkarai
Tel. : Madanamu, Mangara.

Distribution : Sub-tropical Himalayas upto 1,200 m.

Description : Spiny trees possessing ovate, entire leaves. Flowers white or yellow, fragrant. Fruits globose, greenish-purple.

Reproductive Cycle : April-September.

Habitat Ecology : Scrub forests. Shrubberries.

Material Examined : HPU-EL-SKS-5426; Bhauda (1,050 m), Panapari (1,100 m), Rangru (910 m), Thandol (950 m), Tikkar (910 m).

Parts Used : Leaves. Thorns. Fruits. Wood.

Folk Uses : Leaves given as **fodder** for sheep and goats. Ripe fruits **edible**. Bandage of fruit paste applied for checking **headache**. Thorns studded in tail of cattle for curing infection (**'nanaain'**). Powdered 1-2 g of its fruits filled with 'suhaaga' (borax), latex of *Ficus palmata* and subsequently covered with a thin layer of dough given with honey to infants to check **pulmonary ailments**. Wood employed as **fuel**.

Active Constituents : These are dumetoronins A, B, C, D, E, F, β - D - galactopyra - nosyl (1→3) - deanolic acid, lupeol, two new triterpenoid glycosides (*fruit*), α-amyrin, β-sitosterol, oleanolic acid, ursolic acid, mannitol (*heartwood*), and randialic acid-A and randialic acid-B (*bark*).

Uses in Literature : Known earlier as an abortifacient, anodyne, anthelmintic, antispasmodic (for teething ailments of children), astringent, calmative, colour intensifier, diaphoretic, edible (ripe fruit & fruits eaten after roasting or cooking), emetic, expectorant, fish-poison, fodder, insect repellent, nauseant, nervine, sacred tree, sedative, and for biliousness, bobbins, bodyache during fever, boils, cotton-reels, cramps, diarrhoea, dyeing, dysentery, fever, jaundice, lactation, insecticide preparation, malaria, mathematical instruments, pimples, ploughs, wood for walking sticks, rheumatism, sores, umbrella-handles, whooping cough, vomiting and soap industry (Ambasta, 1986; Bennet *et al.*, 1991; Bose, 1984; Chauhan, 1999; Chopra *et al.*, 1956; Das & Agarwal, 1991; Dey & Bahadur, 1973; Jain, 1991; Kirtikar & Basu, 1984; Lalramnghinglova, 1996; Prajapati *et al.*, 2006; Roy *et al.*, 1998; Sharma, 2003; Sharma & Rana, 2005; Singh & Prakash, 1996; Watt, 1972).

***Ranunculus arvensis* L.** (Pl. 43B)

Syn.: *R. tuberculatus* DC.

Family: Ranunculaceae.

Vern. : Jaldhar.

English & Regional Names:

Eng. : Corn crowfoot, Crow's claw, Devil's claw, Devil's coach wheel, Devil's curry comb, Dill cup, English stavesacre, Field crowfoot, Goldweed, Field hog, Hellweed, Horsegold, Hungerweed, Joy, Peagle, Prickle-bags, Starveacre, Urchin Crow Foot, Yellowcup

P. : Mandkolla, Mindhal.

Distribution : Kashmir - Kumaon (W. Himalayas).

Description : Annuals with branched smooth pale green stem and radical, stalked, wedge-shaped leaves. Flowers yellow. Fruits globose. Achenes prickly.

Reproductive Cycle : April-June.

Habitat Ecology : Shady places.

Material Examined : HPU-EL-SKS-5671; Jiun (1,550 m), Ladoh (980 m), Lolani (1,600 m), Nanaun (900 m), Saparu (1,250 m).

Part Used : Leaves.

Folk Use : Leaf-paste applied for healing **boils**.

Uses in Literature : Useful against asthma, intermittent fever, gout; and as antibiotic, febrifuge and vegetable (leaves) (Ambasta, 1986; Chopra *et al.*, 1956; Jain *et al.*, 1991; Kapur & Singh, 1996; Kirtikar & Basu, 1984; Samant *et al.*, 2001; Singh & Kumar, 2000a).

***Ranunculus laetus* Wall. *ex* Don** (Pl. 43C)

Syn.: *R. distans* Royle; *R. brevirostris* Edgew.

Family: Ranunculaceae.

Vern. : Jaldhar.

Distribution : Temperate Himalayas.

Description : Hairy, branched, perennials herbs with woody rootstock and glossy, smooth, leaves which are cut into oblong segments. Flowers yellow. Achenes glabrous.

Reproductive Cycle : March-April.

Habitat Ecology : Marshy grounds. Banks of irrigated channels.

Material Examined : HPU-EL-SKS-5676; Bindraban (1,270 m), Darugnu (1,400 m), Purani-palam (1,100 m), Rajeher (1,050 m), Sungal (1,050 m).

Parts Used : Aerial Plant Parts. Leaves.

Folk Uses : Aerial plant paste applied for suppuration of **boils**. Leaf-paste considered good for **skin diseases**.

Uses in Literature : So far there exists no record of its use in literature.

***Ranunculus muricatus* L.** (Pl. 43D)

Syn.: *R. cabulicus* Boiss.

Family: Ranunculaceae.

Vern. : Changer.

Distribution : Punjab, Kashmir-W. Asia, Europe, Temperate N. America.

Description : Diffuse annuals having erect stem upto 30 cm. Leaves radical, heart-shaped at base, upper ones wedge-shaped.

Flowers yellow, solitary. Fruits a globose head of flat, spinous achenes.

Reproductive Cycle : March-June.

Habitat Ecology : Moist-shady places.

Material Examined : HPU-EL-SKS-5556; Bhagpur (1,600 m), Boda (1,000 m), Dadh (1,100 m), Paror (940 m), Tapa (945 m).

Parts Used : Aerial Plant Parts.

Folk Use : Aerial plant parts consumed as **vegetable** ('saag').

Uses in Literature : Used so far for asthma, edible, gout, intermittent fevers; and as acrid, rubefacient, vesicant (Ambasta, 1986; Caius, 2003; Chandra, 1997; Chopra *et al.*, 1956; Kapur & Srivastava, 1996; Kirtikar & Basu, 1984; Samant *et al.*, 2001; Singh & Kumar, 2000a).

***Raphanus sativus* L.** (Pl. 43E)

Family: Brassicaceae.

Vern. : Mooli.

English, Sanskrit & Regional Names:

Eng. : Garden Raddish, Haripura, Hastidanta, Kunjor

Sans. : Mulaka, Bhumikakshara

B. : Mula

Bo. : Mula, Muro

G. : Muli, Mula, Muro, Muri

H. : Muli, Muro

Kan. : Mullangi

Mar. : Mula, Muri

P. : Muli, Mungra, Taramera

Tam. : Mullangi

Tel. : Mullangi.

Distribution : Throughout India upto 5,000 m in the Himalayas.

Description : Cultivated biennial herbs having fleshy roots and rough pilose leaves. Flowers usually white or lilac with purple veins. Pods indehiscent. Seeds pendulous, globose.

Reproductive Cycle : April-May.

Habitat Ecology : Cultivated land.

Material Examined : HPU-EL-SKS-5679; Bhagpur (1,600 m), Bodhal (970 m), Dehan (1,130 m), Holta (1,200 m), Thandol (950 m).

Parts Used : Leaves. Fruit. Root.

Folk Uses : Leaves and fruits used as **vegetable**, and considered good against **jaundice**. Roots raw eaten as **salad** and for making **soups**.

Active Constituents : These are (–) sulforaphene from (*seeds*). 2-O-(p-coumaroyl)-, 2-O-caffeoyl-and 2-o-feruloyl-malates (*leaves*).

Uses in Literature : Used in India as anthelmintic, carminative, diuretic, emmanegogue, expectorant, laxative, stomachic, tonic; and for bronchial difficulty, cholera, earache, fever, heart ailments, hic-cough, impotency, inflammations, leprosy, paralysis, piles, spleen, tumours, urinary complaints and whooping cough (Ambasta, 1986; Caius, 2003; Chandrasekar & Srivastava, 2003; Chatterjee & Pakrashi, 1991;Khanna & Kumar, 2000; Kirtikar & Basu, 1984).

Reinwardtia indica **Dumort.** (Pl. 43F)

Syn.: *R. trigyna* (Roxb.) Planch.; *R. tetragyna* Planch.; *R. repens* Planch.; *Linum trigynum* Roxb.; *L. repens* Don; *Macrolinum trigynum* Reichb; *Kittelocharis trigyna* Alef.

Family: Linaceae.

Vern. : Basant.

Regional Names:

- **Bo.** : Abai
- **H.** : Basanti
- **P.** : Balbasant, Gudbatal, Karkum
- **U.P.** : Basant.

Distribution : Pakistan - S.W. China. India. Myanmar. S. E. Asia; upto 1,800 m.

Description : Shrubs having entire, ovate-lanceolate leaves. Flowers bright yellow arranged in racemose. Capsule globose.

Reproductive Cycle : November-May.

Habitat Ecology : Edges of terraced fields. Grazing grounds, common.

Material Examined : HPU-EL-SKS-5553; Balu (1,030 m), Bhauda (1,050 m), Chhanchri (930 m), Gharthun (900 m), Rangru (910 m).

Parts Used : Flowers. Leaves.

Folk Uses : Flowers **offered** to appease Goddess Parvati. Mixture of its paste cooked into '**pakodas**'; also applied locally for healing **cuts** and **injuries**. 50 ml of its decoction and leaves taken thrice daily as **heart tonic**.

Uses in Literature : Known to be useful earlier as an edible (flowers), offering, poison; and for maggot infested wounds, 'founder' disease in cattle, paralysis, typhoid, wound (vet.) and whitlow (Ambasta, 1986; Chopra *et al.*, 1956; Jain, 1991; Kirtikar & Basu, 1984; Lal *et al.*, 1996; Manandhar, 1990; Purohit & Samant, 1995; Singh & Kumar, 2000a; Thakur, 2005; Watt, 1972).

***Rhamnus triquetra* Wall.** (Pl. 44A)

Syn.: *Ceanothus triquetra* Wall.

Family: Rhamnaceae.

Vern. : Guldarein.

Regional Name:

P. : Gardhan, Gudlei, Phagora, Phulla, Rangrek.

Distribution : N.W. Himalayas: 1,000-1,700 m.

Description : Shrubs with tomentose shoots and simple, lanceolate leaves. Flowers greenish-white. Fruits obovoid, 3-lobed.

Reproductive Cycle : May-August.

Habitat Ecology : Common; forest areas.

Material Examined : HPU-EL-SKS-5580; Balah (900 m), Boda (1,000 m), Khaira (940 m), Nadli (1,000 m), Tapa (945 m)

Parts Used : Leaves. Wood.

Folk Uses : Leaves given as **fodder**. Wood used as **fuel** and for making **handles** of agricultural implements.

Active Constituents : These are chrysophanol, emodin, physcion, kaempferol, β-sitosterol and its glucoside (*whole plant*), and frangulin (*heartwood*).

Biological Activity : CNS depressant (emodin) and anti-inflammatory (kaempferol 7-methyl ether) activities found positive.

Uses in Literature : Used earlier as an astringent, blood purifier, deobstruent, tonic (vet.); and for boils, fodder and skin diseases (Jain, 1991; Sood & Kaushal, 2008; Watt, 1972).

Rhododendron arboreum **Sm.** (Pl. 44B)

Syn.: *R. puniceum* Roxb.; *R. cinnamomeum* Wall.; *R. album* Sweet; *R. campbelliae* Hook. f.; *R. windsorii* Nutt.; *R. nilagiricum* Zenk.

Family: Ericaceae.

Vern. : Braah, Brass.

English & Regional Names:

- **Eng.** : Tree Rhododendron, Rose Tree
- **B.** : Baras
- **H.** : Burans
- **Mal.** : Kattupuvarasu
- **P.** : Ardawal, Bras, Mandal
- **Tam.** : Alingi, Billi.

Distribution : Temperate Himalayas: 1,500-3,600 m.

Description : Evergreen trees having oblong-lanceolate leaves which are woolly beneath. Flowers large, blood red in head-like corymbs. Capsule scaly, curved, longitudinally ribbed.

Reproductive Cycle : March-May.

Habitat Ecology : Forests. Shrubberies.

Material Examined : HPU-EL-SKS-5700; Bhagpur (1,600 m), Lolani (1,600 m).

Parts Used : Flowers. Wood.

Folk Uses : Flowers raw **eaten** or made into **pakoras**, **chutney**; its sugary decoction considered **refreshing cardiac drink** and highly effective against **nose bleeding**, **burning sensation of soles** and **astringent to bowels**. Wood used for making **agricultural implements**, **tool handles**, and as **fuel**.

Active Constituents : These are taraxerol, ursolic acid, betulinic acid (*bark*) and ursolic acid, quercetin, resin (*flowers*).

Biological Activity : Aq. ethanolic and methanolic extrs. of flowers exhibit significant anti-inflammatory activity against carrageeuin, PG(E2), histamine, and 5HT induced rats hind paw oedema in decreasing order of activity.

Uses in Literature : Known earlier as refreshing drink and vegetable; and for amoebic dysentery, boxes, building, chutneys, diarrhoea, fever, fuel, gum-stocks, headache, jam, jelly, kukri, nose bleeding, park-saddles, plywood, posts, rheumatism, smoking pipe, tool handles, wounds, decorations of temples, blood dysentery, jaundice, clearing fish bone stuck in throat, poisoning rats, and plates (Agarwal, 2003; Ambasta, 1986; Arora & Pandey, 1996; Bennet *et al.*, 1991; Chauhan, 1984; Chhetri, 2005; Chopra *et al.*, 1956; Dash *et al.*, 2003; Jain,

1991; Kirtikar & Basu, 1984; Lal *et al.*, 1996; Mahato & Chaudhary, 2005; Manandhar, 1990; Panda, 1996; Ranjan, 2000; Roy *et al.*, 1998; Sahni, 1998; Samant *et al.*, 2001; Sharma, 2003; Singh, 1996b; Singh & Kumar, 2000a; Sinha, 1996; Watt, 1972).

***Ricinus communis* L. (Pl. 44C)**

Syn.: *R. inermis* Jacq.; *R. lividus* Jacq.; *R. speciosus* Burm.; *R. spectabilis* Blume; *R. viridis* Willd.; *Croton spinosu* L.

Family: Euphorbiaceae.

Vern. : Aieran, Arand, Erand.

English, Sanskrit & Regional Names:

Eng. : Castor Oil Plant, Palma-christi
Sans. : Dirghadantaka, Triputiphala, Vyadatvaka
Ass. : Eri
B. : Bherenda
Bo. : Erendi
G. : Diveligo
H. : Arandi, Erand, Ind
Kan. : Haralu
Mal. : Avanakka, Chittavanaku
Mar. : Yorandicha
Ori. : Bheronta, Tindoko
P. : Aneru, Bedanjir, Harnauli
Tam. : Amanakku, Muttukkottlai
Tel. : Chittamudamu
U.P. : Rend.

Distribution : Naturalized in Himalayas upto 2,400 m.

Description : Tall evergreen shrubs possessing 2-10 palmately-lobed, peltate leaves. Flowers yellowish-red on terminal branches. Capsule prickly. Seeds shining black.

Reproductive Cycle : March-May (intermittent flowering).

Habitat Ecology : Common in hills in the wastelands.

Material Examined : HPU-EL-SKS-5410; Bandla (1,280 m), Chachian (1,300 m), Darugnu (1,400 m), Jiun (1,550 m), Saparu (1,250 m).

Parts Used : Leaves. Seeds. Stem.

Folk Uses : Seed paste prepared in butter good for **cutaneous** diseases. Oil smeared warm leaves applied over furuncles to hasten **suppuration**. Stem used for **scouring teeth**. Seed oil used against **constipation** and also massaged for relief from **rheumatic pains**.

Active Constituents : Various constituents isolated are hyperoside, rutin, ricinine (*flowers*), apigenin, chlorogenic (*endosperm*), β - amyrin, stigmasterol, β-sitosterol, ricinine, quercetin, rutin, hyperoside, n - dimethylricinine, 3-0 - β-d-xylokpyranosides, 3-0-β-d -glucopyranosides and 3-0-β-D- rutinosides of kaempferol and quercetin (*leaves*).

Biological Activity : Anticholestatic and hepatoprotective activities found to be +ve.

Uses in Literature : Described earlier as an anthelmintic (vet.), antidote to food poison, antipyretic (vet.), anti-rheumatic, cathartic, contraceptive jellies, edible, emmanagogue, fish poison, fodder, fuel, insulating materials, lactagogue, laxative, lubricant, lumbago, luminant, manure, psychoactive plant, skin and hair dressing, silkworm host, soothing agent used in ointments, symbolic (wards off evil), vermicide; and for abortion, amenorrhoea, anasarca, asthma, backache, body swelling, boils, bone fracture (vet.), burns, candles, cholera, chronic ulcer, constipation (vet.), cough, culinary purposes, cutaneous affections, dental caries, dermatitis, dropsy, dyeing, dysentery, dysentery (vet.), eczemas, facilitating childbirth, fever, flatulence, fly-paper, gum trouble, hard and dry wax in ears, headache, heat stroke, heat-sealing adhesives, hydrocele, imitation leather, inflammation of breast during lactation, inflammation of eyes, injury, jaundice, loosening of limbs, muscular pain, nervous disorders, night blindness, plastic moulding composition, pneumonia, poultice to boils and sores, preservation of fruits and vegetables, preserving leather and hair oil, producing sterility for one month, rheumatism (vet.), scorpion and snake-bites, soap making, sores, sprain and muscle cramps, swelling, synthetic nylon, transparent soaps, typewriter inks, ulcers, warts, wound (vet.), wounds, and in bone sprain of horses (Ambasta, 1986; Bhandary *et al.*, 1996; Bhatt *et al.*, 2003b; Borthakur, 1993; Dixit & Kumar, 2003; Hembrom, 1996; Hembrom & Goel, 2005; Huidrom, 1996; Jain, 1968, 1991; Jain & Saklani, 1992; Jain *et al.*, 1994; Jha & Kumar, 2003; Jha *et al.*, 1991; Joshi & Joshi, 2005; Katewa & Arora, 1997; Khanna *et al.*, 1996a; Kharkonger & Joseph, 1981; Kumar *et al.*, 2004; Lalramnghinglova, 1996; Mahato & Chaudhary, 2005; Mahato & Mahato, 1996; Manandhar, 1996; Mao, 1993; Masih, 2003; Mishra *et al.*, 1996a, 1997; Mokat & Deokule, 2004; Nayak *et al.*, 2003; Padhy & Dash, 2003; Panda, 1996; Pandey, 2003; Pawar & Patil, 2004; Rajendran *et al.*, 2003; Rama Rao & Henry, 1996; Ranjan, 2003; Rawat & Chowdhury, 1998; Saini, 1996a; Satapathy & Brahmam, 1996; Sharma & Borthakur, 2003; Sharma *et al.*, 1979; Shome *et al.*, 1996; Singh *et al.*, 1996b, 1997; Siwakoti & Varma, 1996; Srivastava *et al.*, 2003b; Tirkey *et al.*, 2003b; Tiwari & Tiwari, 1996; Vihari, 1995; Viswanathan & Singh, 1996; Watt, 1972).

***Rorippa nasturtium-aquaticum* (L.) Hayek** (Pl. 44D)

Syn.: *Nasturtium officinale* R. Br.; *N. fontanum* Asch.; *Sisymbrium nasturtium-aquaticum* L.

Family: Brassicaceae.

Vern. : Chhoo-nali.

English & Regional Names:

Eng. : Water Cress

P. : Piriya halim.

Distribution : Throughout India. Asia. Temperate Europe.

Description : Much-branched herbaceous plant with pinnate, fleshy leaves and white flowers in short racemes. Pods linear. Seeds small.

Reproductive Cycle : April-September.

Habitat Ecology : Bogs. Irrigated Channels, Streamsides, Damp Places.

Material Examined : HPU-EL-SKS-5642; Bandla (1,280 m), Bhagpur (1,600 m), Chachian (1,300 m), Dharaman (1,100 m), Rangru (910 m).

Parts Used : Aerial Plant Parts.

Folk Uses : Aerial plant parts used as a **pot-herb**, and considered good for **improving eyesight**.

Active Constituents : *Seeds* contain gluconasturtiin and a non-drying fatty oil (24%).

Uses in Literature : Used so far as an appetizer, antiscorbutic, blood-purifier, chronic bronchitis, diuretic, pickle, pot-herb, salad, soups, stimulant; and for asthma, chest affections, cold, dry throat, goitre, kidney trouble, polypus of nose, prolonged illness, tuberculosis and stranguary (Ambasta, 1986; Arora, 1997; Arora & Pandey, 1996; Bennet *et al.*, 1991; Bhargava, 1959; Chandrasekar & Srivastava, 2003; Chevallier, 1996; Chhetri, 2005; Chopra *et al.*, 1956; Dash *et al.*, 2003; Devi, 2003; Panda, 1996; Prajapati *et al.*, 2006; Ranjan, 2000; Samant *et al.*, 2001; Sood & Thakur, 2004; Uphof, 1968; Watt, 1972).

***Rosa brunonii* Lindl.** (Pl. 44E)

Syn.: R. lindleyi Wall.; R. moschata auct (non Mill.) Hook. f.; R. pubescens Roxb.; R. glandulifera Roxb.; R. recurva Roxb.

Family: Rosaceae.

Vern. : Galabadi.

English, Sanskrit & Regional Names:

Eng. : Himalayan Musk Rose
Sans. : Sewati
H. : Karer, Kuji, Kunja, Kwiala
Kash. : Phulwari, Chal
P. : Tarni.

Distribution : Temperate Himalayas: 1,350-2,650 m.

Description : Climbing prickly shrubs. Flowers white. Fruits globose, dark brown.

Reproductive Cycle : April-October.

Habitat Ecology : Frequent on slopes. Waste places.

Material Examined : HPU-EL-SKS-5691; Andretta (980 m), Bhuana (1,400 m), Darugnu (1,400 m), Jiun (1,55 m), Lolani (1,600 m).

Part Used : Flowers.

Folk Use : Flowers paste considered a good **astringent** for cuts.

Uses in Literature : Considered useful for eye troubles, skin diseases, headache, opthalmia, making walking sticks, perfumery, wound; as fodder and edible (fruits) (Ambasta, 1986; Aswal & Mehrotra, 1994; Kapur & Srivastava, 1996; Singh & Kumar, 2000a; Sharma, 2003; Sood & Thakur, 2004).

***Rosa chinensis* Jacq.** (Pl. 44F)

Syn.: *R. indica* L.; *R. sinica* L.

Family: Rosaceae.

Vern. : Galaab, Gulaab.

Distribution : Cultivated in India.

Description : Glabrous evergreen plant with narrow stipules which are adnate nearly to the apex. Flowers large on long slender pedicels. Calyx segments reflexed. Fruit a hip.

Reproductive Cycle : Throughout the year.

Habitat Ecology : Temple and house courtyards. Hedges.

Material Examined : HPU-EL-SKS-5504; Banuri (1,150 m), Chauki (900 m), Gharthun (900 m), Pahra (950 m), Sungal (1,050 m).

Part Used : Whole Plant. Flowers.

Folk Uses : Plant grown as an **ornamental** hedge. Flowers **offered** to deities; and its petals used for preparing '**gulakand**' which is considered good against digestive disorders.

Active Constituents : *Flowers* possess linalool acetate, limonene, methoxy-methylbenzene, citronellol, β-caryophyllene, hexanol acetate, linalool and hexanol.

Uses in Literature : Used so far for treating sprains, injuries, foul ulcers and wounds, (Ambasta, 1986; Kirtikar & Basu, 1984).

***Rosa damascena* Mill.** (Pl. 45A)

Family: Rosaceae.

Vern. : Gulabadi.

English, Sanskrit & Regional Names:

Eng. : Bussora Rose, Damask Rose, Persian Rose, Summer Damask Rose

Sans. : Atimanjula, Lakshapushpa, Mahakumari, Shatapatri

B. : Gulab, Sudburg

Mal. : Penimirpushpam

Tam. : Irosa

Tel. : Gulabri, Pannivu, Roja.

Distribution : Throughout India.

Description : Prickly shrubs. Leaves compound with five serrated ovate-oblong leaflets. Stipules dilated. Flowers corymbose, red. Fruits a hip, obovate.

Reproductive Cycle : April-June.

Habitat Ecology : Wild as a hedge plant.

Material Examined : HPU-EL-SKS-5640; Arla (1,100 m), Bhauda (1,050 m), Dadh (1,100 m), Jasun (1,000 m), Tikkar (910 m).

Parts Used : Flowers.

Folk Uses : Flowers **offered** to various deities. Fermented mixture of its petals with sugar (**'Gulakand'**) good for **digestive disorders**. Petal decoction also considered good for **leucorrhoea** and **conspitation**.

Active Constituents : Aromatic constituents of *rose oil* (otto of rose) are rhodinol and geranyl acetate. Cyanidin-3,5-diglucoside isolated from *petals*.

Uses in Literature : Reported so far to be acrid, antipyretic aphrodisiac, appetiser, astringent, bitter, burning sensation, cardiotonic, eye lotion, expectorant, laxative; and for chronic fever, cooling, excessive perspiration, headache, leprosy, stomatitis, tonic, tonsillitis and toothache (Ambasta, 1986; Bhatt *et al.*, 2001; Harborne & Baxter, 2001; Kirtikar & Basu, 1984; Mehtani & Ingle, 1999; Said, 1997; Sanyal, 1994; Sharma, 2003).

***Roylea cinerea* (D. Don) Baill.** (Pl. 45B)

Syn.: *R. elegans* Wall.; *R. calycina* (Roxb.) Briq.; *Ballota cinerea* Don; *Phlomis calycina* Roxb.

Family: Lamiaceae.

Vern. : Kakdoen.

Regional Names:

H. : Patkarru

P. : Kaur, Kauri.

Distribution : Sub-Tropical W. Himalayas: 600-2,400 m.

Description : Hoary shrubs with shortly-stalked, coarsely toothed leaves. Flowers white, tinged with pink in whorls. Nutlets smooth.

Reproductive Cycle : March-November.

Habitat Ecology : Open rocky slopes.

Material Examined : HPU-EL-SKS-5439; Bhaura (940 m), Chimbalhar (1,150 m), Kathena (1,290 m), Purani-palam (1,100 m), Sulah (950 m),

Part Used : Leaves.

Folk Uses : Crushed leaves mixed with wheat flour fed to cattle as **blood purifier**, **stomachache** and for **sores** all over body. Leaf-paste applied for suppuration of **boils**, and its juice good against **lice** (vet).

Active Constituents : *Leaves* yield betulin, β-amyrin, β-sitosterol, stigmasterol, cetyl alcohol, glucose, fructose, arabinose and palmitic, stearic, oleic, oxalic and tartaric acids.

Biological Activity : Plant extr. depresses CNS and relaxed skeletal muscle activities.

Uses in Literature : Known so far as bitter tonic, blood purifier, febrifuge, snuff in tonsil; and for contusion produced by winds, fever, and pimples (Ambasta, 1986; Jain, 1991; Watt, 1972).

Rubus ellipticus **Sm.** (Pl. 45C)

Syn.: *R. rotundifolius* Wall.; *R. gowry-phul* Roxb.; *R. gowree-phul* Roxb.; *R. flavus* Ham.; *R. sessilifolius* Miq.; *R. hirsutus* Roxb.; *R. wallichianus* Wight & Arn.

Family: Rosaceae.

Vern. : Aakhe.

English & Regional Names:

- **Eng.** : Himalayan Yellow-Raspberry
- **Ass.** : Jotelupoka
- **H.** : Anehhu, Hinsalu
- **Kash.** : Gouriphal
- **P.** : Akhi.

Distribution : S.E. Asia: 600-2,300 m.

Description : Straggling shrubs with bristly branches and compound leaves. Leaflets-3. Flowers white, in panicles. Drupes globose, yellow, succulent, aggregate.

Reproductive Cycle : January-May.

Habitat Ecology : Shrubberies. Wastelands.

Material Examined : HPU-EL-SKS-5448; Bhatu (1,040 m), Chornali (1,280 m), Darugnu (1,400 m), Gopalpur (1,350 m), Lolani (1,600 m).

Parts Used : Leaves. Fruits.

Folk Uses : Leaves **relished** by sheep and goats. Fruits **edible**, and **laxative** in nature.

Biological Activity : Effect on CNS and antifertility activity found positive; spasmolytic activity confirmed.

Uses in Literature : So far known for cracked heals, cuts and wounds, dyeing, dysentery, fodder, indigestion, malaria, stomachache, worms; and astringent and edible (ripe fruits) (Ambasta, 1986; Arora & Pandey, 1996; Bennet *et al.*, 1991; Bisht & Jain, 2006; Dash *et al.*, 2003; Jain, 1991; Lal *et al.*, 1996; Mao 1993; Megoneitso & Rao, 1983; Murthy *et al.*, 2003; Nautiyal *et al.*, 2003; Panda, 1996; Purohit & Samant, 1995; Rao, 1990; Ranjan, 2000; Roy *et al.*, 1998; Saklani & Jain, 1994; Samant *et al.*, 2001; Sharma & Rana, 2005; Sood & Thakur, 2004; Sood & Kaushal, 2008; Watt, 1972).

***Rubus niveus* Thunb.** (Pl. 45D)

Syn.: *R. albescens* Roxb.; *R. bijugus* Focke; *R. distans* Don; *R. furfuracens* Wall.; *R. horsefieldii* Miq.; *Rubus lasiocarpus* Sm.; *R. indicus* Heyne; *R. mysorensis* Heyne; *R. rosaeflorus* Roxb.; *R. rotundifolius* Royle

Family: Rosaceae.

Vern. : Kala-Akha.

English and Regional Names:

Eng. : Raspberry, Mysore Raspberry, Mahabaleshwar Raspberry

H. : Kala Hinsalu

Mar. : Gauri Phal

P. : Gunachal.

Distribution : Throughout Himalayas. 1,350-3,500 m.

Description : Prickly shrubs. Leaves compound. Leaflets 5-11, ovate-lanceolate. Flowers dark pink. Drupes succulent, black.

Reproductive Cycle : March-May.

Habitat Ecology : Forests. Shrubberies.

Material Examined : HPU-EL-SKS-5701; Bhagpur (1,600 m), Bhauda (1,050 m), Dadh (1,100 m), Kathena (1,290 m), Molichak (1,020 m).

Parts Used : Whole Plant. Leaves. Fruits.

Folk Uses : Plant grown as **hedges**. Ripe fruits **edible**, and considered **refrigerant**. Leaves given as **fodder** for sheep and goats.

Uses in Literature : Considered useful as an antidote to snakebite, diuretic, edible (berries), fodder, laxative; and for cuts, stomachache, infected wounds, nocturnal micturition and swellings (Agarwal, 2003; Ambasta, 1986; Arora & Pandey, 1996; Bisht & Jain, 2006; Hosagoudar & Henry, 1996c; Jain, 1991; Kapur, 1996b; Kapur & Srivastava, 1996; Mao, 1993; Nautiyal *et al.*, 2003; Negi & Gaur, 1991; Negi *et al.*, 1999; Purohit & Samant, 1995; Rana *et al.*, 1996; Roy *et al.*, 1998; Saklani & Jain, 1994; Samant *et al.*, 1996, Singh & Kumar, 2000a; Sharma & Rana, 2005; Sood & Kaushal, 2008; Watt, 1972).

Rumex hastatus Don (Pl. 45E)

Family: Polygonaceae.

Vern. : Aambi.

Regional Names:

Kum. : Amlora, Chulmora

P. : Amla, Amlora, Ami, Khattimal, Malori-gha.

Distribution : Common in Himalaya: Kashmir - Kumaon.

Description : Erect branched herbs with coriaceous, hastatus, radical, alternate leaves; and pink polygamous flowers in racemes. Nut brown.

Reproductive Cycle : November-April.

Habitat Ecology : Rock crevices. Moist slopes;

Material Examined : HPU-EL-SKS-5678; Arla (1,100 m), Boda (1,000 m), Dharaman (1,100 m), Jiun (1,550 m), Sungal (1,050 m).

Parts Used : Leaves. Inflorescence.

Folk Uses : Leaf decoction (two teaspoons thrice a day) taken to cure **fever**. Leaves and inflorescence eaten raw and considered **refreshing**; also forms one of the ingredient of '**saag**' (pot-herb).

Uses in Literature : Used in India as a pot-herb, coolant, sauce, and condiment; and for burns, cough, cuts, dry throat, headache, fever, food-poisoning, syphilitic ulcer (Ambasta, 1986; Arora & Pandey, 1996; Chandrasekar & Srivastava, 2003; Dash *et al.*, 2003; Gupta, 1962; Lal *et al.*, 1996; Nautiyal *et al.*, 2003; Negi *et al.*, 1999; Purohit & Samant, 1995; Rajwar, 1983; Ranjan, 2000; Samant *et al.*, 1996, 2001; Sharma & Rana, 2005; Singh, 1996b; Singh & Kumar, 2000a; Singh *et al.*, 2001; Sood & Thakur, 2004).

Rumex nepalensis Spreng. (Pl. 45F)

Syn.: *R. roxburghianus* Schult.f.; *R. hamatus* Trevir.; *R. ramulosus* Meisn.; *R. eregrinus* Boiss.; *R. tuberosus* Roxb.; *R. uncinatus* Hort.

Family: Polygonaceae.

Vern. : Jangli-palak.

Regional Names:

B. : Pahari palang

Kas. : Palak

Kum. : Kulii.

Distribution : Temperate Himalayas.

Description : Erect stout herbs, 45-120 cm tall with stiff branches and bisexual flowers in whorls on leafless racemes. Fruiting sepals fringed with hooked teeth. Nuts small.

Reproductive Cycle : May-October.

Habitat Ecology : Frequently along nallahs, water sources, damp places.

Material Examined : HPU-EL-SKS-5594; Bandla (1,280 m), Chachian (1,300 m), Chornali (1,280 m), Nanaun (900 m), Panchrukhi (990 m).

Parts Used : Aerial plant parts. Leaves.

Folk Uses : Vegetable from aerial plant parts taken as **anthelmintic**. Fresh leaves rubbed to give relief from **irritation** caused by stinging hairs of *Urtica dioica.*

Biological Activity : Plant CNS depressant.

Uses in Literature : Already reported as a colic, coolant, diuretic, purgative, pot-herb; and for boils, chutneys, dyeing woollen clothes and leather, hepatitis, indigestion, pickles, rheumatism, scurvy, swelling of muscles and syphilitic ulcers (Ambasta, 1986; Bennet, 1983; Chhetri, 2005; Dash *et al.*, 2003; Devi, 2003; Gangwar & Ramkrishnan, 1990; Jain *et al.*, 1977; Kant & Dutt, 2005; Krishna & Singh, 1987; Lal *et al.*, 1996; Manandhar, 1996; Megoneitso & Rao, 1983; Nautiyal *et al.*, 2003; Negi *et al.*, 1999; Roy *et al.*, 1998; Saklani & Jain, 1994; Samant *et al.*, 1996, 2001; Sood & Thakur, 2004).

Saccharum officinarum L.

Syn.: *S. sinense* Roxb.

Family: Poaceae.

Vern. : Ganna.

English, Sanskrit & Regional Names:

Eng. : Sugar Cane, Noble Cane
Sans. : Adhipatra, Gudakastha, Rasola
Ass. : Pura-kuhiyar
B. : Ak, Kushiar, Puri
Bo. : Gol, Serdi
G. : Naisakar, Sheradi, Uns
H. : Ganna, Ikh, Kumad
Mal. : Kantarakam, Madhutrianam
Mar. : Aos, Kabbo
Ori. : Aku, Gudodaru

P. : Ganna, Khaud, Paunda
Tam. : Angarigai, Madudirunam
Tel. : Arukanupulakranugu, Tellacheraku.

Distribution : Cultivated throughout India.

Description : Stems solid. Upper leaves long, scabrid. Spikelets lanceolate, surrounded by dense, silky white hairs.

Reproductive Cycle : January-April.

Habitat Ecology : Cultivated.

Material Examined : HPU-EL-SKS-5682; Balah (900 m), Bhauda (1,050 m), Chandpur (1,290 m), Gharthun (900 m), Tanda (1,120 m).

Parts Used : Whole Plant. Stem Juice.

Folk Uses : Plant considered **sacred** and used for performing religious rites. Stem juice **refreshing** and considered good against **jaundice** and **anaemia**.

Active Constituents : Six glycans-saccharans A, B, C, D, E and F, neocarlinoside, neoisoschaftoside, orientin, swerti-ajaponin, vicenin-2, p-coumaric, ferulic, caffeic and 3,4-di-0-methylcaffeic acids isolated from *stalk juice*.

Biological Activity : Saccharans A, B, C, D, E and F exhibited hypoglycaemic activity in normal and alloxan produced hyperglycaemic rats.

Uses in Literature : Reported to be useful in India as an anthelmintic, antidote to arsenical poisoning, antiseptic, aphrodisiac, apierent, condiment, detergent, diuretic, emollient, fodder, thatching material for huts, laxative, manure, nutritious, pectoral, refrigerant, sweetener, symbolic in sculptures (bow of Kamdev—god of love made from canes), tonic, torches (dried canes); and for bagasse or fibre, beverages, bronchitis, calculus complaints, carbon paper, catarrhal affections, disorders of bile and wind, fever (postnatal), gall bladder, incipient consumption, jaundice (juice as drink), insanity, lactation (vet.), logs in to rafts, preservation of meat and fish, press-mud, production of lime, rough cordage, shoe and other polishes, soothing and protecting the alimentary tract inflammation, strengthening gum, sugarcane wax, tying, ulcers, wax paper, and wounds (vet.) (Ambasta, 1986; Borthakur, 1996b; Dagar & Dagar, 1996; Gogoi *et al.*, 2003b; Jain, 1991; Kurian, 1995; Lalramnghinglova, 1996; Sahu, 1996; Siwakoti & Varma, 1996; Tirkey *et al.*, 2003b; Watt, 1972).

Saccharum spontaneum **L.** (Pl. 46A)

Syn.: *S. aegyptiacum* Willd.; *S. Balense* Boga.; *S. biforum* Forsk.; *S. canaliculatum* Roxb.; *S. chinense* Nees.; *S. glaza* Reinw.; *S. insulare* Brongn.; *S. propinguum* Steud.; *S. punctatum* Schumach.; *S. semidecumbens* Roxb.; *Imperata spontanea* Beauv.

Family: Poaceae.

Vern. : Sankheda.

English, Sanskrit & Regional Names:

Eng. : Thatch Grass, Wild Sugarcan

Sans. : Kasa, Kasha, Kasalu, Khaggara

B. : Kagara, Kas, Kash, Kashiya

G. : Kans, Kansado, Kansadoghas

H. : Kagara, Kans, Kansi, Kas, Kar, Kosa, Kus

Mal. : Kusha, Nannana

Mar. : Kagara, Kasara

Tam. : Kosangan, Kurbagam, Nanal, Nanarbul, Nanmugappul

Tel. : Billu-gaddi, Kore-gadi, Rellu-guddi, Verri-cheruku.

Distribution : Warmer parts of India, Sri Lanka, and ascending to 2,000 m in the Himalayas.

Description : Tall perennial grasses having slender, solid, smooth, polished stem, linear, rigid, coriaceous leaves. Inflorescence in panicles. Flowers in glumes, lanceolate, ciliate. Grains oblong-subglobose.

Reproductive Cycle : June-October.

Habitat Ecology : Dry slopes.

Material Examined : HPU-EL-SKS-5547; Balah (900 m), Bhuana (1,400 m), Chimbalhar (1,150 m), Gopalpur (1,350 m), Panapari (1,100 m).

Part Used : Stem.

Folk Uses : Stem used widely as **dip-pens**; also hung on main door to keep away **evil spirits**.

Uses in Literature : Reported earlier as aphrodisiac, astringent, bitter, laxative, liver tonic, sweet tonic; and for abdominal worms, biliousness, burning sensation, blood diseases, cooling, diseases of the blood, haemorrhagic diathesis, inflammation of hands and legs, kidney and bladder stones, paper manufacturing and thatching, pthisis, soil binder, strangury, urinary system trouble and vesical calculi (Ambasta, 1986; Bhatt *et al.*, 2003c; Chhetri, 2005; Dwivedi, 2003; Kirtikar & Basu, 1984., Mahato & Chaudhary, 2005; Maliya, 2005; Parrotta, 2001; Srivastava *et al.*, 2003c).

Salix karelini Turcz *ex* Stsch. (Pl. 46B)

Syn.: *S. hastata* auct (non L.) Hook. f.

Family: Salicaceae.

Vern. : Pahadi-Beeuins.

Distribution : Afghanistan-C. Nepal, C. Asia; 2,100-4,500 m.

Description : Small-sized trees with elliptical-ovate, finely toothed leaves. Young leaves woolly haired. Catkins borne on leaflets stems. Fruit a capsule. Seeds with tuft of hairs.

Reproductive Cycle : May-July.

Habitat Ecology : Open shady slopes. Streamside. Wetlands.

Material Examined : HPU-EL-SKS-5589; Chandpur (1,290 m), Dharaman (1,100 m), Gopalpur (1,350 m), Jiun (1,550 m), Tanda (1,120 m).

Parts Used : Leaves. Wood.

Folk Uses : Leaves lopped for **fodder**. Wood suitable for **tool handles** and **fuel**.

Uses in Literature : There exists no report of its use in literature.

Salix tetrasperma Roxb. (Pl. 46C)

Syn.: *S. disperma* Don; *S. horsefieldiana* Miq.; *S. nilagirica* Miq.

Family: Salicaceae.

Vern. : Beeuins.

English, Sanskrit & Regional Names:

Eng. : Indian Willow
Sans. : Varuna
Ass. : Bhi, Veh
B. : Boishaki, Punijami
Bo. : Bacha, Wallunj
H. : Bains, Bainsh, Bilsa
Kan. : Bariche, Niravanji
Kash. : Bins, Yir
M.P. : Dhjanie
Mal. : Arali, Attuphala, Vanti
Mar. : Bitaso, Walunj
Ori. : Baisi, Panijamo
P. : Bakshel, Magsher
Tam. : Niruvanji, Suvedam

Tel. : Etipisinika.

Distribution : Throughout Tropical and Sub-Tropical India, below 1,650 m.

Description : Small trees with silky young branches. Leaves alternate, ovate-lanceolate. Flowers in scattered clusters in catkins. Capsules long-stalked, glabrous. Seeds with tuft of hairs at the base.

Reproductive Cycle : April-July.

Habitat Ecology : Open slopes. Roadsides.

Material Examined : HPU-EL-SKS-5623; Andretta (980 m), Bindraban (1,270 m), Chachian (1,300 m), Ladoh (980 m), Sulah (950 m).

Parts Used : Whole Plant. Wood.

Folk Uses : Plant used as a stock for **grafting** almond. Wood employed for making **agricultural implements**, **arms of bed**, **tool handles**, and as **fuel**.

Active Constituents : *Leaves* and *bark* yield rhamnose and salicin.

Uses in Literature : Used so far for agricultural implements, baskets, bent, brush-backs, cabinet and fancy work, epilepsy, gunpowder, charcoal, house construction, matchboxes, packing cases, pencil and cricket stumps, piles and swellings, ploughs, rheumatism, stone in the bladder, tanning, venereal diseases, wood for posts and planks; and as cattle feed, febrifuge, and pen holders (Ambasta, 1986, Watt, 1972).

***Salvia coccinea* L.** (Pl. 46D)

Syn.: *S. gerardiana* Benth.

Family: Lamiaceae.

Vern. : Phultulsi.

Distribution : More or less throughout India; often planted.

Description : Erect pubescent herbs with simple, opposite, ovate, crenate leaves and scarlet flowers in interrupted terminal racemes. Nutlets 3-quetrous, smooth. Seeds small.

Reproductive Cycle : June-August.

Habitat Ecology : Dry slopes, frequent.

Material Examined : HPU-EL-SKS-5499; Bhuana (1,400 m), Chhanchri (930 m), Dehan (1,130 m), Holta (1,200 m), Saparu (1,250 m).

Part Used : Flowers.

Folk Use : Flowers used as an **offering** to various deities.

Active Constituents : Pelargonidin-3-caffeoylglucosido-5-dimalonylglucuside isolated from *flowers*.

Uses in Literature : Known to be useful for lumbago, renal troubles and tubercular cough; as a sacred plant (Ambasta, 1986; Sood & Thakur, 2004).

***Santalum album* L.** (Pl. 46E)

Syn.: *S. myrtifolium* Roxb.; *S. verum* L.; *Sirium myrtifolium* Roxb.

Family: Santalaceae

Vern. : Chananh.

English, Sanskrit and Regional Names:

Eng. : The White Sandal Wood Tree

Sans. : Ananditam, Chandana, Krishna Chandanam, Taliaparnam

B. : Chandan, Harichandan, Pitchanada, Srikhand, Sufaid, Chundun

Bo. : Chandan, Safed Chandan, Sandal

G. : Sukhet, Sukhud

H. : Chandal, Chandan, Chandoi, Safed-chandan, Sandal

Kan. : Bavanna, Gandada, Gandha, Srigandha

Mal. : Chandanam, Handana-mutti

Mar. : Chandan, Gandha-chakoda

Ori. : Chandno, Gondassaru

P. : Chandan

Tam. : Kulavuri, Sandanam, Shandanak Kattai, Sundel

Tel. : Chandanamu, Chandnapuchettu, Gandhapu-cheeka, Hari Chandanam, Pita Chanadan.

Distribution : Deccan Peninsula.

Description : Small evergreen glabrous trees having opposite, long, thin leaves. Flowers axillary and in terminal paniculate blood-red cymes. Drupes globose, black.

Reproductive Cycle : September-January.

Habitat Ecology : In deciduous forests.

Material Examined : HPU-EL-SKS-5666; Andretta (980 m), Balah (900 m), Boda (1,000 m), Dadh (1,100 m), Tikkar (910 m).

Part Used : Wood.

Folk Uses : Wood paste used as '**tikka**', **face-pack** and for **burning sensation of soles** and **headache**. Pieces of its wood used for performing '**havan**'.

Active Constituents : These are β-sitosterol, glucose, fructose, sucrose (*leaves*) and exo-norbicycloekasantalal (*sandalwood oil*).

Uses in Literature : So far reported as alexeteric, antipyretic, aphrodisiac, antidote to snake venom, astringent to bowels, diaphoretic, diuretic, laxative, tonic to heart; and for biliousness, boils, bronchitis, burning sensation, cold dry, fever, gleets, gonorrhoea, headache, heart and urinary diseases, inflammations, pimples, prickly heat, religious ceremonies, smallpox, strangury, and brain, vaginal discharges and wounds (Ambasta, 1986; Diwakar & Ansari, 2003; Kirtikar & Basu, 1984; Panthi & Chaudhary, 2003; Saikia & Nath, 2003; Sen & Batra, 1997; Sharma, 2003).

Sapindus mukorosii Gaertn. (Pl. 46F)

Syn.: *S. abruptus* Lour.

Family: Sapindaceae.

Vern. : Dodan, Ritha.

English, Sanskrit & Regional Names:

Eng. : Soap Nut Tree of North India

Sans. : Arishta, Phenila

Ass. : Haithaguti

B. : Chhota Ritha, Ritha

Bo. : Kanmar, Ritha

H. : Aritha, Ritha

Ori. : Ita

P. : Dodan.

Distribution : Cultivated throughout N. India, ascending to 1,650 m.

Description : Trees. Leaves alternate, glabrous, paripinnate. Leaflets around 15. Flowers purple, in panicles. Drupes yellow, globose.

Reproductive Cycle : April-September.

Habitat Ecology : Edges of fields. Roadsides.

Material Examined : HPU-EL-SKS-5688; Bhawarna (1,100 m), Chauki (900 m), Dharaman (1,100 m), Paror (940 m), Thandol (950 m).

Parts Used : Fruit. Seeds. Wood.

Folk Uses : Pericarp used as **detergent** for woollen and silken clothes, and shampooing **hair**. Poultice of seeds applied against **dogbite**. Wood employed for making **arms of bed** and **window frames.**

Active Constituents : Comprises mainly sapindoside V, sapindosides A & B, hederagenin as common aglyconel, apindoside V, mukorosside (*fruits*), protein (31.87), carbohydrate (2.21) and starch (11.94%) (*kernel*).

Biological Activity : Saponins show high spermicidal activity in both spot and IPPF tests with human semen, with MED of 0.05 & 0.51, respectively.

Uses in Literature : Reported so far as anthelmintic, cattle feed, contraceptive, detergent, emetic, expectorant, fish poison, hair wash, kernel's flour, purgative; and for charcoal making, chlorosis, epilepsy, polishing jewellery pneumonia, salivation, washing and bleaching cardamoms (Agarwal, 2003; Ambasta, 1986; Banerjee & Pal, 1996; Borthakur, 1993; Chauhan, 1999; Chhetri, 2005; Das & Agarwal, 1991; Jain, 1991; Kirtikar & Basu, 1984; Rao, 1996; Rao *et al.*, 1996; Saklani & Jain, 1994; Samant *et al.*, 2001; Sharma, 2003; Sharma & Rana, 2005; Singh, 1991, 1996b, 1999; Siwakoti & Varma, 1996; Vihari, 1995).

Sapium insigne (Royle) Benth. (Pl. 47A)

Syn.: *Falconeria insignis* Royle; *F. wallichiana* Royle; *Carumbium insigne* Kurz

Family: Euphorbiaceae.

Vern. : Ainkhar.

Regional Names:

Ass. : Mah-Kola
Bo. : Duella
H. : Khina, Khindra
Kan. : Kanuupade, Kurda, Kurudanandi, Nanaka
Mar. : Hura, Ura
P. : Bilodar, Bilodja
Tam. : Garpashola, Tillai
Tel. : Garbhasula, Kadura.

Distribution : Kashmir-Bhutan: 500-1,800 m.

Description : Deciduous trees. Leaves alternate, toothed, oblong-lanceolate. Flowers yellow-green, borne on terminal spikes. Capsule 3-lobed, 3-seeded.

Reproductive Cycle : May-August.

Habitat Ecology : Open slopes. Drier areas.

Material Examined : HPU-EL-SKS-5535; Balah (900 m), Bhaura (940 m), Chauki (900 m), Jasun (1,000 m), Tikkar (910 m).

Part Used : Latex.

Folk Uses : Latex applied for removing **warts**, and as **fish-poison**.

Uses in Literature : Used so far for drums, floats, furniture, matchboxes, packing cases, sandals, toys, wounds; and also as cattle feed, fish-poison, and germicide (Ambasta, 1986; Jain, 1991; Joshi & Joshi, 2005; Kulkarni & Kumbhojkar, 1996; Watt, 1972).

Sapium sebiferum (L.) Roxb. (Pl. 47B)

Syn.: *Exooecaria sebifera* Muell.; *Stillangia sebifera* Michx. *S. sinense* Baill.; *Stillingfleetia sebifera* Bojer; *Carumbium sebiferum* Kurz.; *Croton sebiferus* L.

Family: Euphorbiaceae.

Vern. : Makhan, Pahadi-Taali.

English, Sanskrit & Regional Names:

Eng. : Chinese Tallow Tree
Sans. : Toyapippali
B. : China
Bo. : Pipalyank
H. : Vilayati-Shisham
Kan. : Meenadabattimara
Ori. : Ranojita.

Distribution : Cultivated in various parts of India.

Description : Medium-sized trees having green, long-pointed, petioled, ovate leaves. Capsule coriaceous, green.

Reproductive Cycle : June-August.

Habitat Ecology : Forests. Shrubberies. Roadside.

Material Examined : HPU-EL-SKS-5550; Arla (1,100 m), Bodhal (970 m), Dharaman (1,100 m), Gopalpur (1,350 m), Jiun (1,550 m).

Parts Used : Leaves. Wood.

Folk Uses : Leaves used as **fodder**. Wood employed for making **furniture** and as **fuel**.

Active Constituents : Active principles due to 6,7,8-trimethoxy coumarin (*bark*), gallic acid, astragallin, (–) loliolide, kaempferol, quercetin and β-sitosterol glycoside (*leaves*).

Biological Activity : Antihypertensive activity found to be positive.

Uses in Literature : Known to be used earlier as black dye, block-mount, feed, fertilizer, flour for human consumption, fuel, purgative; and for bedsteads, candles, cosmetics, cricket bats, cutaneous troubles, dressing cloth, dyeing, dyspepsia, furniture, inferior quality pencils, paints, plastics, printing blocks, rearing silkworms, soaps, stillingia oil, tables, toys, and varnishes (Ambasta, 1986; Drury, 1978; Jain, 1991; Watt, 1972).

***Scandix pecten veneris* L.** (Pl. 47C)

Family: Apiaceae.

Vern. : Garhbeenhi, Jangli Dhaniya.

English Name:

Eng. : Venus's comb.

Distribution : W. Himalayas.

Description : Herbs; stem glabrous; leaves pinnately compound; flowers white in simple or compound umbels. Fruits oblong, flattened with long beak and blunt ridges.

Reproductive Cycle : May-September.

Habitat Ecology : Shady localities.

Material Examined : HPU-EL-SKS-5628; Balu (1,030 m), Chachian (1,300 m), Dharaman (1,100 m), Saliana (1,000 m), Sidhpur (1,100 m).

Parts Used : Aerial Parts.

Folk Use : Aerial parts cooked as a v**egetable**; also given as **fodder**.

***Sechium edule* Sw.** (Pl. 47D)

Family: Cucurbitaceae.

Vern. : Lonku.

English & Regional Names:

Eng. : Chow Chow, Chayote

Kan. : Seema Badane

Tam. : Kattirikkai, Seeme.

Distribution : Cultivated throughout India.

Description : Herbaceous climbers having simple, palmately 5- nerved leaves. Flowers creamish. Fruits light-green, hard with ridges and soft spines.

Reproductive Cycle : June-September.

Habitat Ecology : Grown locally for its fruits.

Material Examined : HPU-EL-SKS-5485; Darugnu (1,400 m), Holta (1,200 m), Jasun (1,000 m), Panapari (1,100 m), Saparu (1,250 m).

Part Used : Fruit.

Folk Uses : Fruits cooked as **vegetable**, also considered good against **diabetes**.

Active Constituents : These are α-amino-γ-ureidobutyric acid (*seeds*), and glucosides of sitosterol and stigmasterol (*fruits*).

Biological Activity : β-sitosterol and stigmasterol-β-D-glucosides exhibit anti-inflammatory activity in rat paw oedema.

Uses in Literature : Known to be used so far for its edible fruits, fibre, and fodder (Ambasta, 1986; Negi & Gaur, 1991; Watt, 1972).

Selaginella chrysocaulos (Hook. & Grev.) Spring (Pl. 47E)

Family: Selaginellaceae.

Vern. : Kungu.

Distribution : N. W. Himalayas.

Description : Herbaceous perennials; thallus dorsiventral, prostrate; leaves anisophillous, green, serrated, microphyllous, ligulate; strobilus terminal, bisporangiate, megasporophyll resticted to base and microsporophyll to the upper region.

Reproductive Cycle : July-September.

Habitat Ecology : Shady slopes.

Material Examined : HPU-EL-SKS-5489; Chandpur (1,290 m), Ghugar (1,250 m), Jiun (1,550 m), Nadli (1,000), Saparu (1,250 m).

Part Used : Aerial Plant Parts. Spores.

Folk Uses : Plant extr. considered **stomachic**. Spores applied as vermilion ('**tikka**').

***Sesamum orientale* L.** (Pl. 47F)

Syn.: *S. indicum* DC.; *S. luteum* Retz.; *S. occidentale* Heer & Regel

Family: Pedaliaceae.

Vern. : Til.

English, Sanskrit & Regional Names:

Eng. : Gingelly, Sesame

Sans. : Homadhanya, Jatila, Pavitra, Puraphala

B. : Bhadutil, Kalatil, Kheslatil, Krishnatil, Raktatil, Rasi, Sankitil, Sumsum, Tel, Til, Tilmi

Bom. : Kala katwa, Purbia, Tal, Til

Guj. : Mithutel, Tal, Tel, Til
Hin. : Bariktel, Gingli, Krishnatel, Mithatel, Tir
Kum. : Bhunguru, Til
Mal. : Chitelu, Ellu, Korellu
Mar. : Til, Zilechatil
P. : Kunjad, Til, Tili
Tam. : Ellu, Nuvvulu, Yelluchedi
Tel. : Nuvvu, Pollanuvvulu
Urdu : Til.

Distribution : Hot regions of India.

Description : Erect, hairy much branched annuals with an offensive smell. Leaves variable on the same plant. Flowers axillary, solitary. Capsule tetragonal. Seeds numerous.

Reproductive Cycle : July-September.

Habitat Ecology : Wastelands. Damp places.

Material Examined : HPU-EL-SKS-5481; Balah (900 m), Bodhal (970 m), Nanaun (900 m), Makol (900 m), Rangru (910 m).

Parts Used : Seeds.

Folk Uses : Seeds employed for performing **religious ceremonies**; **chutney** preparation; and chewed to check **loosening of teeth**; and also along with jaggery prescribed to check **night wetting in kids**. Oil from seeds applied locally against **rheumatic** pains and **sprain**; and also utilized for **cooking purposes**.

Active Constituents : Active principles due to globulin, thiamine, niacine (*seeds*), pedaliin (*leaves*), and pinoresinol (*plant*).

Uses in Literature : Used earlier in India as an abortifacient, amenorrhoea, antiaging, antibacterial, antifungal, antiviral, catarrh, demulcent, diuretic, digestive, edible (seed oil), emollient, laxative; and for backache, blood in urine, blurry vision, boils, burns, chronic dysentery, cystitis, dyspepsia, dizziness, flatulence, hair loss, marriage ceremonies, migraines, osteoporosis, receding gums, rejuvenative, tonic, stiff joints, sunburns, tooth decay, ulcers, weakness, wound, and for eye and post-delivery pains, kidney and bladder disorders, spleen complaints and ulcers (Ambasta, 1986; Bhattacharjee *et al.*, 1980; Chopra *et al.*, 1956; Duhoon & Tripathi, 2003; Jain, 1991; Jha & Kumar, 2003; Kshirsagar *et al.*, 2003; Kumar & Rao, 2003; Kumar *et al.*, 2004; Maya *et al.*, 2003; Nadanakunjidam & Abirai, 2003; Nayak *et al.*, 2003; Rajendran *et al.*, 2003; Saikia & Nath, 2003; Sharma, 2003; Singh & Singh, 2003; Thomas & Britto, 2003).

Sida cordifolia L. (Pl. 48A)

Syn.: *S. herbacea* Cav. Diss.; *S. mican* Cav. Diss.; *S. rotundifolia* Cav. Diss.; *S. athaeifolia* Sw.

Family: Malvaceae.

Vern. : Dredain.

English, Sanskrit & Regional Names:

Eng.	: Country-Mallow
Sans.	: Kathorayashtika, Raktatandalu
Ass	: Saru-sonborial
B.	: Bala, Svetabarela
G.	: Baladana, Khareti
H.	: Barial, Kharenti
Kan.	: Hettugigada, Hettuthi, Kisangi
Mal.	: Katturam
Mar.	: Chikana, Khiranti
Ori.	: Beudia Naula, Bisvokapari
P.	: Kharent
Tam.	: Arivalamaippundu
Tel.	: Muttuvapulagamu, Suvarnamu.

Distribution : Throughout Sub-Tropical Parts of India.

Description : Wild undershrubs. Leaves cordate-oblong, obtuse, crenate. Flowers yellow. Fruits reticulated.

Reproductive Cycle : August-September.

Habitat Ecology : Waste places. Ravines.

Material Examined : HPU-EL-SKS-5487; Balah (900 m), Bhuana (1,400 m), Jasun (1,000 m), Ladoh (980 m), Tikkar (910 m).

Part Used : Seeds.

Folk Uses : Powdered seeds alone or mixed in **porridge** considered highly **nutritious** and prescribed for **backache**.

Active Constituents : Isolated constituents are sterculic, malvalic, coronaric acids (*seed oil*).

Biological Activity : Antiprotozoal (*Entamoeba histolytica* strain STA) and CNS depressant activities confirmerd.

Uses in Literature : Described earlier as an antirheumatic, aphrodisiac, astringent, febrifuge, cardiac stimulant, colic, demulcent, diuretic, edible, ingredient of 'chywanprash', opthalmic, tonic; and for boils, bloody flux, cooling, cystitis, epilepsy, dysentery, excessive menstrual flow, eye diseases, facial paralysis, fibres, fits, dog and snakebites, general debility, gonorrhoea, haematuria, heart irregularity, hemiplagia, improving sexual strength, infective viral hepatitis, leucorrhoea, nervous disorders, pregnant micturition, promoting sexual desire, sciatica, shivering (vet.), shivering fits, spermatorrhoea, stomachache, urinary troubles, fever, venereal diseases, wounds, and cordages (Ambasta, 1986; Aminuddin & Girach, 1991, Bhatt *et al.*, 2003b; Borthakur, 1993; Bose, 1984; Chauhan, 1999; Das & Agarwal, 1991; Dey, 1980; Gogoi & Borthakur, 2001; Gogoi & Das, 2003; Gogoi *et al.*, 2003a; Jain, 1968, 1991; Jha *et al.*, 1991; Khanna & Kumar, 2000; Khanna *et al.*, 1996c; Khare, 2004; Kirtikar & Basu, 1984; Kumar & Goel, 1998; Mahato & Chaudhary, 2005; Saini, 1996a, b; Sarin, 1996; Sharma, 2003; Singh, 2003; Srivastava *et al.*, 2003a; Thomas & Britto, 2003; Upadhyay & Singh, 2005; Watt, 1972).

***Silene conoidea* L.** (Pl. 48B)

Family: Caryophyllaceae.

Vern. : Tokala.

Distribution : W. Himalayas. N. India.

Description : Glandular-pubescent dichotomously branched, herbs. Leaves linear, lanceolate. Flowers deep-pink, borne in terminal panicles. Fruits enclosed in inflated calyx.

Reproductive Cycle : March-May.

Habitat Ecology : As weed in wheat fields.

Material Examined : HPU-EL-SKS-5559; Banuri (1,150 m), Chauki (900 m), Dharaman (1,100 m), Tanda (1,120 m), Thandol (950 m).

Parts Used : Aerial Plant Parts.

Folk Uses : Aerial plant parts employed as **vegetable** and **fodder**.

Uses in Literature : Known to be used in India as an edible, sweet, emollient, fumigant, and opthalmia (Ambasta, 1986; Koul *et al.*, 1982; Sarkar *et al.*, 2003).

***Sisymbrium irio* L.** (Pl. 48C)
Family: Brassicaceae.

Vern. : Foonkalan.

English & Regional Names:

Eng. : London Rocket
Bo. : Khakhi
H. : Khubkalan
Mar. : Ran-tikhi
P. : Janglisarson, Maktrusa
Urdu : Khubakalan.

Distribution : Punjab, Peshawar, Baluchistan, Asia, Afghanistan, Arabia, Mediterranean.

Description : Erect glabrous annuals or biennials having stalked, pinnatifid, toothed leaves. Flowers minute, yellow. Pods slender, erect, with curved pedicels.

Reproductive Cycle : December-May.

Habitat Ecology : Moist-shady places. Agricultural fields. Wastelands.

Material Examined : HPU-EL-SKS-5606; Chhanchri (930 m), Gopalpur (1,350 m), Jiun (1,550 m), Malnu (1,000 m), Panapari (1,100 m).

Parts Used : Whole Plant. Inflorescence.

Folk Uses : Plant used for **magico-religious** purposes. Smoke of burning inflorescence good for '**measals**'.

Uses in Literature : Used in India as coolant, edible, expectorant, febrifuge; restorative, rubefacient, stimulant, tonic; and for asthma and soap-making (Ambasta, 1986; Aswal & Mehrotra, 1994; Caius, 2003; Kirtikar & Basu, 1984; Singh & Pandey, 1998).

***Solanum erianthum* Don** (Pl. 48D)
Syn.: *S. verbascifolium* L.; *S. pubescens* Roxb.
Family: Solanaceae.

Vern. : Laah.

English, Sanskrit & Regional Names:

Eng. : Potato Tree
Sans. : Gandhira, Pathi
Kan. : Kadusinde, Sowdangigida
M.P. : Chichora

	Mal.	: Chunta, Erichunta, Melanjunta
	P.	: Kalamewa, Kharawine, Ola, Tiari
	Tam.	: Anaichundai, Malaichundai, Sundai
	Tel.	: Budamu, Rasagadi.

Distribution : Throughout India.

Description : Shrubs. Leaves alternate, woolly, both ends or base sub-rhomboid. Flowers pentamerous, violet. Berries globose-yellow.

Reproductive Cycle : July-October.

Habitat Ecology : Wastelands. Roadsides. Rarely among boulders.

Material Examined : HPU-EL-SKS-5652; Andretta (980 m), Bhaura (940 m), Boda (1,000 m), Jasun (1,000 m), Khaira (940 m).

Part Used : Leaves.

Folk Uses : Leaf extr. used as drops for healing **eye injuries** in cattles. Poultice of leaf-paste good against **cuts** and **wounds**.

Active Constituents : *Leaves* yield solasodine (57.0, 71.0), solasodiene (3.0, 3.5), diosgenin (8.5, 25.0%), solafloridine (8.0%, trt.), tomatidenol (23.0%), vespertilin, 5,16-dehydropregnenolone, solanine, solasonine, solamargine and solanidine.

Uses in Literature : Reported to be used for body pains, burns, diarrhoea, dysentery, fever, flavouring in curries, horse afflicted with glanders, killing worms in stomach, urinary troubles, vaginal discharges; and as vegetable (Ambasta, 1986; Henry *et al.*, 1996; Hosagoudar & Henry, 1996b; Jain, 1991; singh & Kumar, 2000b; Siwakoti & Varma, 1996; Watt, 1972).

Solanum melongena L. (Pl. 48E)

Syn.: *S. incarnum* L.; *S. insanum* L.; *S. zeilanicum* Scop.; *S. undulatum* Lam.; *S. trongum* Lam.; *S. longum* Roxb.; *S. pseudo-undulatum* Blume; *S. ovigerum* Blume; *S*; *esculentum* Dunal; *S. heteracanthum* Dunal; *S. melanocarpum* Dunal; *S. sativum* Dunal; *S. pressum* Dunal; *S. cumingii* Dunal; S. torrum var. *inerme* Dalziel & Gibbs

Family: Solanaceae.

Vern. : Baingan.

English, Sanskrit & Regional Names:

	Eng.	: Brinjal, Egg Plant
	Sans.	: Dirghavartaki, Nilakantaka
	Ass.	: Jati Bani

B. : Bartaku, Kulibegun
Bo. : Waingi
G. : Rihgni
H. : Baigun, Baingan
Kan. : Badanekayi, Doddabadane
Kash. : Tukmibadanjanurmi, Vangun
Mal. : Hinguli, Nilavalutina
Mar. : Vangi
Ori. : Bartakki
P. : Ban, Vataun
Tam. : Kalliyam, Mullukkattari
Tel. : Mettavanga, Nidupuvanga.

Distribution : Widely cultivated for its edible fruits.

Description : Prickly shrubs with woolly, lobed leaves and blue flowers. Berries glabrous, whitish-yellow or deep-purple.

Reproductive Cycle : June-September.

Habitat Ecology : Cultivated, near houses.

Material Examined : HPU-EL-SKS-5519; Balu (1,030 m), Chimbalhar (1,150 m), Ghugar (1,250 m), Saparu (1,250 m), Thandol (950 m)

Part Used : Fruits.

Folk Uses : Fruits cooked as **vegetable**, and also **pickled**.

Active Constituents : *Fruit* yields solasodine and diosegin.

Uses in Literature : Known so far as an alternative, antiasthmatic, cathartic, stomachic, diuretic, hydrogogue, laxative, stimulant, sialagogue, tonic, vegetable; and for anasarca, asthma, bronchitis, fever, diarrhoea, disorders of the sight, dropsy, dysuria, dyspepsia, enlargement of liver, epistaxis (vet.), heart diseases, hydrophobia, inflammation of kidneys and bladder, otitis, producing uncomfortable pregnancy sleep, psoriasis, skin diseases, snakebite, stomatitis, toothache, ulcers, virulent gonorrhoea (Ambasta, 1986; Gill *et al.*, 1993; Jain, 1991; Pandey, 2003; Rajan *et al.*, 2003; Rama Rao & Henry, 1996; Thomas & Britto, 2003; Watt, 1972).

Solanum nigrum L. (Pl. 48F)

Syn.: *S. rubrum* Mill.; *S. triangular* Lam.; *A. villosum* Lam.; *S. incertum* Dunal; *S. nodiflorum* Jacq.; *S. uliginosum* Blume; *S. roxburghii* Dunal; *S. fistulosum* Rich.; *S. miniatum* Bernh.; *S. paludosum* Dunal; *S. erythraeum* Dunal

Family: Solanaceae.

Vern. : Makoi.

English, Sanskrit & Regional Names:

Eng. : Black Nightshade
Sans. : Bahuphala, Bahutikta, Ghanaghana, Kaka, Vayasi
Ass. : Pichkati
B. : Gurkamai, Mako
Bo. : Ghati, Kamuni
G. : Piludi
H. : Gurkamai, Kabaiya, Mekoi
Mar. : Ghati, Kakmachi, Mako
P. : Kachmach, Kambei, Meko
Tel. : Gajuchettu, Kachi, Kasaka
Urdu : Makoya.

Distribution : Throughout India.

Description : Divaricately branched, erect, leafy annuals. Flowers umbellate, white. Berries globose, purplish black.

Reproductive Cycle : January-July.

Habitat Ecology : Edges of cultivated areas. Wastelands.

Material Examined : HPU-EL-SKS-5453; Darugnu (1,400 m), Ghugar (1,250 m), Panapari (1,100 m), Saparu (1,250 m), Tikkar (910 m).

Parts Used : Leaves. Fruits.

Folk Uses : Leaf-juice smeared on tongue and buccal cavity for healing **mouth ulcers**. Ripe fruits **edible** and considered as **heart tonic**.

Active Constituents : Therapeutic properties due to 12-hydroxysolasodine, N-methyl solasodine, 23-O-acetyl-12 β-hydroxysolasodine, solanocapsine and solamargine.

Biological Activity : Plant CNS depressant, hypotensive and spasmolytic.

Uses in Literature : Known so far as an antidote to opium toxic, antiseptic, diaphoretic, diuretic, edible (shoots, leaves, fruits), emollient, expectorant, laxative, psychoactive, sedative, stomachic; and for boils cough, diarrhoea, dysentery, ear complaints, easy expulsion of foetus, eye complaints, fever, goitre, heart ailments, headache, hydragogue, inflammation of scortum, testicles, kidney, abdominal pain, bladder, jaundice, liver complaints, mouth and stomach ulcers, nostril complaints, pain, piles, psoriasis, rat-bite, scrotum swelling, rheumatism, ringworms, skin diseases, sores, sprain, swelling, throat trouble, ulcers in mouth and urinary complaints (Agarwal, 2003; Ambasta, 1986; Arora & Pandey, 1996; Aswal & Mehrotra, 1994; Beigh *et al.*, 2003; Bennet *et al.*, 1991; Bhattacharya, 1996; Binu *et al.*, 2003; Chakraborty *et al.*, 2003; Chatterjee & Pakrashi, 1997; Chauhan, 1999; Chhetri & Shrestha, 2004; Chopra *et al.*, 1956; Chun-Lin & Jieru, 1995; Das & Agarwal, 1991; Dey, 1973; Dhiman, 1976; Dwivedi, 2003; Gogoi & Das, 2003; Gogoi *et al.*, 2003a, b; Hosagoudar & Henry, 1996b; Jain, 1991; Jain *et al.*, 1994; Khare, 2004; Kirtikar & Basu, 1984; Kumar *et al.*, 2003; Kurian, 1995; Mahato & Chaudhary, 2005; Meena *et al.*, 2003; Murthy *et al.*, 2003; Nayak *et al.*, 2003; Nunez & Castro, 1995; Pawar & Patil, 2004; Prajapati *et al.*, 2006; Rajendran *et al.*, 2003; Ranjan, 2003; Rawat & Chowdhury, 1998; Roy *et al.*, 1998; Saklani & Jain, 1994; Samant *et al.*, 2001; Sharma, 2003; Sharma & Rana, 2005; Singh & Kumar, 2000a; Singh & Pandey, 1998; Singh *et al.*, 1999, 2001; Subramaniam, 2000; Srivastava *et al.*, 2003a; Sudhakar & Vedavathy, 2000; Upadhyay & Singh, 2005).

***Solanum torvum* Sw.** (Pl. 49A)

Syn.: *S.indicum* L.; *S. stramoniflium* Roxb.; *S. ferrugineum* Jacq.; *S. pseudosaponaceum* Blume; *S. wightii* Miq.

Family: Solanaceae.

Vern. : Chhota-baingan.

Regional Names:

- **Ass.** : Hathibhekuri, Bhit-Tita
- **B.** : Gotabegun, Titbaigun
- **H.** : Bhurat, Ranabaigan
- **Kan.** : Kadusunde, Sondaegida, Sundakkayi
- **Mal.** : Chunta, Kattuchunta
- **Tam.** : Kotukattarai, Malaichundai, Sundai
- **Tel.** : Chundakai, Kondavuste, Kottuvastu, Ustekaya.

Distribution : Naturalised in Himalayas: 250-750 m.

Description : Prickly shrubs. Leaves ovate with sinuate margins. Flowers white. Berries globose, yellow when ripe.

Reproductive Cycle : April-August.

Habitat Ecology : Wastelands.

Matrial Examined : HPU-EL-SKS-5687; Balah (900 m), Bodhal (970 m), Rangru (910 m), Tapa (945 m), Tikkar (910 m)

Part Used : Fruits.

Folk Uses : Ripe fruits **edible** and unripe ones cooked as **vegetable**.

Uses in Literature : Known earlier as an abortifacient, analgesic, antidote to poison, digestive, diuretic, oral contraceptive, sedative, psychoactive plant, vegetable (leaves, fruit); and also for aches, bleeding, chapped hands and feet, cough, diarrhoea, dog bite, ear pain, eczema, gastric problem, itching and enlargement of spleen, malaria, respiration (vet.), skin diseases, snakebite, spinal cord, stomachache, toothache, tooth decay, and wounds (Ambasta, 1986; Augustine & Sivadasan, 2005; Borthakur, 1996b; Chandra, 1997; Chowdhery, 1996; Dwarakan & Ansari, 1996; Gill *et al.*, 1993, 1997; Gogoi *et al.*, 2003b; Gopal & Chandra, 2003; Jain, 1991; Jain *et al.*, 1994; Kirtikar & Basu, 1984; Lalramnghinglova, 2002; Mao, 1993; Nadanakunjidam & Abirai, 2003; Nath & Maiti, 2003; Negi & Gaur, 1991; Negi *et al.*, 1993; Rama Rao & Henry, 1996; Rawat & Chowdhury, 1998; Roy *et al.*, 1998; Saklani & Jain, 1994; Samant *et al.*, 2001; Singh, 2000; Singh *et al.*, 1999; Sinha, 1996; Srivastava *et al.*, 2003a; Sudhakar & Vedavathy, 2000; Viswanathan & Singh, 1996).

Solanum viarum Dunal (Pl. 49B)

Family: Solanaceae.

Vern. : Bhindi, Kandayai.

Distribution : A naturalized weed.

Description : Erect herbs with grey tomentum. Leaves spiny lobed. Flowers white. Berries yellow when ripe. Seeds compressed.

Reproductive Cycle : May-August.

Habitat Ecology : Wastelands. Roadsides.

Material Examined : HPU-EL-SKS-5689; Bhaura (940 m), Chachian (1,300 m), Dadh (1,100 m), Purani-palam (1,100 m), Tanda (1,120 m).

Part Used : Fruits.

Folk Uses : Crushed fruits used against **dysentery** and **cough**. Ash from fruits massaged over gums as **germicide**.

Uses in Literature : Reported to be useful against menstrual complaints, toothache, worms in teeth, wounds, and as a contraceptive (Chaudhuri *et al.*, 1982; Maheshwari *et al.*, 1980, 1981).

***Sorghum bicolor* (L.) Moench** (Pl. 49C)

Syn. *S. vulgare* Pers.; *Hocus sorghum* L.

Family: Poaceae

Vern. : Chari.

English, Sanskrit and Regional Names:

Eng.	: The Great Millet, Indian Millet
Sans.	: Dirghamala
B.	: Janera, Jondla, Jour, Jowari, Jundri
H.	: Juvar
Kan.	: Jolah
Mar.	: Juari
P.	: Joar, Junri
Tam.	: Cholam
Tel.	: Bonda-janu, Talla.

Distribution : Widely cultivated, chiefly in A.P., M.P., Maharashtra, Rajasthan, U. P. & Plains of Uttarakhand.

Description : Tall robust annuals having decompound panicles which are usually crowded whorls of erect branches and branchlets. Spikes rarely subeffuse. Spikelets pedicelled usually neutar.

Reproductive Cycle : July-August.

Habitat Ecology : Agricultural fields.

Material Examined : HPU-EL-SKS-5482; Bhauda (1,050 m), Bhuana (1,400 m), Holta (1,200 m), Rajeher (1,050 m), Sulah (950 m).

Parts Used : Aerial parts. Seeds.

Folk Uses : Aerial parts used as **fodder**. Seed-flour cooked as '**chapati**'; and useful against **constipation**, **heart** and **kidney ailments**.

Uses in Literature : Considered in India as aphrodisiac, appetizer, cooling, diuretic, demulcent; and for biliousness, blood diseases, constipation, edible, fodder, piles, ulcers, tumours, loose motions, urinary bladder and kidney ailments (Caius, 2003; Drury, 1978; Kirtikar & Basu, 1984; Sen & Batra, 1997; Sood & Kaushal, 2008).

***Spilanthes oleracea* L.** (Pl. 49D)
Syn.: *S. acmella* var. *oleracea* C. B. Clarke
Family: Asteraceae.

Vern. : Karkara.

English, Sanskrit & Regional Names:

Eng. : Brazian Cress, Para Cress
Sans. : Alalkar, Maharashtri, Sarahattika
B. : Roshunia
H. : Pakormul, Pokarmul, Ukra
Kan. : Vana
Mar. : Acharbondi, Pipulka
P. : Pakarmul
Tam. : Ukra
Tel. : Maratimogga, Maratitige.

Distribution : Introduced from Brazil; naturalized in India.

Description : Perennial herbs with triangular, obtuse leaves. Inflorescence heads with reddish brown florets. Achenes scabrid.

Reproductive Cycle : May-October.

Habitat Ecology : Cultivated land.

Material Examined : HPU-EL-SKS-5408; Bhagpur (1,600 m), Bhawarna (1,100 m), Chauki (900 m), Dharaman (1,100 m), Tapa (945 m).

Part Used : Inflorescence.

Folk Uses : Inflorescence chewed or eaten to relieve **mouth ulcers**, **toothache**, **affections of gums**, **throat**, and **rheumatism**.

Active Constituents : Isolated constituents from the *whole plant* are acetic, lauric, myristic, palmitic, linoleic acids, myricyl alcohol, stigmasterol and sitosterol-β-D-glucoside.

Uses in Literature : Described in literature as antiscorbutic, cathartic, diuretic, digestive, laxative, odontalgic, sialagogue, stimulant, tonic; and for toothache, throat and gum affections, eczema, headaches, internal worms, affections of throat and gums, paralysis of the tongue, toothache and stammering (Ambasta, 1986; Caius, 2003; Chakraborty *et al.*, 2003; Gogoi & Das, 2003; Kirtikar & Basu, 1984; Mitaliya & Bhatt, 2003; Nath & Maiti, 2003; Parrotta, 2001; Ranjan, 2003; Rawat & Chowdhury, 1998; Roy *et al.*, 1998; Sharma, 2003; Sinha, 1996).

***Spondias pinnata* (L.f.) Kurz** (Pl. 49E)

Syn.: *S. acuminata* Roxb.; *S. amara* Lam.; *S. macrophylla* Wall *ex* Hook.f.; *S. mangifera* Willd.; *Evia amara* Comm.

Family: Anacardiaceae.

Vern. : Muara.

English, Sanskrit & Regional Names:

Eng. : Bile Tree, Indian Hog Plum, Traveller's Delight, Wild Mango
Sans. : Advagabhjogya, Madhuramlaka
Ass. : Amara
B. : Ambra, Amna
Bo. : Amarah, Jangliam
G. : Ranamba
H. : Amara, Ambodha
Kan. : Amatekaya, Ambattemara
Mal. : Ampalam, Kattampalm, Pulimavu
Mar. : Amb, Ambada, Ranamba
Ori. : Ambula, Amrotoko
P. : Ambara, Bahamb
Tam. : Ambiram, Kattumagirangai
Tel. : Adavimamena, Irurumamidi.

Distribution : Temperate Himalayas upto 900 m.

Description : A medium-sized tree. Leaflets 9-13, elliptic-oblong. Flowers greenish-white. Drupes fleshy-yellow.

Reproductive Cycle : April-June.

Habitat Ecology : Cultivated.

Material Examined : HPU-EL-SKS-5526; Balah (900 m), Chhanchri (930 m), Rangru (910 m), Tikkar (910 m).

Parts Used : Leaves. Fruits.

Folk Uses : Ripe fruits **edible**, and unripe ones used for making '**chutney**' and **pickles**. Tender leaves also consumed as '**chutney**', and **fodder** for cattles.

Active Constituents : Associated properties due to β-amyrin, oleanolic acid (*fruits*), lignoceric acid, 24-methylene cycloartanone, stigmast-4-en-3-one, β-sitosterol and its glucoside (*aerial parts*), glycine, cystine, serine, alanine, and leucine (*fruits*).

Uses in Literature : Recorded in India as an aphrodisiac, aromatic, anticancerous, astringent, edible (fruits), tonic; and for boats, cholera, chutney, diarrhoea, dysentery, dyspepsia, earache, flavouring, floats, fumigation, horses in pneumonia, itching, jams, match splints, night fever, non-ornamental plywood, pickles, regularizing menstruation, reviving taste, rheumatism, ringworm, stomachache, tanning, tetanus, tea-chests, unbleachable, wood-pulp (Ambasta, 1986; Arora & Pandey, 1996; Bennet *et al.*, 1991; Borthakur, 1996b; Chun-Lin & Jieru, 1995; Jain 1991, 1996; Kirtikar & Basu, 1984; Kulkarni & Kumbhojkar, 1993; Lalramnghinglova, 1996; Mohanty & Padhy, 1996; Nazarudeen *et al.*, 1996; Negi & Gaur, 1991; Prajapati *et al.*, 2006; Roy *et al.*, 1998; Samant *et al.*, 2001; Sanyal, 1994; Sharma & Borthakur, 2003; Sharma *et al.*, 2000; Singh & Kumar, 2000a; Singh *et al.*, 1996b; Watt, 1972).

***Sporobolus indicus* R. Br.**

Family: Poaceae.

Vern. : Bonkra.

Distribution : Throughtout India, ascending to 1,700 m.

Description : Perennials upto 1m and glabrous entire leaves. Panicle very narrow tapering to fine point. Spikelets green, crowded. Upper empty glumes much longer than the flowering ones.

Reproductive Cycle : July-September.

Habitat Ecology : Open Slopes.

Material Examined : HPU-EL-SKS-5656; Bhaura (940 m), Chachian (1,300 m), Dehan (1,130 m), Purani-palam (1,100 m), Panapari (1,100 m).

Parts Used : Aerial Plant Parts.

Folk Uses : Aerial plant parts given as **fodder** and used for making **brooms**.

Uses in Literature : Reported as fodder (Ambasta, 1986).

***Stellaria media* (L.) Vill.** (Pl. 49F)

Syn.: *S. monogyna* Don; *Alsine media* L.; *Alsinella wallichiana* Benth.

Family: Caryophyllaceae.

Vern. : Khaukhua.

English & Regional Names:

Eng. : Chickweed, Stichwort

Ass. : Morolia.

Distribution : Temperate Himalayas. N. India.

Description : Much branched, flaccid herbaceous annuals, upto 60 cm with lower leaves stalked and upper sessile. Flowers in terminal cymes. Capsules ovoid. Seeds reniform.

Reproductive Cycle : March-April.

Habitat Ecology : Cultivated areas. Damp shady places.

Material Examined : HPU-EL-SKS-5557; Chandpur (1,290 m), Darugnu (1,400 m), Holta (1,200 m), Molichak (1,020 m), Saparu (1,250 m).

Parts Used : Aerial Plant Parts.

Folk Uses : Aerial plant parts used as **vegetable** and **fodder**.

Biological Activity : Plant anti-inflammatory.

Uses in Literature : Reported in India as a vegetable raw or boiled and for bone fractures, eczema, eye infection, haemorrhoids, ulcers, skin sores, expelling intestinal worms, fever, cough, itching, mouth ulcers, urticaria, rheumatism, sores and swelling (Agarwal, 2003; Ambasta, 1986; Arora, 1981; Aswal & Mehrotra, 1994; Ballabh & Chaurasia, 2006; Chandra, 1997; Chopra *et al.*, 1956; Chun-Lin & Jieru, 1995; Devi, 2003; Kapur & Srivastava, 1996; Nath & Maiti, 2003; Prajapati *et al.*, 2006; Saklani & Jain, 1994; Sharma, 2003; Singh & Kumar, 2000a; Sinha, 1996; Sood & Thakur, 2004; Viswanathan, 2000).

Stephania glabra (Roxb.) Miers (Pl. 50A)

Syn.: *S. rotuda* Hook. f.; *Cissampelos glabra* Roxb.; *C. longa* Miers; *C. japonica* Miers; *C. roxburghiana* Miers; *C. glabra* Miers; *C. glandulifera* Miers; *Cocculus glabra* Roxb.; *C. finlaysoniansus* Wall.; *Clypea wightii* Arn.

Family: Menispermaceae.

Vern. : Bish-Khapar.

Regional Name:

U.P. : Gindaru, Purha.

Distribution : Hilly districts throughout India, ascending to 2,000 m.

Description : Climbers having light-green, broadly-ovate leaves and greenish-yellow flowers. Fruits red, pisiform.

Reproductive Cycle : June-August.

Habitat Ecology : Moist shady places.

Material Examined : HPU-EL-SKS-5502; Bhauda (1,050 m), Gharthun (900 m), Paror (940 m), Rangru (910 m), Tanda (1,120 m).

Part Used : Root-tuber.

Folk Uses : An amulet of root-tuber around neck of livestock checks body **sores** 'bashahar' and also fed raw for **skin diseases** and **purification of blood**.

Active Constituents : *Tubers* yield palmatine, dehydrocorydalmine, palmatrubine stepharanine, tetrahydropalmatine, crydalmine and stepholidine.

Biological Activity : Anticholinesterase, anti-inflammatory and spasmolytic activities found positive.

Uses in Literature : Used earlier for asthma, body and headache, eye diseases, intestinal complaints, pulmonary tuberculosis, skin diseases, tumour in stomach and as psychoactive plant (Ambasta, 1986; Jain, 1991, 2003; Jain *et al.*, 1994)

Syzygium cumini (L.) Skeels (Pl. 50B)

Syn.: S. *jambolanum* DC.; *Calyptranthes jambolana* Willd.; *C. capitellata* Ham.; *Eugenia frondosa* Wall.; *E. jambolana* Lam.; *E. caryophyllifolia* Lam.; *Myrtus cumini* L.

Family: Myrtaceae.

Vern. : Jamun.

English, Sanskrit & Regional Names:

Eng. : Black Plum, Java Plum
Sans. : Jambaram, Rajaphala, Shukapriya
Ass. : Jamu, Kola- jamu
B. : Chotajam, Kalajam
Bo. : Jhambul
G. : Jambudi, Ravanam
H. : Jamni Phalani, Jamun, Phalanda
Kan. : Nerale
Mal. : Naga, Perinnaral
Mar. : Jaman, Rajjambula
Ori. : Bhotajamo, Chuajamo
P. : Jammu
Tam. : Kottainagam, Takkolam
Tel. : Jambuvu, Peddoneredu.

Distribution : Throughout India.

Description : Evergreen trees with opposite, coriaceous leaves; and white, fragrant flowers. Fruits dark purple, globose, succulent.

Reproductive Cycle : April-August.

Habitat Ecology : Fields. Grasslands.

Material Examined : HPU-EL-SKS-5516; Bindraban (1,270 m), Boda (1,000 m), Jiun (1,550 m), Rangru (910 m), Saparu (1,250 m).

Parts Used : Leaves. Bark. Fruits. Seeds. Wood.

Folk Uses : Oil smeared leaves applied to hasten **suppuration** of **furuncle** and **boils**. Fruits **edible**, and also given to induce **rumination** in livestock. One teaspoonful of seed powder taken with cold water every morning for **diabetes**. Bark and tender twigs chewed for **mouth ulcers**. Wood employed for **agricultural tools** and **construction work**.

Active Constituents : These are betulinic acid, friedelin, friedelinol, kaempferol and its 3-O-glucoside, quercetin, sitosterol and its glucoside, sucrose (*stem bark*), heptacosane, nonacosane, triacontane, hentriacontane, octacosanol, triacosanol, dotriacosanol, betulinic acid and crotegolic (masilinic) acid (*leaves*), two new anthocyanins-delphinidin-3-gentiobioside and malvidin-3-laminariboside- along with petunidin-3-gentiobioside (*fruits*), gallic and ellagic acids, corilagin and related ellagitannins, 3,6-hexahydroxyphenoylglucose and its isomer 4,6-hexahydroxydiphenoyl-glucose, 1-alloylglucose, 3-galloylglucose and quercetin (*seeds*).

Biological Activity : Anorexigenic (plant extr.) and hypoglycaemic (seed extr.) activities found (+)ve.

Uses in Literature : Known to be used as antidote to Nux vomica poisoning, astringent, colouring material in Mithila paintings, colic, constuction wood, edible (fruits), fodder, foliage for silkworms, green manure, mouthwash, props for shafts and galleries in mines, sacred (temple sculptures and symbolic abode of gods), tonic, vegetable; and for agricultural implements, anaemia, asthma, blisters in mouth, boat building, branches (festivals and marriages), bronchitis, cancer, cart-wheels, cholera (vet.), colouring fish nets, diabetes, diarrhoea, digestive complaints, dyeing and tanning, dysentery, enlargement of spleen, painful menstruation, expelling placenta after delivery, fermentation for rice beer, furniture, giddiness, gums, indigestion, intestinal worms, jams, jellies, leucorrhoea, liver disorders, natural shiny black hair, piles, pimples, preserves, purifying blood, redness of eyes, sleepers, squashes, stomach pains, stored grain pest, swelling (vet.), thatching, tool handles, troughs, ulcers, vinegar (ripe fruits), well-curbs, and wound maggots (Ambasta, 1986; Arora & Pandey, 1996; Aminuddin & Girach, 1993; Bennet *et al.*, 1991; Bhandary & Chandrashekar, 2002; Bhatt *et al.*, 2003a, b; Bhattacharya *et al.*, 2005; Bondya & Sharma,

2004; Borthakur & Sharma, 1996; Borthakur *et al.*, 2004; Chevallier, 1996; Chopra *et al.*, 1956; Chowdhery, 1996; Chowdhury *et al.*, 1996; Chun-Lin & Jieru, 1995; Dey & Bahadur, 1973; Diwakar & Ansari, 2003; Dixit & Pal, 1996; Duke *et al.*, 2002; Dwivedi, 2003; Gogoi *et al.*, 2003a; Hebber *et al.*, 2003; Henry *et al.*, 1996; Hosagoudar & Henry, 1996a; Jain, 1968, 1991; Jha & Basak, 1994; Karnick, 1994; Kayal *et al.*, 2003; Khanna *et al.*, 1996c; Khare, 2004; Kulkarni & Kumbhojkar, 1996; Lalramnghinglova, 1996; Mandal & Basu, 1996; Masih, 2003; Mishra *et al.*, 1996b; Mitaliya *et al.*, 2003; Panthi & Chaudhary, 2003; Prajapati *et al.*, 2006; Radhakrishnan *et al.*, 1996; Rajendran & Sikarwar, 2003; Rajendran *et al.*, 2003; Rama Rao & Henry, 1996; Ranjan, 2000; Roy *et al.*, 1998; Sahni, 1998; Saklani & Jain, 1994; Saluja, 1996; Sanyal, 1994; Sharma, 2003; Sharma & Rana, 2005; Sharma & Singh, 2001; Sharma *et al.*, 2000; Shome *et al.*, 1996; Sikarwar & Kaushik, 1992; Singh, 1996b, Singh & Kumar, 2000a; Sinha, 1996; Sinha & Sinha, 2001; Siwakoti & Siwakoti, 2000; Siwakoti & Varma, 1996; Sood & Thakur, 2004; Subramani & Goraya, 2003; Sudhakar & Vedavathy, 2000; Tirkey *et al.*, 2003a; Upadhyay & Singh, 2005; Watt, 1972).

***Syzygium operculatum* (Roxb.) Nied.** (Pl. 50C)

Syn.: *Calypatanthes caryophyllifolia* Moon; *Eugenia operculata* Roxb.; *Syzygium nervosum* DC.

Family: Myrtaceae.

Vern. : Kathmanh.

Sanskrit & Regional Names:

- **Sans.** : Bhumijambu
- **Ass.** : Mokara-jam, Godhajan
- **H.** : Deigdugia, Jamava, Piaman, Raijman
- **Mal.** : Maral
- **Mar.** : Naral
- **Ori.** : Moniosiajamo.

Distribution : Sub-Himalayan regions; Yamuna-Assam.

Description : Trees upto 18 m tall. Young leaves pink, older ones yellowish-brown. Flowers pale yellow or white, sessile. Fruits (berries) globose or ellipsoid, purple when ripe.

Reproductive Cycle : April-July.

Habitat Ecology : Wastelands along field borders.

Material Examined : HPU-EL-SKS-5695; Bindraban (1,270 m), Chachian (1,300 m), Kathena (1,290 m), Molichak (1,020 m), Nanaun (900 m).

Parts Used : Leaves. Fruits. Stem.

Folk Uses : Ripe fruits **edible**. Leaves given as **fodder**. Stem good for **agricultural implements** and as **fuelwood**.

Uses in Literature : Described earlier for bronchitis, biliousness, rheumatism, dysentery, ulcers; and as aphrodisiac, astringent, edible (fruits), refrigerant, tonic and wood (Ambasta, 1986; Chandra, 1997; Kirtikar & Basu, 1984; Murthy *et al*., 2003; Parrotta, 2001; Pearson & Brown, 1932).

Tagetes minuta **L.** (Pl. 50D)

Syn.: *T. glandulifera* Schrank

Family: Asteraceae.

Vern. : Gatt.

English Name:

Eng. : Stinking-Roger.

Distribution : N.W. Himalayas: 1,250-2,500 m.

Description : Tall aromatic annuals with pinnatisect leaves and yellow heads in corymbose clusters. Achenes flattened, crowned by pappus.

Reproductive Cycle : September-October.

Habitat Ecology : Moist Places.

Material Examined : HPU-EL-SKS-5525; Bhatu (1,040 m), Bodhal (970 m), Gharthun (900 m), Rajeher (1,050 m), Thandol (950 m).

Parts Used : Flower. Leaves.

Folk Uses : Flowers offered to appease **deities**; and its juice (50 ml two times weakly for six months) taken to treat **piles**. Poultice of its paste in curd good against **burns**. Leaf juice used as drops to check **earache** and its paste applied for **cutaneous disease**.

Active Constituents : Associated properties of the *plant* due to tagetone, phenylethyl alcohol, ocimens, salicylaldehyde, phenylacetaldehyde, eudesmol, linalylacetate, limonene and linalool.

Biological Activity : Plant hypotensive, spasmolytic, bronchodilatory and anti-inflammatory.

Uses in Literature : Known earlier as an aperient, aromatic, diaphoretic, diuretic and insect repellent; and for allergy, bone fractures, earache, insect bite, killing maggots in wounds, ringworm, and tonsillitis (Ambasta, 1986; Baruah & Sarma, 1987; Dash *et al*., 2003; Islam & Hasin, 2003; Jain, 2003b;

Kumar *et al.*, 2003; Sen & Behera, 2003; Sharma, 2003; Singh & Singh, 2003; Upadhye & Kulkarni, 2004).

Terminalia arjuna **Wight & Arn.** (Pl. 50E)

Syn.: *T. berryi* Wight & Arn.; *T. glabra* Wight & Arn.; *T. ovalifolia* Rottl.; *Pentaptera glabra* Roxb.; *P. angustifolia* Roxb.

Family: Combretaceae.

Vern. : Arjan, Arjun.

English, Sanskrit & Regional Names:

Eng. : White Murdah
Sans. : Dhananjaya, Karaviraka, Savya Sachi
Ass. : Orjun
B. : Arjun, Kahu
Bo. : Arjunasadadra, Jamla
G. : Arjunasadada, Dhaulasadr
H. : Arjun, Kangach, Kowa
Kan. : Maddi
Mal. : Vellamarutu
Mar. : Arjunladada, Saumadat
Ori. : Orjuno, Pornoyosahajo
P. : Arjan, Jumla
Tam. : Attumarudu, Kulamatrudu
Tel. : Kakubhamu, Tittumaddi
U.P. : Kauha, Arjun.

Distribution : Commonly along riverian areas in the Sub-Himalayan tracts.

Description : Large-sized trees with smooth bark. Leaves narrowed at the base, often cordate. Spikes usually panicled. Fruits glabrous, ovoid, angled.

Reproductive Cycle : May-September.

Habitat Ecology : Planted along agricultural fields, roadsides.

Material Examined : HPU-EL-SKS-5657; Andretta (980 m), Bindraban (1,270 m), Gopalpur (1,350 m), Sulah (950 m), Tikkar (910 m).

Parts Used : Bark. Fruits. Wood.

Folk Uses : Crushed bark (100 g) with wheat flour (1-2 kg) cooked in desi ghee and taken (two teaspoons, 4 times) to check **excessive bleeding** during menses. Its decoction (two teaspoons daily) prescribed as **cardiac tonic**. Wood used for **agricultural implements** and **building** works. Fruits **pickled**.

Active Constituents : *Bark* contains cevasidin, β-sitosterol, friedelin, methyl deanolate and gallic, ellegic & arjunic acids, β-sitosterol, (+) catechol, (+) gallocateenol, epicateenol, epigalco catechol, arjunolone, arjunglucoside III, arjunoside I and II, arjunflucoside I, arjunetin, leucoganidin, terminoic acid and arjunin.

Uses in Literature : Reported so far as an antidote to poison, antidysenteric, astringent, detergent, deobstruent, expectorant, febrifuge, fish poison, hair wash, psychoactive plant, purgative, sacred, sex stimulant, styptic, tonic; and for agricultural implements, anaemia, apthae, asthma, bilious affections, blood diseases, blood dysentery, blood pressure, boats and canoes, boils, chest pain and heart palpitation, cirrhosis of liver, cuts and wounds, dysuria, earache, edible, electric poles, epilepsy, fatigues, fever, fractures, house construction, hypertension, leprosy, leucoderma, leucorrhoea, masts, menstrual disorders, musical instruments, painful urination, plywood manufacture, pneumonia, sores, skin diseases, tanning, tumours, ulcers, water traps, water trough, wheels of country carts, and wounds (Agarwal, 2003; Ambasta, 1986; Bhatt *et al.*, 2003c; Bhuyan, 2003; Bose, 1984; Chauhan, 1999; Chopra *et al.*, 1956; D'Rozario *et al.*, 1999; Drury, 1978; Das & Agarwal, 1991; Datt, 1996; Dey & Bahadur, 1973; Diwakar & Ansari, 2003; Dixit & Kumar, 2003; Dwivedi, 2003; Gogoi & Borthakur, 2001; Hembrom, 1991; Henry *et al.*, 1996; Hosagoudar & Henry, 1996a; Jain, 1968, 1991; Jain *et al.*, 1994; Kalakoti, 2001; Karnick, 1994; Khare, 2004; Kothari & Moorthy, 1996; Kumar & Lal, 2003; Kumar *et al.*, 2003; Kurian, 1995; Lalramnghinglova, 1996; Masih, 2003; Mitaliya & Bhatt, 2003; Pearson & Brown, 1932; Prasad *et al.*, 2003; Prajapati *et al.*, 2006; Rama Rao & Henry 1996; Ranjan, 2003; Sahni, 1998; Sanyal, 1994; Sarin, 1996; Satapathy & Brahmam, 1996; Sen & Behera, 2003; Sharma, 2003; Sharma & Sood, 1997; Singh & Pandey, 1998; Singh & Prakash, 1996; Singh *et al.*, 1997; Sinha & Sinha, 2001; Srivastava *et al.*, 2003a; Watt, 1972).

Terminalia bellirica (Gaertn.) Roxb. (Pl. 50F)
Syn.: *Myrobalamus bellirica* Gaertn.
Family: Combretaceae.

Vern. : Bahera.

English, Sanskrit & Regional Names:

Eng.	: Bastard Myrobalan, Beddanuts, Belleric Myrobalan
Sans.	: Anilaghnaka, Kaliyugalaya, Karshaphala
Ass.	: Bauri, Hulluch
B.	: Bahera, Buhura
Bo.	: Buhudda, Bherda, Yella
G.	: Bahedamunjad, Bahedo
H.	: Bahera, Bhairach, Sagona
Kan.	: Tare Kai, Shanti Kayi
Kash.	: Babelo, Balali
Mal.	: Tanni, Tusham
Mar.	: Beharda, Bhirda, Goting, Vavara
Ori.	: Horinda, Jonghihorida
P.	: Bahira, Bayrah
Tam.	: Akkam, Kalitturuman
Tel.	: Karakkaya, Nallakaraka.

Distribution : Throughout the plains and Sub-Himalyan tracts.

Description : Tall trees. Leaves simple, entire, alternate. Drupes hard, obovoid, more often pentagonal.

Reproductive Cycle : May-September.

Habitat Ecology : Along agricultural fields and grassy slopes.

Material Examined : HPU-EL-SKS-5441; Bhauda (1,050 m), Boda (1,000 m), Jasun (1,000 m), Thandol (950 m), Tikkar (910 m).

Parts Used : Leaves. Fruits. Seeds. Wood.

Folk Uses : Leaves given as **fodder**. Kernels **edible** and considered **nutritious**. Wood employed for **agricultural tools**, and as **fuel**. Fruits form one of the constituents of '**triphala**' and prescribed for **digestive** disorders, **liver complaints**, **pulmonary ailments** and **constipation**.

Active Constituents : *Fruits* afford β-sitosterol, gallic acid, ellagic acid, ethyl gallate, galloyl glucose and chebulegic acid.

Uses in Literature : Reported to be used as an anthelmintic, antiseptic, aperient, astringent, attenuant, brain tonic, colic, demulcent, diuretic, edible (kernels), expectorant, fish poison, fodder, hallucinogenic, ornamental around temple yards, important constituent of 'triphala', laxative, narcotic, pit-props in coal mines, psychoactive plant, writing inks; and for acidity, ageing, agricultural tools, asthma, black hairs, boards, boats, bronchitis, carts, cholera, coffee boxes, cold, constipation, cooling, cooper's work, cough, cramps, diabetes, diarrhoea (vet.), dropsy, dugouts, dyeing, dysentery (vet.), dysuria, enlargement of spleen, fever, gastric complaints, headache, hoarseness, horses in dysentery, inflammation, inflammatory diseases of eye, intestinal complaints, itching by ticks, leprosy, liver complaint, loose motions, manufacture of molasses, muscular pain, opthalmia, packing cases, piles, ploughs, rafters, respiratory disease, rheumatic, soothing lotion, stomach pain, swellings, skin diseases, slate frames, small crafts, snakebite, soap manufacture, tannin, throat troubles, turning, ulcer (vet.), vigour and vitality, vomiting, and weakness (Agarwal, 2003; Ambasta, 1986; Aminuddin & Girach, 1993; Bennet *et al.*, 1991; Bhatt *et al.*, 2003a; Bhattacharya *et al.*, 2005; Borthakur, 1996b; Borthakur & Gogoi, 1994; Bose, 1984; Chakraborty & Bhattacharjee, 2003; Chauhan, 1999; Chevallier, 1996; Chopra *et al.*, 1956; Das & Agarwal, 1991; Dey & Bahadur, 1973; Dixit & Kumar, 2003; Dwivedi, 2003; Gogoi & Borthakur, 1991; Goud & Pullaiah, 1996b; Jain, 1968, 1991; Jain *et al.*, 1994; Jha & Kumar, 2003; Kalakoti, 2001; Khare, 2004; Kirtikar & Basu, 1984; Kothari & Moorthy, 1996; Kumar & Lal, 2003; Kumar & Pullaiah, 2003; Lalramnghinglova, 1996; Masih, 2003; Maya *et al.*, 2003; Mishra *et al.*, 1997; Mohanty & Padhy, 1996; Murthy *et al.*, 2003; Nadanakunjidam, 2003; Nadanakunjidam & Abirai, 2003; Nayak *et al.*, 2003; Nazarudeen *et al.*, 1996; Panthi & Chaudhary, 2003; Parrotta, 2001; Prajapati *et al.*, 2006; Prakash, 2001; Prasad *et al.*, 1996; Pullaiah & Kumar, 1996; Rama Rao & Henry, 1996; Ranjan, 2000; Rao, 1990, 1996; Roy *et al.*, 1998; Sanyal, 1994; Sen & Batra, 1997; Sharma, 2003; Sharma & Borthakur, 2003; Sharma & Singh, 2001; Sharma & Sood, 1997; Singh, 1991, 1999; Singh & Kumar, 2000a; Singh & Pandey, 1998; Siwakoti & Varma, 1996; Srivastava *et al.*, 2003a; Subramaniam, 2000; Thomas & Britto, 2003; Tirkey *et al.*, 2003a; Verma *et al.*, 1995; Watt, 1972).

Terminalia chebula **Retz.** (Pl. 51A)

Syn.: *T. aruta* Ham.; *T. reticulata* Roth; *Myrobalanus chebula* Gaertn.; *Embryogonia arborea* Teys & Binn.

Family: Combretaceae.

Vern. : Harad.

English, Sanskrit & Regional Names:

Eng. : Chebulic Myrobalan
Sans. : Harra
Ass. : Silikha
B. : Haritaki

G.	: Hardo
H.	: Har, Harar, Harra
Mal.	: Kadukka
Mar.	: Hirda
Ori.	: Haridra
P.	: Har, Harar
Tam.	: Kadukkai
Tel.	: Karaka
U.P.	: Harr.

Distribution : Throughout India.

Description : Small trees with deciduous leaves. Flowers pubescent, deciduous. Fruits green, spindle-shaped.

Reproductive Cycle : April-June.

Habitat Ecology : Barren lands. Scrub forests.

Material Examined : HPU-EL-SKS-5425; Bhuana (1,400 m), Chauki (900 m), Chhanchri (930 m), Khaira (940 m), Rangru (910 m).

Parts Used : Leaves. Fruit. Wood.

Folk Uses : Leaves employed as **fodder**. Roasted fruits or **pickled** ones prescribed for **cough**, **digestive disorders**, cardiac **tonic**, **diabetes**, and **lung ailments**. Powdered fruits massaged for **bleeding gums**. Wood used as **fuel**.

Active Constituents : *Fruits* possess ellagitannin, terchebulin, punicalagin, terflavins A, B, C and D, shilimic, gallic, triacontanoic and palmitic acids, β-sitosterol, daucosterol, triethyl ester of chebulic acid and ethyl ester of gallic acid arjungenin, terminoic acid and arjunolic acid.

Biological Activity : Antibacterial and spasmolytic activities positive.

Uses in Literature : Described earlier as an alterative, antidote to poisons, astringent, blood purifier, cardiac tonic, colic, cooling wash to eyes, dentrifice, dye for hairs, edible (fruits, pickle), gargle in inflammation of mucous membrane of mouth, fish poison, fodder, important constituent of 'triphala', ingredient of 'chywanprash', laxative, ornamental, psychoactive plant, sacred for making writing ink, stomachic, and also for acidity, ageing, agricultural implements, anaemia, anthrax (vet.), ascites, asthma, bile troubles, bleeding, blood pressure, burn, chronic diarrhoea and ulcer, cold, colouration of houses, conjunctivitis, cough, cuts, diabetes, diarrhoea, dysentery (vet.), dysuria, eczema, enlarged spleen and liver, eye diseases, fever, flatulence, gastropathies (vet.), haematuria, headache, hiccups, dysentery in horses, house construction, intestinal worms, liver (vet.), measles, menstrual complaints, muscular pain, musical instruments, ornamental rosaries, piles, pneumonia, preventing the effects of ageing, rheumatic pain, skin diseases, sores, sore throat, stomach complaint, stomachache (vet.), tannin, teeth bleeding and ulceration of gums, urinary diseases, vaginal discharge, vigour and enhances vitality, vomiting, worms and wounds (Agarwal, 2003; Ambasta, 1986; Aminuddin & Girach, 1993; Arinathan *et al.*, 2003; Bhattacharya *et al.*, 2005; Borthakur, 1996a, b; Borthakur & Gogoi, 1994; Chakraborty & Bhattacharjee, 2003; Chauhan, 1999; Chevallier, 1996; Chopra *et al.*, 1956; Chowdhery, 1996; Chhetri, 1994; Das & Agarwal, 1991; Dey & Bahadur, 1973; Diwakar & Ansari, 2003; Dixit & Kumar, 2003; Girach, 1992; Gogoi & Borthakur, 1991, 2001; Hosagoudar & Henry, 1996a; Islam & Jha, 2003; Jain, 1968, 1991; Jain & Patole, 2001; Jain *et al.*, 1994; Khanna & Kumar, 2000; Kirtikar & Basu, 1984; Kshirsagar & Singh, 2000; Kumar, 1996; Kumar & Lal, 2003; Lalramnghinglova, 1996; Masih, 2003; Mehrotra *et al.*, 1995; Mishra *et al.*, 1997; Mohanty & Padhy, 1996; Nayak *et al.*, 2003; Panthi & Chaudhary, 2003; Parrotta, 2001; Prajapati *et al.*, 2006; Prakash, 2001; Rama Rao & Henry, 1996; Ranjan, 2000, 2003; Rawat & Chowdhury, 1998; Said, 1997; Saini, 1996a; Saklani & Jain, 1994; Sarangi & Sahu, 2005; Sarin, 1996; Sen & Batra, 1997; Sharma, 2003; Sharma & Borthakur, 2003; Sharma & Sood, 1997; Singh, 1991, 2000, 2001; Siwakoti & Varma, 1996; Srivastava *et al.*, 2003c; Subramaniam, 2000; Tirkey *et al.*, 2003b; Verma *et al.*, 1995; Watt 1972).

***Terminalia tomentosa* (DC.) Wight & Arn.** (Pl. 51B)

Syn.: *T. alata* Heyne *ex* Roth; *T. crenulata* Wight & Arn.

Family: Combretaceae.

Vern. : Rasaan.

English, Sanskrit & Regional Names:

Eng.	: Black Murdah
Sans.	: Krishnatvaka, Nissaraphalaka, Shyamasa Raka
Ass.	: Amari
B.	: Ashan, Piashel
Bo.	: Ain, Marthi, Sadri
G.	: Hadri, Sadada, Sajadio
H.	: Asluna, Kauha, Sein
Kan.	: Sadada
Mal.	: Karimarutu, Temparu
Mar.	: Madat, Sadada, Yeu
Ori.	: Kolashahko, Pondosahajo
P.	: Aisan, Sein
Tam.	: Aruchananam, Karuppumarudu, Puddavem
Tel.	: Inumaddi, Nalamaddi.

Distribution : Sub-Himalayan tracts. Deccan peninsula.

Description : Tall trees. Leaves subopposite, elliptic. Spikes panicled. Flowers dull yellow. Fruits brown, glabrous.

Reproductive Cycle : August-March.

Habitat Ecology : Along agricultural fields.

Material Examined : HPU-EL-SKS-5667; Balah (900 m), Bhaura (940 m), Jasun (1,000 m), Rangru (910 m), Tikkar (910 m).

Parts Used : Leaves. Bark. Wood.

Folk Uses : Leaves employed as **fodder**. Bark decoction used for **tanning** leather. Wood considered good for **agricultural implements**, **building works**, **tool handles** and as **fuel.**

Active Constituents : Constituents isolated are β-sitosterol, oleanolic acid, arjunolic acid, maslinic lactone and termonolic (*heartwood*), arjunic, arjunolic, betulinic and ellagic acids and β-sitosterol (*bark*).

Uses in Literature : Used earlier for agricultural implements, anaemia, beams, burns, charcoal, cholera, construction of carts, cuts, diarrhoea, doors, dyeing, dysentery, edible purposes (ash of bark), electric casing, engine, brake blocks, fever, fire-proof building, fuel, furniture, haematuria, loose motions, musical instruments, oil plywood manufacturing, rafters, rearing silk worms, rough carpentry, sores, stomachache, tannin, tool handles, toys, ulcers, windows, wounds; and as adhesive, antiemetic, cardio-tonic, fish poison, fodder, purgative, rice pounders, stimulant, symbolic (sacred for idols) and tonic (Ambasta, 1986; Jain, 1991; Manandhar, 1990; Watt, 1972).

***Thuja orientalis* L. (Pl. 51C)**

Syn.: *Biota orientalis* (L.) Endl.

Family: Cupressaceae.

Vern. : Sarua, Morpankh.

English & Hindi Names:

Eng. : Oriental Arbor-vitae

H. : Mayur Pankh, Morpankhi.

Distribution : Throughout world.

Description : Ornamental shrubs possessing opposite, green, scale-like leaves, small male cones and microsporophylls arranged in whorls. Pollen grains wingless. Female cones globose, woody. Ovules erect.

Reproductive Cycle : July-October.

Habitat Ecology : Cultivated often met as an escape.

Material Examined : HPU-EL-SKS-5724; Bhawarna (1,100 m), Ghugar (1,250 m), Gopalpur (1,350 m), Purani-palam (1,100 m), Tapa (945 m).

Parts Used : Whole Plant. Seeds.

Folk Uses : Plant grown as an **ornamental** tree. Powdered seeds (10 g with honey, daily for one year) prescribed for **diabetes**.

Active Constituents : These are palmitic (5.28), stearic (7.30), C-18 acids (36.29), C-20 acids (6.10) and linoleic acid (44.6%) (*seed oil*).

Biological Activity : Quercitin exhibits haemostatic activity in mice. Deoxypodophyllotoxin and ethyl p-methoxy-trans-cinnamate showed cytotoxic activity against HeLa cells.

Uses in Literature : Reported in literature for barrels and casks, dysentery, fencing posts, furniture, house building and tannin; and as antipyretic, diuretic, tonic and psychoactive (Ambasta, 1986; Jain *et al.*, 1994; Singh & Prakash, 1996).

Thysanolaena maxima (Roxb.) Kuntze (Pl. 51D)

Syn.: *T. acarifera* Arn. & Nees; *T. agrostis* Nees; *Agrostis maxima* Roxb.

Family: Poaceae.

Vern. : Garnaal.

English & Regional Names:

Eng. : Bouquet-grass, Tiger-grass

B. : Phuljanta

Bo. : Barucha

M.P. : Chir

Ori. : Phulbadhuni.

Distribution : Throughout India.

Description : Tall handsome glabrous grasses upto 30 cm high with large, linear-lanceolate coriaceous leaves. Sheaths glabrous. Panicles large, soft; branches very numerous, filiform with spikelets small, ovoid-lanceolate.

Reproductive Cycle : July-October.

Habitat Ecology : Moist-shady steep slopes.

Material Examined : HPU-EL-SKS-5730; Chornali (1,280 m), Gharthun (900 m), Holta (1,200 m), Sulah (950 m), Thandol (950 m).

Parts Used : Leaves. Culms.

Folk Uses : Leaves given as **fodder**; culms employed for making **winnowing pans** ('chhaj').

Uses in Literature : Known to be useful as mouthwash during fever, fodder, brooms, reed-pens for writing (Ambasta, 1986; Kirtikar & Basu, 1984; Purohit & Samant, 1995).

Tinospora cordifolia Miers (Pl. 51E)

Syn.: *T. palminervis* Miers; *Coccculus cordifloius* DC.; *C. convolvulaceus* DC.; *C. verrucosus* Wall.; *Menispermum cordifloium* Willd.

Family: Menispermaceae.

Vern. : Gloan, Gulja.

English, Sanskrit & Regional Names:

Eng. : Gulancha Tinospora
Sans. : Amritalata, Jwaranashini, Madhuparnika
Ass. : Amarlata, Siddhilata
B. : Gadancha, Gulancha
Bo. : Gharol, Giroli
G. : Gado, Gulo
H. : Gilol, Gurach
Kan. : Amrutaballi
Kash. : Amrita, Gilo
Mal. : Amrytu, Peyamrytum
Mar. : Gharol, Gulavella
Ori. : Gulancha, Gulochi
P. : Batindu, Gilogularich
Tam. : Amridavalli, Kaippuchindil
Tel. : Duyuntige, Jivantika
U.P. : Gurch.

Distribution : Throughout Tropical India.

Description : Glabrous climbing shrubs with succulent stem and cordate, glabrous, acuminate leaves. Flowers yellow, unisexual; male fascicled, female ones solitary. Drupes deep-red, pea-sized.

Reproductive Cycle : June-July.

Habitat Ecology : Common in hedges.

Material Examined : HPU-EL-SKS-5668; Balah (900 m), Bhauda (1,050 m), Khaira (940 m), Makol (900 m), Tikkar (910 m).

Part Used : Stem.

Folk Uses : Stem forms one of the important ingredients of '**havan**' for relegious ceremonies; and used as **cattle feed** for promoting **lactation**. Also, its extr. after soaking for 10-15 days in water given to relieve **body heat** in livestock whereas its decoction in milk prescribed to check **indigestion**, **persistent fever**, **jaundice** and **diabetes** (one teaspoonful for 6-7 days). One teaspoon of powdered dried stem added to 'halva' (sweet pudding) prescribed to check **backache**.

Biological Activity : Anti-inflammatory, analgesic, antipyretic, CNS depressant and hypoglycaemic activities found to be positive.

Uses in Literature : Reported earliert as an analgesic, anthelmintic, antiarthritic, antiperiodic, acidity, antipyretic, aphrodisiac, appetiser, bitter tonic, blood purifier, cardiac stimulant, carminative, colic (vet.), diuretic (vet.), emetic, expectorant, fibre for tying bundles, fodder, ingredient of 'chywanprash', psychoactive plant, sacred, stimulant, tonic (vet.), vermifuge (vet.); and for acidity of urine, anthrax (vet.), asthma, biliousness, bone fractures, bone fractures (vet.), bronchitis, chronic gonorrhoea, cough, cough (vet.), debility, diabetes, diarrhoea, diptheria, dysentery, dyspepsia, fever (vet.), gastric trouble, general debility, gout, haematuria (vet.), headache, impaction (vet.), improving the health of a lean and thin horse (vet.), indigestion (vet.), insect bite (vet.), intermittent fevers, jaundice, jaundice (vet.), leprosy, leucorrhoea, malaria, mastitis (vet.), nocturnal emission, pthisis, piles, pulmonary tubercular, reducing bile production, rheumatism, scorpion sting, skin disease, snakebite, spermatorrhoea, splenic affections, stomach disorders, stomach ulcers, syphilis, ulcer (vet.), urinary diseases, vein injuries, venereal complaints and visceral obstruction (Ambasta, 1986; Bhatt *et al.*, 2003a, b; Chopra *et al.*, 1956; Diwakar & Ansari, 2003; Dixit & Kumar, 2003; Dwivedi, 2003; Goud & Pullaiah, 1996b; Jain, 1968, 1991; Jain & Saklani, 1992; Jain *et al.*, 1994; Jha & Kumar, 2003; Jha *et al.*, 1991; Katewa & Arora, 1997; Kshirsagar *et al.*, 2003; Kumar *et al.*, 2003; Lalramnghinglova, 1996; Mahato & Mahato, 1996; Mehrotra *et al.*, 1995; Prasad *et al.*, 1996, 2003; Rama Rao & Henry, 1996; Rawat & Chowdhury, 1998; Reddy *et al.*, 1996; Sahoo & Mudgal, 1995; Saini, 1996a; Sensarma, 1991; Sharma, 1996, 2003; Singh, 1991; Singh & Pandey, 1998; Siwakoti & Varma, 1996; Srivastava *et al.*, 2003a; Upadhyay, 2003; Verma *et al.*, 1995; Watt, 1972).

***Toona ciliata* Roem.** (Pl. 51F)

Syn.: *Cedrella toona* Roxb. *ex* Rottl.; *C. febrifuga* Blume; *C. hexandra* Wall.; *C. longifolia* Wall.; *C. australis* F. Muell.

Family: Meliaceae.

Vern. : Tooni.

English, Sanskrit & Regional Names:

Eng. : Happy Tree, Indian Mahogany, Mountain Cedar, Red Cedar, Sandal

Sans. : Kubheraka, Nandivriksha

Ass. : Henduripoma, Jia

B. : Lud, Tunna

Bo. : Deodari, Tundu, Tupa

H. : Tun, Tuna, Tunkajhar

Mal. : Devabaram, Malarveppu

Mar. : Deodari

Ori. : Mona Limba

P. : Bisru, Darab, Khanam, Tun

Tam. : Madagirvembu, Sandavanembu

Tel. : Gali, Nandi.

Distribution : Tropical Himalayas. Throughout the hilly districts of C. & S. India upto 1,000 m.

Description : Trees with even pinnate leaflets varying from 10-20. Panicles longer than leaves. Flowers pink coloured. Fruits ovoid capsule, 5-valved.

Reproductive Cycle : May-June.

Habitat Ecology : Wastelands. Roadsides. Open slopes.

Material Examined : HPU-EL-SKS-5456; Bhaura (940 m), Khaira (940 m), Makol (900 m), Nanaun (900 m), Thandol (950 m).

Parts Used : Leaves. Flowers. Bark. Wood.

Folk Uses : Leaf and flower paste applied to relieve **headache**, and also to treat **skin diseases**. Wood considerd as high quality timber wood and employed for making **tool handles**, **furniture** and **building works**. Bark juice taken to check **dysentery**.

Active Constituents : Constituents isolated are cedrelone (*heartwood*), geranylgeranoil (*trunk wood*), cedrelone, 1, 2-dihydrocedrelone, bergapten and β-sitosterol, a new tetranorterpinoid-6-deoxy cedrelone, a coumarin-siderin (*wood*), two new B-secotetranortiterpenoids- toonacilin and 6-acetoxytoonacilin (*bark*).

Biological Activity : Activities found positive are anticancerous, antiamoebic, hypoglycaemic, spasmolytic and CNS depressant.

Uses in Literature : Used in India as antiperiodic, antipyretic, antiseptic, astringent, cattle feed, dye (yellow), emmenagogue, hoops for sieves, mild febrifuge, tonic; and for bridges, bronchitis, ceilings, cigar cases, diarrhoea, doors, dyeing, dysentery, floors, furniture, gastric trouble, and surgical instrument cases, plywood manufacture, tea boxes, ulcers, windows, wood carvings, and wounds (Ambasta, 1986; Jain, 1991; Kapur & Singh, 1996; Siwakoti & Varma, 1996; Watt, 1972).

Trichosanthes cucumerina L. (Pl. 52A)

Family: Cucurbitaceae.

Vern. : Meis.

Sanskrit & Regional Names:

Sans. : Amrita Phala, Bijagarbha, Jvaranashana, Jyotsna, Kachhughni, Kachhwia, Kadopatola, Karkashadla, Kasabhanjam, Kulaja, Lataphala, Nagaphala, Panduka, Panjina, Rajapatola, Rajimana, Rajiphala, Tiktabhadraka, Tiktaka, Vagimana, Varatika

B. : Banchichanga, Banpatol, Rama Chapadavali

Bo. : Jangalipadavala, Kadupadavala, Pudoli, Ramparul

Mal. : Kaippampatolium, Katupatolum, Patavalum, Pepatolum

Mar. : Janglipadavali, Janglipareta, Kadupadavala, Perula, Ramachapadavali, Rampadavala

P. : Gwalkakri

Tel. : Adavipolla, Chayudpollah, Chetipotla, Patolanu.

Distribution : Throughout India, Sri Lanka, Malay Peninsula-Malaya.

Description : Annuals with long, slender, furrowed stem and long, orbiculary transform, denticulate lobed leaves. Male flowers white in axillary raceme. Female ones axillary. Fruits ovoid, fusiform, scarlet when ripe. Seeds semi-ellipsoidal, compressed, rugulose surrounded with red pulp.

Reproductive Cycle : May-August.

Habitat Ecology : Climber on trees and shrub.

Material Examined : HPU-EL-SKS-5521; Arla (1,100 m), Bindraban (1,270 m), Dehan (1,130 m), Kapain (1200 m), Pahra (950 m).

Part Used : Fruits.

Folk Uses : Unripe fruits cooked as a **vegetable** and considered **purgative** in nature.

Active Constituents : Cucurbitacin B isolated from *fruits*.

Uses in Literature : Used so far as an antipyretic, appetizer, blood purifier, cardiac tonic, cholegogue, edible, emetic, expectorant, febrifuge, laxative; and for impotency, intermittent fever, intestinal worms, malaria, scorpion bite and snakebite, urinary disorders, and gastrointestinal problems (Ambasta, 1986; Brahmam & Saxena, 1990; Chopra *et al.*, 1956; Drury, 1978; Katewa *et al.*, 2001; Khare, 2004; Kirtikar & Basu, 1984; Sanyal, 1994; Sharma & Rana, 2005; Sharma , 1996).

***Trifolium resupinatum* L. (Pl. 52B)**

Family: Fabaceae.

Vern. : Barseen.

English & Regional Names:

Eng. : Persian Clover

H. : Shaftal.

Distribution : Common in Sub-Tropical regions.

Description : Herbaceous annuals. Stipules adnate to the petioles. Flowers purplish, in dense axillary heads. Pods included.

Reproductive Cycle : May-June.

Habitat Ecology : Damp places. Marshy localities.

Material Examined : HPU-EL-SKS-5600; Bhagpur (1,600 m), Chachian (1,300 m), Kapain (1,200 m), Nanaun (900 m), Paror (940 m).

Parts Used : Aerial Plant Parts.

Folk Use : Aerial plant parts used as a green **fodder** for livestock.

Active Constituents : *Leaves* contain arginine, histidine, methionine, tryptophan and total lysine.

Uses in Literature : Known to be used in India as a fodder (Uphof, 1968).

Trigonella foenum-graecum L. (Pl. 52C)

Family: Fabaceae.

Vern. : Maethi.

English, Sanskrit & Regional Names:

Eng. : Fenugreek
Sans. : Methi, Methika
B. : Methi, Methika
G. : Bhaji, Methi
H. : Methi, Muthi
Kan. : Menthya, Mente Soffu
Mal. : Uluva, Venthiam
P. : Methi, Methun
Tam. : Vendayam
Tel. : Mentulu.

Distribution : Cultivated in various parts of India.

Description : Strongly-scented annual. Leaflets three, stipules entire. Flowers yellow, sessile, axillary. Pods many-seeded, beaked.

Reproductive Cycle : March-May.

Habitat Ecology : Cultivated.

Material Examined : HPU-EL-SKS-5669; Bhawarna (1,100 m), Chornali (1,280 m), Dharaman (1,100 m), Panchrukhi (990 m), Tanda (1,120 m).

Parts Used : Leaves. Seeds.

Folk Uses : Leaves used as a **pot-herb**. Seed-powder taken for **flatulence**, **rheumatic pains** and **diabetes**.

Active Constituents : These are grecunins H-N, fenugrin B, sapogenins, trigonelloside C, trigonelline, choline, 7-acetoxy-4-methyl coumarin, kaempferol, luteolin and quercetin (*seeds*), diosgenin, graecunins A-C, gitogenin, tigogenin, kaempferol, quercetin and β-sitosterol (*leaves*).

Biological Activity : Seeds (aq. extr.) exhibit antibiotic activity against *Micrococcus pyogenes* var. *aureus.*

Uses in Literature : Reported earlier for baldness, boils, diabetes, gastric trouble, gonorrhoea, inflammatory tumours, carbuncles, suppuration of acne, painful joints, oedema, piles, rheumatism, sores, syphilis and ulcers; and as an aphrodisiac, aromatic carminative, colic, cooling drink, curry powder, emollient in intestinal inflammation, fodder, galactagogue, pot-herb, salads (Chopra *et al.*, 1956; Jain, 1991; Kumar *et al.*, 2004; Pal & Paul, 2003; Pandey, 2003; Sen & Behera, 2003; Sharma, 2003; Uphof, 1968).

***Triticum aestivum* L.** (Pl. 52D)

Syn.: *T. amyleum* Ser.; *T. compositum* L.; *T. dicoccum* Schrank; *T. durum* Desf.; *T. hybernum* L.; *T. vulgare* Vill.; *T. turgidum* L.

Family: Poaceae.

Vern. : Kanak.

English and Regional Names:

Eng. : Wheat.

Sans. : Arupa, Bahudugdha, Godhuma, Rasala, Saman, Suman, Yava, Yavana

B. : Gam, Giun, Gom

Bo. : Gahu, Ghawn, Ghawut-ghum, Giun

G. : Gawn, Ghavum

H. : Gehun, Giun, Kunak

Kan. : Godhi

Mar. : Gahum, Gahung

P. : Do, Dro, Gehun, Kanak, Rozatt, Zud

Tam. : Godumai, Godumbay-arisi

Tel. : Godumulu.

Distribution : Cultivated in many parts of India, Deccan Peninsula; Himalayas upto 4,350 m.

Description : Annual grasses with tufted culms and glabrous, leaves. Spikes glabrous or hairy, awned or awnless; spikelets compressed, hermaphrodite. Stamens 3. Stigmas short. Grains oblong, swollen.

Reproductive Cycle : October-May.

Habitat Ecology : Cultivated land.

Material Examined : HPU-EL-SKS-5583; Bhatu (1,040 m), Chandpur (1,290 m), Dehan (1,130 m), Kapain (1,200 m), Panchrukhi (990 m), Sulah (950 m).

Part Used : Seeds.

Folk Uses : Chewed grains applied locally against **suppuration** of boils. Extr. of soaked seeds given to livestock as **coolant**.

Active Constituents : *Leaves* afford isoorientin, lutonarin, lucenin-3, vicenin-2, wyomin and tricin.

Uses in Literature : Reported so far as aphrodisiac, coolant, fattening, indigestible, laxative, oleaginous, tonic; and for anaemia, boils, cleaning uterus after childbirth, foul ulcers, increasing appetite and taste, medication, over-abscess to remove pus, plague sores, ringworm, swellings in tuberculosis and tumours (Bhatt *et al.*, 2003c; Caius, 2003; Kirtikar & Basu, 1984; Kumar *et al.*, 2004; Pandey, 2003; Sen & Batra, 1997; Srivastava *et al.*, 2003a).

***Tulipa stellata* Hook.** (Pl. 52E).
Syn.: *T. clusiana* var. *stellata* Regel
Family: Liliaceae.

Vern. : Minmayaaoon.
Regional Names:
P : Bhumphor, Chamoti, Chamuni, Jal-kukar, Paduna, Piperi.
Distribution : Common throughout the greater part of Temperate W. Himalayas.
Description : Bulbous herbs with linear channeled leaves and terminal pink flowers. Capsule loculicidal.
Reproductive Cycle : March-May.
Habitat Ecology : Moist shady slopes.
Material Examined : HPU-EL-SKS-5576; Boda (1,000 m), Gharthun (900 m), Jasun (1,000 m), Tikkar (910 m).
Parts Used : Bulbs. Flowers.
Folk Uses : Flowers **offered** to appease deities. Bulbs **eaten** raw.
Uses in Literature : Known to be used for its edible bulbs, offering, rheumatism; and as a tonic (Ambasta, 1986; Arora & Pandey, 1996; Kapur, 1996b; Koul *et al.*, 1982; Roy *et al.*, 1998; Samant *et al.*, 2001; Singh & Kumar, 2000a; Sharma & Rana, 2005; Sood & Prakash, 2007; Thakur, 2005).

***Urtica ardens* Link**
Syn.: *Urtica parviflora* Roxb.
Family: Urticaceae.

Vern. : Chhoti-Aianh.
Distribution : Temperate Himalayas and Nilgiri Hills.
Description : Slender perennial herbs with stinging hairs. Leaves ovate, acuminate, cordate, serrate. Flowers small in peduncled cymes. Achenes with persistent sepals.
Reproductive Cycle : February-May.
Habitat Ecology : Damp Places.
Material Examined : HPU-EL-SKS-5734; Bhagpur (1,600 m), Darugnu (1,400 m), Jiun (1,550 m), Kathena (1,290 m), Lolani (1,600 m).
Part Used : Leaves.
Folk Use : Leaves cooked as **vegetable**.

Uses in Literature : Recorded earlier for hypertension, fever; and as a blood purifier and edible (leaves, branches, seeds) (Ambasta, 1986; Arora & Pandey, 1996; Bennet *et al.* 1991; Chhetri, 2005; Chopra *et al.*, 1956; Roy *et al.*, 1998; Samant *et al.*, 1996, 2001; Srivastava *et al.*, 2003c).

***Urtica dioica* L.** (Pl. 52F)
Family: Urticaceae.

Vern. : Ainh.

English & Regional Names:

Eng. : Nettle, Stinging Nettle

H. : Bichhu booti.

Distribution : Kashmir- Shimla (N.W. Himalayas).

Description : Erect, grooved herbs with stinging hairs. Leaves cordate, long stalked, toothed. Flowers green, dioecious, clustered on drooping axillary panicles. Achenes flat, included in the persistent perianth.

Reproductive Cycle : June-October.

Habitat Ecology : Moist Shady Slopes. Wastelands.

Material Examined : HPU-EL-SKS-5728; Bhagpur (1,600 m), Chandpur (1,290 m), Jiun (1,550 m), Kapain (1,200 m), Lolani (1,600 m).

Parts Used : Leaves. Root. Tender Shoot.

Folk Uses : Root-juice prescribed for **inflammation** of **urinary tract, liver complaints** and as **blood purifier**. Tender shoot and leaves consumed as '**chutney**' and **vegetable**. Also, these are beaten over joints to check **rheumatic pains**.

Active Constituents : Contains lecithin.

Biological Activity : Plant antiviral.

Uses in Literature : Known earlier as an anthelmintic, antiseptic, astringent, diuretic, emmenagogue, haemostatic, lithotriptic, pot-herb; and for asthma, boils, cough, cuts and wounds, dandruff, diarrhoea, dislocated bones, dog bite, enhanced milk flow (vet.), fibres, gout (nettle tea), haematuria allergies, jaundice, kidney troubles, leocorrhoea, paralysis, rheumatism, sciatica, menorrhagia, nephritis, rheumatism, sprain, swelling, throat diseases, toothache and urticaria (Agarwal, 2003; Ambasta, 1986; Arora & Pandey, 1996; Aswal & Mehrotra, 1994; Chopra *et al.*, 1956; Devi, 2003; Ganai & Nawachoo, 2003; Jain, 1991; Kala & Rawat, 2001; Kaul, 1997; Khare, 2004; Kirtikar & Basu, 1984; Lal *et al.*, 1996; Mahato & Chaudhary, 2005; Manandhar, 1996, 2000; Negi *et al.*,

1999; Panthi & Chaudhary, 2003; Prajapati *et al*., 2006; Said, 1997; Saklani & Jain, 1994; Samant *et al*., 1996, 2001; Sarin, 1990; Sharma, 2003; Singh & Kumar, 2000a; Singh *et al*., 2001; Sood & Thakur, 2004; Uphof, 1968; Yonzone *et al*., 1996; Viswanathan, 2000).

***Vanda tessellata* (Roxb.) G. Don** (Pl. 53A)

Syn.: *Aerides tessellatum* Wight; *Cymbidium alagnata ex* Wall.; *C. tessellatum* Sw.; *C. tesselloides* Roxb.; *Epidendrum tessellatum* Roxb.; *E. Tesselloides* Steud.; *Vanda roxburghii* R. Br.

Family: Orchidaceae.

Vern. : Bhangru.

Sanskrit & Regional Names:

Sans.	: Atirasa, Bhujangakshi, Gandha-nakuli, Rasna, Vandaka
B.	: Nai, Rasna
G.	: Rasno
H.	: Banda, Nai, Persara, Rasana, Vanda
Kan.	: Bandanike, Vandaka, Gidda
Mar.	: Rasana
Tel.	: Chittiveduri, Kanapabadanika, Mardaru, Saga, Vandanika
Urdu	: Banda.

Distribution : Native of Tropical Asia.

Description : Epiphytes with stout stem and linear-oblong, recurved leaves which are unequally lobed at apex. Flowers yellow in racemes. Capsules oblong, winged.

Reproductive Cycle : January-December.

Habitat Ecology : Common in wet, swampy and shady habitat on trees like mango (*Mangifera indica*), 'bargad' (*Ficus benghalensis*), and 'kachnaar' (*Bauhinia variegata*).

Material Examined : HPU-EL-SKS-5454; Balah (900 m), Bhaura (940 m), Chhanchri (930 m), Jasun (1,000 m), Nadli (1,000 m).

Part Used : Roots.

Folk Uses : Crushed roots alongwith wheat flour cooked in to 'halva' (**porridge**), for **internal injuries** and **lung ailments** in case of humans as well as cattle.

Uses in Literature : So far considered alexiteric, antibacterial antipyretic, laxative, skin infections, antiarthritic, heating, fish poison, rheumatic (vet.); and for abdominal diseases, boils on the scalp, bronchitis, dyspepsia, ear cleaning, ear drops, earache, fever, inflammations, lumbago, migraine, nervous disorders, otitis, paralysis, piles, rheumatism, sore eyes, tremors, tonic to liver and brain, toothache, and antidote to snakebite (Ambasta, 1986; Caius, 2003; Chauhan, 1999; Dwivedi, 2003; Girach, 1992; Ramdas *et al.*, 2000; Reddy *et al.*, 2003b; Sharma, 2003; Singh & Kumar, 2000b; Singh & Pandey, 1998; Tirkey, 2004; Upadhyay, 2003; Upadhyay & Chauhan, 2000; Upadhyay & Singh, 2005).

***Verbascum thapsus* L.** (Pl. 53B)

Syn.: *V. indicum* Wall.; *V. blattaria* Schmidt.

Family: Scrophulariaceae.

Vern. : Ban-tambakhoo.

English & Regional Names:

- **Eng.** : Common Mullein, Cow's Lungwort
- **H.** : Gidar tamaku
- **Kan.** : Kaduhogesooppina Gida
- **P.** : Gidar-tamaku, Gurganna, Karathri, Phasruk, Vulr.

Distribution : Kashmir-Bhutan (Temperate Himalayas).

Description : Erect herbs having decurrent, tomentose leaves with yellow-green stellate hairs and winged stems. Flowers sessile, yellow in terminal woolly spikes. Capsule ovoid.

Reproductive Cycle : January-April.

Habitat Ecology : Open slopes. Wastelands. Borders of agricultural fields.

Material Examined : HPU-EL-SKS-5729; Balu (1,030 m), Bhuana (1,400 m), Chimbalhar (1,150 m), Lahla (930 m), Sulah (950 m).

Parts Used : Aerial Plant Parts. Inflorescence. Root.

Folk Uses : Crushed aerial plant parts with wheat dough prescribed for relieving **constipation** in livestock. Root paste applied for hastening suppuration of **boils**. Steam emanating from its inflorescence in boiling water good against '**measles**'.

Active Constituents : Isolated constituents from different parts of plant are crocetin (*flowers*), aucubin, catalpol (*roots*), rotenone and coumarin (*leaves*).

Uses in Literature : Useful earlier as an aphrodisiac, demulcent, emollient, febrifuge, fish poison, heart stimulant and symbolic; and for asthma, cough, diarrhoea, inflammation, leucoderma, pulmonary diseases of cattle, cramps and migraine (Ambasta, 1986, Aswal & Mehrotra, 1994; Chopra *et al.*, 1956; Jain, 1991; Nautiyal *et al.*, 2003; Sharma, 2003).

Vicia faba L. (Pl. 53C)

Syn.: *Fava vulgaris* Moench

Family: Fabaceae.

Vern. : Keon, Keeoon.

Regional Names:

- **H.** : Bakla, Anhuri, Kala Matar
- **Kan.** : Kadu, Huralikayee
- **Kash.** : Katun
- **P.** : Chastang, Kabli-Bakla, Raj-Rawan.

Distribution : N. W. India.

Description : Strong erect annuals upto 120 cm height. Leaves paripinnate, ending in a tendril. Flowers small, white with hairy style. Pods flat, dehiscent. Seeds large.

Reproductive Cycle : December-March.

Habitat Ecology : Cultivated as well as wild.

Material Examined : HPU-EL-SKS-5575; Bhatu (1,040 m), Chandpur (1,290 m), Dharaman (1,100 m), Gopalpur (1,350 m), Panapari (1,100 m).

Parts Used : Aerial Plant Parts. Pods.

Folk Uses : Unripe pods cooked as **vegetable**. Aerial plant parts employed as **fodder**.

Uses in Literature : Reported to be used as vegetable, diuretic, and aphrodisiac (Ambasta, 1986).

Vicia hirsuta Koch (Pl. 53D)

Syn.: *Ervum hirsutum* L.; *E. filiforme* Roxb.; *E. lens* Wall.

Family: Fabaceae.

Vern. : Choti-Rodi.

English & Regional Names:

- **Eng.** : The Hairy Tare, Tiny Vetch
- **B.** : Musur Chuna
- **H.** : Jhunjhuni-ankari, Munmuna
- **Kum.** : Jhanjhaniya-kuri, Masuri, Masur-chana.

Distribution : Temperate and Sub-Tropical Regions.

Description : Hairy annuals with weak stems and leaves ending in tendrils. Flowers white, tinged with blue. Pods flat, 2-seeded.

Reproductive Cycle : March-April.

Habitat Ecology : Cultivated fields. Grassy Meadows. Damp places.

Material Examined : HPU-EL-SKS-5726; Bodhal (970 m), Dadh (1,100 m), Holta (1,200 m), Kapain (1,200 m), Tapa (945 m).

Parts Used : Seeds. Tender aerial plant parts.

Folk Uses : Tender aerial plant parts consumed as **vegetable** and **fodder**. Flour from seeds used for making '**pakoras**'.

Active Constituents : *Seeds* contain HCN.

Biological Activity : Seed extr. exhibits non-specific agglutinating reaction.

Uses in Literature : Reported so far as a pot-herb, pulse, fodder (Ambasta, 1986; Roy *et al.*, 1998; Sharma & Rana, 2005).

Vicia sativa L. (Pl. 53E)

Family: Fabaceae.

Vern. : Badi-Rodi.

English & Regional Names:

- **Eng.** : Common Vetch
- **B.** : Ankari
- **Bihar** : Chirinji
- **H.** : Akra, Ankra
- **Ori.** : Choni, Rothi.

Distribution : Throughout India.

Description : Annuals with slender stems, toothed stipules and solitary red-blue flowers. Pods glabrescent.

Reproductive Cycle : January-February.

Habitat Ecology : Grassy Hill Slopes.

Material Examined : HPU-EL-SKS-5555; Bhaura (940 m), Chachian (1,300 m), Dehan (1,130 m), Molichak (1,020 m), Nadli (1,000 m).

Parts Used : Aerial plant parts. Seeds.

Folk Uses : Aerial plant parts used as a **pot-herb** and **fodder** for livestock. Grounded seeds employed for culinary preparations like '**pakoras**'.

Active Constituents : *Seed* yields vicianine and HCN.

Biological Activity : Seed shows specific haemagglutinating activity.

Uses in Literature : Used earlier as an astringent, edible (seeds), fodder of high protein value, pot-herb, and for diarrhoea (Ambasta, 1986; Roy *et al.*, 1998; Sharma & Rana, 2005; Singh & Kumar, 2000a).

Vigna radiata (L.) Wilez. (Pl. 53F)

Syn.: *Phaseolus mungo* Wall.; *P. radiatus* L.; *P. roxburghii* Wight & Arn.; *P. setulosus* Dalziel; *Vigna opisotricha* A. Rich.

Family: Fabaceae.

Vern. : Maah.

Sanskrit and Regional Names:

Sans. : Daniemasha, Hurita, Masha

B. : Mash-kulai, Tircorai-kalai

G. : Adad, Arad

H. : Dora, Thikiri, Urid, Urud

Kan. : Hasaru, Uddu

Mar. : Maga, Udid

P. : Mah, Mash, Urad

Tam. : Panny-pyre, Patchay-pyre

Tel. : Karu:minumulu, Minumulu, Patasa-pesalu.

Distribution : Cultivated throughout the plains; ascending to 2,000 m in N.W. Himalayas.

Description : Plant herbaceous with trifoliate leaves. Flowers with conspicuous bracts. Keel elongated. Style twisted. Pods cylindrical.

Reproductive Cycle : June-October.

Habitat Ecology : Arid soil.

Material Examined : HPU-EL-SKS-5492; Bhaura (940 m), Chhanchri (930 m), Gharthun (900 m), Gopalpur (1,350 m), Saparu (1,250 m).

Parts Used : Leaves. Seeds.

Folk Uses : Leaves used as **fodder.** Seeds consumed as **pulse**, or for making edible preparation like '**badiyan**' **(nuggets)**, '**bhalla**'; and also **offered** to appease lord 'Shanni Dev' (Saturn).

Viola patrinii Ging. (Pl. 54A)

Family: Violaceae.

Vern. : Banafsha, Banaksa.

Distribution : Temperate Himalayas, Japan.

Description : Glabrous or pubescent plants with very short stems and tufted, triangular, usually narrow elongate leaves. Flowers dark lilac, scented. Fruits 3-valved capsule. Seeds with shining testa.

Reproductive Cycle : April-May.

Habitat Ecology : Moist-Shady Places. Borders of Agricultural Fields.

Material Examined : HPU-EL-SKS-5599; Bhagpur (1,600 m), Chimbalhar (1,150 m), Gopalpur (1,350 m), Lolani (1,600 m), Panchrukhi (990 m).

Parts Used : Aerial Plant Parts.

Folk Use : Decoction from aerial plant parts prescribed for **throat affections** (1 teaspoon daily for a week).

Uses in Literature : Earlier known as blood purifier, pot-herb, purgative; and for biliousness, cancerous growth, cough, cold, foul sores, swellings, syphilis, scrofula, and ulcers (Ambasta, 1986; Chopra *et al.*, 1956; Kirtikar & Basu, 1984).

Viola pilosa Blume (Pl. 54B)

Syn.: *V. serpens* Wall. *ex* Ging.

Family: Violaceae.

Vern. : Banafsha, Banaksha

Regional Names:

- **H.** : Banafsha, Thungtu
- **Kum.** : Thungtu
- **P.** : Banafsha.

Distribution : India, hilly districts.

Description : Small, stoloniferous glabrous herbs with toothed stipules and lilac flowers. Fruit a capsule.

Reproductive Cycle : February-April.

Habitat Ecology : Moist Shady Slopes, frequent.

Material Examined : HPU-EL-SKS-5598; Andretta (980 m), Bhauda (1,050 m), Jasun (1,000 m), Rangru (910 m), Malahu (1,050 m).

Parts Used : Aerial Plant Parts.

Folk Uses : Decoction of aerial plant parts (1 teaspoonful twice a day for one week) cures **cough**, **cold** and **chest affections**; also its paste in goat-milk applied over chest in kids to check **cough**.

Active Constituents : Therapeutic properties due to rutin, violin and salicylic acid (*herb*), and rutin, quercetin, violanthin, violaxanthin, p-hydroxycinnamic acid, delphinidin, C-glycoside-saponarin and 15-cis-violaxanthin (*flowers*).

Biological Activity : Aq. extr. of flower exhibits antibiotic properties.

Uses in Literature : Known so far as an antipyretic, demulcent, diaphoretic, diuretic, emetic, emollient, febrifuge, purgative; and for asthma, biliousness, bleeding piles, cold, constipation, cough, throat cancer, lung diseases and source of 'banafsa' (Ambasta, 1986; Aswal, 1996; Aswal & Mehrotra, 1994; Chauhan, 1999; Chhetri, 2005; Chopra *et al.*, 1956; Kala & Rawat, 2001; Kirtikar & Basu, 1984; Krishna & Singh, 1987; Lal *et al.*, 1996; Nautiyal, 1981; Nautiyal *et al.*, 2003; Prajapati *et al.*, 2006; Samant *et al.*, 1996; Sarin, 1996; Shah & Joshi, 1971; Sharma, 2003; Singh, 1999, 2001; Singh & Kumar, 2000a; Sood & Thakur, 2004).

Vitex negundo L. (Pl. 54C)

Syn.: *V. bicolor* Willd.; *V. arborea* Desf.; *V. paniculata* Lam.; *V. incisa* Lam.

Family: Verbenaceae.

Vern. : Bana.

English, Sanskrit & Regional Names:

Eng. : Common Chaste Tree, Indian Privet, Nigandi
Sans. : Indrani, Nilapushpa, Shepali, Svetasuraso
Ass. : Aggla-Chita, Pasutia
B. : Nishinda, Samalu
Bo. : Katri, Shiwari
G. : Nagaol, Nirgari
H. : Mewri, Nirgunda, Sinduari
Kan. : Lakkigida, Nakkilu, Nekki
M.P. : Nagad
Mal. : Vellanochi
Mar. : Nirgur
Ori. : Begundia, Indrani
P. : Bankahu, Moraun
Tam. : Siduvavam, Tiributamerittan

Tel. : Nellavavilli, Sinduvaramu

U.P. : Tarvan, Kali-nirgundi.

Distribution : Afghanistan-Bhutan; India, China; S.E. Asia upto 2,000 m.

Description : Shrubs. Leaves digitate, 3-5 foliate. Flowers blue purple. Fruits black, ovoid, succulent.

Reproductive Cycle : March-August.

Habitat Ecology : Cultivated Areas. Wastelands.

Material Examined : HPU-EL-SKS-5446; Bhatu (1,040 m), Chachian (1,300 m), Holta (1,200 m), Malnu (1,000 m), Sulah (950 m).

Parts Uses : Leaves. Stem.

Folk Uses : Tender leaf-paste with *Piper nigrum* and jaggery prescribed for **stomachache**. Leaf-juice poured in nasal passage to check **headache** due to cold. Stem used for **scouring teeth** and curing **mouth ulcers**. Fomentation of its leaves in boiling water or bath with it good against **rheumatism**, **sprains**, **internal wounds**, and **post-delivery pains**.

Active Constituents : *Stem bark* exhibits two new leueoanthoeyamidins determined as 6, 8-di-0-methyllencodelphinidin and 3, 4-di-0-methyelencocyanidin-7-0-rhammogluco-side.

Biological Activity : Extr. of leaves exhibits anticancerous activity against *Ehrlich ascites* tumour cells.

Uses in Literature : Known earlier as an abortifacient, alterative, anasarca, anodyne, athelmintic, antifertility in man, antiparasitic, appetizer, astringent, aromatic, cardiac, cephalic, colic, haemostatic, demulcent in dysentery and piles, diaphoretic, diuretic, emmenagogue, expectorant, febrifugal, fuel wood, insect repellent, nervine tonic, nocturnal emission, psychoactive plant, sterilizer, and symbolic to keep evil spirits away and vegetable (seeds); and for basketry, bladder complaints, blisters, boils, bone fractures, catarrhal fever, checking pain, cholera, chronic liver problem associated with loss of appetite, cold, conjunctivitis, cooling effect, cough, diarrhoea, dropsy, dullness of hearing, dyeing, dysentery and piles, dyspepsia, eczema, epilepsy, eye complaints, fomentation of leaves—cold, gastric troubles, gout, gum troubles, haemorrhage of the digestive system, headache, resolving obstructions in kidney and spleen, indigestion, insect bites, intermittent fevers, irritable fevers, irritable bladder, itching, leprosy, lung diseases, making tooth sticks, mental disease, paralysis, paralysis of animals, pneumonia, post-natal, reducing sexual desire and muscular pain of the body, rheumatoid arthritis, scabies, sciatica, scrofulous ulcers, sinuses, sores, stored grain pests, syphilis, testes disease, restoring fertility, tonsillitis, toothache, ulcers, urinary troubles, waist pain, warm bath to ladies after delivery, and wounds (Ambasta, 1986; Bhatt *et al.*, 2003a; Bhandary & Chandrashekar, 2002; Bhandary *et al.*, 1996; Borthakur & Sarma, 1996; Borthakur *et al.*, 2004; Bose, 1984; Chhetri, 2005; Das & Agarwal, 1991; Das & Sharma, 2003; Deshmukh & Rothe, 2003; Geetha *et al.*, 1996a, Girach, 1996; Girach & Aminuddin, 1995; Gogoi *et al.*, 2003a; Gopal & Chandra, 2003; Hembrom, 1996; Hembrom & Goel, 2005; Jain, 1991; Jain & Sharma, 1996; Jain *et al.*, 1994; Katewa & Arora, 1997; Katewa *et al.*, 2000; Khanna *et al.*, 1996a; Khare, 2004; Kirtikar & Basu, 1984; Kothari & Moorthy, 1996; Kshirsagar & Singh, 2000; Kulkarni & Kumbhojkar, 1996; Kurian, 1995; Lal *et al.*, 1996; Mahato & Mahato, 1996; Mishra *et al.*, 1997; Mitaliya *et al.*, 2003; Nayak *et al.*, 2003; Painuli & Maheshwari, 1996; Pal *et al.*, 2000; Panthi & Chaudhary, 2003; Pawar & Patil, 2004; Prajapati *et al.*, 2006; Ragupathy & Mahadevan, 1991; Rajendran *et al.*, 2003; Rama Rao & Henry, 1996; Ranjan, 2003; Ravikumar & Shankar, 2003; Saini, 1996b; Saklani & Jain, 1994; Sarin, 1996; Satapathy & Brahmam, 1996; Sharma, 1996; Sharma & Sood, 1997; Singh, 1991, 2001; Singh & Kumar, 2000a; Singh *et al.*, 1996b; Siwakoti & Varma, 1996; Srivastava *et al.*, 2003b; Subramani & Goraya, 2003; Thomas & Britto, 2003; Upadhyay & Chauhan, 2000; Upadhyay, 2003; Watt, 1972).

Vitis repanda **Wight & Arn.** (Pl. 54D)

Syn.: *V. laeta* Wall.; *V. rosea* Royle; *Cissus acuminata* Thwaites; *C. aquosa* Wall.; *C. repanda* Vahl; *C. repens* Thwaites; *C. riparia* Wall.; *C. vitiginea* L.

Family: Vitaceae.

Vern. : Maraingu.

Regional Names:

- **Ass.** : Medmedia-lop
- **H.** : Pani-bel.

Distribution : N. W. Himalayas.

Description : Climbers having stem and leaves appressed with woolly tomentum. Leaves broadly-cordate, coarsely crenate. Fruits pyriform.

Reproductive Cycle : April-June.

Habitat Ecology : Commonly growing on trees.

Material Examined : HPU-EL-SKS-5745; Bhagpur (1,600 m), Bhuana (1,400 m), Jiun (1,550 m), Kathena (1,290 m), Purani-palam (1,100 m).

Part Used : Fruits.

Folk Use : Ripe fruits **edible** and considered **refrigerant**.

Uses in Literature : Reported for its edible shoots, boils, fracture ulceration, potable water (Ambasta, 1986; Bennet *et al.*, 1991; Chandra, 1997; Kirtikar & Basu, 1984).

***Vitis vinifera* L.** (Pl. 54E)
Syn.: *V. labrusa* L.; *V. vulpina* L.
Family: Vitaceae.

Vern. : Angoor.

English, Sanskrit and Regional Names:

Eng. : Grapes
Sans. : Draksha, Mridvika
B. : Angurphal, Drakhyaluta, Kismis, Manakka
Bo. : Abai, Drakh, Kishmish
G. : Darakh, Drakasha
H. : Angur, Dakh, Drakh, Kismis, Manakka
Kan. : Drakshi
Mar. : Draksha
P. : Angur, Buri, Dakh, Dakki, Dehla, Gandeli, Mitha, Newala, Tanaur
Tam. : Kodi-mun-dirrippazham, Kodrimundrie
Tel. : Draksha-pondu, Gostani Draksha.

Distribution : Cultivated extensively in N. W. India.

Description : Glabrous climbers having cordate 3-7- angled or lobed leaves. Petals united at the apex. Flowers small, reddish-brown. Dried fruits sweet.

Reproductive Cycle : March-April.

Habitat Ecology : Hedges.

Material Examined : HPU-EL-SKS-5615; Andretta (980 m), Bhawarna (1,100 m), Darugnu (1,400 m), Jasun (1,000 m), Rajeher (1,050 m).

Part Used : Fruits.

Folk Uses : Ripe fruits **edible** and considered **refrigerant**.

Active Constituents : *Young fruits* contain jasmonic acids, cholesterol, β-sitosterol, and ergosterol.

Uses in Literature : Considered earlier as aphrodisiac, appetizer; coolant, diuretic, demulcent, fattening, laxative, tonic; good for asthma, bad effects of drinking, blood diseases, cough, eyes and throat, fever, headache, inflammation of spleen, jaundice, joint pain, piles, scabies, skin affections, stops bleeding from the mouth, stones in bladder, swelling of testicals and vomiting (Ambasta, 1986; Kirtikar & Basu, 1984; Mehrotra *et al.*, 1995).

***Woodfordia fruticosa* (L.) Kurz** (Pl. 54F)

Syn.: *W. floribunda* Salisb. *W. tomentosa* Bedd.; *Lythrum fruticosum* L.; *Grislea tomentosa* Roxb.; *G. micropetala* Hochst. *ex* Steud.; *G. punctuata* Ham.; *G. uniflora* Rich.

Family: Lythraceae.

Vern. : Dhawain, Dhaun.

English, Sanskrit & Regional Names:

Eng. : Fire Flame Bush, Red Bell Bush, Shirantitea, Woodfordia
Sans. : Bahupushpika, Rodhrapushpini, Reyya-manu
Ass. : Eda, Era
B. : Dawa, Dhadki, Dhawanji
Bo. : Dhavri, Dowari
G. : Dhawadina
H. : Dhai, Dhaura, Thawi
Kan. : Tamrapushpi, Dhataaki Huvvu
Kash. : Thawi
Mal. : Taativipu, Tatire
Mar. : Dhaiphal, Phulsatti
Ori. : Dhatubiphala, Horibadi
P. : Dawi, Khurd
Tam. : Thatarijargi, Velakkai
Tel. : Gaddapisniki, Reyyamanu
U.P. : Phoolchhari.

Distribution : Pakistan-S.W. China; India; Myanmar; W. Asia; Africa: 300-1,800 m.

Description : Shrubs. Leaves opposite, lanceolate, pointed. Flowers red. Capsule ovate.

Reproductive Cycle : March-June.

Habitat Ecology : Forests, Steep Slopes.

Material Examined : HPU-EL-SKS-5571; Darugnu (1,400 m), Gopalpur (1,350 m), Panapari (1,100 m), Rajeher (1,050 m), Tikkar (910 m).

Parts Used : Leaves. Flowers. Wood.

Folk Uses : Leaves given as **fodder** to livestock. Flowers serve as source of **nectar** for rearing honey-bees. Its decoction good against **dysentery**, **stomachache** and **leucorrhoea**. Powdered flowers along with curd prescribed to check **dysentery**; also it forms one of the important ingredients of local wine, '**sur**'. Wood considered as good **fuel**.

Active Constituents : Isolated ingredients are ellagic acid, polystachoside, myricetin-3-galactoside and pelar gonidin-3,5-dighnoside (*leaves, flowers*), cyaniding-3, 5-diglucoside, deta coganol, β-sitosterol, chrysophanol-8-D-β-D-glucopyranoside, quercetin and tannins (20%) (*flowers*).

Biological Activity : Extr. of flowers exhibits significant abortifacient activity in mice.

Uses in Literature : Recorded to be used as an astringent, cool drink, edible (flowers, fruits, tender shoots), fodder, fuelwood, ornamental, mosquito and insect repellent, stimulant, styptic, supportive, tonic, vegetable, vermillion; and for axe handles, bleeding in pregnancy, body pains, bone fractures, bowel complaints, burns, cancer, carbuncle, cholera, complaints of liver, cough, diarrhoea, dropsy, dysentery, dyeing and tanning material, expelling worms from the body, fever, fermentations for most of 'aristas' and 'asvas' of ayurveda, foul ulcers, haemorrhages, haemorrhoids, injuries, leprosy, leucorrhoea, loss of appetite in pregnancy, mucous disorders, muscular pain, nausea, night blindness, pain, pimples, diabetes, rheumatism, sore and wounds (vet.), safe stimulant in pregnancy, sores, seminal weakness, skin diseases, smallpox, spleen complaints, sprain, tanning, vesicular eruptions and enhancing taste and intoxication in toddy (Agarwal, 2003; Ambasta, 1986; Aminuddin & Girach, 1996; Arora & Pandey, 1996; Bennet *et al.*, 1991; Bhatt *et al.*, 2003a; Borthakur, 1993; Bose, 1984; Chauhan, 1999; Chhetri, 2005; Jain, 1991; Khare, 2004; Kirtikar & Basu, 1984; Kothari & Moorthy, 1996; Kothari *et al.*, 2003; Kulkarni & Kumbhojkar, 1996; Kumar, 2003; Kurian, 1995; Masih, 2003; Murthy *et al.*, 2003; Nayak *et al.*, 2003; Panthi & Chaudhary, 2003; Parrotta, 2001; Patil, 2004; Patole & Jain, 2002; Purohit & Samant, 1995; Rai & Upadhyay, 1996; Rama Rao & Henry, 1996; Saini, 1996a; Samant *et al.*, 2001; Sanyal, 1994; Sarangi & Sahu, 2005; Shah & Singh, 1990; Sharma, 2003; Sharma & Sood, 1997; Singh, 1991, 1999; Singh & Kumar, 2000a; Singh & Pandey, 1998; Siwakoti & Siwakoti, 2000; Siwakoti & Varma, 1996; Sood & Thakur, 2004; Sood & Kaushal, 2008; Srivastava *et al.*, 2003a, b; Upadhyay & Chauhan, 2000; Verma & Chauhan, 2006; Watt, 1972).

***Wrightia arborea* (Dennst.) Mabb.** (Pl. 55A)
Syn.: *W. tomentosa* (Roxb.) Roem. & Sch. Bip.
Family: Apocynaceae.

Vern. : Khaloa.

Regional Names:

Ass.	: Atkuri, Garo Dudhkuri
B.	: Dudhkoraiya
Bo.	: Kala Inderjau, Tambada Kuda
G.	: Dudhlo, Kala Inderjau
H.	: Dudhi, Dharauli, Daira
Kan.	: Bilikudegidda, Billiganagalugidda
Mal.	: Maiden Pala, Nilampala
Mar.	: Kalainderjau
Ori.	: Pal Kurwan, Harido
Tam.	: Pala
Tel.	: Tellapaala, Peddapaala, Koilamukhri.

Distribution : Throughtout Tropical India.

Description : Trees upto 9 m high, and with yellow milky juice, opposite divaricate scabrous branches, and smooth yellowish grey bark. Leaves acuminate, tomentose on both sides. Flowers malodourous; fruit cylindrical. Seeds long, slender, attenuated at the apex.

Reproductive Cycle : May-January.

Habitat Ecology : Borders of agricultural fields. Open spaces.

Material Examined : HPU-EL-SKS-5579; Boda (1,000 m), Jasun (1,000 m), Malahu (1,050 m), Rangru (910 m), Thandol (950 m).

Parts Used : Leaves. Fruits. Stem.

Folk Uses : Leaves lopped for **fodder**. Stem as **fuelwood**. Crushed fruits (2-3 kg) mixed with wheat flour (10-15 kg) given to check **dysentery** in cattle (500 g-1 kg daily for a week).

Active Constituents : These are conkurchine, conessine, conessidine, holarrhine, burchine (*fruits*), tritriacent-16-one and hexacosan-3, 6-diol-12oic acid (*stem bark*).

Biological Activity : Wrightiadione exhibited cytotoxicity against murine P388 lymphocytic leukaemia cell line.

Uses in Literature : Reported earlier as adulterant, dye (yellow), edible, floss for stuffing; and for antidote to snakebite, scorpion sting, diarrhoea, fodder, menstrual and renal complaints (Ambasta, 1986; Bennet *et al.*, 1991; Bhatt *et al.*, 2003b; Kirtikar & Basu, 1984; Rastogi & Mehrotra, 1999; Sood & Kaushal, 2008; Srivastava *et al.*, 2003a, b).

Zanthoxylum armatum DC. (Pl. 55B)

Syn.: *Z. alatum* Roxb.; *Z. hostile* Wall.

Family: Rutaceae.

Vern. : Tirmir.

Sanskrit & Regional Names:

Sans. : Mahamuni, Sauravanaja, Tumburu
B. : Gaira, Nepali-dhamia
G. : Tejabala, Tejbal
H. : Tejbal, Timur
Kan. : Jimmi, Tejovanti, Tejpatri, Tumbura, Tumburudra
Mal. : Thumboonal, Thumbooni, Valiyavaluzhavam
Mar. : Tejabal
Ori. : Tundapoda
P. : Tezbal, Timmal
Tam. : Thejyovathi
Tel. : Tejovathi.

Distribution : Kashmir-S.W.China; S.E.Asia: 1,100-2,500 m.

Description : Shrubs having stems and branches armed with long prickles. Leaves imparipinnate, aromatic. Flowers yellow. Drupes pale-red, globose.

Reproductive Cycle : April-June.

Habitat Ecology : Shrubberies. Cultivated Areas. Scrub Forests.

Material Examined : HPU-EL-SKS-5725; Bindraban (1,270 m), Chauki (900 m), Holta (1,200 m), Jiun (1,550 m), Kapain (1,200 m).

Parts Used : Leaves. Fruits. Stem. Wood.

Folk Uses : Fruits chewed against **mouth ulcers**, **foul breath**, and as **blood purifier**. Wood suitable for making **walking sticks** and **pestles**. Stem used for **scouring teeth**. Leaves relished as '**chutney**'.

Active Constituents : Principle constituents are linalool, limonene, methyl cinnamate, lineole, trans-cinnamic acid, nevadensin, umbelliferone, β-sitosterol and its glucoside *(fruits)*, β-sitosterol, pinoresionol dimethyl ether and a mixture of alkaloids (*aerial parts*), linalool (58.30), limonene

(24.46), alloaroma dendrene (8.98), methyl linnamate (8.92), myrcene (3.55), α-thujene (1.65), α-fenchol (0.68), p-cymene (0.65), β-caryophyllane (0.50), α-terpineol (0.31), camphor (0.25), 1,8-lineole (0.25), carvone (0.20), tagetonol (0.16), lis-linnamic (0.12) and γ-terpinene (0.08%) (*fruit oil*).

Biological Activity : Stem and bark exhibit hypoglycaemic and anthelmintic activities, respectively.

Uses in Literature : Used in India as an anthelmintic, antiseptic, aromatic, astringent, carminative, condiment, deodorant, disinfectant, edible (seeds, leaves, fruits), febrifuge, flavouring agent in *paan*, stimulant, stomachic, tonic in fever; and also for boils, cholera, clubs, cough, dental diseases, diarrhoea, dyspepsia, eczema, fencing, headache, gum troubles, hair lotions, indigestion, intoxicating fish, itching, leucoderma, mouth blisters, serving furs, paralysis, pestles, piles, purification of water, religious ceremonies, rheumatism, round worms, scabies, smallpox, snakebite, soap making, toothache, warding off house flies, and wood for house building (Agarwal, 2003; Ambasta, 1986; Ansari, 1991; Borthakur & Sarma, 1996; Chakraborty, 2003; Chandra, 1997; Chauhan, 1999; Chhetri, 2005; Chopra *et al.*, 1956; Chowdhery, 1996; Das & Agarwal, 1991; Dey & Bahadur, 1973; Duke *et al.*, 2002; Girach, 2001; Harborne & Baxter, 2001; Jain, 1991, 2003; Joshi & Joshi, 2005; Kapur & Singh, 1996; Kirtikar & Basu, 1984; Lal *et al.*, 1996; Lalramnghinglova, 2000, 2002; Manandhar, 1996; Mao, 1993; Negi & Gaur, 1991; Pal *et al.*, 2000; Panthi & Chaudhary, 2003; Prajapati *et al.*, 2006; Purohit & Samant, 1995; Rana *et al.*, 1996; Ranjan, 2000; Ravishankar & Henry, 1992; Rawat & Chowdhury, 1998; Saklani & Jain, 1994; Sharma, 2003; Singh, 1991, 1999; Singh & Kumar, 2000a; Sinha & Sinha, 2001; Siwakoti & Varma, 1996; Sood & Thakur, 2004; Verma & Chauhan, 2006; Watt, 1972).

Zea mays L. (Pl. 55C)

Syn.: *Mays zea* Gaertn.; *Zea alba* Mill. Gard.; *Z. americana* Mill. Gard.; *Z. segetalis* Salisb.; *Z. vulgaris* Mill. Gard.

Family: Poaceae.

Vern. : Chhali.

English, Sanskrit and Regional Names:

Eng. : Corn, Indian-corn, Maize

Sans. : Yavanala

Ass. : Godhan, Makoi

B. : Bhutta, Janar, Jonar

Bo. : Buta, Makai

	G.	: Makkai, Makkari
	H.	: Barajuar, Bhutta, Junri, Kukri, Makka, Makkai
	Kan.	: Goin-jol, Mekkejola, Musuku-jola
	Mar.	: Buta, Maka, Makai
	Ori.	: Buta, Maka, Mokka
	P.	: Bara-juar, Chale, Kukri, Kuthi, Makkei, Makki
	Tam.	: Mukka-sholam, Turakacholam
	Tel.	: Makka-zonnalu, Mokka-jonna
	Urdu	: Makai.

Distribution : Cultivated throughout N. India.

Description : Cultivated annuals having solid stems showing well-developed nodes and internodes and protandrous flowers; monoecious. Male inflorescence consists of a terminal panicle and female ones—a modified spike, the 'ear', that develops from the axil of one of the largest leaves. Grains large, subglobose, shining and hard when ripe.

Reproductive Cycle : June-September.

Habitat Ecology : Agricultural Land.

Material Examined : HPU-EL-SKS-5520; Balah (900 m), Chauki (900 m), Gharthun (900 m), Nadli (1,000 m), Rajpur (1,100 m).

Parts Used : Aerial Plant Parts. Peduncle. Spathe. Style. Stem. Root.

Folk Uses : Decoction of style (50 ml thrice a day for 15 days) considered good **diuretic** and **refrigerant**. Green as well as dried aerial plant parts given as **fodder**. Dried stem, underground parts and peduncle used as **fuel**. Spathes woven as **mats**.

Active Constituents : Zeanin isolated from sweet immature *kernels*; di-O- (indole-3-acetyl)-myo-inositol and tri-O-(indole-3-acetyl).

Biological Activity : When grape cluster were dipped before blossom, in solution of zeatin, the fruit set increased about three-fold.

Uses in Literature : Known so far in India as an appetiser, astringent, contraceptive, cystitis, demulcent, diuretic, easily digestable, edible, emollient, chloretic, diseases of bladder and kidney, dysmenorrhoea, lithiasis, ointments, pains, piles, poultice, puludism, resolvent, toilet powders, tuberculosis and urethritis (Ambasta, 1986; Caius, 2003; Gill *et al.*, 1997; Hembrom & Goel, 2005; Kirtikar & Basu, 1984; Parrotta, 2001; Singh & Pandey, 1998; Srivastava *et al.*, 2003b).

Zephyranthes carinata Herb. (Pl. 55D)

Family: Amaryllidaceae.

Vern. : Jangli-payaaj.

English Name:

Eng. : Pink lily.

Distribution : Himalayas.

Description : Bulbous herbs. Leaves entire, radical. Flowers pink. Petals lanceolate, spreading. Corolla tube-like green.

Reproductive Cycle : April-July.

Habitat Ecology : Along Water Streams. Borders of Rice Fields.

Material Examined : HPU-EL-SKS-5738; Bhagpur (1,600 m), Chachian (1,300 m), Ghugar (1,250 m), Saparu (1,250 m), Jiun (1,550 m).

Parts Used : Bulbs. Flowers.

Folk Uses : Boiled bulbs **edible**. Flowers offered to **appease** deities.

Active Constituents : Pancratistatin isolated (*bulbs*).

Uses in Literature : Known for its edible bulbs (Agarwal, 2003; Singh & Kumar, 2000a).

Zizyphus jujuba Mill. (Pl. 55E)

Syn.: *Z. mauritiana* Herb.; *Z, soraria* Schult.; *Z. trinervis* Roth; *Z. sativa* Gaertn. DC.; *Z. vulgaris* Lam.; *Rhamnus jujuba* L.

Family: Rhamnaceae.

Vern. : Ber.

English, Sanskrit & Regional Names:

Eng. : Chinese Date, Indian Cherry, Indian Jujuba, Indian Plum

Sans. : Dridhabija, Prithukoli, Svaccha

B. : Bogri

Bo. : Bhurmi, Bordi

G. : Ber, Boreli, Boyedi

H. : Baer, Ber, Beri

Kan. : Elanji, Yalachi

Mal. : Badaram, Badari, Perintulali

Mar. : Baher, Bor

Ori. : Barkoli, Bodokoli

P. : Beri, Unab

Tam. : Iratti, Kulvali, Sivagam

Tel. : Gangaregu, Karkhanduvu.

Distribution : Throughout India.

Description : A moderate-sized evergreen tree having simple leaves which are glabrous above with tomentose underside. Flowers greenish-yellow in axillary clusters. Drupes ellipsoid, red, shining with furrowed stones.

Reproductive Cycle : July-August.

Habitat Ecology : Wastelands. Rocky slopes.

Material Examined : HPU-EL-SKS-5494; Balah (900 m), Bhuana (1,400 m), Chauki (900 m), Jasun (1,000 m), Tikkar (910 m).

Parts Used : Leaves. Fruit. Roots. Wood.

Folk Uses : Ripe fruits **edible**, and considered **refrigerant**. Root-juice taken to check **dysentery** (10 ml in 100 ml of water twice a day). Leaves considered an excellent **fodder** for cattle. Extr. of roots soaked in water overnight in earthen pot prescribed for relieving **body heat**. Wood especially used for making **arms of bed**.

Uses in Literature : Known earlier as antipyretic, aphrodisiac, appetizer (fruits), coolant, edible (ripe and dried fruits, seed kernels), laxative, tonic; and good for abdominal pain, agricultural implements, asthma, diabetes, diarrhoea (roots), digestion, dysentery, eye, house construction, and stomach complaints, giddiness, gingivitis, good for liver, labour pain, leucorrhoea, malaria, measles, obesity, piles, restlessness, stomachache, toothache (leaves), typhoid fever, ulcer and wounds (Agarwal, 2003; Ambasta, 1986; Bennet *et al.*, 1991; Bhatt *et al.*, 2003a, b; Bhattacharya *et al.*, 2005; Chevallier, 1996; Chopra *et al.*, 1956; Das, 2000; Devi, 2003; Dey & Bahadur, 1973; Diwakar & Ansari, 2003; Duke *et al.*, 2002; Ganesan & Kesavan, 2003; Hebber *et al.*, 2003; Khare, 2004; Khanna *et al.*, 1996c; Kirtikar & Basu, 1984; Kumar, 2003; Mitaliya & Bhatt, 2003; Murthy *et al.*, 2003; Nayak *et al.*, 2003; Parrotta, 2001; Pawar & Patil, 2004; Pearson & Brown, 1932; Prajapati *et al.*, 2006; Ranjan, 2000; Roy *et al.*, 1998; Sahni, 1998; Saklani & Jain, 1994; Samant *et al.*, 2001; Sarma *et al.*, 2001; Sen & Batra, 1997; Sharma, 2003; Sharma & Rana, 2005; Singh, 2000; Singh & Pandey, 1998; Sinha & Sinha, 2001; Siwakoti & Siwakoti, 2000; Siwakoti & Varma, 1996; Sood & Thakur, 2004; Sudhakar & Vedavathy, 2000; Upadhyay & Chauhan, 2000; Upadhye & Kulkarni, 2004).

***Zizyphus nummularia* (Burm.f.) Wight & Arn.** (Pl. 55F)
Syn.: *Rhamnus nummularia* Burm.; *Zizyphus lotus* Lam.;
Z. microphylla Roxb.; *Z. rotundifolia* Lam.
Family: Rhamnaceae

Vern. : Jareid, Breid.

English, Sanskrit & Regional Names:

Eng. : Wild Jujuba
Sans. : Ajapriya, Bhubadari, Bhukamtaka, Sukshmaphala
Bo. : Charubor
G. : Chanyabor, Adbaubordi
H. : Jar-Beri, Jhadiaber, Jhabrberi
Kan. : Gidda, Parpalli
Mar. : Gangar, Janglar
P. : Bal, Biras, Jangra
Tam. : Korgodi
Tel. : Neelareegu.

Distribution : Panjab ascending to 1,000 m, Gujarat, W. Peninsula.

Description : Thorny tomentose shrubs with widely-divaricating flexuous branches of which young ones are puberulous. Leaves serrated, dark-green; velvety above, pale and more felted beneath. Flowers greenish-yellow in cymes. Drupes globose, red.

Reproductive Cycle : June-September.

Habitat Ecology : Wastelands. Drier Areas. Borders of Agricultural Fields.

Material Examined : HPU-EL-SKS-5423; Boda (1,000 m), Chhanchri (930 m), Jasun (1,000 m), Panapari (1,100 m), Rangru (910 m).

Parts Used : Whole Plant. Fruits. Leaves. Root.

Folk Uses : Plant good for **fencing** agricultural fields. Ripe fruits **edible** and also considered **refrigerant**. Roots used for imparting typical **aroma** to local local wine '**sur**'. Extr. of its roots in earthen pot prescribed to **relieve body heat** (100 ml twice for a week). Leaves given as **fodder** for goat and its poultice good for suppuration of **boils**.

Active Constituents : Nummularine P-and nummularin S-isolated from *stem-bark*.

Uses in Literature : So far used in India as an appetizer, astringent, charcoal, coolant, edible, fodder, fuel, stomachic; and good for biliousness, bleeding gums, blood motions, boils, fencing fields and bone fractures, joint pains, nasal discharge, scabies, sore throat, stomatitis, toothache, toothbrush and symbol of good rain and good crops (Agarwal, 2003; Ambasta, 1986; Arora

& Pandey, 1996; Bennet *et al.*, 1991; Bhatt *et al.*, 2003c; Chopra *et al.*, 1956; Das & Agarwal, 1991; Dwivedi, 2003; Kirtikar & Basu, 1984; Kumar, 2003; Kumar *et al.*, 2004; Maliya, 2005; Mitaliya & Bhatt, 2003; Murthy *et al.*, 2003; Parrotta, 2001; Prasad *et al.*, 1999; Roy *et al.*, 1998; Samant *et al.*, 2001; Sharma, 2003; Singh & Kumar, 2000a; Singh & Pandey, 1998).

Epilogue

ETHNOBOTANICAL SPECTRUM

Analysis of the data brought to light the use of 335 ethnobotanically important plants belonging to 254 genera under 88 diverse families by the rural populace of tehsil Palampur, district Kangra, (H.P.) (Table IX), of which 156 spp. are herbaceous (herbs 120; herbaceous climbers 30; tender undershrubs 6) and the remaining 174 are woody (Tables III-VIII). The overall diversity can be grouped into 273 ethnobotanically important wild spp., 41 cultivated ones and 21 growing as wild as well as under cultivation (Table II). Based on their growth habit these spp. are distributed over 120 herbs, 105 trees, 54 shrubs, 18 undershrubs, 30 climbers, 3 lianas, and 5 ferns (Tables III-VIII; Figs. 5, 6).

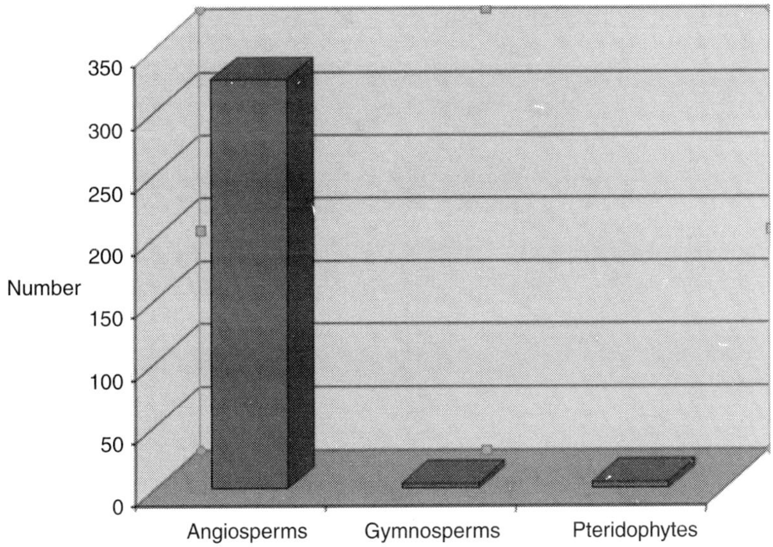

Fig. 4 Histogram Showing Total Number of Plants Used by the Rural Populace of Tehsil Palampur Under Various Divisions of Plant Kingdom.

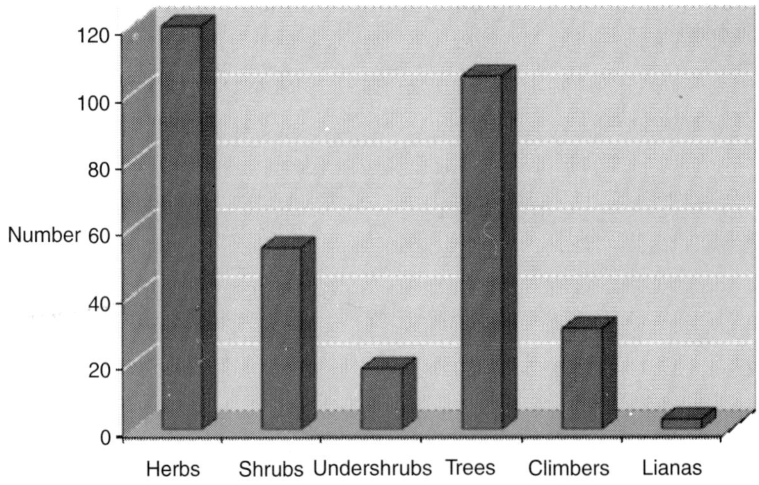

Fig. 5 Histogram Showing Total Number of Herbs, Shrubs, Undershrubs, Trees, Climbers and Lianas Used by the Rural Populace of Tehsil Palampur.

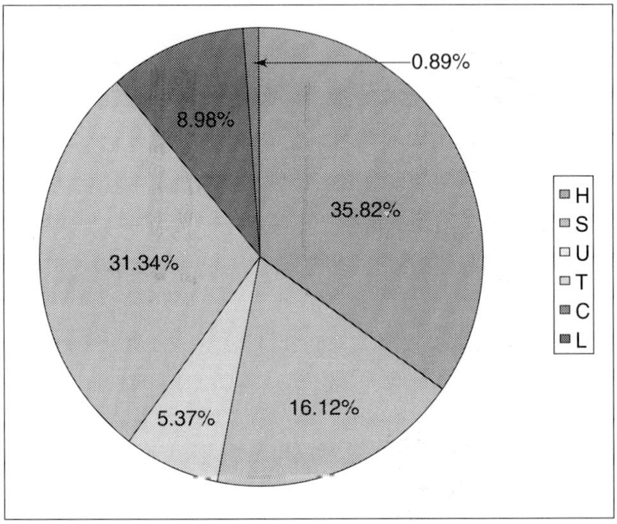

Fig. 6 Relative Percentage of Ethnobotanically Collected Herbs, Shrubs, Undershrubs, Trees, Climbers and Lianas of Tehsil Palampur of District Kangra (H.P.) (H-Herbs, T-Trees, S-Shrubs, U- Undershrubs, C- Climbers, L- Lianas).

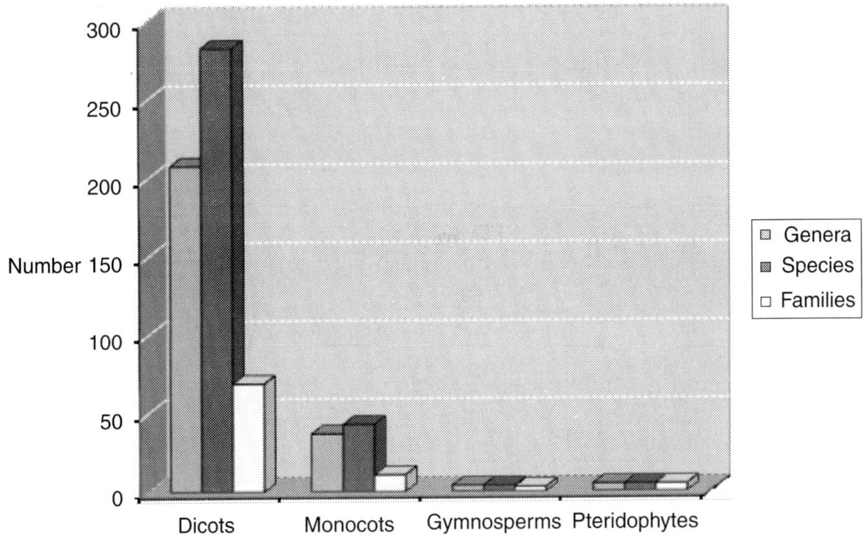

Fig. 7 Total Number of Ethnobotanically Collected Species, Genera and Families Under Various Divisions of Plant Kingdom.

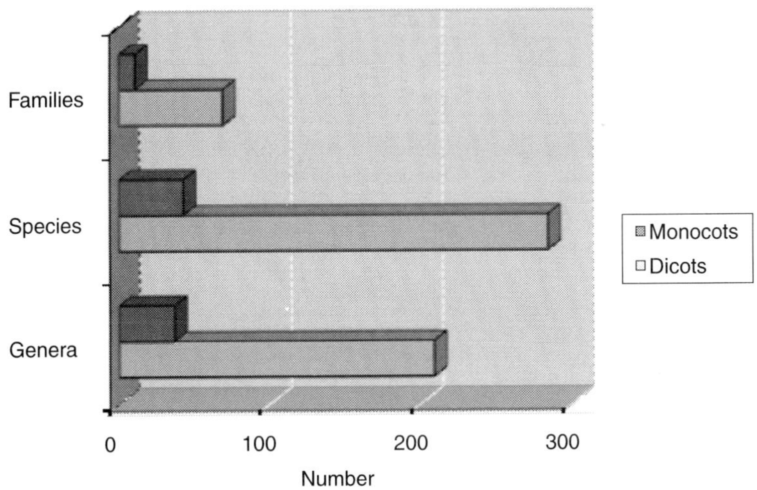

Fig. 8 Histogram Showing the Comparative Number of Genera, Species and Families of Monocots and Dicots Used by the Rural Populace of Tehsil Palampur.

Table II Plant Resource Categories of Tehsil Palampur, District Kangra (H.P.)

S.No.	Wild	Cultivated	Wild as well as Cultivated
1.	*Abrus precatorius*	*Abelmoschus esculentus*	*Acacia catechu*
2.	*Achyranthes aspera*	*Allium cepa*	*Aloe vera*
3.	*Achyranthes bidentata*	*Allium sativum*	*Artocarpus lacucha*
4.	*Acorus calamus*	*Amaranthus paniculatus*	*Basella alba*
5.	*Adiantum capillus veneris*	*Araucaria columnaris*	*Cinnamomum tamala*
6.	*Aegle marmelos*	*Brassica nigra*	*Citrus grandis*
7.	*Aerva lanata*	*Brassica tournefortii*	*Citrus limon*
8.	*Agave cantula*	*Camellia sinensis*	*Citrus medica var medica*
9.	*Agave wightii*	*Capsicum annuum*	*Coccinia grandis*
10.	*Ageratum conyzoides*	*Carica papaya*	*Curcuma longa*
11.	*Ajuga parviflora*	*Carya illinoensis*	*Cyclanthera pedata*
12.	*Albizia chinensis*	*Cestrum nocturnum*	*Cyphomandra betacea*
13.	*Albizia julibrissin*	*Cinnamomum camphora*	*Ipomoea batatas*
14.	*Albizia lebbeck*	*Citrus medica*	*Mangifera indica*
15.	*Alternanthera sessilis*	*Cucumis sativus*	*Musa sapientum*
16.	*Amaranthus viridis*	*Cucurbita hispida*	*Ocimum kilimandscharicum*
17.	*Apium leptophyllum*	*Cucurbita pepo*	*Phoenix sylvestris*
18.	*Aristida cyanantha*	*Diospyros kaki*	*Prunus persica*
19.	*Artemisia capillaris*	*Eriobotrya japonica*	*Punica granatum*
20.	*Artemisia japonica*	*Ficus carica*	*Raphanus sativus*
21.	*Arundinaria falcata*	*Foeniculum vulgare*	*Trigonella foenum-graecum*
22.	*Asparagus adscendens*	*Glycine max*	
23.	*Avena sativa*	*Gossypium arboreum*	
24.	*Bambusa nutans*	*Grevillea robusta*	
25.	*Bauhinia vahlii*	*Juglans regia*	
26.	*Bauhinia variegata*	*Litchi chinensis*	
27.	*Berberis lycium*	*Luffa cylindrica*	
28.	*Bergenia ciliata*	*Nerium indicum*	
29.	*Bidens biternata*	*Ocimum basilicum*	
30.	*Boehmeria macrophylla*	*Ocimum sanctum*	
31.	*Boerhaavia diffusa*	*Prunus domestica*	
32.	*Bombax ceiba*	*Psidium guajava*	
33.	*Butea monosperma*	*Pyrus communis*	
34.	*Caesalpinia decapetala*	*Saccharum officinarum*	
35.	*Cajanus cajan*	*Sechium edule*	

Contd...

Table II Contd...

S.No.	Wild	Cultivated	Wild as well as Cultivated
36.	*Callicarpa macrophylla*	*Solanum melongena*	
37.	*Calotropis procera*	*Tagetes minuta*	
38.	*Cannabis sativa*	*Triticum aestivum*	
39.	*Capillipedium assimile*	*Vicia faba*	
40.	*Carissa congesta*	*Vitis vinifera*	
41.	*Caryopteris bicolor*	*Zea mays*	
42.	*Cassia fistula*		
43.	*Cassia floribunda*		
44.	*Cassia obtusifolia*		
45.	*Cassia occidentalis*		
46.	*Catharanthus roseus*		
47.	*Cedrus deodara*		
48.	*Celastrus paniculatus*		
49.	*Celtis australis*		
50.	*Centella asiatica*		
51.	*Centipeda minima*		
52.	*Centratherum anthelminticum*		
53.	*Cheilanthes farinosa*		
54.	*Chenopodium album*		
55.	*Chenopodium ambrosioides*		
56.	*Cicer arietinum*		
57.	*Cirsium arvense*		
58.	*Cissampelos parriera*		
59.	*Citrus jambhiri*		
60.	*Clematis gauriana*		
61.	*Cocculus laurifolius*		
62.	*Colebrookea oppositifolia*		
63.	*Colocasia esculenta*		
64.	*Commelina paludosa*		
65.	*Cordia dichotoma*		
66.	*Coriandrum sativum*		
67.	*Coronopus didymus*		
68.	*Costus speciosus*		
69.	*Cryptolepis buchanani*		
70.	*Cucumis melo* var. *agrestis*		

Contd...

Table II Contd...

S.No.	Wild	Cultivated	Wild as well as Cultivated
71.	*Cuscuta reflexa*		
72.	*Cymbopogon citratus*		
73.	*Cynodon dactylon*		
74.	*Dalbergia sissoo*		
75.	*Datura innoxia*		
76.	*Datura stramonium*		
77.	*Debregeasia hypoleuca*		
78.	*Deeringia amaranthoides*		
79.	*Dendrocalamus hamiltonii*		
80.	*Didymocarpus pedicellata*		
81.	*Dioscorea bellophylla*		
82.	*Dioscorea bulbifera*		
83.	*Dioscorea oppositifolia*		
84.	*Diplazium esculentum*		
85.	*Diplocyclos palmatus*		
86.	*Dodonaea viscosa*		
87.	*Dolichos biflorus*		
88.	*Drimia indica*		
89.	*Duranta repens*		
90.	*Ehretia acuminata*		
91.	*Ehretia laevis*		
92.	*Eleusine coracana*		
93.	*Emblica officinalis*		
94.	*Engelhardia colebrookeana*		
95.	*Equisetum debile*		
96.	*Eriophorum comosum*		
97.	*Erythrina suberosa*		
98.	*Eucalyptus tereticornis*		
99.	*Eulaliopsis binata*		
100.	*Euonymus hamiltonianus*		
101.	*Euphorbia royleana*		
102.	*Fagopyrum cymosum*		
103.	*Fagopyrum esculentum*		
104.	*Ficus benghalensis*		
105.	*Ficus clavata*		
106.	*Ficus foveolata*		

Contd...

Table II Contd...

S.No.	Wild	Cultivated	Wild as well as Cultivated
107.	*Ficus glomerata*		
108.	*Ficus hispida*		
109.	*Ficus infectoria*		
110.	*Ficus palmata*		
111.	*Ficus religiosa*		
112.	*Ficus roxburghii*		
113.	*Ficus rumphii*		
114.	*Ficus semicordata*		
115.	*Flacourtia indica*		
116.	*Fragaria indica*		
117.	*Fumaria indica*		
118.	*Galium aparine*		
119.	*Geranium nepalense*		
120.	*Gnaphalium affine*		
121.	*Gonatanthus pumilus*		
122.	*Grewia disperma*		
123.	*Grewia elastica*		
124.	*Grewia optiva*		
125.	*Hamiltonia suaveolens*		
126.	*Helicteres isora*		
127.	*Heteropogon contortus*		
128.	*Hibiscus rosa-sinensis*		
129.	*Holarrhena antidysentrica*		
130.	*Hypericum oblogifolium*		
131.	*Impatiens balsamina*		
132.	*Indigofera cassioides*		
133.	*Ipomoea turbinata*		
134.	*Jasminum dispermum*		
135.	*Jasminum grandiflorum*		
136.	*Jasminum humile*		
137.	*Jasminum multiflorum*		
138.	*Jatropha curcas*		
139.	*Justicia adhatoda*		
140.	*Kalanchoe pinnata*		
141.	*Lannea coromandelica*		

Contd...

Table II Contd...

S.No.	Wild	Cultivated	Wild as well as Cultivated
142.	*Lantana camara*		
143.	*Lathyrus aphaca*		
144.	*Lathyrus sativus*		
145.	*Lawsonia inermis*		
146.	*Lepidagathis cuspidata*		
147.	*Lepidium sativum*		
148.	*Leucaena glauca*		
149.	*Leucas cephalotes*		
150.	*Linum usitatissimum*		
151.	*Litsea monopetala*		
152.	*Litsea sebifera*		
153.	*Litsea semicarpifolia*		
154.	*Loranthus falcatus*		
155.	*Lotus corniculatus*		
156.	*Luffa acutangula*		
157.	*Mallotus philippinensis*		
158.	*Medicago polymorpha*		
159.	*Melia azedarach*		
160.	*Mentha longifolia*		
161.	*Mentha piperita*		
162.	*Michelia champaca*		
163.	*Micromeria biflora*		
164.	*Milletia extensa*		
165.	*Mirabilis jalapa*		
166.	*Momordica charantia*		
167.	*Momordica dioica*		
168.	*Moringa oleifera*		
169.	*Morus alba*		
170.	*Mucuna pruriens*		
171.	*Murraya koenigii*		
172.	*Murraya paniculata*		
173.	*Neolitsea umbrosa*		
174.	*Ocimum gratissimum*		
175.	*Opuntia dillenii*		
176.	*Origanum vulgare*		
177.	*Oroxylum indicum*		

Contd...

Table II Contd...

S.No.	Wild	Cultivated	Wild as well as Cultivated
178.	*Ougeinia dalbergioides*		
179.	*Oxalis corniculata*		
180.	*Panicum antidontale*		
181.	*Pennisetum americanum*		
182.	*Pentanema indicum*		
183.	*Perilla frutescens*		
184.	*Persea odoratissima*		
185.	*Phoenix acaulis*		
186.	*Phyllanthus urinaria*		
187.	*Pinus roxburghii*		
188.	*Pistacia khinjuk*		
189.	*Plantago lanceolata*		
190.	*Plumbago zeylanica*		
191.	*Pogostemon benghalense*		
192.	*Polygala chinensis*		
193.	*Polygonum barbatum*		
194.	*Populus ciliata*		
195.	*Populus nigra*		
196.	*Porana paniculata*		
197.	*Portulaca grandiflora*		
198.	*Prinsepia utilis*		
199.	*Prunus cerasoides*		
200.	*Pyrus pashia*		
201.	*Pyrus serotina*		
202.	*Quercus leucotrichophora*		
203.	*Randia dumetorum*		
204.	*Ranunculus arvensis*		
205.	*Ranunculus laetus*		
206.	*Ranunculus muricatus*		
207.	*Reinwardtia indica*		
208.	*Rhamnus triquetra*		
209.	*Rhododendron arboreum*		
210.	*Ricinus communis*		
211.	*Rorippa nasturtium -aquaticum*		
212.	*Rosa brunonii*		

Contd...

Table II Contd...

S.No.	Wild	Cultivated	Wild as well as Cultivated
213.	*Rosa chinensis*		
214.	*Rosa damascena*		
215.	*Roylea cinerea*		
216.	*Rubus ellipticus*		
217.	*Rubus niveus*		
218.	*Rumex hastatus*		
219.	*Rumex nepalensis*		
220.	*Saccharum spontaneum*		
221.	*Salix karelini*		
222.	*Salix tetrasperma*		
223.	*Salvia coccinea*		
224.	*Santalum album*		
225.	*Sapindus mukorosii*		
226.	*Sapium insigne*		
227.	*Sapium sebiferum*		
228.	*Scandix pecten veneris*		
229.	*Selaginella chrysocaulos*		
230.	*Sesamum orientale*		
231.	*Sida cordifolia*		
232.	*Silene conoidea*		
233.	*Sisymbrium irio*		
234.	*Solanum erianthum*		
235.	*Solanum nigrum*		
236.	*Solanum torvum*		
237.	*Solanum viarum*		
238.	*Sorghum bicolor*		
239.	*Spilanthes oleracea*		
240.	*Spondias pinnata*		
241.	*Sporobolus indicus*		
242.	*Stellaria media*		
243.	*Stephania glabra*		
244.	*Syzygium cumini*		
245.	*Syzygium operculatum*		
246.	*Terminalia arjuna*		
247.	*Terminalia bellirica*		

Contd...

Table II Contd...

S.No.	Wild	Cultivated	Wild as well as Cultivated
248.	*Terminalia chebula*		
249.	*Terminalia tomentosa*		
250.	*Thuja orientalis*		
251.	*Thysanolaena maxima*		
252.	*Tinospora cordifolia*		
253.	*Toona ciliata*		
254.	*Trichosanthes cucumerina*		
255.	*Trifolium resupinatum*		
256.	*Tulipa stellata*		
257.	*Urtica ardens*		
258.	*Urtica dioica*		
259.	*Vanda tessellata*		
260.	*Verbascum thapsus*		
261.	*Vicia hirsuta*		
262.	*Vicia sativa*		
263.	*Vigna radiata*		
264.	*Viola patrinii*		
265.	*Viola pilosa*		
266.	*Vitex negundo*		
267.	*Vitis repanda*		
268.	*Woodfordia fruticosa*		
269.	*Wrightia arborea*		
270.	*Zanthoxylum armatum*		
271.	*Zephyranthes carinata*		
272.	*Zizyphus jujuba*		
273.	*Zizyphus nummularia*		

Table III Ethnobotanically Used Herbs of Tehsil Palampur, District Kangra (H.P.)

S.No.	Name	Local Name/s
1.	*Abelmoschus esculentus*	Ghandoli
2.	*Achyranthes aspera*	Puthkanda
3.	*Achyranthes bidentata*	Puthkanda
4.	*Acorus calamus*	Barain, Brein.
5.	*Aerva lanata*	Aung
6.	*Ageratum conyzoides*	Jaihar, Phulnu

Contd...

Table III Contd...

S.No.	*Name*	*Local Name/s*
7.	*Ajuga parviflora*	Neelkanthi
8.	*Allium cepa*	Payaj
9.	*Allium sativum*	Lahsun, Lassan
10.	*Alternanthera sessilis*	Rubadi-Khubadi
11.	*Amaranthus paniculatus*	Seul
12.	*Amaranthus viridis*	Chalairi, Chaulai
13.	*Apium leptophyllum*	Juanhe
14.	*Aristida cyanantha*	Kaduli-Tuli
15.	*Arundinaria falcata*	Banji
16.	*Avena sativa*	Luanhe
17.	*Bambusa nutans*	Ddang
18.	*Bergenia ciliata*	Saprotri
19.	*Bidens biternata*	Lumb
20.	*Boerhaavia diffusa*	Eat-Seat
21.	*Brassica nigra*	Kali-Rai
22.	*Brassica tournefortii*	Rai
23.	*Calotropis procera*	Aak, Ak
24.	*Capillipedium assimile*	Dholu
25.	*Cassia obtusifolia*	Allu, Rellu
26.	*Catharanthus roseus*	Sadabahar
27.	*Centella asiatica*	Meenaki
28.	*Centipeda minima*	Nakk-Chhikannu
29.	*Centratherum anthelminticum*	Brahmjeeri
30.	*Chenopodium album*	Kanah
31.	*Chenopodium ambrosioides*	Bathu
32.	*Cicer arietinum*	Chana
33.	*Cirsium arvense*	Bhruce
34.	*Colocasia esculenta*	Janglee Kachalu
35.	*Commelina paludosa*	Chhura
36.	*Coriandrum sativum*	Beenh, Dhania
37.	*Coronopus didymus*	Garhbeenhi
38.	*Costus speciosus*	Jangli-Haldi
39.	*Curcuma longa*	Haldar
40.	*Cymbopogon citratus*	Makoda
41.	*Cynodon dactylon*	Dharub, Daroov
42.	*Dendrocalamus hamiltonii*	Maggar

Contd...

Table III Contd...

S.No.	*Name*	*Local Name/s*
43.	*Didymocarpus pedicellata*	Musakani
44.	*Dolichos biflorus*	Kulath
45.	*Drimia indica*	Ban-Ganda
46.	*Eleusine coracana*	Mandal
47.	*Eriophorum comosum*	Bagdella
48.	*Eulaliopsis binata*	Baggad
49.	*Fagopyrum cymosum*	Fafru
50.	*Fagopyrum esculentum*	Bhresh
51.	*Foeniculum vulgare*	Meethi-Saunf
52.	*Fragaria indica*	Bhoo-Akha
53.	*Fumaria indica*	Papda
54.	*Galium aparine*	Jhanjheedi
55.	*Geranium nepalense*	Laljari
56.	*Glycine max*	Soyabean
57.	*Gnaphalium affine*	Gadua
58.	*Gonatanthus pumilus*	Aala
59.	*Heteropogon contortus*	Lambia Ghaa
60.	*Impatiens balsamina*	Teur
61.	*Ipomoea batatas*	Shakarkanni
62.	*Kalanchoe pinnata*	Gand-Patar
63.	*Lathyrus aphaca*	Mater-Phali, Khinnu, Sudu
64.	*Lathyrus sativus*	Kalan
65.	*Lepidium sativum*	Halaein
66.	*Leucas cephalotes*	Bhedu
67.	*Linum usitatissimum*	Allain
68.	*Lotus corniculatus*	Chidi
69.	*Medicago polymorpha*	Khukhani
70.	*Mentha longifolia*	Jungli-Pudina
71.	*Mentha piperita*	Bhura-Pudina, Pipermint
72.	*Micromeria biflora*	Janglee-Ajwain
73.	*Musa sapientum*	Kela
74.	*Origanum vulgare*	Kah-Ajwain
75.	*Oxalis corniculata*	Malori
76.	*Panicum antidontale*	Barua
77.	*Pennisetum americanum*	Bajara
78.	*Pentanema indicum*	Jugaali-Di-Dua

Contd...

Table III Contd...

S.No.	Name	Local Name/s
79.	*Perilla frutescens*	Ban-Ajwain
80.	*Phyllanthus urinaria*	Amli
81.	*Plantago lanceolata*	Safed Isbagol
82.	*Polygala chinensis*	Tone-Beidu
83.	*Polygonum barbatum*	Ghaneeri
84.	*Portulaca grandiflora*	Kulfa
85.	*Ranunculus arvensis*	Jaldhar
86.	*Ranunculus laetus*	Jaldhar
87.	*Ranunculus muricatus*	Changer
88.	*Raphanus sativus*	Mooli
89.	*Reinwardtia indica*	Basant
90.	*Rorippa nasturtium- officinale*	Chhoo-Nali
91.	*Rumex hastatus*	Aambi
92.	*Rumex nepalensis*	Jangli-Palak
93.	*Saccharum officinarum*	Ganna
94.	*Saccharum spontaneum*	Ganna
95.	*Scandix pecten veneris*	Jangli Dhaniya
96.	*Sesamum orientale*	Til
97.	*Silene conoidea*	Tokala
98.	*Sisymbrium irio*	Foonkalan
99.	*Solanum nigrum*	Makoi
100.	*Sorghum bicolor*	Chari
101.	*Spilanthes oleracea*	Karkara
102.	*Sporobolus indicus*	Bonkra
103.	*Stellaria media*	Khaukhua
104.	*Thysanolaena maxima*	Garnaal
105.	*Trifolium resupinatum*	Barseen
106.	*Trigonella foenum- graecum*	Maethi
107.	*Triticum aestivum*	Kanak
108.	*Tulipa stellata*	Minmayaaoon
109.	*Urtica ardens*	Chhoti-Ainh
110.	*Urtica dioica*	Ainh
111.	*Vanda tessellata*	Bhangru
112.	*Verbascum thapsus*	Ban-Tambakhoo
113.	*Vicia faba*	Keon, Keeoon
114.	*Vicia hirsuta*	Choti-Rodi

Contd...

Table III Contd...

S.No.	Name	Local Name/s
115.	*Vicia sativa*	Badi-Rodi
116.	*Vigna radiata*	Maah
117.	*Viola patrinii*	Banafsha, Banaksa
118.	*Viola pilosa*	Banafsha, Banaksa
119.	*Zea mays*	Chhali, Chhaliyan
120.	*Zephyranthes carinata*	Jangli-Payaas

Table IV Ethnobotanically Used Shrubs of Tehsil Palampur

S.No.	Name	Local Name/s
1.	*Berberis lycium*	Kashmal, Kasmalein
2.	*Boehmeria macrophylla*	Handa
3.	*Caesalpinia decapetala*	Drogaadi, Taranghadi
4.	*Cajanus cajan*	Dhinger
5.	*Callicarpa macrophylla*	Danther
6.	*Camellia sinensis*	Cha, Chai
7.	*Capsicum annuum*	Mirach, Pipal
8.	*Carissa congesta*	Garna, Garnae
9.	*Caryopteris bicolor*	Ban-Dhavein
10.	*Cassia floribunda*	Bada-Rellu
11.	*Cestrum nocturnum*	Rat-Ki-Rani
12.	*Colebrookea oppositifolia*	Doos
13.	*Datura innoxia*	Dhatura
14.	*Datura stramonium*	Dhatura
15.	*Debregeasia hypoleuca*	Siaru, Syarru
16.	*Dodonaea viscosa*	Mehendu, Mehendadu
17.	*Duranta repens*	Neel-Kanta
18.	*Euphorbia royleana*	Chhoo
19.	*Ficus foveolata*	Phaegdi
20.	*Hamiltonia suaveolens*	Padaar
21.	*Hibiscus rosa-sinensis*	Dunaj-Phul
22.	*Hypericum oblogifolium*	Peelan
23.	*Indigofera cassioides*	Kanthi
24.	*Jasminum humile*	Chamaeli
25.	*Jasminum multiflorum*	Chamaeli
26.	*Justicia adhatoda*	Basunti, Basuti

Contd...

Table IV Contd...

S.No.	Name	Local Name/s
27.	*Lantana camara*	Zahar, Phulanu
28.	*Lawsonia inermis*	Mehendi
29.	*Lepidagathis cuspidata*	Billi
30.	*Loranthus falcatus*	Parand
31.	*Mirabilis jalapa*	Gulbansi
32.	*Mucuna pruriens*	Gazal Bel
33.	*Murraya koenigii*	Ghandla, Kadi Patta
34.	*Murraya paniculata*	Gandhala
35.	*Opuntia dillenii*	Daandi-Chhoo
36.	*Phoenix acaulis*	Khajoori
37.	*Pogostemon benghalense*	Kali Basunti
38.	*Porana paniculata*	Chhube-Beil
39.	*Prinsepia utilis*	Bhekhal
40.	*Rhododendron arboreum*	Braah, Brass
41.	*Ricinus communis*	Arand, Erand
42.	*Rosa brunonii*	Gulabadi
43.	*Rosa chinensis*	Galaab, Gulaab
44.	*Rosa damascena*	Gulabadi
45.	*Roylea cinerea*	Kakdoen
46.	*Rubus ellipticus*	Aakha
47.	*Rubus niveus*	Kala-Akha
48.	*Solanum erianthum*	Laah
49.	*Solanum melongena*	Baingan
50.	*Solanum torvum*	Chhota-Baingan
51.	*Thuja orientalis*	Sarua, Morpankh
52.	*Vitex negundo*	Bana
53.	*Woodfordia fruticosa*	Dhawain, Dhaun
54.	*Zizyphus nummularia*	Jareid, Breid

Table V Ethnobotanically Used Trees of Tehsil Palampur

S.No.	Name	Local Name/s
1.	*Acacia catechu*	Khair
2.	*Aegle marmelos*	Bael, Bil
3.	*Albizia chinensis*	Oyee
4.	*Albizia julibrissin*	Karmarua

Contd...

Table V Contd...

S.No.	Name	Local Name/s
5.	*Albizia lebbeck*	Sarin, Sirin
6.	*Araucaria columnaris*	Araucaria
7.	*Artocarpus lacucha*	Dehu
8.	*Bauhinia variegata*	Karale, Karaalein
9.	*Bombax ceiba*	Simbal
10.	*Butea monosperma*	Palah, Palash
11.	*Carica papaya*	Kharbuja, Papita
12.	*Carya illinoensis*	Kagzi-Khod
13.	*Cassia fistula*	Alia, Amaltas
14.	*Cedrus deodara*	Deyar, Devdar
15.	*Celtis australis*	Khidak, Khirak
16.	*Cinnamomum tamala*	Gudpatraj,Mithpatar, Tejpatar
17.	*Cinnamomum camphora*	Kapur
18.	*Citrus grandis*	Chakotra
19.	*Citrus jambhiri*	Ghamiridi, Jhambhirdi, Khateerali
20.	*Citrus limon*	Khatta
21.	*Citrus medica*	Nimbu
22.	*Citrus medica* var. *medica*	Bajoorein
23.	*Cocculus laurifolius*	Parauda
24.	*Cordia dichotoma*	Lasura, Lasyada
25.	*Cyphomandra betacea*	Desi Tamatar
26.	*Dalbergia sissoo*	Shissham, Tahli
27.	*Diospyros kaki*	Japani-phal
28.	*Ehretia acuminata*	Punna
29.	*Ehretia laevis*	Bhankar
30.	*Emblica officinalis*	Ambla, Amla
31.	*Engelhardia colebrookeana*	Samma
32.	*Eriobotrya japonica*	Galaat
33.	*Erythrina suberosa*	Paryara
34.	*Eucalyptus tereticornis*	Safeda
35.	*Euonymus hamiltonianus*	Chamardein
36.	*Ficus benghalensis*	Badein, Bargad
37.	*Ficus carica*	Anjeer
38.	*Ficus clavata*	Ban-Feguda
39.	*Ficus glomerata*	Umbrein
40.	*Ficus hispida*	Dagrein

Contd...

Table V Contd...

S.No.	Name	Local Name/s
41.	*Ficus infectoria*	Pallakharein
42.	*Ficus palmata*	Dhuda, Khasra
43.	*Ficus religiosa*	Bad, Peepal
44.	*Ficus roxburghii*	Traimbal
45.	*Ficus rumphii*	Badarein
46.	*Ficus semicordata*	Khanurein, Khasarkhurein
47.	*Grevillea robusta*	Silver Oak
48.	*Grewia disperma*	Byulee
49.	*Grewia elastica*	Dhaman
50.	*Grewia optiva*	Beul
51.	*Helicteres isora*	Ban-Beuli
52.	*Holarrhena antidysentrica*	Keor, Kurchi
53.	*Jatropha curcas*	Jamalgota, Rattanjot
54.	*Juglans regia*	Akhrot, Khod
55.	*Lannea coromandelica*	Kahemble
56.	*Leucaena glauca*	Alseenia
57.	*Litchi chinensis*	Litchi
58.	*Litsea monoptela*	Guan
59.	*Litsea sebifera*	Rhaanh
60.	*Litsea semicarpifolia*	Pahadi-Sannanh
61.	*Mallotus philippinensis*	Kaamal
62.	*Mangifera indica*	Aam, Amb
63.	*Melia azedarach*	Drek
64.	*Michelia champaca*	Chamba, Champa
65.	*Moringa oleifera*	Sunanha
66.	*Morus alba*	Toot
67.	*Neolitsea umbrosa*	Gulab- Jamun
68.	*Nerium indicum*	Ghaneera
69.	*Oroxylum indicum*	Tat-Palanga
70.	*Ougeinia dalbergioides*	Sannanh
71.	*Persea odoratissima*	Bhadrol
72.	*Phoenix sylvestris*	Khajoor
73.	*Pinus roxburghii*	Chir, Cheel
74.	*Pistacia khinjuk*	Kakad Singhi
75.	*Populus ciliata*	Phalsa
76.	*Populus nigra*	Poplar

Contd...

Table V Contd...

S.No.	Name	Local Name/s
77.	*Prunus cerasoides*	Paaja, Pajja
78.	*Prunus domestica*	Aloo Bhukara, Palam
79.	*Prunus persica*	Aadu, Aarnei, Aru
80.	*Psidium guajava*	Amrood
81.	*Punica granatum*	Daadu
82.	*Pyrus communis*	Nashpati
83.	*Pyrus pashia*	Kainth
84.	*Pyrus serotina*	Bada-Kainth
85.	*Quercus leucotrichophora*	Baan
86.	*Randia dumetorum*	Rada
87.	*Rhamnus triquetra*	Guldarein
88.	*Rhododendron arboreum*	Braah, Brass
89.	*Salix karelini*	Pahadi-Beeuins
90.	*Salix tetrasperma*	Beeuins
91.	*Santalum album*	Chananh
92.	*Sapindus mukorosii*	Dodan, Ritha
93.	*Sapium insigne*	Ainkhar
94.	*Sapium sebiferum*	Makhan, Pahadi-Taali
95.	*Spondias pinnata*	Muara
96.	*Syzygium cumini*	Jamun
97.	*Syzygium operculatum*	Kathmanh
98.	*Terminalia arjuna*	Arjan, Arjun
99.	*Terminalia bellirica*	Bahera
100.	*Terminalia chebula*	Harad
101.	*Terminalia tomentosa*	Rasaan
102.	*Toona ciliata*	Tooni
103.	*Wrightia arborea*	Khaloa
104.	*Zanthoxylum armatum*	Tirmir
105.	*Zizyphus jujuba*	Ber

Table VI Ethnobotanically Used Undershrubs of Tehsil Palampur

S.No.	Botanical Name	Local Name/s
1.	*Agave cantula*	Ranhbaan, Ramban
2.	*Agave wightii*	Kyor
3.	*Aloe vera*	Kuaarein

Contd...

Table VI Contd...

S.No.	Botanical Name	Local Name/s
4.	*Artemisia capillaris*	Jhauncha
5.	*Artemisia japonica*	Chharmara
6.	*Asparagus adscendens*	Sanspain
7.	*Cannabis sativa*	Bhang
8.	*Cassia occidentalis*	Chhota-Rellu
9.	*Deeringia amaranthoides*	Bhringya
10.	*Ocimum basilicum*	Bhabhri
11.	*Ocimum gratissimum*	Ban-Ajwain
12.	*Ocimum kilimandscharicum*	Tulsi
13.	*Ocimum sanctum*	Sailli
14.	*Plumbago zeylanica*	Chitra
15.	*Salvia coccinea*	Phultulsi
16.	*Sida cordifolia*	Dredain
17.	*Solanum viarum*	Bhindi, Kandayai
18.	*Tagetes minuta*	Gutt

Table VII Ethnobotanically Used Lianas of Tehsil Palampur

S.No.	Botanical Name	Local Name/s
1.	*Bauhinia vahlii*	Taur, Tor
2.	*Celastrus paniculatus*	Sankheeru
3.	*Milletia extensa*	Salod

Table VIII Ethnobotanically Used Climbers of Tehsil Palampur

S.No.	Botanical Name	Local Name/s
1.	*Abrus precatorius*	Chadannu, Rati
2.	*Basella alba*	Poe
3.	*Cissampelos parriera*	Bhatindu
4.	*Clematis gauriana*	Baaker-Bel
5.	*Coccinia grandis*	Kalakadu
6.	*Cucumis melo* var. *agrestis*	Phaphanu
7.	*Cucumis sativus*	Kheera
8.	*Cucurbita hispida*	Dhuda Petha
9.	*Cucurbita pepo*	Kaddu
10.	*Cuscuta reflexa*	Akashbel, Gaasbel

Contd...

Table VIII Contd...

S.No.	Botanical Name	Local Name/s
11.	*Cyclanthera pedata*	Meetha Karela
12.	*Dioscorea bellophylla*	Tardi
13.	*Dioscorea bulbifera*	Gangaradi
14.	*Dioscorea oppositifolia*	Daregal
15.	*Diplocyclos palmatus*	Shivlingi
16.	*Ipomoea turbinata*	Ghondali
17.	*Jasminum dispermum*	Chamaeli
18.	*Jasminum grandiflorum*	Malti
19.	*Luffa acutangula*	Jangli Ghangher
20.	*Luffa cylindrica*	Ghi-Ghangher
21.	*Momordica charantia*	Karela
22.	*Momordica dioica*	Meetha karela
23.	*Mucuna pruriens*	Gazal Bel
24.	*Sechium edule*	Lonku
25.	*Stephania glabra*	Bis-Khapar
26.	*Thuja orientalis*	Sarua, Morpankh
27.	*Tinospora cordifolia*	Gloan, Gulja
28.	*Trichosanthes cucumerina*	Meis
29.	*Vitis repanda*	Maraingu
30.	*Vitis vinifera*	Angoor

For their sustenance and healthcare needs, the various ethnobotanical categories and number of plants used respectively are: various ailments–224 spp; edible–166 spp; fodder–100 spp; magico-religious–60 spp; fuel–48 spp; wood–35 spp; household articles–16 spp; aesthetic–12 spp; beverage–10 spp; fibre–10 spp; condiments–11 spp; hedges–9 spp; local wine ('sur'–9 spp; *Achyranthes bidentata, Albizia chinensis, Callicarpa macrophylla, Celastrus paniculatus, Cymbopogon citratus, Justicia adhatoda, Panicum antidontale, Woodfordia fruticosa, Zizyphus nummularia*); dye–7 spp; walking sticks–6 spp; insecticidal–4; thatching–4 spp; vermillion–4 spp; bedding–3 spp; insect repellent–3 spp; nectar–3 spp; fish poison–2 spp; and for miscellaneous purposes–88 spp. (Appendix-II). Organ-wise analysis reveals that inhabitants of this part of the Himalayas use aerial plant parts (594, 94.14%) more than the underground parts (37, 5.86%) (Figs. 9, 10). Of all the plant parts, leaves predominate in usage (176, 24%) followed by stem/wood (117, 17%), fruits (101, 15%), seeds (60, 15%), aerial plant parts (59, 9%), flowers/inflorescence (56, 8%), whole plant (45, 7%), underground parts (37, 5%), bark (25, 4%), gum/latex (15, 3%) (Figs. 11, 12). Evidently, dependence on a great diversity of plant species reflects their vast knowledge of the local wealth ensuring a year-round supply of food and medicine with least pressure on the surrounding plant diversity.

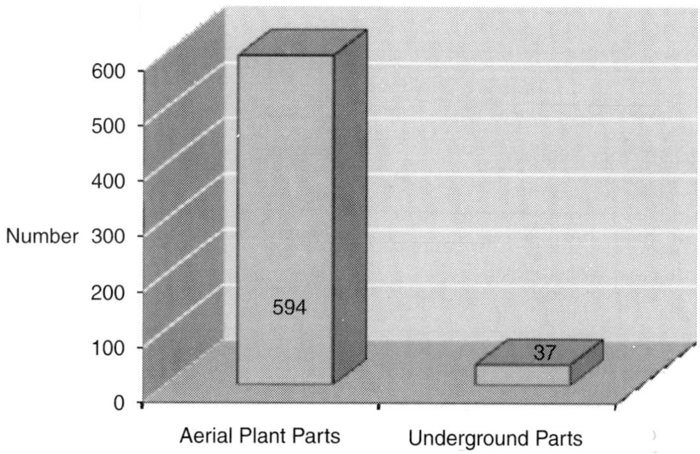

Fig. 9 Histogram Showing Relative Number of Aerial and Underground Plant Parts Used by the Rural Populace of Tehsil Palampur.

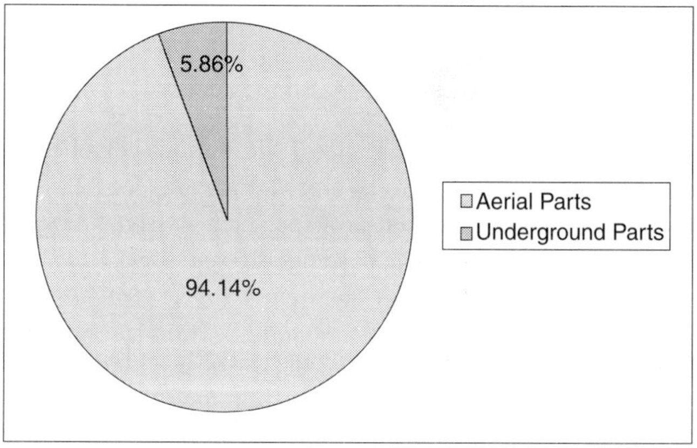

Fig. 10 Pie Diagram Showing Relative Disposition of Underground and Aerial Plant Parts Consumed by the Rural Populace of Tehsil Palampur

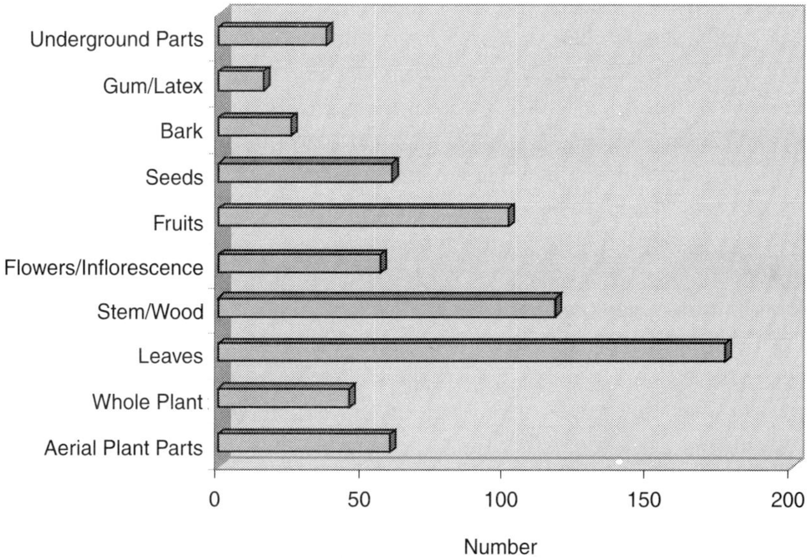

Fig. 11 Histogram Showing the Quantitative Reflection of Various Plant Parts Consumed by the Rural Populace of Tehsil Palampur.

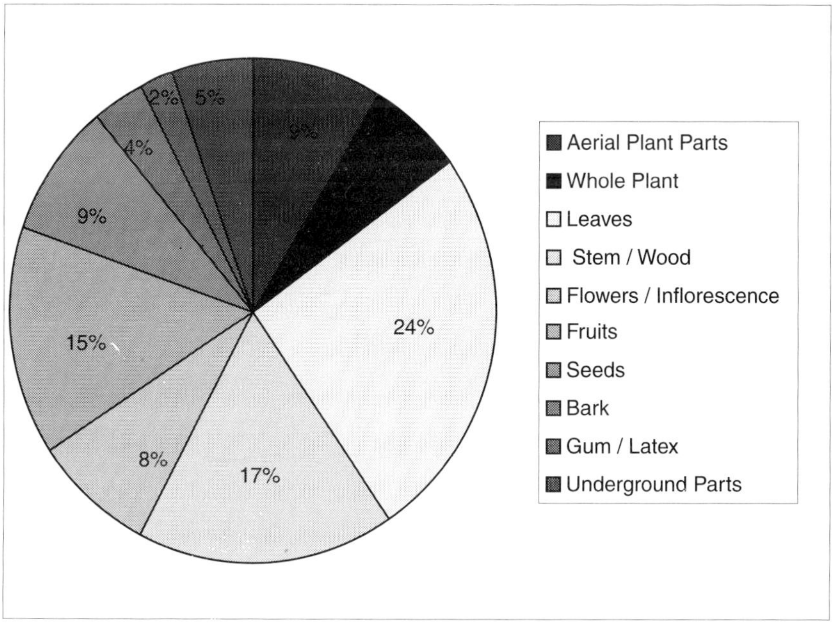

Fig. 12 Relative Per Cent Use of the Various Plant Parts by the Rural Populace of Tehsil Palampur.

Of the plants delineated presently, the floristic composition includes 5 pteridophytes (*Adiantum capillus veneris, Cheilanthes farinosa, Diplazium esculentum, Equisetum debile, Selaginella chrysocaulos*-1 genus each), 4 gymnospermous taxa (*Araucaria columnaris, Cedrus deodara, Pinus roxburghii, Thuja orientalis*-1 genus each) and 326 angiospermous taxa belonging to 245 genera under 80 families of which 43 spp. (37 genera, 11 families) are the monocots and 283 as dicots (208 genera, 69 families) (Figs. 4,7,8; Table IX) with predominance of polypetalous families (36), followed by gamopetalous (21) and monochlamydous (12) ones (Table X). Relatively, dicotyledonous spp. represent 84.48% of the total plants collected from the region with proportion of spp. belonging to monocots to dicots is 1: 6.58, of genera 1:6.62, and of families 1: 6.27. In overall perspective, dicots represent 86.25% of the total angiospermous families and 84.89% of the total angiospermous genera reported from the region. Phenologically, a large majority of the collected taxa flowers and fruits during summer (33%) and rainy (36%) seasons (Figs. 13, 14).

Table IX Families, Genera and Species of Ethnobotanically Important Plants of Tehsil Palampur Under Various Divisions of Plant Kingdom

S.No.	*Division*	*Total No. of Families*	*Genera*	*Species*
1.	Dicots	69	208	283
2.	Monocots	11	37	43
3.	Gymnosperms	03	04	04
4.	Pteridophytes	05	05	05

Table X Layout of Ethnobotanical Families of Palampur Tehsil as per Bentham & Hooker System of Classification (1862-1883)

Polypetalous Families	*Gamopetalous Families*	*Monochlamydous Families*	*Monocot Families*	*Gymnospermous Families*
Anacardiaceae	Acanthaceae	Chenopodiaceae	Amaryllidaceae	Araucariaceae
Apiaceae	Amaranthaceae	Euphorbiaceae	Araceae	Cupressaceae
Berberidaceae	Asclepiadaceae	Fagaceae	Arecaceae	Pinaceae
Brassicaceae	Apocynaceae	Juglandaceae	Commelinaceae	
Cactaceae	Asteraceae	Lauraceae	Cyperaceae	
Caricaceae	Bignoniaceae	Loranthaceae	Dioscoreaceae	
Caryophyllaceae	Boraginaceae	Nyctaginaceae	Liliaceae	
Celastraceae	Convolvulaceae	Polygonaceae	Musaceae	
Combretaceae	Ebenaceae	Proteaceae	Orchidaceae	
Crassulaceae	Ericaceae	Salicaceae	Poaceae	
Cucurbitaceae	Gesneriaceae	Santalaceae	Zingiberaceae	
Fabaceae	Hypericaceae	Urticaceae		
Flacourtiaceae	Lamiaceae			
Fumariaceae	Oleaceae			

Contd...

Table X Contd...

Polypetalous Families	*Gamopetalous Families*	*Monochlamydous Families*	*Monocot Families*	*Gymnosperm-ous Families*
Geraniaceae	Pedaliaceae			
Linaceae	Plantaginaceae			
Lythraceae	Plumbaginaceae			
Magnoliaceae	Scrophulariaceae			
Malvaceae	Rubiaceae			
Meliaceae	Solanaceae			
Menispermaceae	Verbenaceae			
Moringaceae				
Myrtaceae				
Portulacaceae				
Punicaceae				
Ranunculaceae				
Rhamnaceae				
Rosaceae				
Rutaceae				
Sapindaceae				
Saxifragaceae				
Sterculiaceae				
Theaceae				
Tiliaceae				
Violaceae				
Vitaceae				

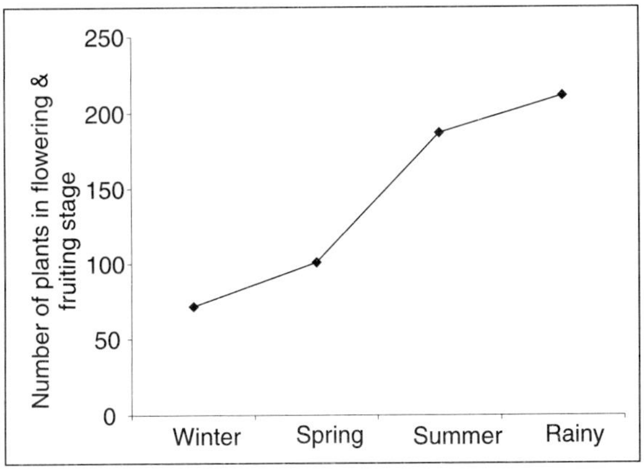

Fig. 13 Flowering and Fruiting Seasons of the Ethnobotanical Plants Used by the Rural Populace of Tehsil Palampur, District Kangra (H.P.).

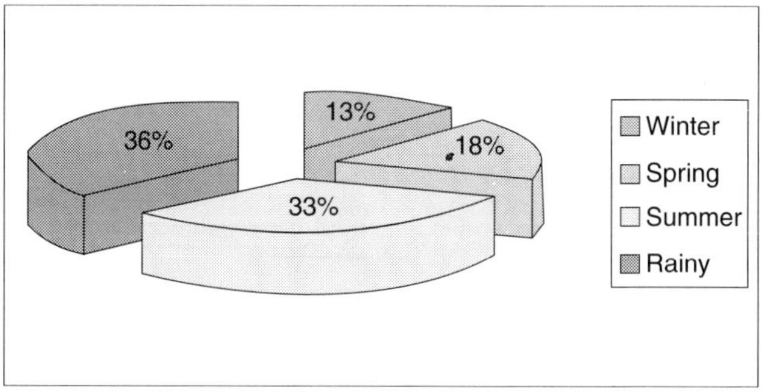

Fig.14 Pie Diagram Showing Relative Percentage of Flowering and Fruiting Seasons of the Collected Ethnobotanical Plants (Present Study).

Family-wise analysis indicates that the maximum diversity of ethnobotanical plants is represented by Fabaceae (34 spp, 25 genera; 10.17%), Urticaceae (20 spp, 8 genera; 6.00%), Poaceae (20 spp, 19 genera; 6.00%), followed by Lamiaceae (15 spp, 11 genera; 4.49%), Rosaceae (14 spp, 7 genera; 4.48%), Cucurbitaceae (13 spp, 9 genera; 3.89%), Asteraceae (11 spp, 10 genera; 3.29%), Solanaceae (10 spp, 5 genera; 2.99%), Rutaceae (9 spp, 4 genera; 2.69%), Euphorbiaceae (8 spp, 7 genera; 2.40%), Brassicaceae (7 spp, 6 genera; 2.09%), Amaranthaceae (7 spp, 5 genera; 2.09%), Lauraceae (7 spp, 4 genera; 2.09%), Liliaceae (6 spp, 5 genera; 1.79%) and Polygonaceae (6 spp, 4 genera; 1.79%) (Figs. 15, 16; Tables XI-XIII). Appraisal of pooled data in Tables XI-XIII further revealed that out of 88 families, 50 families are with 1 or 2 spp; 12 families comprise 3 spp; 7 families with 4 spp each; 4 families with 5 spp each; and 15 families represented by more than five spp. The top ten families, however, contribute to 45.97 % of the total taxa from the region (Tables XII, XIII). With regard to high used ethnobotanical genera (present study), *Ficus* with 12 spp showed highest spp diversity, followed by *Citrus, Solanum* (5 spp each), *Cassia, Jasminum, Ocimum, Terminalia* (4 spp each), *Albizia, Prunus, Pyrus, Litsea, Ranunculus, Rosa, Grewia, Vicia* (3 spp each), etc. (Figs. 17, 18; Tables XIV, XV). Likewise, commonly employed monocots with maximum number of spp are those of *Dioscorea, Agave, Phoenix* and *Saccharum* (Fig. 18). Statistically, the fifteen dominating genera constitute 21.91% of the total taxa.

Table XI Respective Percentage Distribution of Taxa Under Predominant Ethnobotanical Families of Tehsil Palampur (H.P.)

Families	*Total Number of Spp*	*Total % in Taxa*
Fabaceae	34	10.17
Urticaceae	20	6.00
Poaceae	20	6.00
Lamiaceae	15	4.49

Contd...

Table XI Contd...

Families	*Total Number of Spp*	*Total % in Taxa*
Rosaceae	14	4.18
Cucurbitaceae	13	3.89
Asteraceae	11	3.29
Solanaceae	10	2.99
Rutaceae	9	2.69
Euphorbiaceae	8	2.40
Amaranthaceae	7	2.09
Brassicaceae	7	2.09
Lauraceae	7	2.09
Liliaceae	6	1.79
Polygonaceae	6	1.79
Apiaceae	5	1.50
Apocynaceae	5	1.50
Verbenaceae	5	1.50
Malvaceae	5	1.50
Combretaceae	4	1.20
Convolvulaceae	4	0.90
Menispermaceae	4	1.20
Myrtaceae	4	1.20
Oleaceae	4	1.20
Ranunculaceae	4	1.20
Salicaceae	4	1.20
Anacardiaceae	3	0.90
Araceae	3	0.90
Amaryllidaceae	3	0.90
Boraginaceae	3	0.90
Dioscoreaceae	3	0.90
Juglandaceae	3	0.90
Rhamnaceae	3	0.90
Rubiaceae	3	0.90
Sapindaceae	3	0.90
Tiliaceae	3	0.90
Chenopodiaceae	3	0.90
Geraniaceae	3	0.90
Acanthaceae	2	0.60

Contd...

Table XI Contd...

Families	*Total Number of Spp*	*Total % in Taxa*
Arecaceae	2	0.60
Asclepiadaceae	2	0.60
Caryophyllaceae	2	0.60
Celastraceae	2	0.60
Linaceae	2	0.60
Lythraceae	2	0.60
Meliaceae	2	0.60
Nyctaginaceae	2	0.60
Pinaceae	2	0.60
Violaceae	2	0.60
Vitaceae	2	0.60
Zingiberaceae	2	0.60
Araucariaceae	1	0.30
Athyriaceae	1	0.30
Berberidaceae	1	0.30
Bignoniaceae	1	0.30
Cactaceae	1	0.30
Caricaceae	1	0.30
Cheilanthaceae	1	0.30
Commelinaceae	1	0.30
Crassulaceae	1	0.30
Cupressaceae	1	0.30
Cyperaceae	1	0.30
Ebenaceae	1	0.30
Equisetaceae	1	0.30
Ericaceae	1	0.30
Fagaceae	1	0.30
Flacourtiaceae	1	0.30
Fumariaceae	1	0.30
Gesneriaceae	1	0.30
Hypericaceae	1	0.30
Loranthaceae	1	0.30
Magnoliaceae	1	0.30
Moringaceae	1	0.30

Contd...

Table XI Contd...

Families	*Total Number of Spp*	*Total % in Taxa*
Musaceae	1	0.30
Orchidaceae	1	0.30
Pedaliaceae	1	0.30
Plantaginaceae	1	0.30
Plumbaginaceae	1	0.30
Polypodiaceae	1	0.30
Portulacaceae	1	0.30
Proteaceae	1	0.30
Punicaceae	1	0.30
Santalaceae	1	0.30
Saxifragaceae	1	0.30
Scrophulariaceae	1	0.30
Selaginellaceae	1	0.30
Sterculiaceae	1	0.30
Theaceae	1	0.30

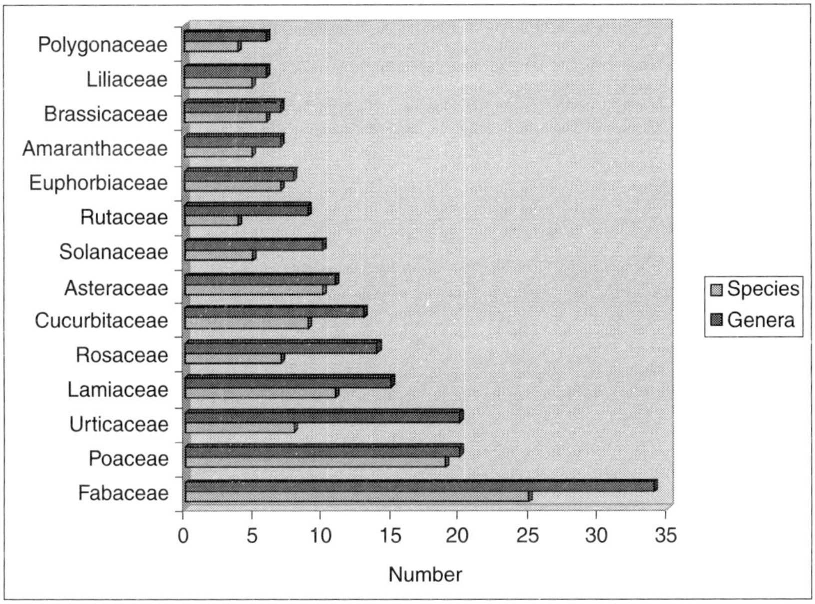

Fig. 15 Relative Disposition of Species and Genera Under Predominant Ethnobotanical Families of Tehsil Palampur.

Table XII Relative Disposition of Taxa Under Predominant Ethnobotanically Employed Dicot Families of Tehsil Palampur (H.P.)

Families	*Total Number of Spp*	*Total % in Taxa*
Fabaceae	34	10.17
Urticaceae	20	6.00
Lamiaceae	15	4.49
Rosaceae	14	4.18
Cucurbitaceae	13	3.89
Asteraceae	11	3.29
Solanaceae	10	2.99
Rutaceae	9	2.69
Euphorbiaceae	8	2.40
Amaranthaceae	7	2.09
Brassicaceae	7	2.09
Lauraceae	7	2.09
Polygonaceae	6	1.79
Apiaceae	5	1.50
Apocynaceae	5	1.50
Malvaceae	5	`1.50
Verbenaceae	5	1.50
Combretaceae	4	1.20
Menispermaceae	4	1.20
Myrtaceae	4	1.20
Oleaceae	4	1.20
Ranunculaceae	4	1.20
Salicaceae	4	1.20
Anacardiaceae	4	1.20
Convolvulaceae	3	0.90
Boraginaceae	3	0.90
Geraniaceae	3	0.90
Juglandaceae	3	0.90
Rhamnaceae	3	0.90
Rubiaceae	3	0.90
Sapindaceae	3	0.90
Tiliaceae	3	0.90
Acanthaceae	2	0.60
Asclepiadaceae	2	0.60

Contd...

Table XII Contd...

Families	*Total Number of Spp*	*Total % in Taxa*
Caryophyllaceae	2	0.60
Celastraceae	2	0.60
Chenopodiaceae	3	0.90
Linaceae	2	0.60
Lythraceae	2	0.60
Meliaceae	2	0.60
Nyctaginaceae	2	0.60
Violaceae	2	0.60
Vitaceae	2	0.60

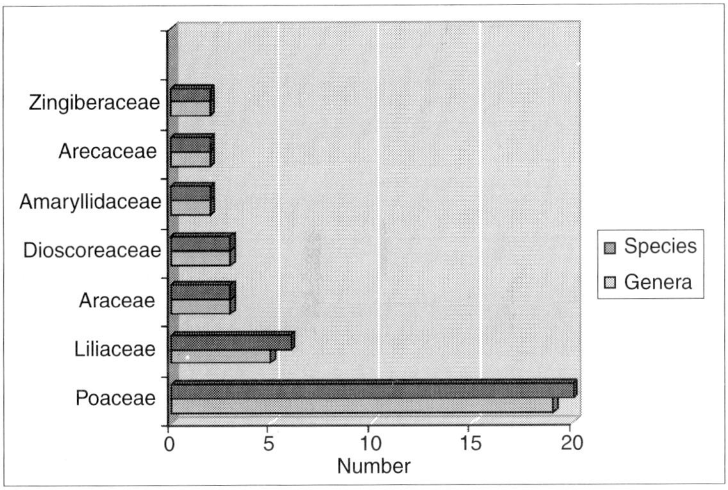

Fig. 16 Predominant Ethnobotanical Monocot Families of Tehsil Palampur.

Table XIII Relative Percentage Disposition of Taxa Under Predominant Ethnobotanical Monocot Families of Tehsil Palampur

Family	*Spp Used*	*Percentage in the Taxa*
Poaceae	20	5.97
Liliaceae	06	1.79
Araceae	03	0.90
Dioscoreaceae	03	0.90
Amaryllidaceae	03	0.90
Arecaceae	02	0.60
Zingiberaceae	02	0.60

Table XIV Predominant Ethnobotanical Dicot Genera of Tehsil Palampur

Genera	*Families*	*No. of Spp*
Ficus	Urticaceae	12
Citrus	Rutaceae	05
Solanum	Solanaceae	05
Cassia	Fabaceae	04
Jasminum	Oleaceae	04
Ocimum	Lamiaceae	04
Terminalia	Combretaceae	04
Albizia	Fabaceae	03
Grewia	Tiliaceae	03
Litsea	Lauraceae	03
Prunus	Rosaceae	03
Pyrus	Rosaceae	03
Rosa	Rosaceae	03
Vicia	Fabaceae	03
Achyranthes	Amaranthaceae	02
Artemisia	Asteraceae	02
Bauhinia	Fabaceae	02
Brassica	Brassicaceae	02
Chenopodium	Chenopodiaceae	02
Cinnamomum	Lauraceae	02
Cucumis	Cucurbitaceae	02
Cucurbita	Cucurbitaceae	02
Datura	Solanaceae	02
Ehretia	Boraginaceae	02
Fagopyrum	Polygonaceae	02
Ipomoea	Convolvulaceae	02
Lathyrus	Fabaceae	02
Luffa	Cucurbitaceae	02
Mentha	Lamiaceae	02
Momordica	Cucurbitaceae	02
Murraya	Rutaceae	02
Populus	Salicaceae	02
Rubus	Rosaceae	02
Rumex	Polygonaceae	02

Contd...

Table XIV Contd...

Genera	*Families*	*No. of Spp*
Salix	Salicaceae	02
Sapium	Euphorbiaceae	02
Urtica	Urticaceae	02
Viola	Violaceae	02
Vitis	Vitaceae	02
Zizyphus	Rhamnaceae	02

Table XV Predominant Ethnobotanical Monocot Genera of Tehsil Palampur

Genera	*Families*	*No. of Species*
Dioscorea	Dioscoreaceae	03
Agave	Amaryllidaceae	02
Phoenix	Araceae	02
Saccharum	Poaceae	02

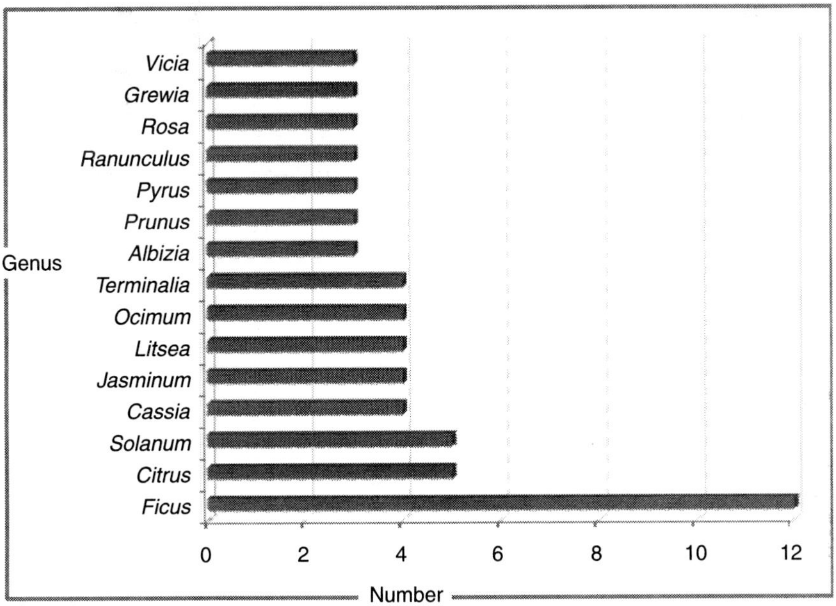

Fig. 17 Predominant Ethnobotanical Dicot Genera of Tehsil Palampur.

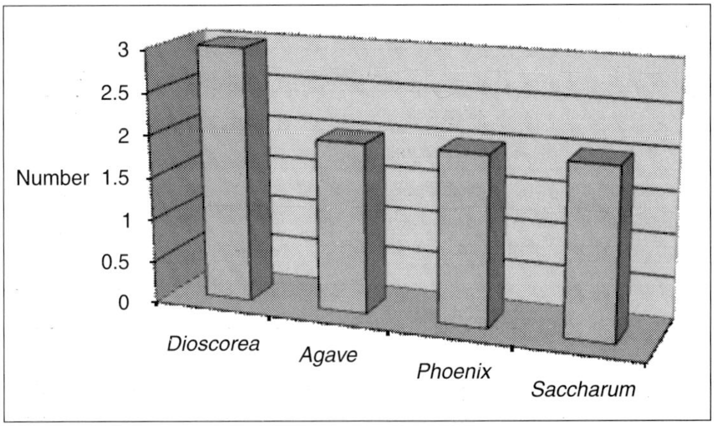

Fig. 18 Predominant Ethnobotanical Monocot Genera of Tehsil Palampur.

For supplementing their diet the local populace use a large number of edible plants (166 spp, 117 genera under 55 families) (Table XVI). Generally, the food habits are developed according to the availability of food and their nutritional value. Based on consumption pattern, these edibles can be categorised into various groups like **fruits** (94 spp: *Abelmoschus esculentus, Aegle marmelos, Artocarpus lacucha, Berberis lycium, Bombax ceiba, Callicarpa macrophylla, Capsicum annuum, Carica papaya, Carissa congesta, Carya illinoensis, Cassia floribunda, Cassia occidentalis, Celastrus paniculatus, Citrus grandis, Citrus jambhiri, Citrus limon, Citrus medica, Citrus medica* var. *medica, Coccinia grandis, Cordia dichotoma, Coriandrum sativum, Cucumis melo* var. *agrestis, Cucumis sativus, Cucurbita hispida, Cucurbita pepo, Cyclanthera pedata, Cyphomandra betacea, Debregeasia hypoleuca, Diospyros kaki, Eleusine coracana, Emblica officinalis, Eriobotrya japonica, Ficus benghalensis, Ficus carica, Ficus clavata, Ficus foveolata, Ficus hispida, Ficus palmata, Ficus roxburghii, Ficus semicordata, Flacourtia indica, Foeniculum vulgare, Fragaria indica, Grewia disperma, Grewia elastica, Grewia optiva, Juglans regia, Litchi chinensis, Litsea monopetala, Luffa acutangula, Luffa cylindrica, Mangifera indica, Momordica charantia, Momordica dioica, Moringa oleifera, Morus alba, Murraya koenigii, Murraya paniculata, Musa sapientum, Neolitsea umbrosa, Opuntia dillenii, Oroxylum indicum, Pennisetum americanum, Phoenix acaulis, Phoenix sylvestris, Prunus cerasoides, Prunus domestica, Prunus persica, Psidium guajava, Punica granatum, Pyrus communis, Pyrus pashia, Pyrus serotina, Randia dumetorum, Raphanus sativus, Rubus ellipticus, Rubus niveus, Sechium edule, Solanum melongena, Solanum nigrum, Solanum torvum, Spondias pinnata, Sorghum bicolor, Syzygium cumini, Syzygium operculatum, Terminalia arjuna, Terminalia bellirica, Terminalia chebula, Trichosanthes cucumerina, Vicia faba, Vitis repanda, Vitis vinifera, Zizyphus jujuba and Zizyphus nummularia*), **leaves** (29 spp: *Aloe vera, Artocarpus lacucha, Basella alba, Boerhaavia diffusa, Brassica nigra, Brassica tournefortii, Colocasia esculenta, Coriandrum sativum, Deeringia amaranthoides, Diplazium esculentum,*

Fagopyrum cymosum, Fagopyrum esculentum, Ficus palmata, Foeniculum vulgare, Gonatanthus pumilus, Lepidium sativum, Medicago polymorpha, Mentha longifolia, Mentha piperita, Morus alba, Murraya koenigii, Oxalis corniculata, Ranunculus muricatus, Raphanus sativus, Spondias pinnata, Trigonella foenum-graecum, Urtica ardens, Urtica dioica, Zanthoxylum armatum), **aerial parts** (21 spp: *Allium cepa, Allium sativum, Alternanthera sessilis, Amaranthus paniculatus, Amaranthus viridis, Apium leptophyllum, Chenopodium ambrosioides, Chenopodium album, Commelina paludosa, Coronopus didymus, Galium aparine, Lathyrus aphaca, Portulaca grandiflora, Rorippa nasturtium-aquaticum, Rumex hastatus, Rumex nepalensis, Scandix pecten veneris, Silene conoidea, Stellaria media, Vicia hirsuta, Vicia sativa*), **seeds** (17 spp: *Bauhinia vahlii, Brassica nigra, Brassica tournefortii, Butea monosperma, Cajanus cajan, Cassia obtusifolia, Cicer arietinum, Dolichos biflorus, Glycine max, Lathyrus aphaca, Lathyrus sativus, Linum usitatissimum, Pinus roxburghii, Sesamum orientale, Trigonella foenum-graecum, Vicia sativa, Vigna radiata*), **flowers/inflorescence** (13 spp: *Agave cantula, Agave wightii, Artocarpus lacucha, Bauhinia variegata, Bergenia ciliata, Bombax ceiba, Cordia dichotoma, Indigofera cassioides, Ipomoea turbinata, Moringa oleifera, Ougeinia dalbergioides, Reinwardtia indica, Rhododendron arboreum*), **underground parts** (11 spp: *Allium cepa, Allium sativum, Colocasia esculenta, Dioscorea bellophylla, Dioscorea bulbifera, Dioscorea oppositifolia, Gonatanthus pumilus, Ipomoea batatas, Raphanus sativus, Tulipa stellata, Zephyranthes carinata*), and **nuts** (2 spp: *Carya illinoensis, Juglans regia*) (Fig.19). For edibility, as many as 84 spp belonging to 63 genera are being used in this part of the Himalayas as pot herbs, 29 spp for pickling, 19 spp for culinaries, 14 spp for chutney, 10 spp for beverages (Table XVIII), 11 spp for flavouring (Table XVII), 7 spp for jams, jellies and sweet puddings, 3 spp for chapaati, 2 spp each for soups and salads etc. (Appendix III). Nevertheless, the maximum diversity of edibles is represented by Fabaceae (19 spp, 13 genera) followed by Urticaceae (13 spp, 5 genera), Cucurbitaceae (12 spp, 8 genera), Rosaceae (9 spp, 4 genera) Rutaceae (8 spp, 3genera), Poaceae (6 spp, 6 genera), Brassicaceae (5 spp, 4 genera) and Solanaceae (5 spp, 3 genera) with top 5 families accounting for 36.75% of the total edible taxa from the region.

Table XVI Edible Plants of Tehsil Palampur

S.No.	*Name*	*Edible Part/s*
1.	*Abelmoschus esculentus*	Fruits
2.	*Aegle marmelos*	Fruits
3.	*Agave cantula**	Peduncle
4.	*Agave wightii*	Peduncle
5.	*Allium cepa*	Aerial Parts, Bulbs
6.	*Allium sativum*	Aerial Parts, Bulbs
7.	*Aloe vera**	Leaves

Contd...

Table XVI Contd...

S.No.	Name	Edible Part/s
8.	*Alternanthera sessilis*	Aerial Parts
9.	*Amaranthus paniculatus*	Aerial Parts
10.	*Amaranthus viridis*	Aerial Parts
11.	*Apium leptophyllum**	Aerial Parts
12.	*Artocarpus lacucha**	Leaves, Flower Buds, Fruits
13.	*Basella alba*	Leaves
14.	*Bauhinia vahlii*	Seeds
15.	*Bauhinia variegata*	Flower Buds, Flowers
16.	*Berberis lycium*	Fruits
17.	*Bergenia ciliata*	Flowers
18.	*Boerhaavia diffusa*	Leaves
19.	*Bombax ceiba*	Flower Buds, Fruits
20.	*Brassica nigra*	Leaves, Seeds
21.	*Brassica tournefortii*	Leaves, Seeds
22.	*Butea monosperma*	Seeds
23.	*Cajanus cajan*	Seeds
24.	*Callicarpa macrophylla*	Fruits
25.	*Capsicum annuum*	Fruits
26.	*Carica papaya*	Fruits
27.	*Carissa congesta*	Fruits
28.	*Carya illinoensis*†	Nut
29.	*Cassia floribunda*	Pods
30.	*Cassia obtusifolia*	Seeds
31.	*Cassia occidentalis*	Pods
32.	*Celastrus paniculatus*	Fruits, Seeds
33.	*Chenopodium album*	Aerial Plant Parts
34.	*Chenopodium ambrosioides*	Aerial Plant Parts
35.	*Cicer arietinum*	Seeds
36.	*Citrus grandis*	Fruits
37.	*Citrus jambhiri*	Fruits
38.	*Citrus limon*	Fruits
39.	*Citrus medica*	Fruits
40.	*Citrus medica* var. *medica*	Fruits
41.	*Coccinia grandis*	Fruits
42.	*Colocasia esculenta*	Corm, Leaves
43.	*Commelina paludosa*	Aerial Plant Parts

Contd...

Table XVI Contd...

S.No.	Name	Edible Part/s
44.	*Cordia dichotoma*	Fruits, Flower buds
45.	*Coriandrum sativum*	Leaves, Fruits
46.	*Coronopus didymus*	Aerial Plant Parts
47.	*Cucumis melo* var. *agrestis*	Fruits
48.	*Cucumis sativus*	Fruits
49.	*Cyclanthera pedata*	Fruits
50.	*Cucurbita hispida**	Fruits
51.	*Cucurbita pepo**	Fruits
52.	*Cyphomandra betacea*	Fruits
53.	*Debregeasia hypoleuca*	Fruits
54.	*Deeringia amaranthoides*	Leaves
55.	*Dendrocalamus hamiltonii*	Young Culms
56.	*Dioscorea bellophylla*	Root Tubers
57.	*Dioscorea bulbifera*	Root Tubers
58.	*Dioscorea oppositifolia*	Root Tubers
59.	*Diospyros kaki*	Fruits
60.	*Diplazium esculentum*	Young Fronds
61.	*Dolichos biflorus*	Seeds
62.	*Eleusine coracana*	Grains
63.	*Emblica officinalis*	Fruits
64.	*Eriobotrya japonica*	Fruits
65.	*Euphorbia royleana*	Stem
66.	*Fagopyrum cymosum*	Leaves
67.	*Fagopyrum esculentum*	Leaves
68.	*Ficus benghalensis*	Fruits
69.	*Ficus carica*	Fruits
70.	*Ficus clavata*	Fruits
71.	*Ficus foveolata*	Fruits
72.	*Ficus hispida*	Fruits
73.	*Ficus palmata*	Leaves, Fruits
74.	*Ficus roxburghii*	Fruits
75.	*Ficus semicordata*	Fruits
76.	*Flacourtia indica*	Fruits
77.	*Foeniculum vulgare*	Leaves, Fruits
78.	*Fragaria indica*	Fruits
79.	*Galium aparine*	Aerial Plant Parts

Contd...

Table XVI Contd...

S.No.	Name	Edible Part/s
80.	*Gonatanthus pumilus**	Leaves, Tubers
81.	*Glycine max*	Seeds
82.	*Grewia disperma*	Fruits
83.	*Grewia elastic*	Fruits
84.	*Grewia optiva*	Fruits
85.	*Indigofera cassioides**	Flowers
86.	*Ipomoea batatas*	Root Tubers
87.	*Ipomoea turbinata*	Pedicels
88.	*Juglans regia*	Nut
89.	*Lathyrus aphaca*	Aerial Plant Parts, Seeds
90.	*Lathyrus sativus*	Seeds
91.	*Lepidium sativum*	Leaves
92.	*Linum usitatissimum*	Seeds
93.	*Litchi chinensis*	Aril
94.	*Litsea monopetala*	Fruits
95.	*Luffa acutangula*	Fruits
96.	*Luffa cylindrica*	Fruits
97.	*Mangifera indica*	Fruits
98.	*Medicago polymorpha*	Leaves
99.	*Mentha longifolia*	Leaves
100.	*Mentha piperita*	Leaves
101.	*Momordica charantia*	Fruits
102.	*Momordica dioica*	Fruits
103.	*Moringa oleifera*	Flower Buds, Pods
104.	*Morus alba**	Leaves, Fruits
105.	*Murraya koenigii*	Leaves, Fruits
106.	*Murraya paniculata**	Fruits
107.	*Musa sapientum*	Fruits
108.	*Neolitsea umbrosa*	Fruits
109.	*Opuntia dillenii*	Fruits
110.	*Oroxylum indicum*	Pods
111.	*Ougeinia dalbergioides*	Flowers
112.	*Oxalis corniculata*	Leaves
113.	*Pennisetum americanum*	Grains
114.	*Phoenix acaulis*	Fruits, Stem
115.	*Phoenix sylvestris*	Fruits, Stem

Contd...

Table XVI Contd...

S.No.	Name	Edible Part/s
116.	*Pinus roxburghii*	Seeds
117.	*Portulaca grandiflora*	Aerial Plant Parts
118.	*Prunus cerasoides*	Fruits
119.	*Prunus domestica*	Fruits
120.	*Prunus persica*	Fruits
121.	*Psidium guajava*	Fruits
122.	*Punica granatum*	Fruits
123.	*Pyrus communis*	Fruits
124.	*Pyrus pashia*	Fruits
125.	*Pyrus serotina*	Fruits
126.	*Randia dumetorum*	Fruits
127.	*Ranunculus muricatus*	Leaves
128.	*Raphanus sativus*	Root, Leaves, Fruits
129.	*Reinwardtia indica*	Flowers
130.	*Rhododendron arboreum*	Flowers
131.	*Rorippa nasturtium-aquaticum*	Aerial Plant Parts
132.	*Rubus ellipticus*	Fruits
133.	*Rubus niveus*	Fruits
134.	*Rumex hastatus*	Aerial Plant Parts
135.	*Rumex nepalensis*	Aerial Plant Parts
136.	*Saccharum officinarum*	Stem
137.	*Scandix pecten veneris*	Aerial Plant Parts
138.	*Sechium edule*	Fruits
139.	*Sesamum orientale*	Seeds
140.	*Silene conoidea*	Aerial Plant Parts
141.	*Solanum melongena*	Fruits
142.	*Solanum nigrum*	Fruits
143.	*Solanum torvum*	Fruits
144.	*Spondias pinnata*	Fruits, Leaves
145.	*Sorghum bicolor*	Fruits
146.	*Stellaria media*	Aerial Plant Parts
147.	*Syzygium cumini*	Fruits
148.	*Syzygium operculatum*	Fruits
149.	*Terminalia arjuna*	Fruits
150.	*Terminalia bellirica*	Fruits
151.	*Terminalia chebula*	Fruits

Contd...

Table XVI Contd...

S.No.	Name	Edible Part/s
152.	*Trichosanthes cucumerina*	Fruits
153.	*Trigonella foenum-graecum*	Leaves, Seeds
154.	*Tulipa stellata*	Bulbs
155.	*Urtica ardens*	Leaves
156.	*Urtica dioica*	Leaves
157.	*Vicia faba*	Pods
158.	*Vicia hirsuta*	Aerial Plant Parts
159.	*Vicia sativa*	Aerial Plant Parts, Seeds
160.	*Vigna radiata*	Seeds
161.	*Vitis repanda*	Fruits
162.	*Vitis vinifera*	Fruits
163.	*Zanthoxylum armatum*	Leaves
164.	*Zephyranthes carinata*	Bulbs
165.	*Zizyphus jujuba*	Fruits
166.	*Zizyphus nummularia*	Fruits

* Additional Uses

† New Record

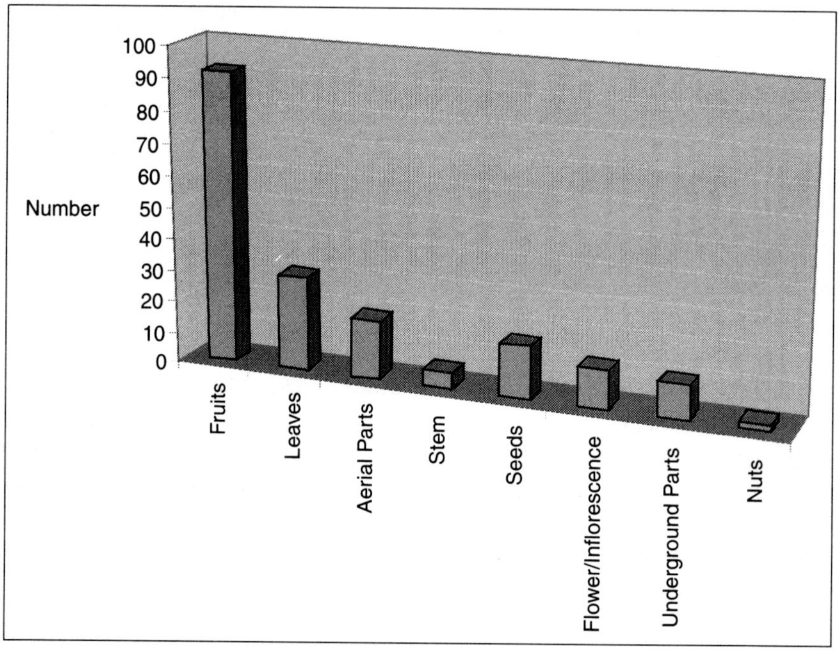

Fig. 19 Histogram Showing Relative Number of Plant Parts Used for Edible Purposes by the Rural Populace of Tehsil Palampur.

Table XVII Flavouring Plants of Tehsil Palampur

Plant Species	*Family*	*Part/s Used*
Allium sativum	Liliaceae	Bulbs
*Artemisia capillaris**	Asteraceae	Aerial Plant Parts
Capsicum annuum	Solanaceae	Fruits
*Cinnamomum tamala**	Lauraceae	Leaves, Stem, Bark
Coriandrum sativum	Apiaceae	Aerial Plant Parts, Seeds
Foeniculum vulgare	Apiaceae	Aerial Plant Parts, Seeds
Murraya koenigii	Rutaceae	Leaves
Murraya paniculata	Rutaceae	Leaves
Ocimum basilicum	Lamiaceae	Leaves, Seeds
*Ocimum kilimandscharicum**	Lamiaceae	Leaves, Seeds
Origanum vulgare	Lamiaceae	Aerial Plant Parts

* Additional Uses

Table XVIII Beverage Plants of Tehsil Palampur

Plant Species	*Family*	*Part Used*
*Achyranthes bidendata**	Amaranthaceae	Plant
*Albizia chinensis**	Fabaceae	Roots
*Callicarpa macrophylla**	Verbenaceae	Inflorescence
Camellia sinensis	Theaceae	Leaf buds
Celastrus paniculatus	Celastraceae	Fruits
Cymbopogon citratus	Poaceae	Stem
Justicia adhatoda	Acanthaceae	Flowers
*Panicum antidontale**	Poaceae	Roots
*Woodfordia fruticosa**	Lythraceae	Flowers
Zizyphus nummularia	Rhamnaceae	Roots

* Additional Uses

Of the total edible diversity recorded presently, notably 17 new uses of as many as 10 plant species hitherto unknown to the science of ethnobotany have been recorded from the region thereby extending the base of our food security system. All these plants hold great promises for shaping the future of mankind. Interestingly, 20 plant species are rich sources of carbohydrates (*Aegle marmelos, Amaranthus paniculatus, Bauhinia variegata, Colocasia esculenta, Dioscorea oppositifolia, Ficus carica, F. benghalensis, F. glomerata, F. palmata, F. reliogiosa, F. semicordata, Fagopyrum esculenta, Flacourtia indica, Ipomaea batatas, Lathyrus sativus, Mangifera indica, Morus alba, Musa paradisiaca, Opuntia dillenii, Phoenix sylvestris*), 8 in proteins (*Cannabis sativa, Cassia obtusifolia, Chenopodium album, Fagopyrum esculentum,*

Lathyrus sativus, Morus alba, Perilla frutescence, Urtica dioica), 4 in fats (*Cannabis sativa, Chenopodium album, Ficus palmata, Juglans* regia), 11 in calcium (*Amaranthus viridis, Ficus roxburghii, Flacourtia indica, Juglans regia, Moringa oleifera, Morus alba, Murraya koenigii, Oxalis corrniculata, Punica granatum, Rorippa nasturtium-aquaticum, Terminalia bellirica*), 7 in phosphorus (*Amaranthus paniculatus, Amaranthus viridis, Juglans regia, Moringa oleifera, Morus alba, Punica granatum, Rorippa nasturtium-aquaticum*) and 3 in iron (*Cannabis sativa, Moringa oleifera, Solanum torvum*) (Tables XIX, XX; Arora, 1981, 1985a,b, 1991, 1997; Arora & Pandey, 1996; Singh & Arora, 1978; Vaidya & Dhumal, 2000, 2001). Use of such a large diversity helps the local populace to overcome their nutritional deficiencies.

Table XIX Nutritive Subsidiary Food Plants of Tehsil Palampur, District Kangra (H.P.)

Carbohydrate Rich Resources	*Protein Rich Resources*	*Fat Rich Resources*
Aegle marmelos	*Cannabis sativa*	*Cannabis sativa*
Amaranthus paniculatus	*Cassia obtusifolia*	*Chenopodium album*
Bauhinia variegata	*Chenopodium album*	*Ficus palmata*
Colocasia esculenta	*Fagopyrum esculentum*	*Juglans regia*
Dioscorea oppositifolia	*Lathyrus sativus*	
Ficus carica	*Morus alba*	
F. benghalensis	*Perilla frutescens*	
F. glomerata	*Urtica dioica*	
F. palmata		
F. reliogiosa		
F. semicordata		
Fagopyrum esculentum		
Flacourtia indica		
Ipomoea batatas		
Lathyrus sativus		
Mangifera indica		
Morus alba		
Musa paradisiaca		
Opuntia dillenii		
Phoenix sylvestris		

For healthcare management, these people (rural populace of tehsil Palampur) use as many as 224 plant spp belonging to 183 genera under 76 families, singly (209) or in combination (61) against 48 types of ailments prevailing in the region. As a matter of fact, the locals use maximum number of plants for **gastrointestinal disorders** (45 spp)

Table XX Nutritive Subsidiary Food Plants of Tehsil Palampur

Calcium Rich Resources	*Phosphorus Rich Resources*	*Iron Rich Resources*
Amaranthus viridis	*Amaranthus paniculatus*	*Ficus roxburghii*
Ficus roxburghii	*Amaranthus viridis*	*Moringa oleifera*
Flacourtia indica	*Juglans regia*	*Solanum torvum*
Juglans regia	*Moringa oleifera*	
Moringa oleifera	*Morus alba*	
Morus alba	*Punica granatum*	
Murraya koenigii	*Rorippa nasturtium-aquaticum*	
Oxalis corniculata		
Punica granatum		
Rorippa nasturtium-aquaticum		
Terminalia bellirica		

viz., *Aegle marmelos, Albizia lebbeck, Allium sativum, Aloe vera, Apium leptophyllum, Artemisia capillaris, Asparagus adscendens, Bauhinia variegata, Boerhaavia diffusa, Cannabis sativa, Carica papaya, Cassia fistula, Centella asiatica, Centratherum anthelminticum, Cinnamomum tamala, Didymocarpus pedicellata, Euphorbia royleana, Ficus benghalensis, Ficus glomerata, Flacourtia indica, Foeniculum vulgare, Indigofera cassioides, Mallotus philippinensis, Mentha longifolia, Mentha piperita, Micromeria biflora, Moringa oleifera, Ocimum basilicum, Ocimum gratissimum, Ocimum sanctum, Origanum vulgare, Pentanema indicum, Perilla frutescens, Rosa chinensis, Rosa damascena, Selaginella chrysocaulos, Solanum viarum, Terminalia bellirica, Terminalia chebula, Tinospora cordifolia, Toona ciliata, Trigonella foenum-graecum, Vitex negundo, Woodfordia fruticosa, Zizyphus jujuba* followed by **dental problems** (32 spp: *Albizia lebbeck , Boehmeria macrophylla, Butea monosperma, Calotropis procera, Carya illinoensis, Cassia occidentalis, Centipeda minima, Cinnamomum tamala, Citrus limon, Citrus medica, Clematis gauriana, Engelhardia colebrookeana, Ficus benghalensis, Heteropogon contortus, Jasminum grandiflorum, Juglans regia, Lantana camara, Lepidagathis cuspidata, Mangifera indica, Murraya koenigii, Murraya paniculata, Ocimum basilicum, Plumbago zeylanica, Prunus cerasoides, Psidium guajava, Ricinus communis, Sesamum orientale, Solanum viarum, Spilanthes oleracea, Terminalia chebula, Vitex negundo, Zanthoxylum armatum*), **lung diseases** (31 spp: *Acacia catechu, Achyranthes aspera, Acorus calamus, Bombax ceiba, Cassia obtusifolia, Cassia occidentalis, Cinnamomum camphora, Cinnamomum tamala, Citrus limon, Cordia dichotoma, Curcuma longa, Datura stramonium, Ficus palmata, Ficus religiosa, Justicia adhatoda, Lawsonia inermis, Lepidium sativum, Micromeria biflora, Momordica charantia, Ocimum kilimandscharicum, Ocimum sanctum, Pistacia khinjuk, Pogostemon benghalense, Punica granatum, Randia dumetorum, Solanum viarum,*

Terminalia bellirica, Terminalia chebula, Vanda tessellata, Viola patrinii, Viola pilosa), **boils** (31 spp: *Ajuga parviflora, Albizia chinensis, Allium cepa, Aloe vera, Alternanthera sessilis, Berberis lycium, Bombax ceiba, Kalanchoe pinnata, Butea monosperma, Centella asiatica, Costus speciosus, Curcuma longa, Cymbopogon citratus, Dalbergia sissoo, Dodonaea viscosa, Euphorbia royleana, Ficus hispida, Ficus roxburghii, Hibiscus rosa sinensis, Jasminum multiflorum, Mirabilis jalapa, Plantago lanceolata, Plumbago zeylanica, Prinsepia utilis, Ranunculus arvensis, Ranunculus laetus, Ricinus communis, Syzygium cumini, Triticum aestivum, Verbascum thapsus, Zizyphus nummularia*), **refrigerant** (30 spp: *Aegle marmelos, Aloe vera, Caesalpinia decapetala, Centella asiatica, Cirsium arvense, Coriandrum sativum, Cucumis sativus, Cymbopogon citratus, Cynodon dactylon, Emblica officinalis, Euphorbia royleana, Holarrhena antidysentrica, Lawsonia inermis, Lepidagathis cuspidata, Morus alba, Musa sapientum, Phoenix acaulis, Phoenix sylvestris, Portulaca grandiflora, Rhododendron arboreum, Rubus niveus, Rumex hastatus, Saccharum officinarum, Santalum album, Triticum aestivum, Vitis repanda, Vitis vinifera, Zea mays, Zizyphus jujuba, Zizyphus nummularia*), **skin diseases** (26 spp: *Abrus precatorius, Aerva lanata, Ajuga parviflora, Allium sativum, Aloe vera, Butea monosperma, Carica papaya, Carissa congesta, Cassia fistula, Centratherum anthelminticum, Coriandrum sativum, Cucumis sativus, Ehretia laevis, Ficus benghalensis, Jatropha curcas, Mallotus philippinensis, Michelia champaca, Nerium indicum, Pinus roxburghii, Polygonum barbatum, Ranunculus laetus, Ricinus communis, Santalum album, Sapium insigne, Tagetes minuta, Toona ciliata*), **rheumatism** (22 spp: *Allium sativum, Artemisia capillaris, Basella alba, Boerhaavia diffusa, Calotropis procera, Cannabis sativa, Cinnamomum camphora, Citrus limon, Cuscuta reflexa, Datura innoxia, Eucalyptus tereticornis, Litsea sebifera, Michelia champaca, Murraya koenigii, Oroxylum indicum, Prinsepia utilis, Ricinus communis, Sesamum orientale, Spilanthes oleracea, Trigonella foenum-graecum, Urtica dioica, Vitex negundo*), **diabetes** (19 spp: *Ajuga parviflora, Carica papaya, Cassia obtusifolia, Catharanthus roseus, Clematis gauriana, Dodonaea viscosa, Eleusine coracana, Kalanchoe pinnata, Ficus palmata, Glycine max, Mangifera indica, Momordica charantia, Psidium guajava, Sechium edule, Syzygium cumini, Terminalia chebula, Thuja orientalis, Tinospora cordifolia, Trigonella foenum-graecum*), **liver complaints** (16 spp: *Aerva lanata, Aloe vera, Cajanus cajan, Carica papaya, Centella asiatica, Cuscuta reflexa, Emblica officinalis, Melia azedarach, Ocimum sanctum, Phyllanthus urinaria, Portulaca grandiflora, Raphanus sativus, Saccharum officinarum, Terminalia bellirica, Tinospora cordifolia, Urtica dioica*), **diuretic** (15 spp: *Achyranthes aspera, Aerva lanata, Amaranthus viridis, Butea monosperma, Cassia occidentalis, Chenopodium album, Cynodon dactylon, Didymocarpus pedicellata, Emblica officinalis, Grewia disperma, Grewia optiva, Heteropogon contortus, Phyllanthus urinaria, Portulaca grandiflora, Zea mays*), **kidney disorders** (15 spp: *Achyranthes aspera, Achyranthes bidentata, Aerva lanata, Bergenia ciliata, Butea monosperma, Catharanthus roseus, Cucumis sativus, Debregeasia hypoleuca,*

Dolichos biflorus, Kalanchoe pinnata, Lawsonia inermis, Portulaca grandiflora, Sesamum orientale, Sorghum bicolor, Urtica dioica), **constipation** (15 spp: *Aegle marmelos, Chenopodium album, Eleusine coracana, Euphorbia royleana, Ficus palmata, Mallotus philippinensis, Opuntia dillenii, Oroxylum indicum, Psidium guajava, Pyrus communis, Ricinus communis, Rosa damascena, Rubus ellipticus, Sorghum bicolor, Terminalia bellirica*), **mouth sores/ulcers** (14 spp: *Acacia catechu, Bidens biternata, Cajanus cajan, Callicarpa macrophylla, Cassia occidentalis, Musa sapientum, Plantago lanceolata, Psidium guajava, Pyrus pashia, Solanum nigrum, Spi!anthes oleracea, Syzygium cumini, Vitex negundo, Zanthoxylum armatum*), **body-tones** (Table XXI; 13 spp: *Coccinia grandis, Celastrus paniculatus, Dendrocalamus hamiltonii, Dioscorea bellophylla, Dioscorea oppositifolia, Ficus glomerata, Geranium nepalense, Gossypium arboreum, Linum usitatissimum, Mucuna pruriens, Oxalis corniculata, Sida cordifolia, Terminalia bellirica*), **heart complaints** (13 spp: *Allium sativum, Camellia sinensis, Cinnamomum tamala, Eleusine coracana, Momordica charantia, Pennisetum americanum, Portulaca grandiflora, Reinwardtia indica, Rhododendron arboreum, Solanum nigrum, Sorghum bicolor, Terminalia arjuna, Terminalia chebula*), **eye diseases** (12 spp: *Abrus precatorius, Berberis lycium, Boerhaavia diffusa, Butea monosperma, Coriandrum sativum, Lepidium sativum, Linum usitatissimum, Litsea sebifera, Mangifera indica, Oxalis corniculata, Rorippa nasturtium-aquaticum, Solanum erianthum*), **hair care** (12 spp: *Adiantum capillus veneris, Ajuga parviflora, Brassica tournefortii, Carya illinoensis, Emblica officinalis, Grewia disperma, Grewia optiva, Hibiscus rosa sinensis, Mangifera indica, Melia azedarach, Pyrus pashia, Sapindus mukorosii*), **astringent** (11 spp: *Aegle marmelos, Asparagus adscendens, Bauhinia variegata, Cordia dichotoma, Diospyros kaki, Foeniculum vulgare, Litchi chinensis, Opuntia dillenii, Pyrus communis, Rhododendron arboreum, Rosa brunonii*), **pains/aches** (11 spp): *Albizia julibrissin, Butea monosperma, Carissa congesta, Celastrus paniculatus, Cuscuta reflexa, Eucalyptus tereticornis, Justicia adhatoda, Moringa oleifera, Sida cordifolia, Tinospora cordifolia, Vitex negundo*), **blood purifier** (10 spp: *Berberis lycium, Brassica nigra, Brassica tournefortii, Carissa congesta, Centella asiatica, Plantago lanceolata, Moringa oleifera, Urtica ardens, Urtica dioica, Zanthoxylum armatum*), **headache** (10 spp: *Abrus precatorius, Centipeda minima, Justicia adhatoda, Lawsonia inermis, Melia azadarach, Michelia champaca, Randia dumetorum, Santalum album, Toona ciliata, Vitex negundo*), **piles** (10 spp: *Aegle marmelos, Butea monosperma, Dioscorea bellophylla, Dioscorea bulbifera, Holarrhena antidysentrica, Jatropha curcas, Momordica charantia, Pennisetum americanum, Punica granatum, Tagetes minuta*), **purgative** (10 spp: *Amaranthus viridis, Cassia floribunda, Cassia occidentalis, Chenopodium album, Emblica officinalis, Ficus carica, Jatropha curcas, Mirabilis jalapa, Terminalia bellirica, Trichosanthes cucumerina*), **wounds** (10 spp: *Agave cantula, Callicarpa macrophylla, Cannabis sativa, Cassia fistula, Commelina paludosa, Emblica officinalis, Jasminum dispermum, Jatropha curcas, Lannea coromandelica, Solanum erianthum*), **sprain** (9 spp: *Artemisia japonica, Cassia floribunda, Dodonaea viscosa, Lantana camara, Litsea sebifera, Murraya koenigii,*

Prunus cerasoides, Sesamum orientale, Vitex negundo), **injury** (8 spp: *Cannabis sativa, Colebrookea oppositifolia, Curcuma longa, Cynodon dactylon, Dodonaea viscosa, Justicia adhatoda, Reinwardtia indica, Solanum erianthum*), **leucorrhoea** (8 spp: *Capillipedium assimile, Cissampelos parriera, Cymbopogon citratus, Dalbergia sissoo, Hibiscus rosa sinensis, Opuntia dillenii, Rosa damascena, Woodfordia fruticosa*), **bone fractures** (7 spp: *Artemisia japonica, Cissampelos parriera, Dodonaea viscosa, Litsea sebifera, Oroxylum indicum, Pentanema indicum, Prunus cerasoides*), **burns** (7 spp: *Aloe vera, Caesalpinia decapetala, Chenopodium album, Dodonaea viscosa, Impatiens balsamina, Lawsonia inermis, Tagetes minuta*), **ear complaints** (7 spp: *Allium cepa, Allium sativum, Cheilanthes farinosa, Costus speciosus, Erythrina suberosa, Panicum antidontale, Tagetes minuta*, **anthelmintic** (6 spp: *Butea monosperma, Centratherum anthelminticum, Chenopodium album, Cordia dichotoma, Momordica charantia, Rumex nepalensis*), **febrifuge** (6 spp: *Apium leptophyllum, Dendrocalamus hamiltonii, Fumaria indica, Momordica charantia, Rumex hastatus, Tinospora cordifolia*), **internal injuries** (6 spp: *Curcuma longa, Eulaliopsis binata, Oroxylum indicum, Prunus cerasoides, Vanda tessellata, Vitex negundo*), **menstrual complaints** (6 spp: *Aloe vera, Amaranthus viridis, Boehmeria macrophylla, Butea monosperma, Holarrhena antidysentrica, Terminalia arjuna*), **antidote** (6 spp: *Achyranthes bidentata, Cannabis sativa, Capsicum annuum, Catharanthus roseus, Rumex nepalensis, Sapindus mukorosii*), **antiseptic** (5 spp: *Agave cantula, Berberis lycium, Chenopodium album, Cymbopogon citratus, Solanum viarum*), **measles** (4 spp: *Aerva lanata, Musa sapientum, Sisymbrium irio, Verbascum Thapsus*), **mouth odour** (4 spp: *Camellia sinensis, Citrus limon, Citrus medica, Zanthoxylum armatum*), **sore-toes** (4 spp: *Carya illinoensis, Cassia fistula, Dalbergia sissoo, Impatiens balsamina*), **antiemetic** (3 spp: *Allium cepa, Citrus grandis, Mentha longifolia*), **massages** (3 spp: *Acorus calamus, Brassica tournefortii, Prunus persica*), **anaemia** (2 spp: *Carissa congesta, Saccharum officinarum*), **fertility** (2 spp: *Diplocyclos palmatus, Mucuna pruriens*), **bed wetting** (2 spp: *Carya illinoensis, Sesamum orientale*, **herpes** (2 spp: *Aerva lanata, Ficus benghalensis*), **nasal problems** (2 spp: *Cynodon dactylon, Rhododendron arboreum*); **paralysis** (2 spp: *Abrus precatorius, Justicia adhatoda*). Only one taxa has been recommended for **swellings** (*Artemisia capillaris*) thereby indicating their less prevalence in the area (Appendix II). Various plant parts such as leaves, stem, flowers, fruits, seeds and underground parts are used orally for their health problems. Notably, top ranking families are: Fabaceae (20 spp), Lamiaceae (12 spp), Poaceae (12 spp), Asteraceae (10 spp), Rosaceae (10 spp), Urticaceae (9 spp), Euphorbiaceae (7 spp) and Rutaceae (6 spp). Of the information recorded presently, additional uses of 127 plants (marked with asterisk) and 4 hitherto unreported plants (marked [†)] for healthcare needs have been reported (Appendix II). It is needless to say that the generated data on perspective healthcare plants, if screened on scientific lines, will hopefully be useful to pharmaceuticals to establish their utility in the formation of new drugs for alleviating sufferings of mankind and usher in the Pink Revolution.

Table XXI Body Tones of Tehsil Palampur

Plant Species	*Family*	*Part Used*
Coccinia grandis	Cucurbitaceae	Fruits
Celastrus paniculatus	Celastraceae	Fruits
*Dendrocalamus hamiltoni**	Poaceae	Culm
Dioscorea bellophylla	Dioscoreaceae	Root tubers
Dioscorea oppositifolia	Dioscoreaceae	Root tubers
Ficus glomerata	Urticaceae	Latex
*Geranium nepalense**	Geraniaceae	Aerial Parts
Gossypium arboreum	Malvaceae	Seeds
Linum usitatissimum	Linaceae	Seeds
Mucuna pruriens	Fabaceae	Seeds
Oxalis corniculata	Geraniaceae	Aerial Parts
Sida cordifolia	Malvaceae	Seeds
Terminalia bellirica	Combretaceae	Seeds

* Additional Uses

For the well-being of their animals, the traditional veterinary herbal remedies highlight the use of as many as 41 plants spp under 40 genera and 27 families for prevailing 21 ailments in the region. In this regard, plants used predominantly are for **digestive problems** (11 spp, 11 genera: *Apium leptophyllum, Bergenia ciliata, Bombax ceiba, Cajanus cajan, Drimia indica, Cissampelos parriera, Leucas cephalotes, Origanum vulgare, Roylea cinerea, Verbascum thapsus, Wrightia arborea*) followed by **lactation** (10 spp, 9 genera: *Acacia catechu, Asparagus adscendens, Cirsium arvense, Cryptolepis buchanani, Grewia optiva, Helicteres isora, Phoenix acaulis, Phoenix sylvestris, Quercus leucotrichophora, Tinospora cordifolia*), **rumination** (8 spp, 8 genera: *Aerva lanata, Carica papaya, Equisetum debile, Grewia optiva, Leucas cephalotes, Pentanema indicum, Pogostemon benghalense, Syzygium cumini*), **fever** (4 spp, 4 genera: *Apium leptophyllum, Drimia indica, Jasminum grandiflorum, Pogostemon benghalense*), **suppuration of boils** (4 spp, 4 genera: *Ajuga parviflora, Curcuma longa, Roylea cinerea, Stephania glabra*), **abortion** (3 spp, 3 genera: *Curcuma longa, Euphorbia royleana, Ficus hispida*), **cough** (3 spp, 3 genera: *Drimia indica, Leucas cephalotes, Vanda tessellata*); **internal injuries** (2 spp, 2 genera: *Oroxylum indicum, Vanda tessellata*), **fracture** (2 spp, 2 genera: *Litsea sebifera, Oroxylum indicum*), **tail-infection** (2 spp, 2 genera: *Euphorbia royleana, Randia dumetorum*), **coolant** (2 spp, 2 genera: *Euphorbia royleana, Tinospora cordifolia*), **cutaneous diseases** (2 spp, 2 genera: *Ajuga parviflora, Stephania glabra*), blood-purifier (2 spp, 2 genera: *Stephania glabra, Roylea cinerea*), **prolapsed uterus** (1 sp: *Holarrhena antidysentrica*), **ovulation** (*Euphorbia royleana*), **urination** (*Drimia indica*), **opthalmic ailments** (*Albizia lebbeck*), **expelling flies** (*Artemisia japonica*) and **leeches** (*Bauhinia vahlii*), **hair-care** (*Brassica tournefortii*) and **killing lice** (*Roylea* *cinerea*). Statistically, appraisal of Table XXII reveals that high used families for this purpose are Lamiaceae (5 spp), Fabaceae (4 spp), Asteraceae and Menispermaceae (3 spp each), Arecaceae, Liliaceae and Apocynaceae (2 spp each). Interestingly, the veterinary families Menispermaceae, Arecaceae, Liliaceae and Apocynaceae do not figure among the largest 7 families for humans. Another significant feature of this data is the report of new uses of 31 plants (marked with asterisk) and new record of 1 plant (marked with†) for ethnoveterinary purposes from the region (Table XXII) whereas for other plants similar ethnoveterinary practices lend support to the claims and authenticity of earlier reports from other parts of the country. Furthermore, plants reported here can provide an opportunity of new drug research for human use also.

Table XXII Ethnoveterinary Plants of Tehsil Palampur

S.No.	Botanical Name	Family
1.	*Acacia catechu*	Fabaceae
2.	*Aerva lanata**	Amaranthaceae
3.	*Ajuga parviflora*†	Lamiaceae
4.	*Albizia lebbeck*	Fabaceae
5.	*Apium leptophyllum**	Apiaceae
6.	*Artemisia japonica**	Asteraceae
7.	*Asparagus adscendens*	Liliaceae
8.	*Bauhinia vahlii**	Fabaceae
9.	*Bergenia ciliata*	Saxifragaceae
10.	*Bombax ceiba**	Malvaceae
11.	*Brassica tournefortii**	Brassicaceae
12.	*Cajanus cajan**	Fabaceae
13.	*Carica papaya**	Caricaceae
14.	*Cirsium arvense**	Asteraceae
15.	*Cissampelos parriera*	Menispermaceae
16.	*Cryptolepis buchanani*	Asclepiadaceae
17.	*Curcuma longa**	Zingiberaceae
18.	*Drimia indica**	Liliaceae
19.	*Equisetum debile**	Equisetaceae
20.	*Euphorbia royleana**	Euphorbiaceae
21.	*Ficus hispida*	Urticaceae
22.	*Grewia optiva**	Tiliaceae
23.	*Helicteres isora**	Sterculiaceae
24.	*Holarrhena antidysentrica**	Apocynaceae
25.	*Jasminum grandiflorum*	Oleaceae
26.	*Leucas cephalotes**	Lamiaceae

Contd...

Table XXII Contd...

S.No.	Botanical Name	Family
27.	*Origanum vulgare**	Lamiaceae
28.	*Oroxylum indicum**	Bignoniaceae
29.	*Pentanema indicum**	Asteraceae
30.	*Phoenix acaulis**	Arecaceae
31.	*Phoenix sylvestris**	Arecaceae
32.	*Pogostemon benghalense**	Lamiaceae
33.	*Quercus leucotrichophora**	Fagaceae
34.	*Randia dumetorum**	Rubiaceae
35.	*Roylea cinerea**	Lamiaceae
36.	*Stephania glabra**	Menispermaceae
37.	*Syzygium cumini**	Myrtaceae
38.	*Tinospora cordifolia**	Menispermaceae
39.	*Vanda tessellata*	Orchidaceae
40.	*Verbascum thapsus**	Scrophulariaceae
41.	*Wrightia arborea**	Apocynaceae

* Additional Uses

† New Record

For cordage requirements, 10 plant spp (Table XXIII) belonging to Tiliaceae (*Grewia disperma, Gewia elastica Grewia optiva*), Malvaceae (*Abelmoschus esculentus, Bombax ceiba*), Fabaceae (*Bauhinia vahlii, Milletia extensa*), Amaryllidaceae (*Agave cantula*), Poaceae (*Eulaliopsis binata*), and Convolvulaceae (*Porana paniculata*) are mainly used in tehsil Palampur. Of these, the use of *Porana paniculata* is new addition to the existing record of cordages for India (Sood & Kaushal, 2008). Of the reported use of 500 natural dye-yielding plant species (Mahanta & Tiwari, 2005), the present study highlights the use of 7 spp belonging to 7 genera under 6 families for dye purposes (Table XXIV); and 48 spp belonging to 40 genera under 26 families with predominance of Fabaceae (9 spp), Tiliaceae, Combretaceae and Salicaceae (3 spp each) for fuelwood (Table XXV) by the local inhabitants. Undoubtedly, the resulted information on the germplasm materials / wild genetic resources hold great potential and required to be brought to the attention of Government to initiate and develop these into a cottage industry.

Table XXIII Fibre Yielding Plants of Tehsil Palampur

Plant Species	Family	Part/s Used
Abelmoschus esculentus	Malvaceae	Stem
Agave cantula	Amaryllidaceae	Leaves
Bauhinia vahlii	Fabaceae	Stem

Contd...

Table XXIII Contd...

Plant Species	*Family*	*Part/s Used*
Bombax ceiba	Malvaceae	Epidermal Seed Hairs
Eulaliopsis binata	Poaceae	Aerial Plant Parts
Grewia disperma	Tiliaceae	Stem
Grewia elastica	Tiliaceae	Stem
Grewia optiva	Tiliaceae	Stem
Milletia extensa	Fabaceae	Stem
*Porana paniculata**	Convolvulaceae	Stem

* Additional Uses

Table XXIV Dye Yielding Plants of Tehsil Palampur

Plant Spp	*Family*	*Part/s Used*
Bauhinia variegata	Fabaceae	Bark
Camellia sinensis	Theaceae	Leaves
Carya illinoensis†	Juglandaceae	Stem Bark, Fruits
Emblica officinalis	Euphorbiaceae	Stem Bark
*Impatiens balsamina**	Geraniaceae	Leaves
Juglans regia	Juglandaceae	Stem Bark, Leaves, Fruits
*Terminalia tomentosa**	Combretaceae	Stem Bark

* Additional Uses
† New Record

Table XXV Fuel Resources of Tehsil Palampur

S.No.	*Botanical Name*	*Local Name/s*	*Family*
1.	*Albizia chinensis*	Oyee	Fabaceae
2.	*Albizia julibrissin*	Karmarua	Fabaceae
3.	*Albizia lebbeck*	Sarin, Sirin	Fabaceae
4.	*Bauhinia variegata**	Karale, Karaalein	Fabaceae
5.	*Cajanus cajan**	Dhinger	Fabaceae
6.	*Carissa congesta*	Garna, Garnae	Apocynaceae
7.	*Caryopteris bicolor**	Ban-dhavein	Verbenaceae
8.	*Cassia fistula**	Alia, Amaltas	Fabaceae
9.	*Celtis australis**	Khidak, Khirak	Urticaceae
10.	*Colebrookea oppositifolia*	Doos	Lamiaceae
11.	*Cordia dichotoma**	Lasura, Lasyada	Boraginaceae
12.	*Dalbergia sissoo*	Shissham, Tahli	Fabaceae
13.	*Debregeasia hypoleuca**	Siaru, Syarru	Urticaceae
14.	*Dendrocalamus hamiltonii*	Maggar	Poaceae

Contd...

Table XXV Contd...

S.No.	Botanical Name	Local Name/s	Family
15.	*Dodonaea viscosa*	Mehendu, Mehendadu	Sapindaceae
16.	*Duranta repens**	Neel-Kanta	Verbenaceae
17.	*Ehretia laevis**	Bhankar	Boraginaceae
18.	*Erythrina suberosa**	Paryara	Fabaceae
19.	*Euonymus hamiltonianus**	Chamardein	Celastraceae
20.	*Ficus glomerata*	Umbrein	Urticaceae
21.	*Flacourtia indica**	Kaangu	Flacourtiaceae
22.	*Grewia disperma**	Byulee	Tiliaceae
23.	*Grewia elastica**	Dhaman	Tiliaceae
24.	*Grewia optiva*	Beul	Tiliaceae
25.	*Hamiltonia suaveolens*	Padaar	Rubiaceae
26.	*Helicteres isora*	Ban-beuli	Sterculiaceae
27.	*Lannea coromandelica**	Kahemble	Anacardiaceae
28.	*Leucaena glauca**	Alseenia	Fabaceae
29.	*Mallotus philippinensis*	Kaamal	Euphorbiaceae
30.	*Moringa oleifera**	Sunanha	Moringaceae
31.	*Murraya koenigii**	Gandhla	Rutaceae
32.	*Oroxylum indicum**	Tat-palanga	Bignoniaceae
33.	*Pinus roxburghii*	Chir, Cheel	Pinaceae
34.	*Populus nigra*	Poplar	Salicaceae
35.	*Prunus domestica**	Aloo Bhukara, Palam	Rosaceae
36.	*Randia dumetorum**	Rada	Rubiaceae
37.	*Rhamnus triquetra**	Guldarein	Rhamnaceae
38.	*Rhododendron arboreum*	Braah, Brass	Ericaceae
39.	*Salix karelini*†	Pahadi-Beeuins	Salicaceae
40.	*Salix tetrasperma*	Beeuins	Salicaceae
41.	*Sapium sebiferum*	Makhan, Pahadi-Taali	Euphorbiaceae
42.	*Syzygium operculatum**	Kathmanh	Myrtaceae
43.	*Terminalia bellirica**	Bahera	Combretaceae
44.	*Terminalia chebula**	Harad	Combretaceae
45.	*Terminalia tomentosa*	Rasaan	Combretaceae
46.	*Woodfordia fruticosa*	Dhawain, Dhaun	Lythraceae
47.	*Wrightia arborea* *	Khaloa	Apocynaceae
48.	*Zea mays**	Chhali, Chhaliyan	Poaceae

* - Additional uses

† - New Record

The region holds rich diversity in forages as evident from the utilization of 100 plants belonging to 78 genera under 35 families, of which Fabaceae (18 spp: *Albizia chinensis, Albizia julibrissin, Albizia lebbeck, Bauhinia vahlii, Bauhinia variegata, Butea monosperma, Caesalpinia decapetala, Indigofera cassioides, Lathyrus aphaca, Leucaena glauca, Medicago polymorpha, Milletia extensa, Ougeinia dalbergioides, Trifolium resupinatum, Vicia faba, Vicia hirsuta, Vicia sativa* and *Vigna radiata*), Poaceae (14 spp: *Arundinaria falcata, Avena sativa, Bambusa nutans, Capillipedium assimile, Cymbopogon citratus, Cynodon dactylon, Dendrocalamus hamiltoni, Heteropogon contortus, Panicum antidontale, Pennisetum americanum, Sorghum vulgare, Sporobolus indicus, Thysanolaena maxima* and *Zea mays*) and Urticaceae (12 spp: *Artocarpus lacucha, Celtis australis, Debregeasia hypoleuca, Ficus clavata, Ficus foveolata, Ficus glomerata, Ficus hispida, Ficus infectoria, Ficus roxburghii, Ficus rumphii, Ficus semicordata* and *Morus alba*) are the high used families. Likewise, high used forage genera are: *Ficus* (8 spp), *Terminalia* (4 spp), *Albizia* (3 spp), *Grewia* (3 spp) and *Vicia* (3 spp). Most of these plants are given to livestock either singly or in combination or after mixing thoroughly with crushed straw of wheat and/or maize for improving the overall quality/quantity of milk in this part of the Himalayas. Of these, species marked with asterisk in table XXVI are not enumerated earlier in the information provided so far on Indian forages (Anonymous, 1994; Sood & Kaushal, 2008).

Table XXVI Fodder Resources of Tehsil Palampur

S.No.	Botanical Name	Local Name/s	Family
1.	*Albizia chinensis*	Oyee	Fabaceae
2.	*Albizia julibrissin*	Karmarua	Fabaceae
3.	*Albizia lebbeck*	Sarin, Sirin	Fabaceae
4.	*Alternanthera sessilis*	Rubadi-Khubadi	Amaranthaceae
5.	*Artocarpus lacucha*	Dehu	Urticaceae
6.	*Arundinaria falcata*	Banji	Poaceae
7.	*Avena sativa*	Luanhe	Poaceae
8.	*Bambusa nutans*	Ddang	Poaceae
9.	*Bauhinia vahlii* *	Taur, Tor	Fabaceae
10.	*Bauhinia variegata*	Karale, Karaalein	Fabaceae
11.	*Berberis lyceum*	Kashmal, Kasmalein	Berberidaceae
12.	*Butea monosperma*	Palah, Palash	Fabaceae
13.	*Caesalpinia decapetala*	Drogaadi, Taranghadi	Fabaceae
14.	*Capillipedium assimile*	Dholu	Poaceae
15.	*Carissa congesta*	Garna, Garnae	Apocynaceae
16.	*Celtis australis*	Khidak, Khirak	Urticaceae
17.	*Cordia dichotoma*	Lasura, Lasyada	Boraginaceae

Contd...

Table XXVI Contd...

S.No.	Botanical Name	Local Name/s	Family
18.	*Cryptolepis buchanani**	Dudhi-Bel	Asclepiadaceae
19.	*Cymbopogon citrates**	Makoda	Poaceae
20.	*Cynodon dactylon*	Doob	Poaceae
21.	*Debregeasia hypoleuca*	Siaru, Syarru	Urticaceae
22.	*Dendrocalamus hamiltonii**	Maggar	Poaceae
23.	*Ehretia acuminata*	Punna	Boraginaceae
24.	*Ehretia laevis*	Bhankar	Boraginaceae
25.	*Eriophorum comosum**	Bagdella	Cyperaceae
26.	*Euonymus hamiltonianus*	Chamardein	Celastraceae
27.	*Ficus clavata*	Ban Feguda	Urticaceae
28.	*Ficus foveolata*	Phaegdi	Urticaceae
29.	*Ficus glomerata*	Umbrein	Urticaceae
30.	*Ficus hispida*	Dagrein	Urticaceae
31.	*Ficus infectoria*	Pallakharein	Urticaceae
32.	*Ficus roxburghii*	Traimbal	Urticaceae
33.	*Ficus rumphii*	Badarein	Urticaceae
34.	*Ficus semicordata*	Khanurein, Khasarkhurein	Urticaceae
35.	*Flacourtia indica*	Kaangu	Flacourtiaceae
36.	*Fumaria indica**	Pitpapra	Fumariaceae
37.	*Galium aparine**	Jhanjheedi	Rubiaceae
38.	*Grewia disperma**	Byulee	Tiliaceae
39.	*Grewia elastica**	Dhaman	Tiliaceae
40.	*Grewia optiva*	Beul	Tiliaceae
41.	*Helicteres isora*	Ban-Beuli	Sterculiaceae
42.	*Heteropogon contortus**	Lambia Ghaa	Poaceae
43.	*Hypericum oblogifolium**	Peelan	Hypericaceae
44.	*Indigofera cassioides*	Kanthi	Fabaceae
45.	*Lannea coromandelica*	Kahemble	Anacardiaceae
46.	*Lathyrus aphaca*	Mater-Phali, Khinnu, Sudu	Fabaceae
47.	*Leucaena glauca*	Alseenia	Fabaceae
48.	*Litsea monopetala*	Guan	Lauraceae
49.	*Litsea semicarpifolia*†	Pahadi-Sannanh	Lauraceae
50.	*Loranthus falcatus**	Parand	Loranthaceae
51.	*Mallotus philippinensis*	Kaamal	Euphorbiaceae
52.	*Medicago polymorpha*	Khukhani	Fabaceae
53.	*Milletia extensa*	Salod	Fabaceae

Contd...

Table XXVI Contd...

S.No.	Botanical Name	Local Name/s	Family
54.	*Moringa oleifera*	Sunanha	Moringaceae
55.	*Morus alba*	Toot	Urticaceae
56.	*Musa sapientum*	Kela	Musaceae
57.	*Ougeinia dalbergioides*	Sannanh	Fabaceae
58.	*Panicum antidontale*	Barua	Poaceae
59.	*Pennisetum americanum*	Bajara	Poaceae
60.	*Persea odoratissima*	Bhadrol	Lauraceae
61.	*Phoenix acaulis*	Khajoori	Arecaceae
62.	*Phoenix sylvestris*	Khajoor	Arecaceae
63.	*Pistacia khinjuk*	Kakad Singhi	Anacardiaceae
64.	*Polygala chinensis**	Tone-Beidu	Polygonaceae
65.	*Polygonum barbatum*	Ghaneeri	Polygonaceae
66.	*Populus ciliata*	Phalsa	Salicaceae
67.	*Populus nigra*	Poplar	Salicaceae
68.	*Porana paniculata**	Chhube-Beil	Convolvulaceae
69.	*Prunus cerasoides*	Paza	Rosaceae
70.	*Pyrus pashia*	Kainth	Rosaceae
71.	*Quercus leucotrichophora*	Baan	Fagaceae
72.	*Randia dumetorum*	Rada	Rubiaceae
73.	*Rhamnus triquetra*	Guldarein	Rhamnaceae
74.	*Rubus ellipticus*	Akha	Rosaceae
75.	*Rubus niveus*	Kala-Akha	Rosaceae
76.	*Salix karelini*†	Pahadi-Beeuins	Salicaceae
77.	*Sapium sebiferum**	Makhan, Pahadi-Taali	Euphorbiaceae
78.	*Scandix pecten veneris*†	Jangli Dhaniya	Apiaceae
79.	*Silene conoidea**	Tokala	Caryophyllaceae
80.	*Sorghum bicolor*	Chari	Poaceae
81.	*Spondias pinnata**	Muara	Anacardiaceae
82.	*Sporobolus indicus*	Bonkra	Poaceae
83.	*Stellaria media**	Khaukhua	Caryophyllaceae
84.	*Syzygium operculatum**	Kathmanh	Myrtaceae
85.	*Terminalia arjuna*	Arjan, Arjun	Combretaceae
86.	*Terminalia bellirica*	Bahera	Combretaceae
87.	*Terminalia chebula*	Harad	Combretaceae
88.	*Terminalia tomentosa*	Rasaan	Combretaceae
89.	*Thysanolaena maxima*	Garnaal	Poaceae

Contd...

Table XXVI Contd...

S.No.	Botanical Name	Local Name/s	Family
90.	*Tinospora cordifolia*	Gloan, Gulja	Menispermaceae
91.	*Trifolium resupinatum*	Barseen	Fabaceae
92.	*Vicia faba**	Keon, Keeoon	Fabaceae
93.	*Vicia hirsuta*	Choti Rodi	Fabaceae
94.	*Vicia sativa*	Badi-Rodi	Fabaceae
95.	*Vigna radiata*†	Maah	Fabaceae
96.	*Woodfordia fruticosa*	Dhawain, Dhaun	Lythraceae
97.	*Wrightia arborea*	Khaloa	Apocynaceae
98.	*Zea mays*	Chhali, Chhaliyan	Poaceae
99.	*Zizyphus jujuba**	Ber	Rhamnaceae
100.	*Zizyphus nummularia*	Jareid, Breid	Rhamnaceae

* Additional Uses

† New Record

The local inhabitants are deeply religious as evident by the reported use of 60 taxa under 49 genera and 31 families for magico-religious purposes. These are: *Abrus precatorius, Achyranthes aspera, Aegle marmelos, Aerva lanata, Asparagus adscendens, Berberis lycium, Bombax ceiba, Butea monosperma, Calotropis procera, Capsicum annuum, Cassia fistula, Cestrum nocturnum, Cinnamomum camphora, Citrus limon, Citrus medica* var. *medica, Cucurbita hispida, Curcuma longa, Cynodon dactylon, Datura innoxia, Datura stramonium, Emblica officinalis, Erythrina suberosa, Euphorbia royleana, Ficus benghalensis, Ficus glomerata, Ficus infectoria, Ficus religiosa, Gnaphalium affine, Gossypium arboreum, Grewia optiva, Hibiscus rosa sinensis, Holarrhena antidysentrica, Jasminum dispermum, Jasminum grandiflorum, Jasminum humile, Jasminum multiflorum, Lotus corniculatus, Mallotus philippinensis, Mangifera indica, Mirabilis jalapa, Musa sapientum, Ocimum kilimandscharicum, Ocimum sanctum, Pistacia khinjuk, Prinsepia utilis, Punica granatum, Reinwardtia indica, Rosa chinensis, Rosa damascena, Saccharum officinarum, Saccharum spontaneum, Salvia coccinea, Santalum album, Sesamum orientale, Sisymbrium irio, Tagetes minuta, Tinospora cordifolia, Tulipa stellata, Vigna radiata* and *Zephyranthes carinata.* For this purpose, Fabaceae (4 spp) and Oleaceae (3 spp) followed by Asteraceae, Rutaceae, Euphorbiaceae and Solanaceae (2 spp each) are the high used families whereas *Jasminum* (3 spp) and *Ficus* (2 spp) are the high used genera. Some of the magico-religious practices associated with the use of *Achyranthes aspera, Aegle marmelos, Aerva lanata, Berberis lycium, Calotropis procera, Cassia fistula, Cestrum nocturnum, Cinnamomum camphora, Citrus medica* var. *medica, Cucurbita hispida, Emblica officinalis, Erythrina suberosa, Euphorbia royleana, Ficus glomerata, Ficus infectoria, Holarrhena antidysentrica, Jasminum dispermum, Jasminum humile, Lotus corniculatus, Musa sapientum, Ocimum kilimandscharicum, Pistacia khinjuk, Prinsepia utilis, Saccharum spontaneum, Sisymbrium irio, Tagetes minuta and Zephyranthes carinata* are unique to this place and hitherto unknown to the science of ethnobotany.

On comparing upon the present ethnobotanical findings with those of earlier ones, additional uses of 127 plants for human ailments, 42 spp for 28 miscellaneous purposes, 21 spp for fodder, 27 spp for fuelwood, 27 spp for magico-religious, 10 spp for edibility, 4 spp for building works, and 7 spp for household articles have been recorded (Table XXVII; Appendix IV). Moreover, uses of 10 plants, viz. *Ajuga parviflora* (diabetes, skin ailments, hair growth, lice repellent), *Carya illinoensis* (edible, bed wetting, dental problems, dye, sore toes), *Citrus jambhiri* (edible, vermillion), *Gnaphalium affine* (sacred), *Litsea semicarpifolia* (fodder, household construction), *Ranunculus laetus* (skin diseases), *Salix karelini* (fodder, fuel), *Scandix pectin veneris* (vegetable, fodder), *Selaginella chrysocaulos* (stomachache, vermillion) and *Vigna radiata* (magico-religious) form the first report of its type as these plants have not been described earlier for any of the ethnobotanical use.

Table XXVII Ethnobotanical Plants of Palampur with Additional Uses

S.No.	*Botanical Name*
1.	*Abrus precatorius*
2.	*Acacia catechu*
3.	*Achyranthes aspera*
4.	*Achyranthes bidentata*
5.	*Adiantum capillus veneris*
6.	*Aegle marmelos*
7.	*Aerva lanata*
8.	*Agave cantula*
9.	*Ageratum conyzoides*
10.	*Albizia chinensis*
11.	*Albizia julibrissin*
12.	*Albizia lebbeck*
13.	*Aloe vera*
14.	*Alternanthera sessilis*
15.	*Apium leptophyllum*
16.	*Aristida cyanantha*
17.	*Artemisia capillaris*
18.	*Artemisia japonica*
19.	*Artocarpus lacucha*
20.	*Arundinaria falcata*
21.	*Asparagus adscendens*

Contd...

Table XXVII Contd...

S.No.	Botanical Name
22.	*Bambusa nutans*
23.	*Basella alba*
24.	*Bauhinia vahlii*
25.	*Bauhinia variegata*
26.	*Berberis lycium*
27.	*Bergenia ciliata*
28.	*Boehmeria macrophylla*
29.	*Bombax ceiba*
30.	*Brassica tournefortii*
31.	*Butea monosperma*
32.	*Cajanus cajan*
33.	*Callicarpa macrophylla*
34.	*Calotropis procera*
35.	*Cannabis sativa*
36.	*Capillipedium assimile*
37.	*Capsicum annuum*
38.	*Carica papaya*
39.	*Carissa congesta*
40.	*Caryopteris bicolor*
41.	*Cassia fistula*
42.	*Cassia floribunda*
43.	*Cassia obtusifolia*
44.	*Cassia occidentalis*
45.	*Catharanthus roseus*
46.	*Celastrus paniculatus*
47.	*Celtis australis*
48.	*Cestrum nocturnum*
49.	*Cheilanthes farinosa*
50.	*Chenopodium album*
51.	*Cinnamomum camphora*
52.	*Cinnamomum tamala*
53.	*Cirsium arvense*
54.	*Cissampelos parriera*
55.	*Citrus medica*

Contd...

Table XXVII Contd...

S.No.	Botanical Name
56.	*Citrus medica* var. *medica*
57.	*Clematis gauriana*
58.	*Cocculus laurifolius*
59.	*Cordia dichotoma*
60.	*Coriandrum sativum*
61.	*Cryptolepis buchanani*
62.	*Cucurbita hispida*
63.	*Cucurbita pepo*
64.	*Curcuma longa*
65.	*Cuscuta reflexa*
66.	*Cymbopogon citratus*
67.	*Dalbergia sissoo*
68.	*Debregeasia hypoleuca*
69.	*Dendrocalamus hamiltonii*
70.	*Didymocarpus pedicellata*
71.	*Dodonaea viscosa*
72.	*Drimia indica*
73.	*Duranta repens*
74.	*Ehretia laevis*
75.	*Eleusine coracana*
76.	*Emblica officinalis*
77.	*Engelhardia colebrookeana*
78.	*Equisetum debile*
79.	*Eriophorum comosum*
80.	*Erythrina suberosa*
81.	*Eucalyptus tereticornis*
82.	*Eulaliopsis binata*
83.	*Euonymus hamiltonianus*
84.	*Euphorbia royleana*
85.	*Ficus benghalensis*
86.	*Ficus glomerata*
87.	*Ficus hispida*
88.	*Ficus infectoria*
89.	*Ficus palmata*
90.	*Ficus religiosa*

Contd...

Table XXVII Contd...

S.No.	Botanical Name
91.	*Ficus roxburghii*
92.	*Flacourtia indica*
93.	*Fumaria indica*
94.	*Galium aparine*
95.	*Geranium nepalense*
96.	*Glycine max*
97.	*Gonatanthus pumilus*
98.	*Grevillea robusta*
99.	*Grewia disperma*
100.	*Grewia elastica*
101.	*Grewia optiva*
102.	*Helicteres isora*
103.	*Heteropogon contortus*
104.	*Hibiscus rosa-sinensis*
105.	*Holarrhena antidysentrica*
106.	*Hypericum oblogifolium*
107.	*Impatiens balsamina*
108.	*Indigofera cassioides*
109.	*Jasminum dispermum*
110.	*Jasminum grandiflorum*
111.	*Jasminum humile*
112.	*Jasminum multiflorum*
113.	*Jatropha curcas*
114.	*Justicia adhatoda*
115.	*Kalanchoe pinnata*
116.	*Lannea coromandelica*
117.	*Lantana camara*
118.	*Lepidagathis cuspidata*
119.	*Lepidium sativum*
120.	*Leucaena glauca*
121.	*Leucas cephalotes*
122.	*Linum usitatissimum*
123.	*Litchi chinensis*
124.	*Loranthus falcatus*
125.	*Lotus corniculatus*

Contd...

Table XXVII Contd...

S.No.	Botanical Name
126.	*Luffa acutangula*
127.	*Luffa cylindrica*
128.	*Mallotus philippinensis*
129.	*Mangifera indica*
130.	*Melia azedarach*
131.	*Michelia champaca*
132.	*Momordica charantia*
133.	*Moringa oleifera*
134.	*Morus alba*
135.	*Murraya koenigii*
136.	*Murraya paniculata*
137.	*Musa sapientum*
138.	*Ocimum basilicum*
139.	*Ocimum kilimandscharicum*
140.	*Ocimum sanctum*
141.	*Opuntia dillenii*
142.	*Origanum vulgare*
143.	*Oroxylum indicum*
144.	*Oxalis corniculata*
145.	*Panicum antidontale*
146.	*Pennisetum americanum*
147.	*Pentanema indicum*
148.	*Perilla frutescens*
149.	*Phoenix acaulis*
150.	*Phoenix sylvestris*
151.	*Phyllanthus urinaria*
152.	*Pinus roxburghii*
153.	*Pistacia khinjuk*
154.	*Plantago lanceolata*
155.	*Plumbago zeylanica*
156.	*Pogostemon benghalense*
157.	*Polygala chinensis*
158.	*Polygonum barbatum*
159.	*Porana paniculata*
160.	*Prinsepia utilis*

Contd...

Table XXVII Contd...

S.No.	Botanical Name
161.	*Prunus cerasoides*
162.	*Prunus domestica*
163.	*Punica granatum*
164.	*Pyrus communis*
165.	*Pyrus pashia*
166.	*Quercus leucotrichophora*
167.	*Randia dumetorum*
168.	*Ranunculus arvensis*
169.	*Reinwardtia indica*
170.	*Rhamnus triquetra*
171.	*Rhododendron arboreum*
172.	*Ricinus communis*
173.	*Rorippa nasturtium-aquaticum*
174.	*Rosa brunonii*
175.	*Rosa chinensis*
176.	*Rosa damascena*
177.	*Roylea cinerea*
178.	*Rubus ellipticus*
179.	*Rubus niveus*
180.	*Rumex nepalensis*
181.	*Saccharum officinarum*
182.	*Salix tetrasperma*
183.	*Sapindus mukorosii*
184.	*Sapium insigne*
185.	*Sechium edule*
186.	*Sesamum orientale*
187.	*Sida cordifolia*
188.	*Silene conoidea*
189.	*Sisymbrium irio*
190.	*Solanum erianthum*
191.	*Solanum viarum*
192.	*Sorghum bicolor*
193.	*Spilanthes oleracea*
194.	*Spondias pinnata*
195.	*Sporobolus indicus*

Contd...

Table XXVII Contd...

S.No.	Botanical Name
196.	*Stellaria media*
197.	*Stephania glabra*
198.	*Syzygium cumini*
199.	*Syzygium operculatum*
200.	*Tagetes minuta*
201.	*Terminalia arjuna*
202.	*Terminalia bellirica*
203.	*Terminalia chebula*
204.	*Terminalia tomentosa*
205.	*Thuja orientalis*
206.	*Tinospora cordifolia*
207.	*Toona ciliata*
208.	*Urtica dioica*
209.	*Vanda tessellata*
210.	*Verbascum thapsus*
211.	*Vicia faba*
212.	*Vitex negundo*
213.	*Woodfordia fruticosa*
214.	*Wrightia arborea*
215.	*Zea mays*
216.	*Zephyranthes carinata*
217.	*Zizyphus jujuba*

OBSERVED DENSITY AND AVAILABILITY OF UNDERUTILIZED PLANT RESOURCES

Results of the collected data on observed density and availability of ethnobotanically important plants in the region indicate high availability of 109 spp, abundant 175 spp and 51 rare spp (Table XXVIII). In India, out of 8,000 medicinal plants only 200 spp fall under rare, endangered and threatened categories (Nayar & Sastri, 1988). In 1994, CITES notified 11 Indian medicinal plants and Government of India banned the export of more than 50 species, which are nearly threatened in the wild state. In the present study, *Aloe vera* and *Vanda tessellata* fall under the category of negative list of exports due to their threatened status, and need to be conserved to prevent their exploitation and extinction with the help of local inhabitants. The information generated holds great promises for policy planners for shaping plans for sustainable use and conservation of biodiversity and socio-economic upliftment of the inhabitants of the region.

Table XXVIII Observed Density and Availability (ODA) of Ethnobotanically Important Plants of Tehsil Palampur, District Kangra (H.P.)

S.No.	Botanical Name	Local Name/s	Observed Density and Availability
1.	*Abelmoschus esculentus*	Ghandoli	+ +
2.	*Abrus precatorius*	Chadannu, Rati	+
3.	*Acacia catechu*	Khair	+ + +
4.	*Achyranthes aspera*	Puthkanda	+ + +
5.	*Achyranthes bidentata*	Puthkanda	+ + +
6.	*Acorus calamus*	Barain, Brein.	+ +
7.	*Adiantum capillus veneris*	Dum-tuli	+ +
8.	*Aegle marmelos*	Bael, Bil	+ +
9.	*Aerva lanata*	Aung	+ + +
10.	*Agave cantula*	Ranhbaan, Ramban	+ +
11.	*Agave wightii*	Kyor	+ + +
12.	*Ageratum conyzoides*	Jaihar, Phulnu	+ + +
13.	*Ajuga parviflora*	Neelkanthi	+ +
14.	*Albizia chinensis*	Oyee	+ +
15.	*Albizia julibrissin*	Karmarua	+ +
16.	*Albizia lebbeck*	Sarin, Sirin	+ +
17.	*Allium cepa*	Payaj	+ + +
18.	*Allium sativum*	Lahsun, Lassan	+ + +
19.	*Aloe vera*	Ghi Kanwar, Kuaarein	+ +
20.	*Alternanthera sessilis*	Rubadi-Khubadi	+ +
21.	*Amaranthus paniculatus*	Seul	+ + +
22.	*Amaranthus viridis*	Chalairi, Chaulai	+ + +
23.	*Apium leptophyllum*	Jungli ajwain, Pahadi ajwain, Juanhe	+
24.	*Araucaria columnaris*	Araucaria	+
25.	*Aristida cyanantha*	Kaduli-Tuli	+ +
26.	*Artemisia capillaris*	Jhauncha	+ +
27.	*Artemisia japonica*	Chharmara	+ +
28.	*Artocarpus lacucha*	Dehu	+ +
29.	*Arundinaria falcata*	Banji	+ + +
30.	*Asparagus adscendens*	Sanspain	+ +
31.	*Avena sativa*	Luanhe	+ + +
32.	*Bambusa nutans*	Dang	+ + +
33.	*Basella alba*	Poe	+

Contd...

Table XXVIII Contd...

S.No.	*Botanical Name*	*Local Name/s*	*Observed Density and Availability*
34.	*Bauhinia vahlii*	Taur, Tor	+ +
35.	*Bauhinia variegata*	Karale, Karaalein	+ + +
36.	*Berberis lycium*	Kashmal, Kasmalein	+ + +
37.	*Bergenia ciliata*	Saprotri	+
38.	*Bidens biternata*	Lumb	+ + +
39.	*Boehmeria macrophylla*	Handa	+ +
40.	*Boerhaavia diffusa*	Eat-seat	+ +
41.	*Bombax ceiba*	Simbal	+ +
42.	*Brassica nigra*	Kali-Rai	+ + +
43.	*Brassica tournefortii*	Rai	+ +
44.	*Butea monosperma*	Palah, Palash	+ +
45.	*Caesalpinia decapetala*	Drogaadi, Taranghadi	+ +
46.	*Cajanus cajan*	Dhinger	+
47.	*Callicarpa macrophylla*	Danther	+
48.	*Calotropis procera*	Aak, Ak	+
49.	*Camellia sinensis*	Cha, Chai	+ + +
50.	*Cannabis sativa*	Bhang	+ + +
51.	*Capillipedium assimile*	Dholu	+ + +
52.	*Capsicum annuum*	Mirach, Pipali	+ + +
53.	*Carica papaya*	Kharbuja, Papita	+ +
54.	*Carissa congesta*	Garna, Garnae	+ + +
55.	*Carya illinoensis*	Kagzi Khod	+
56.	*Caryopteris bicolor*	Ban-Dhavein	+ +
57.	*Cassia fistula*	Aelia, Amaltas	+ +
58.	*Cassia floribunda*	Bada-Rellu	+ +
59.	*Cassia obtusifolia*	Aellu, Rellu	+ +
60.	*Cassia occidentalis*	Chhota-Rellu	+ +
61.	*Catharanthus roseus*	Sadabahar	+
62.	*Cedrus deodara*	Deyar, Devdar	+ +
63.	*Celastrus paniculatus*	Sankheeru	+
64.	*Celtis australis*	Khidak, Khirak	+ + +
65.	*Centella asiatica*	Brahmi, Meenaki	+ +
66.	*Centipeda minima*	Nakk-Chhikannu	+ +
67.	*Centratherum anthelminticum*	Brahmjeeri	+ +
68.	*Cestrum nocturnum*	Rat-Ki-Rani	+ +

Contd...

Table XXVIII Contd...

S.No.	Botanical Name	Local Name/s	Observed Density and Availability
69.	*Cheilanthes farinosa*	Duman-Tuli	+ + +
70.	*Chenopodium album*	Kanah	+ + +
71.	*Chenopodium ambrosioides*	Bathu	+ +
72.	*Cicer arietinum*	Chana	+ +
73.	*Cinnamomum camphora*	Kapur	+
74.	*Cinnamomum tamala*	Gudpatraj, Mithpatra, Tejpatta	+ +
75.	*Cirsium arvense*	Bhruce	+ + +
76.	*Cissampelos parriera*	Bhatindu	+ +
77.	*Citrus grandis*	Chakotra	+ +
78.	*Citrus jhambiri*	Ghamidi, Jhamirdi, Khatirali	+ +
79.	*Citrus limon*	Nimbu	+ +
80.	*Citrus medica*	Baranimbu	+ +
81.	*Citrus medica* var. *medica*	Bajoorein	+
82.	*Clematis gauriana*	Baaker-Bel	+ + +
83.	*Coccinia grandis*	Kalakadu	+
84.	*Cocculus laurifolius*	Parauda	+ +
85.	*Colebrookea oppositifolia*	Doos	+ + +
86.	*Colocasia esculenta*	Janglee Kachalu	+ +
87.	*Commelina paludosa*	Chhura	+ + +
88.	*Cordia dichotoma*	Lasura, Lasyada	+ +
89.	*Coriandrum sativum*	Beenh, Dhania	+ + +
90.	*Coronopus didymus*	Garhbeenhi	+ +
91.	*Costus speciosus*	Jangli-Haldi	+ +
92.	*Cryptolepis buchanani*	Dudhi-Bel	+ +
93.	*Cucumis melo* var. *agrestis*	Phaphanu	+ + +
94.	*Cucumis sativus*	Kheera	+ + +
95.	*Cucurbita hispida*	Dhuda Petha	+
96.	*Cucurbita pepo*	Kaddu	+ + +
97.	*Curcuma longa*	Haldee	+ + +
98.	*Cuscuta reflexa*	Akashbel, Amarbel, Gaasbel	+ + +
99.	*Cyclanthera pedata*	Meetha Karela	+ + +
100.	*Cymbopogon citratus*	Makoda	+ + +
101.	*Cynodon dactylon*	Dharub, Daroov	+ + +
102.	*Cyphomandra betacea*	Desi Tamatar	+
103.	*Dalbergia sissoo*	Shissham, Tahli	+ +

Contd...

Table XXVIII Contd...

S.No.	Botanical Name	Local Name/s	Observed Density and Availability
104.	*Datura innoxia*	Dhatura	+ +
105.	*Datura stramonium*	Dhatura	+ +
106.	*Debregeasia hypoleuca*	Siaru, Syaru	+ +
107.	*Deeringia amaranthoides*	Bhringya	+
108.	*Dendrocalamus hamiltonii*	Maggar	+ + +
109.	*Didymocarpus pedicellata*	Musakani	+ +
110.	*Dioscorea bellophylla*	Tardi	+ +
111.	*Dioscorea bulbifera*	Gangaradi	+ +
112.	*Dioscorea oppositifolia*	Daregal	+
113.	*Diospyros kaki*	Japani-Phal	+
114.	*Diplazium esculentum*	Lungadu	+ +
115.	*Diplocyclos palmatus*	Shivlingi	+ +
116.	*Dodonaea viscosa*	Mehendu, Mehendadu	+ + +
117.	*Dolichos biflorus*	Kulath	+ +
118.	*Drimia indica*	Ban-Ganda	+
119.	*Duranta repens*	Neel-Kanta	+ +
120.	*Ehretia acuminata*	Punna	+ +
121.	*Ehretia laevis*	Bhankar	+ +
122.	*Eleusine coracana*	Mandal	+
123.	*Emblica officinalis*	Ambla, Amla	+ + +
124.	*Engelhardia colebrookeana*	Samma	+
125.	*Equisetum debile*	Sau-Greendi	+ +
126.	*Eriobotrya japonica*	Galaat	+ +
127.	*Eriophorum comosum*	Bagdella	+ + +
128.	*Erythrina suberosa*	Paryara	+ +
129.	*Eucalyptus tereticornis*	Safeda	+ +
130.	*Eulaliopsis binata*	Baggad	+ +
131.	*Euonymus hamiltonianus*	Chamardei	+
132.	*Euphorbia royleana*	Chhoo	+ + +
133.	*Fagopyrum cymosum*	Fafru	+ + +
134.	*Fagopyrum esculentum*	Bhresh	+ ++
135.	*Ficus benghalensis*	Badein, Bargad	+ + +
136.	*Ficus carica*	Anjeer	+
137.	*Ficus clavata*	Ban Feguda	+ +
138.	*Ficus foveolata*	Phaegdi	+ +
139.	*Ficus glomerata*	Umbrein	+ +

Contd...

Table XXVIII Contd...

S.No.	Botanical Name	Local Name/s	Observed Density and Availability
140.	*Ficus hispida*	Dagrein	+ + +
141.	*Ficus infectoria*	Pallakharein	+ + +
142.	*Ficus palmata*	Dhuda, Khasra	+ +
143.	*Ficus religiosa*	Bad, Peepal	+ + +
144.	*Ficus roxburghii*	Traimbal	+ + +
145.	*Ficus rumphii*	Badarein	+ +
146.	*Ficus semicordata*	Khanurein, Khasarkhurein	+ +
147.	*Flacourtia indica*	Kaangu	+ +
148.	*Foeniculum vulgare*	Meethi-Saunf	+ +
149.	*Fragaria indica*	Bhoo-Akha	+ + +
150.	*Fumaria indica*	Papda	+ + +
151.	*Galium aparine*	Jhanjheedi	+ + +
152.	*Geranium nepalense*	Laljari	+ + +
153.	*Glycine max*	Soyabean	+ + +
154.	*Gnaphalium affine*	Gadua	+ +
155.	*Gonatanthus pumilus*	Aala	+ +
156.	*Gossypium arboreum*	Kapah	+
157.	*Grevillea robusta*	Silver Oak	+ +
158.	*Grewia disperma*	Byulee	+ +
159.	*Grewia elastica*	Dhaman	+ +
160.	*Grewia optiva*	Beul	+ +
161.	*Hamiltonia suaveolens*	Padaar	+ +
162.	*Helicteres isora*	Ban-Beuli	+
163.	*Heteropogon contortus*	Lambia Ghaa	+ + +
164.	*Hibiscus rosa-sinensis*	Dunaj-Phul	+ +
165.	*Holarrhena antidysentrica*	Keor, Inderjaun	+ +
166.	*Hypericum oblogifolium*	Peelan	+ +
167.	*Impatiens balsamina*	Teur	+ +
168.	*Indigofera cassioides*	Kanthi	+ +
169.	*Ipomoea batatas*	Shakarkanni	+
170.	*Ipomoea turbinata*	Ghondali	+ +
171.	*Jasminum dispermum*	Chamaeli	+ +
172.	*Jasminum grandiflorum*	Malti	+ +
173.	*Jasminum humile*	Chamaeli	+ +
174.	*Jasminum multiflorum*	Chamaeli	+ +

Contd...

Table XXVIII Contd...

S.No.	Botanical Name	Local Name/s	Observed Density and Availability
175.	*Jatropha curcas*	Jamalghota, Rattanjot	+ +
176.	*Juglans regia*	Akhrot, Khod	+ +
177.	*Justicia adhatoda*	Basunti, Basuti	+ + +
178.	*Kalanchoe pinnata*	Gand-Patar	+ +
179.	*Lannea coromandelica*	Kahemble	+ +
180.	*Lantana camara*	Zahar, Phulanu	+ + +
181.	*Lathyrus aphaca*	Mater-Phali, Khinnu, Sudu	+ + +
182.	*Lathyrus sativus*	Kalan	+ + +
183.	*Lawsonia inermis*	Mehendi	+
184.	*Lepidagathis cuspidata*	Billi	+ +
185.	*Lepidium sativum*	Halaein	+ +
186.	*Leucaena glauca*	Alseenia	+ +
187.	*Leucas cephalotes*	Bhedu	+ +
188.	*Linum usitatissimum*	Allain	+ +
189.	*Litchi chinensis*	Litchi	+ +
190.	*Litsea monopetala*	Guaan	+ +
191.	*Litsea sebifera*	Rhaanh	+
192.	*Litsea semicarpifolia*	Pahadi-Sannanh	+
193.	*Loranthus falcatus*	Parand	+ +
194.	*Lotus corniculatus*	Chidi	+ + +
195.	*Luffa acutangula*	Jangli Ghangher	+ +
196.	*Luffa cylindrica*	Ghi-Ghangher	+ +
197.	*Mallotus philippinensis*	Kaamal	+ +
198.	*Mangifera indica*	Aam, Amb	+ +
199.	*Medicago polymorpha*	Khukhani	+ + +
200.	*Melia azedarach*	Drek	+ +
201.	*Mentha longifolia*	Jungli-Pudina	+ + +
202.	*Mentha piperita*	Bhura-Pudina, Pipermint	+ + +
203.	*Michelia champaca*	Chamba, Champa	+
204.	*Micromeria biflora*	Janglee Ajwain	+ +
205.	*Milletia extensa*	Salod	+
206.	*Mirabilis jalapa*	Gulbansi	+ +
207.	*Momordica charantia*	Karela	+ + +
208.	*Momordica dioica*	Meetha Karela	+ +
209.	*Moringa oleifera*	Sunanha	+

Contd...

Table XXVIII Contd...

S.No.	*Botanical Name*	*Local Name/s*	*Observed Density and Availability*
210.	*Morus alba*	Toot	+ + +
211.	*Mucuna pruriens*	Gazal Bel	+ +
212.	*Murraya koenigii*	Ghandhla, Kadi-Patta	+ +
213.	*Murraya paniculata*	Gandhala	+
214.	*Musa sapientum*	Kela	+ +
215.	*Neolitsea umbrosa*	Gulab-Jamun	+
216.	*Nerium indicum*	Ghaneera	+ +
217.	*Ocimum basilicum*	Bhabhri	+ +
218.	*Ocimum gratissimum*	Ban-Ajwain	+ +
219.	*Ocimum kilimandscharicum*	Tulsi	+ +
220.	*Ocimum sanctum*	Sailli	+ +
221.	*Opuntia dillenii*	Daandi-Chhoo	+ +
222.	*Origanum vulgare*	Ban-Ajwain, Kah-Juainhe	+ +
223.	*Oroxylum indicum*	Tat-Palanga	+ +
224.	*Ougeinia dalbergioides*	Sannanh	+
225.	*Oxalis corniculata*	Malori	+ + +
226.	*Panicum antidontale*	Barua	+ +
227.	*Pennisetum americanum*	Bajara	+
228.	*Pentanema indicum*	Jugaali -Di-Dua	+ +
229.	*Perilla frutescens*	Ban-Juainhe	+ +
230.	*Persea odoratissima*	Bhadrol	+ + +
231.	*Phoenix acaulis*	Khajoori	+ +
232.	*Phoenix sylvestris*	Khajoor	+ +
233.	*Phyllanthus urinaria*	Amli	+ +
234.	*Pinus roxburghii*	Chir, Cheel	+ + +
235.	*Pistacia khinjuk*	Kakad Singhi	+ +
236.	*Plantago lanceolata*	Safed Isbagol	+ +
237.	*Plumbago zeylanica*	Chitra	+ +
238.	*Pogostemon benghalense*	Kali Basunti	+ +
239.	*Polygala chinensis*	Tone-Beidu	+
240.	*Polygonum barbatum*	Ghaneeri	+ +
241.	*Populus ciliata*	Poplar	+ +
242.	*Populus nigra*	Poplar	+ +
243.	*Porana paniculata*	Chhube-Beil	+ +
244.	*Portulaca grandiflora*	Kulfa	+ +

Contd...

Table XXVIII Contd...

S.No.	*Botanical Name*	*Local Name/s*	*Observed Density and Availability*
245.	*Prinsepia utilis*	Bhekhal	+ +
246.	*Prunus cerasoides*	Paaja, Pajja	+ + +
247.	*Prunus domestica*	Aloo Bukhara, Palam	+ +
248.	*Prunus persica*	Aadu, Aarnei, Aru	+ + +
249.	*Psidium guajava*	Amrood	+ + +
250.	*Punica granatum*	Daadu	+ +
251.	*Pyrus communis*	Nashpati	+ +
252.	*Pyrus pashia*	Kainth	+ + +
253.	*Pyrus serotina*	Bada-kainth	+
254.	*Quercus leucotrichophora*	Baan	+ +
255.	*Randia dumetorum*	Rada	+ +
256.	*Ranunculus arvensis*	Jaldhar	+ +
257.	*Ranunculus laetus*	Jaldhar	+ + +
258.	*Ranunculus muricatus*	Changer	+ +
259.	*Raphanus sativus*	Mooli	+ + +
260.	*Reinwardtia indica*	Basant	+ +
261.	*Rhamnus triquetra*	Guldarein	+ +
262.	*Rhododendron arboreum*	Braah, Brass	+ +
263.	*Ricinus communis*	Arand, Erand	+ + +
264.	*Rorippa nasturtium-aquaticum*	Chhoo-Nali	+ + +
265.	*Rosa brunonii*	Gulabadi	+ + +
266.	*Rosa chinensis*	Galaab, Gulaab	+ + +
267.	*Rosa damascena*	Gulabadi	+ +
268.	*Roylea cinerea*	Kakdoen	+ +
269.	*Rubus ellipticus*	Aakhe	+ + +
270.	*Rubus niveus*	Kala-Akha	+ +
271.	*Rumex hastatus*	Aambi	+ + +
272.	*Rumex nepalensis*	Jangli-Palak	+ + +
273.	*Saccharum officinarum*	Ganna	+
274.	*Saccharum spontaneum*	Sankheda	+
275.	*Salix karelini*	Pahadi-Beeuins	+ + +
276.	*Salix tetrasperma*	Beeuins	+ + +
277.	*Salvia coccinea*	Phultulsi	+ + +
278.	*Santalum album*	Chananh	+
279.	*Sapindus mukorosii*	Dodan, Ritha	+ + +

Contd...

Table XXVIII Contd...

S.No.	Botanical Name	Local Name/s	Observed Density and Availability
280.	*Sapium insigne*	Ainkhar	+ +
281.	*Sapium sebiferum*	Makhan, Pahadi-Taali	+ + +
282.	*Scandix pectin veneris*	Jangli Dhaniya	+ + +
283.	*Sechium edule*	Lonku	+ +
284.	*Selaginella chrysocaulos*	Kungu	+ + +
285.	*Sesamum orientale*	Til	+
286.	*Sida cordifolia*	Dredain	+ +
287.	*Silene conoidea*	Tokala	+ + +
288.	*Sisymbrium irio*	Foonkalan	+ + +
289.	*Solanum erianthum*	Laah	+
290.	*Solanum melongena*	Baingan	+ + +
291.	*Solanum nigrum*	Makoi	+ + +
292.	*Solanum torvum*	Chhota-Baingan	+ + +
293.	*Solanum viarum*	Bhindi, Kandayai	+
294.	*Sorghum bicolor*	Chari	+ +
295.	*Spilanthes oleracea*	Karkara	+ +
296.	*Spondias pinnata*	Muara	+ +
297.	*Sporobolus indicus*	Bonkra	+ + +
298.	*Stellaria media*	Khaukhua	+ +
299.	*Stephania glabra*	Bis-Khapar	+ + +
300.	*Syzygium cumini*	Jamun	+ + +
301.	*Syzygium operculatum*	Kathmanh	+ + +
302.	*Tagetes minuta*	Gatt	+ +
303.	*Terminalia arjuna*	Arjan, Arjun	+
304.	*Terminalia bellirica*	Bahera	+
305.	*Terminalia chebula*	Harad	+
306.	*Terminalia tomentosa*	Rasaan	+
307.	*Thuja orientalis*	Sarua, Morpankh	+ + +
308.	*Thysanolaena maxima*	Garnaal	+ +
309.	*Tinospora cordifolia*	Gloan, Gulja	+ +
310.	*Toona ciliata*	Tooni	+ +
311.	*Trichosanthes cucumerina*	Meis	+ +
312.	*Trifolium resupinatum*	Barseen	+ + +
313.	*Trigonella foenum-graecum*	Maethi	+ + +
314.	*Triticum aestivum*	Kanak	+ + +

Contd...

Table XXVIII Contd...

S.No.	Botanical Name	Local Name/s	Observed Density and Availability
315.	*Tulipa stellata*	Minmayaaoon	+ +
316.	*Urtica ardens*	Chhoti-Ainh	+ +
317.	*Urtica dioica*	Ainh	+ +
318.	*Vanda tessellata*	Bhangru	+ + +
319.	*Verbascum thapsus*	Ban-Tambakhoo	+ + +
320.	*Vicia faba*	Keon, Keeoon	+ + +
321.	*Vicia hirsuta*	Choti Rodi	+ + +
322.	*Vicia sativa*	Badi-Rodi	+ + +
323.	*Vigna radiata*	Maah	+ +
324.	*Viola patrinii*	Banafsha, Banaksa	+ + +
325.	*Viola pilosa*	Banafsha, Banaksha	+ +
326.	*Vitex negundo*	Bana	+ +
327.	*Vitis repanda*	Maraingu	+
328.	*Vitis vinifera*	Angoor	+ +
329.	*Woodfordia fruticosa*	Dhawain, Dhaun	+ +
330.	*Wrightia arborea*	Khaloa	+
331.	*Zanthoxylum armatum*	Tirmir	+ +
332.	*Zea mays*	Chhali	+ +
333.	*Zephyranthes carinata*	Jangli-Payaas	+ + +
334.	*Zizyphus jujuba*	Ber	+ +
335.	*Zizyphus nummularia*	Jareid, Breid	+ + +

(+ rare extent; + + considerable extent; + + + abundant)

NEUTRACEUTICAL POTENTIAL OF SOME HIMALAYAN EDIBLES

With regard to the nutritional analysis of four of the presently investigated non-traditional edible taxa, *Eleusine coracana* (grains), *Ficus roxburghii* (fruits), *Emblica officinalis* (unripe fruits) and *Pyrus pashia* (unripe fruits), the analysis reveals high contents of carbohydrates in grains of *Eleusine coracana* (73.39%), fruits of *Ficus roxburghii* (51.36%), and fruits of *Emblica officinalis* (45.89%); high content of proteins in unripe fruits of *Pyrus pashia* (22.92%); high contents of fibres in young fruits of *Pyrus pashia* (55.37%) and *Emblica officinalis* (31.09%); and high content of minerals in fruits of *Ficus roxburghii* (9.78%) (Fig. 20; Table XXIX). Regarding nutritional evaluation of vitamins and minerals in lesser known wild edible fruits of *Ficus roxburghii* (present study), the investigation brought to light that these are fairly rich in vit. C (6.37mg/g) (Fig. 21), iron (15.63 mg/g) and calcium (12.02 mg/g) (Figs. 22; Table XXXI). Interestingly, on comparing upon the nutraceutical potential of the above investigated taxa with those reported earlier in literature (Table XXXI), it is observed that some of the presently investigated edible plants (*Eleusine coracana, Ficus roxburghii, Emblica officinalis* and *Pyrus pashia* are rich in micro-nutrients thereby indicating their potentiality for incorporation as a part of human diet in one form or other to overcome nutritional deficiency and maintenance of good health in this part of Himalaya.

Table XXIX Nutritional Analysis of Some Edibles of Palampur Tehsil (on Dry Weight Basis) (present study)

Name of Plant	*Carbohydrates (%)*	*Proteins (%)*	*Fats (%)*	*Ash (%)*	*Fibres (%)*	*Minerals (%)*
Eleusine coracana (Grains)	73.39	9.94	1.52	3.28	3.18	3.28
Ficus roxburghii (Fruits)	51.36	10.70	7.15		13.40	9.78
Emblica officinalis (Fruits)	45.89	10.71	1.02	3.82	31.09	4.01
Pyrus pashia (Unripe Fruits)	9.94	22.92	3. 90	2.44	55.37	3.82

Analysis Based on: Pearson's Chemical Analysis of Foods (1987)

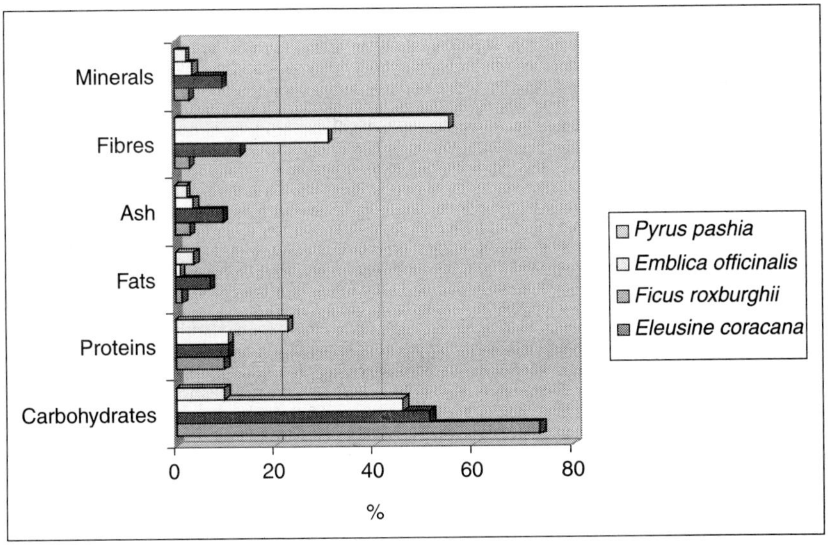

Fig. 20 Nutritional Components of *Eleusine coracana* - (grains); *Ficus roxburghii* (young fruits); *Emblica officinalis* (fruits) and *Pyrus pashia* (fruits).

Table XXX Mineral and Vitamin Components of Fruit Powder of *Ficus roxburghii* (on Dry Weight Basis) (present study)

Phosphorous (P) (%)	*Calcium (Ca) (ppm)*	*Iron (Fe) (ppm)*	*Magnesium (Mg) (ppm)*	*Vitamin A (mg/g)*	*Vitamin B-1(mg/g)*	*Vitamin C (mg/g)*	*Vitamin E (mg/g)*
0.25	12.02	15.63	2.13	0.013	0.026	6.37	0.167
(Standard Method of Chemical Analysis)	(CLL-TP-0040)	(CLL-TP-0040)	(CLL-TP-0040)	(CLL-TP-0068)	(CLL-TP-0109)	(CLL-TP-0109)	(CLL-TP-0068)

Analysis Based on: Pearson's Chemical Analysis of Foods (1987)

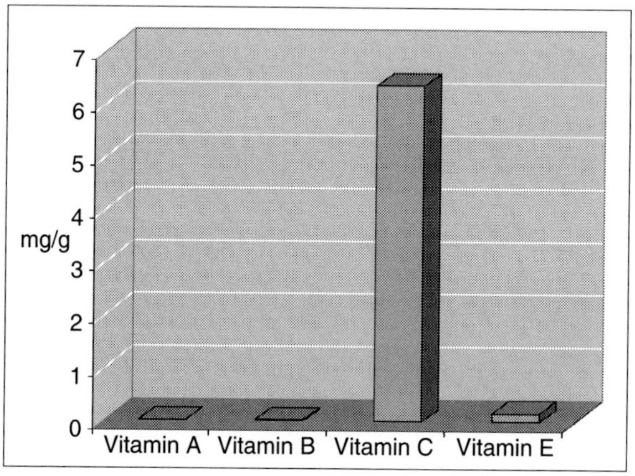

Fig. 21 Histogram Showing Quantitative Reflection of Vitamins (mg/g) in Fruit Powder of *Ficus roxburghii.*

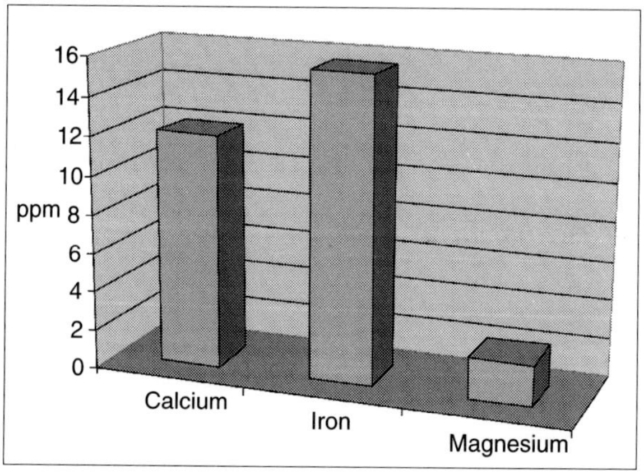

Fig. 22 Histogram Showing Quantitative Reflection of Minerals (mg/g) in Fruit Powder of *Ficus roxburghii.*

Table XXXI Nutritional Components of Some Edible Plants of Tehsil Palampur (on Dry Weight Basis)

Name of Plant	*Carbo-hydrates (%)*	*Protein-s (%)*	*Fats (%)*	*Ash (%)*	*Fibres (%)*	*Starch (%)*	*Minerals (mg/100 gm)*	*Vitamins (I.U. mg / 100 m);*
Amaranthus paniculatus (Aerial Parts)	A, 434; B, 40	-	-	-	-	-	-	-
Cannabis sativa (Seeds)	-	21.51	30.4	4.60	18.64	-	-	A, 434; B, 40
Cynodon dactylon (Aerial Parts)	10.47	-	28.2	11.8	-	-	-	-
Cassia obtusifolia (Seeds)	-	21.12		6.63	13.16	-	0.5%; Ca, 0.4; P, 0.03, Fe, 1.4	A, 260; C, 28
Chenopodium album (Seeds)	-	15.4 -16.8	5.8 -8.1	-	18.4 21.5	-	-	C, 66-96
Dioscorea bulbifera (Root tuber)	75.11-81.39	-	0.8-1.2	-	3.3-9.6		-	-
Dioscorea oppositifolia (Root tuber)	68.54	-	1.52	8.69	6.55	-	-	-
Flacourtia indica (Fruits)	24.20	0.37	-	-	-	-	Ca, 24.1	-
Morus alba (Leaves)	7.0-26.0	16.39	0.4	8.17	1.4	-	0.09%; Ca, 80; P, 40; Fe, 1.9	-
Oxalis corniculata (Leaves)	-	-	-	-	-	-	Ca 5.6%	C, 125; carotene, 3.6
Perilla frutescens (Seeds)	-	23.12	-	-	-	-	-	-
Plantago lanceolata (Leaves)		8.5					Ca,1.6P 0.8-8	
Terminalia bellirica	-	8.63	-	11.7	16-18.59	-	Ca, 2.86; N, 1.36- 2.05	-
Urtica dioica (Leaves)	-	30.4	3.4	16.4	-	-	-	-
Zizyphus jujuba (Fruits)	-	-	-	-	-	-	-	Carotene, 70 mg/100g.

Sources: Anonymous, 1948-1976; Arora & Pandey, 1996; Bennet *et al.,* 1991; Dwivedi, 1993; Rana, 2004; Sharma & Rana, 2005.

NATIVE STATUS OF ETHNOBOTANICAL FLORA

Perusal of Table XXXII and Figs. 23 & 24 revealed that Central, North and South American elements (Brazil, Caribbean Territories, Chile, Mexico, Venezuela, United States), represent the greatest percentage of species in tehsil Palampur of district Kangra (H.P.) (40.00%), followed by Asia (Afghanistan, Bangladesh, Bhutan, China, Japan, Myanmar, Nepal, Pakistan, Sri Lanka, Thailand, etc.) (22.07%), Common Origin (Asia, America, Africa, Europe, Australia) (21.38%), Europe (Austria, Belgium, Bulgaria, France, Germany, Italy, Romania, Spain, Ukraine, etc.) (5.52%) and Africa (Algeria, Egypt, Ethiopia, Ghana, Kenya, South-Africa, Tunisia, Zimbabwe, etc) (8.97%) and Australia (2.07%) (Fig. 24, Table XXXV). These findings are more or less in accordance with those of Nayar (1977) who reported that the alien flora of whole of India is constituted predominantly from American elements (55%). Drake *et al.* (1989), Kowarik (2003), and Perrings *et al.* (2005) opined that various factors like increasing global trade and transport have attributed for the translocation of species either deliberately or by accident, and ultimately in the homogenisation of earth's biota; and are a serious but underestimated problem in biodiversity conservation (Cronk & Fuller, 1995).

Table XXXII Native Status of Aliens of Tehsil Palampur, District Kangra (H.P.)

S.No.	*Continent/Country*	*Species*	*Genera*	*Families*
1.	Asia (Afghanistan, Bangladesh, Bhutan, China, Japan, Myanmar, Nepal, Pakistan, Sri Lanka, Thailand, etc.)	32	31	18
2.	Central, North and South American Continent (Brazil, Caribbean Territories, Chile, Mexico, Venezuela, United States, etc.)	58	51	34
3.	African Continent (Algeria, Egypt, Ethiopia, Ghana, Kenya, South Africa, Tunisia, Zimbabwe, etc.	13	13	07
4.	Europe (Austria, Belgium, Bulgaria, France, Germany, Italy, Romania, Spain, Ukraine, etc.)	08	08	07
5.	Common Origin (Asia, America, Africa, Europe,)	31	31	21
6.	Australian	03	03	03

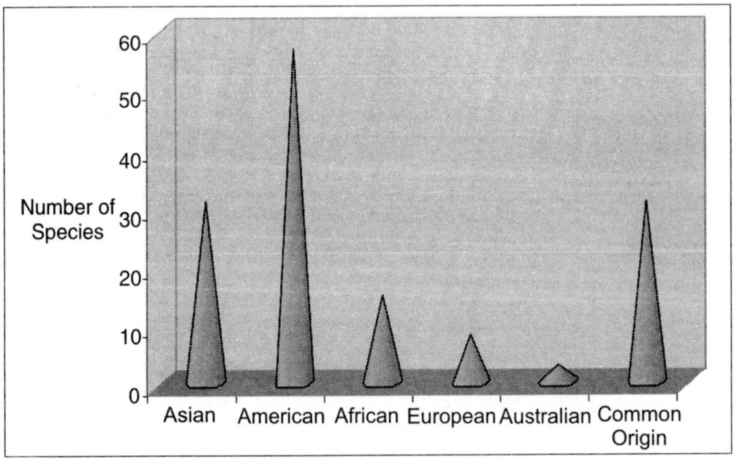

Fig. 23 Histogram Showing Native Status of Ethnobotanically Important Aliens of Tehsil Palampur.

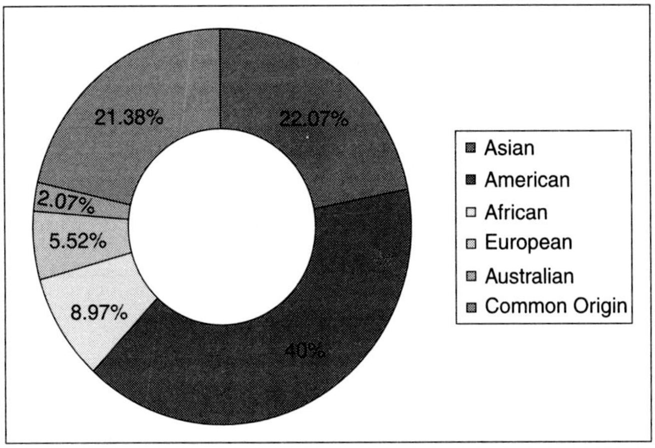

Fig. 24 Doughnut Showing Native Status of Ethnobotanically Important Aliens of Tehsil Palampur.

In comparison, the use of indigenous spp for ethnobotanical purposes is more than the exotic ones by the rural populace of tehsil Palampur (Tables XXXIV, XXXV). Of the 145 exotic aliens belonging to 133 genera under 59 families, use of herbs (51.30%) predominate followed by trees (21.38%), shrubs (12.41%), climbers (8.28%) and undershrubs (6.90%) (Table XXXIII; Fig. 25). Dicotyledonous taxa with 123 spp grouped under 111 genera and 50 families constitute maximum (83.45%) proportion of aliens of the region, followed by Monocots (17 spp) distributed in 17 genera under 4 families, viz., Amaryllidaceae (*Agave americana, Zephyranthes carinata*), Commelinaceae (*Commelina paludosa*), Liliaceae (*Allium cepa, Aloe vera*), Poaceae (*Aristida cyanantha, Arundinaria falcata, Cynodon dactylon, Eleusine coracana, Heteropogon contortus, Panicum antidontale, Pennisetum americanum, Saccharum officinarum, Sorghum bicolor, Sporobolus indicus, Triticum aestivum, Zea mays*), Pteridophytes (3 spp: *Adiantum capillus veneris, Cheilanthes farinosa, Equisetum debile*) and Gymnosperms (2 spp) represented by *Araucaria columnaris* (Araucaricaceae) and *Thuja orientalis* (Cupressaceae) (Tables XXXVI-XXXVIII). Family-wise analysis reveals that out of the 59 alien ethnobotanical families, Fabaceae (15 spp) are at the top followed by Poaceae (12 spp), Solanaceae (8 spp), Asteraceae (7 spp), Cucurbitaceae (6 spp), Lamiaceae (6 spp), Brassicaceae (5 spp), Polygonaceae (5 spp), Rosaceae (5 spp), Convolvulaceae (4 spp), Amaranthaceae, Apiaceae, Caryophyllaceae, Euphorbiaceae, Juglandaceae, Lauraceae and Urticaceae (3 spp each), Chenopodiaceae, Geraniaceae, Liliaceae, Linaceae, Myrtaceae, Ranunculaceae, Salicaceae, Sapindaceae, Verbenaceae with 2 spp each (Tables XXXVII, XXXVIII). The remaining 28 families are with one spp each (Table XL).

Relatively, the top ten ethnobotanical predominant families Fabaceae, Poaceae, Solanaceae, Asteraceae, Cucurbitaceae, Lamiaceae, Brassicaceae, Polygonaceae, Rosaceae and Convolvulaceae with 73 spp account for 50.34 per cent (Table XL). Moreover, the predominant genera with highest number of aliens are *Solanum*

(4 spp), *Vicia* (3 spp), *Chenopodium, Fagopyrum, Ipomoea, Lathyrus, Mentha, Prunus, Ranunculus* (2 spp each) and the remaining 124 genera are represented by one sp. (Table XLI). As a matter of fact, alien crops used for ethnobotanical purposes are: *Allium cepa, Brassica nigra, Camellia sinensis, Capsicum annum, Cucurbita melo* var. *agrestis, C. pepo, Fagopyrum cymosum, F. esculentum, Ipomoea batatas, Linum usitatissimum, Litchi sinensis, Momordica charantia, Pennisetum americanum, Saccharum officinarum, Sechium edule, Sesamum indicum, Solanum melongena, Trifolium resupinatum, Triticum aestivum* and *Vitis vinifera* (Table XXXIX)

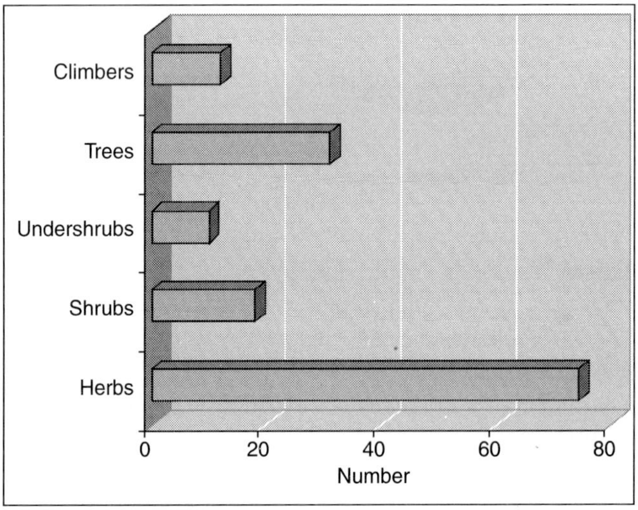

Fig. 25 Histogram Showing Ethnobotanical Aliens (Herbs, Shrubs, Undershrubs, Trees and Climbers) of Tehsil Palampur.

Table XXXIII Percentage of Aliens Belonging to Different Categories

Categories	*Sub-Categories*	*Percentage of Spp*
Origin	Asian	22.07%
	American (Central, South & North)	40.00%
	African	8.97%
	European	5.52%
	Australian	2.07%
	Common Origin	21.38%
Habit	Herbs	51.03%
	Shrubs	12.41%
	Undershrubs	6.90%
	Trees	21.38%
	Climbers	8.28%
	Lianas	0.00%

Table XXXIV Exotic Ethnobotanical Resources of Tehsil Palampur

S.No.	Botanical Name	Family	Origin
1.	*Achyranthes aspera*	Amaranthaceae	Mediterranean, N. Africa, Europe
2.	*Adiantum capillus veneris*	Polypodiaceae	Caribbean Territories, N. America
3.	*Agave cantula*	Amaryllidaceae	Mexican Highlands
4.	*Ageratum conyzoides*	Asteraceae	Caribbean Territories, C. America
5.	*Albizia julibrissin*	Fabaceae	Continental USA
6.	*Allium cepa*	Liliaceae	Central Asia
7.	*Aloe vera*	Liliaceae	E. Africa & S. Africa
8.	*Alternanthera sessilis*	Amaranthaceae	Caribbean Territories
9.	*Amaranthus viridis*	Amaranthaceae	Caribbean Territories
10.	*Apium leptophyllum*	Apiaceae	U.S.A.
11.	*Araucaria columnaris*	Araucariaceae	New Caledonia
12.	*Aristida cyanantha*	Poaceae	Mexico, Zambia
13.	*Artemisia capillaris*	Asteraceae	Euresia
14.	*Arundinaria falcata*	Poaceae	Chinese-Japanese Region
15.	*Bidens biternata*	Asteraceae	Java, South Africa
16.	*Brassica nigra*	Brassicaceae	N. America
17.	*Camellia sinensis*	Theaceae	China
18.	*Capsicum annuum*	Solanaceae	Tropical America
19.	*Carica papaya*	Caricaceae	Southern Mexico and C. America
20.	*Carya illinoensis*	Juglandaceae	Central USA
21.	*Cassia occidentalis*	Fabaceae	Tropical America
22.	*Centipeda minima*	Asteraceae	Euresia
23.	*Cestrum nocturnum*	Solanaceae	North and South America
24.	*Cheilanthes farinosa*	Cheilanthaceae	Malay Peninsula
25.	*Chenopodium album*	Chenopodiaceae	South America, Central America, Mexico
26.	*Chenopodium ambrosioides*	Chenopodiaceae	Central America
27.	*Cicer arietinum*	Fabaceae	Ethopia, China
28.	*Cinnamomum camphora*	Lauraceae	China and Japan
29.	*Cirsium arvense*	Asteraceae	Mediterranean Region
30.	*Cissampelos parriera*	Menispermaceae	Caribbean Territories, N. America
31.	*Commelina paludosa*	Commelinaceae	U.S.A.
32.	*Coronopus didymus*	Brassicaceae	Europe, S. America
33.	*Cryptolepis buchanani*	Asclepiadaceae	Continental USA

Contd...

Table XXXIV Contd...

S.No.	Botanical Name	Family	Origin
34.	*Cucumis melo* var. *agrestis*	Cucurbitaceae	Africa
35.	*Cucurbita pepo*	Cucurbitaceae	Mexico
36.	*Cuscuta reflexa*	Convolvulaceae	USA
37.	*Cyclanthera pedata*	Cucurbitaceae	S. America
38.	*Cynodon dactylon*	Poaceae	Tropical Africa
39.	*Cyphomandra betacea*	Solanaceae	S. America
40.	*Datura innoxia*	Solanaceae	Central and South America
41.	*Diospyros kaki*	Ebenaceae	China, Turkey
42.	*Diplocyclos palmatus*	Cucurbitaceae	Africa
43.	*Dodonaea viscosa*	Sapindaceae	Australia
44.	*Duranta repens*	Verbenaceae	C. America
45.	*Eleusine coracana*	Poaceae	Africa, Ethopia
46.	*Engelhartia colebrookeana*	Juglandaceae	East Asia, Japan, Vietnam
47.	*Equisetum debile*	Equisetaceae	Europe, Asia
48.	*Eriobotrya japonica*	Rosaceae	China and Japan
49.	*Erythrina suberosa*	Fabaceae	C. & S. America, C. Asia
50.	*Eucalyptus tereticornis*	Myrtaceae	Australia
51.	*Euonymus hamiltonianus*	Celastraceae	USA
52.	*Fagopyrum cymosum*	Polygonaceae	Chinese Center
53.	*Fagopyrum esculentum*	Polygonaceae	China
54.	*Ficus clavata*	Urticaceae	Tibet
55.	*Flacourtia indica*	Flacourtiaceae	Caribbean Territories
56.	*Foeniculum vulgare*	Apiaceae	Mediterranean Region
57.	*Fumaria indica*	Fumariaceae	Turkey
58.	*Galium aparine*	Rubiaceae	Alaska, Bulgaria
59.	*Glycine max*	Fabaceae	Caribbean Territories, China
60.	*Gnaphalium affine*	Asteraceae	China, Korea, Japan, Taiban
61.	*Grevillea robusta*	Proteaceae	Queensland and New South Wales
62.	*Heteropogon contortus*	Poaceae	Caribbean Territories, N. America, Oceania
63.	*Hibiscus rosa-sinensis*	Malvaceae	China
64.	*Impatiens balsamina*	Geraniaceae	China
65.	*Ipomoea batatas*	Convolvulaceae	Central Mexican Center
66.	*Ipomoea turbinata*	Convolvulaceae	N. America

Contd...

Table XXXIV Contd...

S.No.	Botanical Name	Family	Origin
67.	*Jasminum dispermum*	Oleaceae	China
68.	*Juglans regia*	Juglandaceae	Asia-Temperate-Afghanistan, erbaijan, Iran, Iraq, Turkey, Tajikistan, Turkmenistan, Asia-Tropical- Pakistan, Europe – Bulgaria, former Yugoslavia, Greece
69.	*Kalanchoe pinnata*	Crassulaceae	Madagascar
70.	*Lantana camara*	Verbenaceae	Caribbean Territories, Central U.S., Hawaii, Australia
71.	*Lathyrus aphaca*	Fabaceae	Continental U.S., Caribbean Territories
72.	*Lathyrus sativus*	Fabaceae	Abyssinian Center
73.	*Lepidium sativum*	Brassicaceae	Latin America
74.	*Leucaena glauca*	Fabaceae	C. America
75.	*Leucas cephalotes*	Lamiaceae	C. Asia
76.	*Linum usitatissimum*	Linaceae	Continental US
77.	*Litchi chinensis*	Sapindaceae	China
78.	*Loranthus falcatus*	Loranthaceae	C. Asia
79.	*Lotus corniculatus*	Fabaceae	Europe
80.	*Medicago polymorpha*	Fabaceae	Medditerranean
81.	*Mentha longifolia*	Lamiaceae	Africa-Algeria, Ethiopia, Kenya, Lesotho, Morocco, Namibia, S. Africa, Swaziland, Tunisia, Zimbabwe, ASIA Temperate-Afghanistan, Cyprus, Egypt, Iran, Iraq, Israel, Lebanon, Syria, Turkey, Armenia, Azerbaijan, Georgia, Russian Federation, Ukraine, Albania, Bulgaria, Former Yugoslavia, Italy, Greece, Romania, France, Portugal, Spain
82.	*Mentha piperita*	Lamiaceae	Europe
83.	*Mirabilis jalapa*	Nyctaginaceae	Mexico
84.	*Morus alba*	Urticaceae	China, Europe and N. America
85.	*Neolitsea umbrosa*	Lauraceae	Afghanistan, China, Japan, Java
86.	*Ocimum kilimandscharicum*	Lamiaceae	East Africa
87.	*Opuntia dillenii*	Cactaceae	U.S.A. to Ecuador, Caribbean Territories
88.	*Origanum vulgare*	Lamiaceae	Mediterranean basin to Central Asia
89.	*Ougeinia dalbergioides*	Fabaceae	East Asia, Japan, Vietnam
90.	*Oxalis corniculata*	Geraniaceae	Caribbean Territories
91.	*Panicum antidontale*	Poaceae	Continental USA

Contd...

Table XXXIV Contd...

S.No.	*Botanical Name*	*Family*	*Origin*
92.	*Pennisetum americanum*	Poaceae	Africa
93.	*Persea odoratissima*	Lauraceae	Mexico, Vietnam
94.	*Phyllanthus urinaria*	Euphorbiaceae	Caribbean Territories, N. America
95.	*Plantago lanceolata*	Plantaginaceae	Africa
96.	*Plumbago zeylanica*	Plumbaginaceae	Caribbean Territories
97.	*Polygala chinensis*	Polygonaceae	China
98.	*Polygonum barbatum*	Polygonaceae	S. America
99.	*Populus nigra*	Salicaceae	Africa- Algeria, Morocco, Tunisia, Asia-temperate- Armenia, Azerbaijan, China, Georgia, Iran, Kazakhstan, Russian Federation (Ciscaucasia, Dagestan, Eastern Siberia, Western Siberia), Tajikistan, Turkey, EUROPE-Albania, Austria, Belarus, Belgium, Bulgaria, Czechoslovakia, Former Yugoslavia, France, Germany, Hungary, Italy, Netherlands, Poland, Romania, Russian Federation (European part), Spain, Switzerland, Ukraine
100.	*Porana paniculata*	Convolvulaceae	Caribbean Territories
101.	*Portulaca grandiflora*	Portulacaceae	Caribbean Territories, Persia, China
102.	*Prunus domestica*	Rosaceae	Eurasia
103.	*Prunus persica*	Rosaceae	China
104.	*Psidium guajava*	Myrtaceae	Tropical America
105.	*Punica granatum*	Punicaceae	Iran
106.	*Pyrus communis*	Rosaceae	Eurasia
107.	*Ranunculus arvensis*	Ranunculaceae	North America
108.	*Ranunculus muricatus*	Ranunculaceae	N. America, Oceania
109.	*Raphanus sativus*	Brassicaceae	China, Middle Asia
110.	*Reinwardtia indica*	Linaceae	Euresia
111.	*Rhamnus triquetra*	Rhamnaceae	Continental U.S.
112.	*Rhododendron arboreum*	Ericaceae	Eurasia, Europe, Armenia
113.	*Ricinus communis*	Euphorbiaceae	N. America
114.	*Rorippa nasturtium-aquaticum*	Brassicaceae	Latin America
115.	*Rosa chinensis*	Rosaceae	China
116.	*Rumex nepalensis*	Polygonaceae	Ethopia

Contd...

Table XXXIV Contd...

S.No.	Botanical Name	Family	Origin
117.	*Saccharum officinarum*	Poaceae	South Pacific (probably New Guinea)
118.	*Salix karelini*	Salicaceae	Turkmenistan, China
119.	*Salvia coccinea*	Lamiaceae	Mexico
120.	*Sapium sebiferum*	Euphorbiaceae	China
121.	*Scandix pecten veneris*	Apiaceae	N. Africa, Mediterranean Origin
122.	*Sechium edule*	Cucurbitaceae	S. Mexico
123.	*Sesamum orientale*	Pedaliaceae	Africa
124.	*Silene conoidea*	Caryophyllaceae	Continental US
125.	*Sisymbrium irio*	Brassicaceae	USA
126.	*Solanum erianthum*	Solanaceae	Caribbean Territories, N. America
127.	*Solanum melongena*	Solanaceae	S. America
128.	*Solanum torvum*	Solanaceae	Caribbean Territories
129.	*Solanum viarum*	Solanaceae	S. America
130.	*Sorghum bicolor*	Poaceae	Ethopia
131.	*Sporobolus indicus*	Poaceae	Caribbean Territories, Africa
132.	*Stellaria media*	Caryophyllaceae	Croatia, Armenia
133.	*Tagetes minuta*	Asteraceae	Continental US
134.	*Thuja orientalis*	Cupressaceae	Mediterranean, China, S. Korea
135.	*Trifolium resupinatum*	Fabaceae	China, Mediterranean
136.	*Triticum aestivum*	Poaceae	C. Asia, USA
137.	*Urtica dioica*	Urticaceae	Europe
138.	*Verbascum thapsus*	Scrophulariaceae	Europe
139.	*Vicia faba*	Fabaceae	Continental US
140.	*Vicia hirsuta*	Fabaceae	East Asia
141.	*Vicia sativa*	Fabaceae	China, Mediterranean Region
142.	*Viola patrinii*	Violaceae	N. America
143.	*Vitis vinifera*	Vitaceae	Turkey, Europe, Caribbean Territories, Tunisia, Afghanistan
144.	*Zea mays*	Poaceae	Tropical America
145.	*Zephyranthes carinata*	Amaryllidaceae	S. America

Table XXXV Indigenous Ethnobotanical Resources of Tehsil Palampur

S.No.	Botanical Name	Family	Origin
1.	*Abelmoschus esculentus*	Malvaceae	India, Asia, Caribbean Territories
2.	*Abrus precatorius*	Fabaceae	India
3.	*Acacia catechu*	Fabaceae	Bangladesh, Bhutan, China, India, Nepal, Pakistan, Sri Lanka, Thailand

Contd...

Table XXXV Contd...

S.No.	Botanical Name	Family	Origin
4.	*Achyranthes bidentata*	Amaranthaceae	Tropical Asia
5.	*Acorus calamus*	Araceae	India
6.	*Aegle marmelos*	Rutaceae	India
7.	*Aerva lanata*	Amaranthaceae	India
8.	*Agave wightii*	Amaryllidaceae	India, East Asia
9.	*Ajuga parviflora*	Lamiaceae	China, Japan, India
10.	*Albizia chinensis*	Fabaceae	Bangladesh, Bhutan, India, Myanmar, Nepal, New Guinea, Philippines, Pakistan, Malaysia, Sri Lanka, Thailand, Vietnam
11.	*Albizia lebbeck*	Fabaceae	Bangladesh, Bhutan, India, Java, Myanmar, Nepal, Pakistan, Malaysia, Sri Lanka
12.	*Allium sativum*	Liliaceae	N.W. India, Afghanistan, Tajikistan, USA
13.	*Amaranthus paniculatus*	Amaranthaceae	S. Asia, Central America
14.	*Artemisia japonica*	Asteraceae	Africa, Asia, Caribbean Territories
15.	*Artocarpus lacucha*	Urticaceae	India, Bangladesh
16.	*Asparagus adscendens*	Liliaceae	India
17.	*Avena sativa*	Poaceae	Asia Minor, South-East Europe
18.	*Bambusa nutans*	Poaceae	Indo-Malayan
19.	*Basella alba*	Chenopodiaceae	Africa, S.E. Asia
20.	*Bauhinia vahlii*	Fabaceae	India
21.	*Bauhinia variegata*	Fabaceae	India, China, Myanmar, Nepal, Pakistan, Thailand
22.	*Berberis lycium*	Berberidaceae	India, Nepal, Pakistan
23.	*Bergenia ciliata*	Saxifragaceae	Afghanistan, Bhutan, India
24.	*Boehmeria macrophylla*	Urticaceae	Africa, Oman, Arabia, Yemen, China, Japan, Vietnam, India, Nepal, Indonesia, Malaysia, Philippines, U.S., Mexico, Fiji, Panama, Cuba, Jamaica, Bolivia, Ecuador, Peru, Argentina, Chile, Paraguay and Uruguay
25.	*Boerhaavia diffusa*	Nyctaginaceae	India, Himalayas
26.	*Bombax ceiba*	Malvaceae	Australia, Bhutan, India, Cambodia, Indonesia, Malaysia, Myanmar, Nepal, Sri Lanka, Taiwan, Thailand, Vietnam
27.	*Brassica tournefortii*	Brassicaceae	N. America , Africa, Asia

Contd...

Table XXXV Contd...

S.No.	*Botanical Name*	*Family*	*Origin*
28.	*Butea monosperma*	Fabaceae	India
29.	*Caesalpinia decapetala*	Fabaceae	Asia
30.	*Cajanus cajan*	Fabaceae	Asia, India
31.	*Callicarpa macrophylla*	Verbenaceae	India
32.	*Calotropis procera*	Asclepiadaceae	Africa-cape Verde, Algeria, Egypt, Libya, Morocco, Eritrea, Ethiopia, Somalia, Sudan, Kenya, Tanzania, Uganda, Cameroon, Equatorial Guinea, Gambia, Ghana, Guinea Bissau, Mauritania, Nigeria, Senegal, Sierra Leone, Asia- Temperate- Oman, Saudi Arabia, Yemen, Afghanistan, Egypt, Iran, Israel, Jordan, Asia-tropical- India, Nepal, Pakistan, Myanmar, Thailand, Vietnam
33.	*Cannabis sativa*	Urticaceae	Central Asia and Himalayas
34.	*Capillipedium assimile*	Poaceae	Indo-Malaya
35.	*Carissa congesta*	Apocynaceae	Asia, India
36.	*Caryopteris bicolor*	Verbenaceae	Hindustan, Himalayas, Asia, China
37.	*Cassia fistula*	Fabaceae	Bangladesh, Java, Laos, India, Malaysia, Sri Lanka, Sumatra
38.	*Cassia floribunda*	Fabaceae	Bangladesh, Java, India, Malaysia, Myanmar, Nepal, Moluccas, Kalimantan, Sri Lanka, Sumatra
39.	*Cassia obtusifolia*	Fabaceae	India- Tamilnadu, Satpura Range
40.	*Catharanthus roseus*	Apocynaceae	Madagascar and India
41.	*Cedrus deodara*	Pinaceae	Himalayas, USA
42.	*Celastrus paniculatus*	Celastraceae	India
43.	*Celtis australis*	Urticaceae	N. India and Temperate Himalayas
44.	*Centella asiatica*	Apiaceae	India
45.	*Centratherum anthelminticum*	Asteraceae	India
46.	*Cinnamomum tamala*	Lauraceae	South Asia, Sri Lanka
47.	*Citrus grandis*	Rutaceae	S.E.Asia & Caribbean Territories
48.	*Citrus jambhiri*	Rutaceae	Hindustani Center
49.	*Citrus limon*	Rutaceae	India, China, Australia, New Zealand, Mexico, U.S.A., Italy, Sicily, France, Spain
50.	*Citrus medica*	Rutaceae	India, China

Contd...

Table XXXV Contd...

S.No.	Botanical Name	Family	Origin
51.	*Citrus medica var. medica*	Rutaceae	India
52.	*Clematis gauriana*	Ranunculaceae	N.India, Himalayas, Nepal
53.	*Coccinia grandis*	Cucurbitaceae	India
54.	*Cocculus laurifolius*	Menispermaceae	Himalayas, Japan
55.	*Colebrookea oppositifolia*	Lamiaceae	Asia
56.	*Colocasia esculenta*	Araceae	India, China, Tropical East Asia
57.	*Cordia dichotoma*	Boraginaceae	India, China, Indonesia, Japan, Malaysia, Myanmar, Pakistan, Sri Lanka, Vietnam, Thailand, Laos
58.	*Coriandrum sativum*	Apiaceae	Asia Minor
59.	*Costus speciosus*	Zingiberaceae	India
60.	*Cucumis sativus*	Cucurbitaceae	India
61.	*Cucurbita hispida*	Cucurbitaceae	Indo-China
62.	*Curcuma longa*	Zingiberaceae	India
63.	*Cymbopogon citratus*	Poaceae	India, Sri Lanka, Malaysia
64.	*Dalbergia sissoo*	Fabaceae	Afghanistan, Bangladesh, Bhutan, India, Iran, Iraq, Myanmar, Nepal, Oman, Pakistan
65.	*Datura stramonium*	Solanaceae	Caspian regions of Europe, N. America, Indian subcontinent
66.	*Debregeasia hypoleuca*	Urticaceae	W. Asia- W. Himalayas
67.	*Deeringia amaranthoides*	Amaranthaceae	Australia, China, Bhutan, India, Indonesia, Malaysia, Laos, Taiwan, Thailand, Vietnam
68.	*Dendrocalamus hamiltonii*	Poaceae	NE India, Nepal, Bhutan
69.	*Didymocarpus pedicellata*	Gesneriaceae	India
70.	*Dioscorea bellophylla*	Dioscoreaceae	Hindustani Center
71.	*Dioscorea bulbifera*	Dioscoreaceae	Asia
72.	*Dioscorea oppositifolia*	Dioscoreaceae	China, Asia
73.	*Diplazium esculentum*	Athyriaceae	India, Polynesia
74.	*Dolichos biflorus*	Fabaceae	Hindustan center
75.	*Drimia indica*	Liliaceae	India

Contd...

Table XXXV Contd...

S.No.	Botanical Name	Family	Origin
76.	*Ehretia acuminata*	Boraginaceae	Australia, Bhutan, China, India, Indonesia, Taiwan
77.	*Ehretia laevis*	Boraginaceae	Assam, Burma, Bengladesh
78.	*Emblica officinalis*	Euphorbiaceae	Cambodia, China, Taiwan, Bangladesh, Bhutan, Fiji, India, Malaysia, New Caledonia, Papua New Guinea, Myanmar
79.	*Eriophorum comosum*	Cyperaceae	Asia
80.	*Eulaliopsis binata*	Poaceae	Asia
81.	*Euphorbia royleana*	Euphorbiaceae	India-Himalayas, Bhutan, Nepal, Pakistan, Myanmar
82.	*Ficus benghalensis*	Urticaceae	India, Pakistan
83.	*Ficus carica*	Urticaceae	Asia Minor
84.	*Ficus foveolata*	Urticaceae	Asia
85.	*Ficus glomerata*	Urticaceae	Australia, China, India, Indonesia, Malaysia, Myanmar, Nepal, Pakistan, Papua New Guinea, Sri Lanka, Thailand, Vietnam
86.	*Ficus hispida*	Urticaceae	India, Asia
87.	*Ficus infectoria*	Urticaceae	India
88.	*Ficus palmata*	Urticaceae	Afghanistan, Egypt, Ethiopia, Iran, India, Israel, Nepal, Oman, Pakistan, Somalia, Sudan
89.	*Ficus religiosa*	Urticaceae	Bangladesh, India, Nepal, Pakistan, Thailand, Vietnam
90.	*Ficus roxburghii*	Urticaceae	China, Bhutan, India, Nepal, Malaysia, Myanmar, Thailand
91.	*Ficus rumphii*	Urticaceae	India, Myanmar, Indo-China, Thailand, Malaya, Java
92.	*Ficus semicordata*	Urticaceae	India
93.	*Fragaria indica*	Rosaceae	India
94.	*Geranium nepalense*	Geraniaceae	Himalayas, Nepal
95.	*Gonatanthus pumilus*	Araceae	India
96.	*Gossypium arboreum*	Malvaceae	S.E. Asia
97.	*Grewia disperma*	Tiliaceae	Iran, India
98.	*Grewia elastica*	Tiliaceae	Iran, India
99.	*Grewia optiva*	Tiliaceae	Bhutan, India - Himachal Pradesh, Jammu and Kashmir, Uttar Pradesh, Pakistan

Contd...

Table XXXV Contd...

S.No.	Botanical Name	Family	Origin
100.	*Hamiltonia suaveolens*	Rubiaceae	Indo-China
101.	*Helicteres isora*	Sterculiaceae	India
102.	*Holarrhena antidysentrica*	Apocynaceae	India
103.	*Hypericum oblogifolium*	Hypericaceae	India- Himachal Pradesh, Jammu and Kashmir, Uttar Pradesh, Nepal, Pakistan
104.	*Indigofera cassioides*	Fabaceae	Himalayas
105.	*Jasminum grandiflorum*	Oleaceae	India
106.	*Jasminum humile*	Oleaceae	Afghanistan, Bhutan, China, India, Nepal, Pakistan, Tajikistan
107.	*Jasminum multiflorum*	Oleaceae	India
108.	*Jatropha curcas*	Euphorbiaceae	Asia
109.	*Justicia adhatoda*	Acanthaceae	India
110.	*Lannea coromandelica*	Anacardiaceae	Bangladesh, Bhutan, Cambodia, China, India, Laos, Myanmar, Nepal, Pakistan, Sri Lanka, Thailand, Vietnam
111.	*Lawsonia inermis*	Lythraceae	W. Asia, N. Africa, India
112.	*Lepidagathis cuspidata*	Acanthaceae	S.W. Himalayas
113.	*Litsea monopetala*	Lauraceae	Asia, India
114.	*Litsea sebifera*	Lauraceae	India and South China, Malaysia, Australia, Western Pacific Island
115.	*Litsea semicarpifolia*	Lauraceae	Tropical Asia
116.	*Luffa acutangula*	Cucurbitaceae	India
117.	*Luffa cylindrica*	Cucurbitaceae	India
118.	*Mallotus philippinensis*	Euphorbiaceae	Australia, Bangladesh, Bhutan, China, India, Japan, Malaysia, Myanmar, Nepal, Philippines, Sri Lanka, Taiwan, Vietnam
119.	*Mangifera indica*	Anacardiaceae	South - East Asia, Indomalayan
120.	*Melia azedarach*	Meliaceae	Australia, Bhutan, China, India, Indonesia, Japan, Laos, Nepal, Papua Guinea, Pacific Islands, Philippines, Sri Lanka, Thailand, Vietnam
121.	*Michelia champaca*	Magnoliaceae	E. Himalaya, Assam, Western Ghats
122.	*Micromeria biflora*	Lamiaceae	India- Himalayas
123.	*Milletia extensa*	Fabaceae	S.E. Asia

Contd...

Table XXXV Contd...

S.No.	Botanical Name	Family	Origin
124.	*Momordica charantia*	Cucurbitaceae	Southern Asia
125.	*Momordica dioica*	Cucurbitaceae	India
126.	*Moringa oleifera*	Moringaceae	India
127.	*Mucuna pruriens*	Fabaceae	India
128.	*Murraya koenigii*	Rutaceae	India, Subtropical forests in much of southern Asia
129.	*Murraya paniculata*	Rutaceae	China, S.E. Asia, Malay Peninsula
130.	*Musa sapientum*	Musaceae	Tropical Asia
131.	*Nerium indicum*	Apocynaceae	Africa- Algeria, Ethiopia, Libya, Morocco, Tunisia, Asia temperate - Afghanistan, china Cyprus, Iran, Iraq, Israel, Jordan, Lebanon, Oman, United Arab Emirates, Syria, Turkey, Asia Tropical- India, Nepal, Pakistan, Europe- Albania, Former Yugoslavia, Greece, Italy, Malta, France, Portugal, Spain
132.	*Ocimum basilicum*	Lamiaceae	Tropical Asia
133.	*Ocimum gratissimum*	Lamiaceae	Africa, Angola, Ghana, Guinea, India, Kenya, Malawi, Malaysia, Namibia, Nepal, Rwanda, Somalia, South Madagascar, Sri Lanka, Tanzania, Thailand, Uganda
134.	*Ocimum sanctum*	Lamiaceae	India and other tropical regions of Asia
135.	*Oroxylum indicum*	Bignoniaceae	India
136.	*Pentanema indicum*	Asteraceae	Asia
137.	*Perilla frutescens*	Lamiaceae	Asia
138.	*Phoenix acaulis*	Arecaceae	India, Myanmar
139.	*Phoenix sylvestris*	Arecaceae	India, Nepal
140.	*Pinus roxburghii*	Pinaceae	Afghanistan, China, Pakistan, Bhutan, India, Myanmar, Nepal
141.	*Pistacia khinjuk*	Anacardiaceae	Afghanistan, India, Nepal, Pakistan, Myanmar
142.	*Pogostemon benghalense*	Lamiaceae	India, Afghanistan
143.	*Populus ciliata*	Salicaceae	China, Bhutan, India, Nepal, Myanmar
144.	*Prinsepia utilis*	Rosaceae	Temperate Himalaya
145.	*Prunus cerasoides*	Rosaceae	India

Contd...

Table XXXV Contd...

S.No.	Botanical Name	Family	Origin
146.	*Pyrus pashia*	Rosaceae	Afghanistan, Bhutan, China, India, Nepal, Myanmar, Pakistan, Thailand, Vietnam
147.	*Pyrus serotina*	Rosaceae	Asia Minor
148.	*Quercus leucotrichophora*	Fagaceae	Himalayas, India
149.	*Randia dumetorum*	Rubiaceae	India
150.	*Ranunculus laetus*	Ranunculaceae	India
151.	*Rosa brunonii*	Rosaceae	India
152.	*Rosa damascena*	Rosaceae	India
153.	*Roylea cinerea*	Lamiaceae	India, Nepal
154.	*Rubus ellipticus*	Rosaceae	S. E. Asia, Himalayas
155.	*Rubus niveus*	Rosaceae	Asia, China, India
156.	*Rumex hastatus*	Polygonaceae	Asia, India
157.	*Saccharum spontaneum*	Poaceae	India, Asia
158.	*Salix tetrasperma*	Salicaceae	India
159.	*Santalum album*	Santalaceae	India
160.	*Sapindus mukorosii*	Sapindaceae	India
161.	*Sapium insigne*	Euphorbiaceae	India, China, Pen. Malaysia
162.	*Selaginella chrysocaulos*	Selaginellaceae	S. W. China, N. Myanmar, N.E. India, Himalaya
163.	*Sida cordifolia*	Malvaceae	India
164.	*Solanum nigrum*	Solanaceae	India, Asia,Africa
165.	*Spilanthes oleracea*	Asteraceae	South Asia, India
166.	*Spondias pinnata*	Anacardiaceae	S.E. Asia, India
167.	*Stephania glabra*	Menispermaceae	Himalaya, China, Caucasian
168.	*Syzygium cumini*	Myrtaceae	India
169.	*Syzygium operculatum*	Myrtaceae	Asia, India
170.	*Terminalia arjuna*	Combretaceae	India
171.	*Terminalia bellirica*	Combretaceae	India, Malaysia, Philippines
172.	*Terminalia chebula*	Combretaceae	India, Tropical Asia
173.	*Terminalia tomentosa*	Combretaceae	India
174.	*Thysanolaena maxima*	Poaceae	S.E. Asia
175.	*Tinospora cordifolia*	Menispermaceae	India, Mayanmar, Sri Lanka

Contd...

Table XXXV Contd...

S.No.	Botanical Name	Family	Origin
176.	*Toona ciliata*	Meliaceae	Afghanistan, Australia, Bangladesh, Bhutan, Cambodia, China, India, Indonesia, Laos, Malaysia, Myanmar, Nepal, Pakistan, Papua New Guinea, Philippines, Sri Lanka, Thailand
177.	*Trichosanthes cucumerina*	Cucurbitaceae	Africa, Asia, Australia
178.	*Trigonella foenum-graecum*	Fabaceae	India, Ethiopia
179.	*Tulipa stellata*	Liliaceae	East Asia, Himalayas, Central and N. America
180.	*Urtica ardens*	Urticaceae	India
181.	*Vanda tessellata*	Orchidaceae	Indo-China
182.	*Vigna radiata*	Fabaceae	India, Japan
183.	*Viola pilosa*	Violaceae	Bhutan, China, India, Myanmar, Nepal, Pakistan
184.	*Vitex negundo*	Verbenaceae	India, Asia, Sri Lanka
185.	*Vitis repanda*	Vitaceae	India
186.	*Woodfordia fruticosa*	Lythraceae	India
187.	*Wrightia arborea*	Apocynaceae	Bihar, Assam, Punjab, Rajasthan and Western Peninsula
188.	*Zanthoxylum armatum*	Rutaceae	Bangladesh, Bhutan, China, India, Indonesia, Japan, Korea, Myanmar, Nepal, Pakistan, Taiwan, Thailand, Vietnam
189.	*Zizyphus jujuba*	Rhamnaceae	India, Myanmar
190.	*Zizyphus nummularia*	Rhamnaceae	India

Table XXXVI Taxonomic Components of Ethnobotanical Aliens of Tehsil Palampur

Plant Group	No. of Families	No. of Genera	No. of Species
Dicotyledons	50	111	123
Monocotyledons	04	17	17
Gymnosperms	02	02	02
Pteridophytes	03	03	03
Total	59	133	145

Table XXXVII Exotic Dicot Families of Tehsil Palampur

Family	*Botanical Name*
Amaranthaceae	*Achyranthes aspera, Alternanthera sessilis, Amaranthus viridis*
Apiaceae	*Apium leptophyllum, Foeniculum vulgare, Scandix pecten veneris*
Asclepiadaceae	*Cryptolepis buchanani*
Asteraceae	*Ageratum conyzoides, Artemisia capillaris, Bidens biternata, Centipeda minima, Cirsium arvense, Gnaphalium affine, Tagetes minuta*
Brassicaceae	*Brassica nigra, Coronopus didymus, Lepidium sativum, Rorippa nasturtium-aquaticum, Sisymbrium irio*
Cactaceae	*Opuntia dillenii*
Caryophyllaceae	*Silene conoidea, Stellaria media*
Caricaceae	*Carica papaya*
Celastraceae	*Euonymus hamiltonianus*
Chenopodiaceae	*Chenopodium album, Chenopodium ambrosioides*
Crassulaceae	*Kalanchoe pinnata*
Cucurbitaceae	*Cucumis melo* var. *agrestis, Cucurbita pepo, Cyclanthera pedata, Diplocyclos palmatus, Momordica charantia, Sechium edule*
Ebenaceae	*Diospyros kaki*
Ericaceae	*Rhododendron arboretum*
Euphorbiaceae	*Phyllanthus urinaria, Ricinus communis, Sapium sebiferum*
Fabaceae	*Albizia julibrissin, Cassia occidentalis, Cicer arietinum, Erythrina suberosa, Glycine max, Lathyrus aphaca, Lathyrus sativus, Leucaena glauca, Lotus corniculatus, Medicago polymorpha, Ougeinia dalbergioides, Trifolium resupinatum, Vicia faba, Vicia hirsuta, Vicia sativa*
Flacourtiaceae	*Flacourtia indica*
Fumariaceae	*Fumaria indica*
Geraniaceae	*Impatiens balsamina, Oxalis corniculata*
Juglandaceae	*Carya illinoensis, Engelhardia colebrookeana, Juglans regia*
Lamiaceae	*Leucas cephalotes, Mentha longifolia, Mentha piperita, Ocimum kilimandscharicum, Origanum vulgare, Salvia coccinea*
Lauraceae	*Cinnamomum camphora, Neolitsea umbrosa, Persea odoratissima*
Linaceae	*Linum usitatissimum, Reinwardtia indica*
Loranthaceae	*Loranthus falcatus*
Malvaceae	*Hibiscus rosa-sinensis*
Menispermaceae	*Cissampelos parriera*
Myrtaceae	*Eucalyptus tereticornis, Psidium guajava*
Nyctaginaceae	*Mirabilis jalapa*
Oleaceae	*Jasminum dispermum*

Contd...

Table XXXVII Contd...

Family	*Botanical Name*
Pedaliaceae	*Sesamum orientale*
Plantaginaceae	*Plantago lanceolata*
Plumbaginaceae	*Plumbago zeylanica*
Polygonaceae	*Fagopyrum cymosum, Fagopyrum esculentum, Polygala chinensis, Polygonum barbatum, Rumex nepalensis*
Portulacaceae	*Portulaca grandiflora*
Proteaceae	*Grevillea robusta*
Punicaceae	*Punica granatum*
Ranunculaceae	*Ranunculus arvensis, Ranunculus muricatus*
Rhamnaceae	*Rhamnus triquetra*
Rosaceae	*Eriobotrya japonica, Prunus domestica, Prunus persica, Pyrus communis, Rosa chinensis*
Rubiaceae	*Galium aparine*
Salicaceae	*Populus nigra, Salix karelini*
Sapindaceae	*Dodonaea viscosa, Litchi chinensis*
Scrophulariaceae	*Verbascum thapsus*
Solanaceae	*Capsicum annuum, Cestrum nocturnum, Cyphomandra betacea, Datura innoxia, Solanum erianthum, Solanum melongena, Solanum torvum, Solanum viarum*
Theaceae	*Camellia sinensis*
Urticaceae	*Ficus clavata, Morus alba, Urtica dioica*
Verbenaceae	*Duranta repens, Lantana camara*
Violaceae	*Viola patrinii*
Vitaceae	*Vitis vinifera*

Table XXXVIII Exotic Monocot Families of Tehsil Palampur

Family	*Name*
Amaryllidaceae	*Agave cantula, Zephyranthes carinata*
Commelinaceae	*Commelina paludosa*
Liliaceae	*Allium cepa, Aloe vera*
Poaceae	*Aristida cyanantha, Arundinaria falcata, Cynodon dactylon, Eleusine coracana, Heteropogon contortus, Panicum antidontale, Pennisetum americanum, Saccharum officinarum, Sorghum bicolor, Sporobolus indicus, Triticum aestivum, Zea mays*

Table XXXIX Alien Crops of Tehsil Palampur

S.No.	Name	Family
1.	*Allium cepa*	Liliaceae
2.	*Brassica nigra*	Brassicaceae
3.	*Camellia sinensis*	Theaceae
4.	*Capsicum annuum*	Solanaceae
5.	*Cucurbita melo* var. *agrestis*	Cucurbitaceae
6.	*C. pepo*	Cucurbitaceae
7.	*Fagopyrum cymosum*	Polygonaceae
8.	*F. esculentum*	Polygonaceae
9.	*Ipomoea batatas*	Convolvulaceae
10.	*Linum usitatissimum*	Linaceae
11.	*Litchi sinensis*	Sapindaceae
12.	*Momordica charantia*	Cucurbitaceae
13.	*Pennisetum americanum*	Poaceae
14.	*Saccharum officinarum*	Poaceae
15.	*Sechium edule*	Cucurbitaceae
16.	*Sesamum indicum*	Pedaliaceae
17.	*Solanum melongena*	Solanaceae
18.	*Trifolium resupinatum*	Fabaceae
19.	*Triticum aestivum*	Poaceae
20.	*Vitis vinifera*	Vitaceae

Table XL Predominant Ethnobotanical Alien Families of Tehsil Palampur

S.No.	Family	Total Number of Spp
1.	Fabaceae	15
2.	Poaceae	12
3.	Solanaceae	8
4.	Asteraceae	7
5.	Cucurbitaceae	6
6.	Lamiaceae	6
7.	Brassicaceae	5
8.	Polygonaceae	5
9.	Rosaceae	5
10.	Convolvulaceae	4
11.	Amaranthaceae	3
12.	Apiaceae	3

Contd...

Table XL Contd...

S.No.	*Family*	*Total Number of Spp*
13.	Caryophyllaceae	3
14.	Euphorbiaceae	3
15.	Juglandaceae	3
16.	Lauraceae	3
17.	Urticaceae	3
18.	Liliaceae	2
19.	Chenopodiaceae	2
20.	Geraniaceae	2
21.	Linaceae	2
22.	Myrtaceae	2
23.	Ranunculaceae	2
24.	Salicaceae	2
25.	Sapindaceae	2
26.	Verbenaceae	2
27.	Amaryllidaceae	1
28.	Asclepiadaceae	1
29.	Cactaceae	1
30.	Caricaceae	1
31.	Celastraceae	1
32.	Commelinaceae	1
33.	Crassulaceae	1
34.	Ebenaceae	1
35.	Ericaceae	1
36.	Flacourtiaceae	1
37.	Fumariaceae	1
38.	Loranthaceae	1
39.	Malvaceae	1
40.	Menispermaceae	1
41.	Nyctaginaceae	1
42.	Oleaceae	1
43.	Pedaliaceae	1
44.	Plantaginaceae	1
45.	Plumbaginaceae	1
46.	Portulacaceae	1
47.	Proteaceae	1

Contd...

Table XL Contd...

S.No.	Family	Total Number of Spp
48.	Punicaceae	1
49.	Rhamnaceae	1
50.	Rubiaceae	1
51.	Scrophulariaceae	1
52.	Theaceae	1
53.	Violaceae	1
54.	Vitaceae	1

Table XLI Predominant Alien Genera of Tehsil Palampur

S.No.	Genus	Number of Spp
1.	*Solanum*	4
2.	*Vicia*	3
3.	*Chenopodium*	2
4.	*Fagopyrum*	2
5.	*Ipomoea*	2
6.	*Lathyrus*	2
7.	*Mentha*	2
8.	*Prunus*	2
9.	*Ranunculus*	2
10.	*Achyranthes*	1
11.	*Agave*	1
12.	*Ageratum*	1
13.	*Albizia*	1
14.	*Allium*	1
15.	*Aloe*	1
16.	*Alternanthera*	1
17.	*Amaranthus*	1
18.	*Apium*	1
19.	*Aristida*	1
20.	*Artemisia*	1
21.	*Arundinaria*	1
22.	*Bidens*	1
23.	*Brassica*	1
24.	*Camellia*	1
25.	*Capsicum*	1

Contd...

Table XLI Contd...

S.No.	Genus	Number of Spp
26.	*Carica*	1
27.	*Carya*	1
28.	*Cassia*	1
29.	*Centipeda*	1
30.	*Cestrum*	1
31.	*Cicer*	1
32.	*Cinnamomum*	1
33.	*Cirsium*	1
34.	*Cissampelos*	1
35.	*Commelina*	1
36.	*Coronopus*	1
37.	*Cryptolepis*	1
38.	*Cucumis*	1
39.	*Cucurbita*	1
40.	*Cuscuta*	1
41.	*Cyclanthera*	1
42.	*Cynodon*	1
43.	*Cyphomandra*	1
44.	*Datura*	1
45.	*Diospyros*	1
46.	*Diplocyclos*	1
47.	*Dodonaea*	1
48.	*Duranta*	1
49.	*Eleusine*	1
50.	*Engelhardia*	1
51.	*Eriobotrya*	1
52.	*Erythrina*	1
53.	*Eucalyptus*	1
54.	*Euonymus*	1
55.	*Ficus*	1
56.	*Flacourtia*	1
57.	*Foeniculum*	1
58.	*Fumaria*	1
59.	*Galium*	1
60.	*Glycine*	1

Contd...

Table XLI Contd...

S.No.	Genus	Number of Spp
61.	*Gnaphalium*	1
62.	*Grevillea*	1
63.	*Heteropogon*	1
64.	*Hibiscus*	1
65.	*Impatiens*	1
66.	*Jasminum*	1
67.	*Juglans*	1
68.	*Kalanchoe*	1
69.	*Lantana*	1
70.	*Lepidium*	1
71.	*Leucaena*	1
72.	*Leucas*	1
73.	*Linum*	1
74.	*Litchi*	1
75.	*Loranthus*	1
76.	*Lotus*	1
77.	*Medicago*	1
78.	*Mirabilis*	1
79.	*Momordica*	1
80.	*Morus*	1
81.	*Neolitsea*	1
82.	*Ocimum*	1
83.	*Opuntia*	1
84.	*Origanum*	1
85.	*Ougeinia*	1
86.	*Oxalis*	1
87.	*Panicum*	1
88.	*Pennisetum*	1
89.	*Persea*	1
90.	*Phyllanthus*	1
91.	*Plantago*	1
92.	*Plumbago*	1
93.	*Polygala*	1
94.	*Polygonum*	1
95.	*Populus*	1
96.	*Porana*	1

Contd...

Table XLI Contd...

S.No.	Genus	Number of Spp
97.	*Portulaca*	1
98.	*Psidium*	1
99.	*Punica*	1
100.	*Pyrus*	1
101.	*Reinwardtia*	1
102.	*Rhamnus*	1
103.	*Rhododendron*	1
104.	*Ricinus*	1
105.	*Rorippa*	1
106.	*Rosa*	1
107.	*Rumex*	1
108.	*Saccharum*	1
109.	*Salix*	1
110.	*Salvia*	1
111.	*Sapium*	1
112.	*Scandix*	1
113.	*Sechium*	1
114.	*Sesamum*	1
115.	*Silene*	1
116.	*Sisymbrium*	1
117.	*Sorghum*	1
118.	*Sporobolus*	1
119.	*Stellaria*	1
120.	*Tagetes*	1
121.	*Trifolium*	1
122.	*Triticum*	1
123.	*Urtica*	1
124.	*Verbascum*	1
125.	*Viola*	1
126.	*Vitis*	1
127.	*Zea*	1
128.	*Zephyranthes*	1

ECONOMIC VALUATION OF SOURCE FLORA

Of the data on total importance value (TIV) of 335 plants worked out presently (Table XLII), *Dendrocalamus hamiltonii* (Poaceae) tops the list with highest value

Table XLII Total Importance Values (TIV) of Ethnobotanical Important Plants of Tehsil Palampur

Botanical Name	Dye	Fibre	Fodder	Fuel	Medi-cine	Oil	Tannin /Gum	Timber	Others	Edible	TIV%
Abelmoschus esculentus	-	1	-	-	-	-	-	-	1	2	20
Abrus precatorius	-	-	-	-	2	-	-	-	2	-	20
Acacia catechu	-	-	-	-	2	-	-	2	2	-	30
Achyranthes aspera	-	-	-	-	2	-	-	-	1	-	15
Achyranthes bidentata	-	-	-	-	2	-	-	-	1	-	15
Acorus calamus	-	-	-	-	2	-	-	-	2	-	20
Adiantum capillus veneris	-	-	-	-	-	-	-	-	-	2	10
Aegle marmelos	-	-	-	-	2	-	-	-	2	2	30
Aerva lanata	-	-	-	-	2	-	-	-	2	-	20
Agave cantula	-	-	-	-	2	-	-	-	2	1	25
Agave wightii	-	-	-	-	2	-	-	-	-	1	15
Ageratum conyzoides	-	-	-	-	1	-	-	-	1	-	10
Ajuga parviflora	-	-	-	-	2	-	-	-	2	-	20
Albizia chinensis	-	-	2	2	2	-	-	2	2	-	50
Albizia julibrissin	-	-	2	2	2	-	-	2	-	-	40
Albizia lebbeck	-	-	2	2	2	-	-	2	-	-	40
Allium cepa	-	-	-	-	2	-	-	-	-	2	20
Allium sativum	-	-	-	-	2	-	-	-	-	2	20
Aloe vera	-	-	-	-	2	-	-	-	-	2	20
Alternanthera sessilis	-	-	2	-	1	-	-	-	-	2	25
Amaranthus paniculatus	-	-	-	-	-	-	-	-	-	2	10

Contd...

Table XLII Contd...

Botanical Name	Dye	Fibre	Fodder	Fuel	Medi-cine	Oil	Tannin /Gum	Timber	Others	Edible	TIV%
Amaranthus viridis	-	-	-	-	2	-	-	-	-	2	20
Apium leptophyllum	-	-	-	-	2	-	-	-	-	1	15
Araucaria columnaris	-	-	-	-	-	-	-	1	1	-	10
Aristida cyanantha	-	-	-	-	-	-	-	-	2	-	10
Artemisia capillaris	-	-	-	-	2	-	-	-	-	-	10
Artemisia japonica	-	-	-	-	2	-	-	-	2	-	20
Artocarpus lacucha	-	-	2	-	-	-	-	-	-	2	20
Arundinaria falcata	-	-	-	-	-	-	-	2	2	-	30
Asparagus adscendens	-	-	-	-	2	-	-	-	2	-	20
Avena sativa	-	-	2	-	-	-	-	-	-	-	10
Bambusa nutans	-	-	2	-	-	-	-	2	2	-	30
Basella alba	-	-	-	-	2	-	-	-	-	2	20
Bauhinia vahlii	-	2	2	-	1	-	-	-	2	1	40
Bauhinia variegata	2	-	2	2	2	-	-	-	-	2	40
Berberis lycium	-	-	2	-	2	-	-	-	-	2	30
Bergenia ciliata	-	-	-	-	2	-	-	-	-	2	20
Bidens biternata	-	-	-	-	2	-	-	-	-	-	10
Boehmeria macrophylla	-	-	-	-	-	-	-	-	2	-	10
Boerhaavia diffusa	-	-	-	-	2	-	-	-	-	2	20
Bombax ceiba	-	-	-	-	2	-	-	2	2	2	40
Brassica nigra	-	-	-	-	-	2	-	-	1	2	25

Contd...

Table XLII Contd...

Botanical Name	Dye	Fibre	Fodder	Fuel	Medi-cine	Oil	Tannin /Gum	Timber	Others	Edible	TIV%
Brassica tournefortii	-	-	-	-	2	2	-	-	2	2	40
Butea monosperma	-	-	2	-	2	-	-	-	2	1	35
Caesalpinia decapetala	-	-	2	-	1	-	-	-	-	-	15
Cajanus cajan	-	-	-	1	2	-	-	-	-	2	20
Callicarpa macrophylla	-	-	-	-	2	-	-	-	-	2	20
Calotropis procera	-	-	-	-	2	-	-	-	2	-	20
Camellia sinensis	2	-	-	-	2	-	-	-	2	-	30
Cannabis sativa	-	-	-	-	2	-	-	-	1	-	15
Capillipedium assimile	-	-	2	-	1	-	-	-	2	-	25
Capsicum annuum	-	-	-	-	1	-	-	-	2	2	25
Carica papaya	-	-	-	-	2	-	-	-	-	2	20
Carissa congesta	-	-	1	2	2	-	-	-	1	2	40
Carya illinoensis	2	-	-	-	2	-	-	-	2	2	40
Caryopteris bicolor	-	-	-	2	-	-	-	-	1	-	15
Cassia fistula	-	-	-	2	2	-	-	-	2	-	30
Cassia floribunda	-	-	-	-	2	-	-	-	-	2	20
Cassia obtusifolia	-	-	-	-	2	-	-	-	1	1	20
Cassia occidentalis	-	-	-	-	2	-	-	-	1	1	20
Catharanthus roseus	-	-	-	-	2	-	-	-	2	-	20
Cedrus deodara	-	-	-	-	-	-	-	2	-	-	10
Celastrus paniculatus	-	-	-	-	1	-	-	-	-	2	15

Contd...

Table XLII Contd...

Botanical Name	Dye	Fibre	Fodder	Fuel	Medi-cine	Oil	Tannin /Gum	Timber	Others	Edible	TIV%
Celtis australis	-	-	2	2	-	-	-	2	2	-	40
Centella asiatica	-	-	-	-	2	-	-	-	-	-	10
Centipeda minima	-	-	-	-	2	-	-	-	1	-	15
Centratherum anthelminticum	-	-	-	-	2	-	-	-	-	-	10
Cestrum nocturnum	-	-	-	-	-		-	-	2	-	10
Cheilanthes farinosa	-	-	-	-	2	-	-	-	1	-	15
Chenopodium album	-	-	-	-	2	-	-	-	-	2	20
Chenopodium ambrosioides	-	-	-	-	-	-	-	-	-	2	10
Cicer arietinum	-	-	-	-	-	-	-	-	-	2	10
Cinnamomum camphora	-	-	-	-	2	-	-	-	2	-	20
Cinnamomum tamala	-	-	-	-	2	-	-	-	2	-	20
Cirsium arvense	-	-	-	-	2	-	-	-	2		20
Cissampelos parriera	-	-	-	-	2	-	-	-	-		10
Citrus grandis	-	-	-	-	1	-	-	-	-	2	15
Citrus jambhiri	-	-	-	-	-	-	-	-	2	2	20
Citrus limon	-	-	-	-	2	-	-	-	2	2	30
Citrus medica	-	-	-	-	1	-	-	-	2	2	25
Citrus medica var. medica	-	-	-	-	-	-	-	-	2	2	20
Clematis gauriana	-	-	-	-	2	-	-	-	1	-	15
Coccinia grandis	-	-	-	-	-	-	-	-	-	2	10
Cocculus laurifolius	-	-	-	-		-	-	-	2	-	10

Contd...

Table XLII Contd...

Botanical Name	Dye	Fibre	Fodder	Fuel	Medi-cine	Oil	Tannin /Gum	Timber	Others	Edible	TIV%
Colobrookia oppositifolia	-	-	-	2	1	-	-	-	-	-	15
Colocasia esculenta	-	-	-	-		-	-	-	-	2	10
Commelina paludosa	-	-	-	-	1	-	-	-	-	2	15
Cordia dichotoma	-	-	2	2	2	-	-	-	-	2	40
Coriandrum sativum	-	-	-	-	1	-	-	-	1	2	20
Coronopus didymus	-	-	-	-		-	-	-	-	2	10
Costus speciosus	-	-	-	-	2	-	-	-	-	-	10
Cryptolepis buchanani	-	1	1	-	1	-	-	-	-	-	15
Cucumis melo var. agrestis	-	-	-	-	-	-	-	-	-	2	10
Cucumis sativus	-	-	-	-	2	-	-	-	2	2	30
Cucurbita hispida	-	-	-	-	-	-	-	-	2	2	20
Cucurbita pepo	-	-	-	-	-	-	-	-	2	2	20
Curcuma longa	-	-	-	-	2	-	-	-	2	-	20
Cuscuta reflexa	-	-	-	-	2	-	-	-	2	-	20
Cyclanthera pedata	-	-	-	-	-	-	-	-	-	2	10
Cymbopogon citratus	-	-	2	-	2	-	-	-	1	-	25
Cynodon dactylon	-	-	2	-	2	-	-	-	2	-	30
Cyphomandra betacea	-	-	-	-	2	-	-	-	-	2	20
Dalbergia sissoo	-	-	-	2	2	-	-	2	2	-	40
Datura innoxia	-	-	-	-	2	-	-	-	2	-	20
Datura stramonium	-	-	-	-	1	-	-	-	2	-	15

Contd...

Table XLII Contd...

Botanical Name	Dye	Fibre	Fodder	Fuel	Medi-cine	Oil	Tannin /Gum	Timber	Others	Edible	TIV%
Debregeasia hypoleuca	-	-	2	2	1	-	-	-	-	1	30
Deeringia amaranthoides	-	-	-	-	1	-	-	-	-	2	15
Dendrocalamus hamiltonii	-	-	2	2	2	-	-	2	2	2	60
Didymocarpus pedicellata	-	-	-	-	2	-	-	-	-	-	10
Dioscorea bellophylla	-	-	-	-	2	-	-	-	-	2	20
Dioscorea bulbifera	-	-	-	-	2	-	-	-	-	2	20
Dioscorea oppositifolia	-	-	-	-	1	-	-	-	-	2	15
Diospyros kaki	-	-	-		1	-	-	-	-	2	15
Diplazium esculentum	-	-	-	-	-	-	-	-	-	2	10
Diplocyclos palmatus	-	-	-	-	2	-	-	-	2	-	20
Dodonaea viscosa	-	-	-	2	2	-	-	-	2	-	30
Dolichos biflorus	-	-	-	-	2	-	-	-	-	2	20
Drimia indica	-	-	-	-	2	-	-	-	-	-	10
Duranta repens	-	-	-	2	-	-	-	-	2	-	20
Ehretia acuminata	-	-	2	-	-	-	-	2	2	-	30
Ehretia laevis	-	-	2	1	1	-	-	2	2	-	40
Eleusine coracana	-	-	-	-	2	-	-	-	-	2	20
Emblica officinalis	1	-	-	-	2	-	2	-	1	2	40
Engelhardia colebrookeana	-	-	-	-	1	-	-	2	2	-	25
Equisetum debile	-	-	-	-	2	-	-	-	-	-	10
Eriobotrya japonica	-	-	-	-	-	-	-	-	-	2	10

Contd...

Table XLII Contd...

Botanical Name	Dye	Fibre	Fodder	Fuel	Medi-cine	Oil	Tannin /Gum	Timber	Others	Edible	TIV%
Eriophorum comosum	-	-	2	-	-	-	-	-	2	-	20
Erythrina suberosa	-	-	-	2	2	-	-	-	2	-	30
Eucalyptus tereticornis	-	-	-	-	2	-	-	-	2	-	20
Eulaliopsis binata	-	-	2	-	1	-	-	-	2	-	25
Euonymus hamiltonianus	-	-	1	1	-	-	-	-	1	-	15
Euphorbia royleana	-	-	-	-	2	-	-	-	2	1	25
Fagopyrum cymosum	-	-	-	-	-	-	-	-	-	2	10
Fagopyrum esculentum	-	-	-	-	-	-	-	-	-	2	10
Ficus benghalensis	-	-	-	-	2	-	-	-	2	1	25
Ficus carica	-	-	-	-	1		-	-	-	2	15
Ficus clavata	-	-	2	-	-	-	-	-	-	2	20
Ficus foveolata	-	-	2	-	-	-	-	-	-	1	15
Ficus glomerata	-	-	2	2	1	-	-	-	2	-	35
Ficus hispida	-	-	1	-	2	-	-	-	-	2	25
Ficus infectoria	-	-	2	-	-	-	-	-	2	-	20
Ficus palmata	-	-	-	-	2	-	-	-	2	2	30
Ficus religiosa	-	-	-	-	2	-	-	-	2		20
Ficus roxburghii	-	-	2	-	-	-	-	-	2	2	30
Ficus rumphii	-	-	2	-	-	-	-	-	-	-	10
Ficus semicordata	-	-	2	-	-	-	-	-	2	2	30
Flacourtia indica	-	-	1	1	2	-	-	-	1	2	35

Contd...

Table XLII Contd...

Botanical Name	Dye	Fibre	Fodder	Fuel	Medi-cine	Oil	Tannin /Gum	Timber	Others	Edible	TIV%
Foeniculum vulgare	-	-	-	-	2	-	-	-	-	2	20
Fragaria indica	-	-	-	-	-	-	-	-	-	2	10
Fumaria indica	-	-	2	-	1	-	-	-	-	-	15
Galium aparine	-	-	1	-	-	-	-	-	-	1	10
Geranium nepalense	-	-	-	-	1	-	-	-	-	-	5
Glycine max	-	-	-	-	1	-	-	-	-	2	15
Gnaphalium affine	-	-	-	-	-	-	-	-	1	-	5
Gonatanthus pumilus	-	-	-	-	-	-	-	-	-	2	10
Gossypium arboreum	-	-	-	-	-	-	-	-	2	-	10
Grevillea robusta	-	-	-	-	-	-	-	-	2	-	10
Grewia disperma	-	2	2	2	1	-	-	-	2	1	50
Grewia elastica	-	2	2	2	-	-	-	-	2	1	45
Grewia optiva	-	2	2	2	1	-	-	-	2	1	50
Hamiltonia suaveolens	-	-	-	2	-	-	-	-	-	-	10
Helicteres isora	-	-	2	2	-	-	-	-	-	-	20
Heteropogon contortus	-	-	2	-	2	-	-	-	-	-	20
Hibiscus rosa-sinensis	-	-	-	-	2	-	-	-	2	-	20
Holarrhena antidysentrica	-	-	-	-	2	-	-	-	2	-	20
Hypericum oblogifolium	-	-	1	-	1	-	-	-	-	-	10
Impatiens balsamina	-	-	-	-	1	-	-	-	1	-	10
Indigofera cassioides	-	-	1	-	1	-	-	-	-	2	20

Contd...

Table XLII Contd...

Botanical Name	Dye	Fibre	Fodder	Fuel	Medi-cine	Oil	Tannin /Gum	Timber	Others	Edible	TIV%
Ipomoea batatas	-	-	-	-	-	-	-	-	-	2	10
Ipomoea turbinata	-	-	-	-	-	-	-	-	-	2	10
Jasminum dispermum	-	-	-	-	1	-	-	-	1	-	10
Jasminum grandiflorum	-	-	-	-	2	-	-	-	1	-	15
Jasminum humile	-	-	-	-	-	-	-	-	1	-	5
Jasminum multiflorum	-	-	-	-	1	-	-	-	1	-	10
Jatropha curcas	-	-	-	-	2	-	-	-	2	-	20
Juglans regia	1	-	-	-	1	-	-	1	1	2	30
Justicia adhatoda	-	-	-	-	2	-	-	-	2	-	20
Kalanchoe pinnata	-	-	-	-	2	-	-	-	1	-	15
Lannea coromandelica	-	-	1	1	1	-	-	-	-	-	15
Lantana camara	-	-	-	-	1	-	-	-	2	-	15
Lathyrus aphaca	-	-	2	-	-	-	-	-	-	2	20
Lathyrus sativus	-	-	-	-	-	-	-	-	-	2	10
Lawsonia inermis	-	-	-	-	2	-	-	-	-	-	10
Lepidagathis cuspidata	-	-	-	-	2	-	-	-	-	-	10
Lepidium sativum	-	-	-	-	2	-	-	-	-	2	20
Leucaena glauca	-	-	2	2	-	-	-	-	-	-	20
Leucas cephalotes	-	-	-	-	2	-	-	-	-	-	10
Linum usitatissimum	-	-	-	-	-	2	-	-	1	2	25
Litchi chinensis	-	-	-	-	1	-	-	-	-	2	15

Contd...

Table XLII Contd...

Botanical Name	Dye	Fibre	Fodder	Fuel	Medi-cine	Oil	Tannin /Gum	Timber	Others	Edible	TIV%
Litsea monopetala	-	-	2	-	-	-	-	1	2	1	30
Litsea sebifera	-	-	-	-	2	-	-	1	-	-	15
Litsea semicarpifolia	-	-	2	-	-	-	-	1	-	-	15
Loranthus falcatus	-	-	1	-	1	-	-	-	-	-	10
Lotus corniculatus	-	-	-	-	-	-	-	-	1	-	5
Luffa acutangula	-	-	-	-	-	-	-	-	1	2	15
Luffa cylindrica	-	-	-	-	-	-	-	-	1	2	15
Mallotus philippinensis	-	-	1	1	2	-	-	-	2	-	30
Mangifera indica	-	-	-	-	2	-	-	2	2	2	40
Medicago polymorpha	-	-	2	-	-	-	-	-	-	2	20
Melia azedarach	-	-	-	-	2	-	-	2	2	-	30
Mentha longifolia	-	-	-	-	2	-	-	-	-	2	20
Mentha piperita	-	-	-	-	2	-	-	-	-	2	20
Michelia champaca	-	-	-	-	2	-	-	-	2	-	20
Micromeria biflora	-	-	-	-	2	-	-	-	-	-	10
Milletia extensa	-	2	2	-	-	-	-	-	1	-	25
Mirabilis jalapa	-	-	-	-	1	-	-	-	1	-	10
Momordica charantia	-	-	-	-	2	-	-	-	-	2	20
Momordica dioica	-	-	-	-	-	-	-	-	-	2	10
Moringa oleifera	-	-	2	2	2	-	-	-	-	2	40
Morus alba	-	-	2	-	-	-	-	-	2	2	30

Contd...

Table XLII Contd...

Botanical Name	Dye	Fibre	Fodder	Fuel	Medi-cine	Oil	Tannin /Gum	Timber	Others	Edible	TIV%
Mucuna pruriens	-	-	-	-	2	-	-	-	2	-	20
Murraya koenigii	-	-	-	2	2	-	-	-	2	2	40
Murraya paniculata	-	-	-	-	-	-	-	-	2	2	20
Musa sapientum	-	1	2	-	2	-	-	-	2	2	45
Neolitsea umbrosa	-	-	-	-	-	-	-	1	-	1	10
Nerium indicum	-	-	-	-	2	-	-	-	2	-	20
Ocimum basilicum	-	-	-	-	2	-	-	-	1	-	15
Ocimum gratissimum	-	-	-	-	2	-	-	-	-	-	10
Ocimum kilimandscharicum	-	-	-	-	2	-	-	-	2	-	20
Ocimum sanctum	-	-	-	-	2	-	-	-	2	-	20
Opuntia stricta	-	-	-	-	2	-	-	-	1	1	20
Origanum vulgare	-	-	-	-	2	-	-	-	-	-	10
Oroxylum indicum	-	-	-	2	2	-	-	-	-	1	25
Ougeinia dalbergioides	-	-	2	-	-	-	-	-	1	1	20
Oxalis corniculata	-	-	-	-	2	-	-	-	2	2	30
Panicum antidontale	-	-	2	-	1	-	-	-	1	-	20
Pennisetum americanum	-	-	2	-	1	-	-	-	-	2	25
Pentanema indicum	-	-	-	-	2	-	-	-	-	-	10
Perilla frutescens	-	-	-	-	2	-	-	-	-	-	10
Persea odoratissima	-	-	2	-	-	-	-	2	-	-	20
Phoenix acaulis	-	-	2	-	1	-	-	-	2	2	35

Contd...

Table XLII Contd...

Botanical Name	Dye	Fibre	Fodder	Fuel	Medi-cine	Oil	Tannin /Gum	Timber	Others	Edible	TIV%
Phoenix sylvestris	-	-	2	-	1	-	-	-	2	2	35
Phyllanthus urinaria	-	-	-	-	2	-	-	-	-	-	10
Pinus roxburghii	-	-	-	1	1	-	-	1	2	1	30
Pistacia khinjuk	-	-	1	-	2	-	-	-	1	-	20
Plantago lanceolata	-	-	-	-	2	-	-	-	-	-	10
Plumbago zeylanica	-	-	-	-	2	-	-	-	1	-	15
Pogostemon benghalense	-	-	-	-	2	-	-	-	-	-	10
Polygala chinensis	-	-	1	-	-	-	-	-	1	-	10
Polygonum barbatum	-	-	1	-	1	-	-	-	-	-	10
Populus ciliata	-	-	1	-	-		-	2	2	-	20
Populus nigra	-	-	1	1	-	-	-	1	2	-	25
Porana paniculata	-	1	1	-	-	-	-	-	-	-	10
Portulaca grandiflora	-	-	-	-	2	-	-	-	-	2	20
Prinsepia utilis	-	-	-	-	2	-	-	-	1	-	15
Prunus cerasoides	-	-	1	-	2	-	-	-	2	1	30
Prunus domestica	-	-	-	1	-	-	-	-	-	2	15
Prunus persica	-	-	-	-	2	1	-	-	1	2	30
Psidium guajava	-	-	-	-	2	-	-	-	1	2	25
Punica granatum	-	-	-	-	2	-	-	-	1	2	25
Pyrus communis	-	-	-	-	2	-	-	-	-	2	20
Pyrus pashia	-	-	1	-	2	-	-	-	2	2	35

Contd...

Table XLII Contd...

Botanical Name	Dye	Fibre	Fodder	Fuel	Medi-cine	Oil	Tannin /Gum	Timber	Others	Edible	TIV%
Pyrus serotina	-	-	-	-	-	-	-	-	-	2	10
Quercus leucotrichophora	-	-	2	-	-	-	-	2	-	-	20
Randia dumetorum	-	-	1	2	2	-	-	-	-	1	30
Ranunculus arvensis	-	-	-	-	1	-	-	-	-	-	05
Ranunculus laetus	-	-	-	-	1	-	-	-	-	-	05
Ranunculus muricatus	-	-	-	-	-	-	-	-	-	2	10
Raphanus sativus	-	-	-	-	1	-	-	-	-	2	15
Reinwardtia indica	-	-	-	-	2	-	-	-	1	1	20
Rhamnus triquetra	-	-	1	2	-	-	-	-	1	-	20
Rhododendron arboreum	-	-	-	2	2	-	-	-	2	2	40
Ricinus communis	-	--	-	-	2	-	-	-	1	-	15
Rorippa nasturtium-aquaticum	-	-	-	-	1	-	-	-	-	2	15
Rosa brunonii	-	-	-	-	1	-	-	-	-	-	05
Rosa chinensis	-	-	-	-	2	-	-	-	2	-	20
Rosa damascena	-	-	-	-	2	-	-	-	1	-	15
Roylea cinerea	-	-	-	-	2	-	-	-	1	-	15
Rubus ellipticus	-	-	1	-	1	-	-	-	-	2	20
Rubus niveus	-	-	1	-	1	-	-	-	1	2	25
Rumex hastatus	-	-	-	-	2	-	-	-	-	1	15
Rumex nepalensis	-	-	-	-	2	-	-	-	-	1	15
Saccharum officinarum	-	-	-	-	2	-	-	-	2	2	30

Contd...

Table XLII Contd...

Botanical Name	Dye	Fibre	Fodder	Fuel	Medi-cine	Oil	Tannin /Gum	Timber	Others	Edible	TIV%
Saccharum spontaneum	-	-	-	-	-	-	-	-	2	-	10
Salix karelini	-	-	1	2	-	-	-	-	2	-	25
Salix tetrasperma	-	-	-	2	-	-	-	1	2	-	25
Salvia coccinea	-	-	-	-	-	-	-	-	2	-	10
Santalum album	-	-	-	-	2	-	-	-	2	-	20
Sapindus mukorosii	-	-	-	-	1	-	-	2	2	-	25
Sapium insigne	-	-	-	-	1	-	-	-	1	-	10
Sapium sebiferum	-	-	2	1	-	-	-	1	2	-	30
Scandix pecten veneris	-	-	1	-	-	-	-	-	-	2	15
Sechium edule	-	-	-	-	1	-	-	-	-	2	15
Selaginella chrysocaulos	-	-	-	-	2	-	-	-	1	-	15
Sesamum orientale	-	-	-	-	2	2	-	-	2	2	40
Sida cordifolia	-	-	-	-	2	-	-	-	-	-	10
Silene conoidea	-	-	2	-	-	-	-	-	-	2	20
Sisymbrium irio	-	-	-	-	1	-	-	-	1	-	10
Solanum erianthum	-	-	-	-	2	-	-	-	-	-	10
Solanum melongena	-	-	-	-	-	-	-	-	-	2	10
Solanum nigrum	-	-	-	-	2	-	-	-	-	1	15
Solanum torvum	-	-	-	-	-	-	-	-	-	2	10
Solanum viarum	-	-	-	-	2	-	-	-	-	-	10
Sorghum bicolor	-	-	2	-	1	-	-	-	-	2	25

Contd...

Table XLII Contd...

Botanical Name	Dye	Fibre	Fodder	Fuel	Medi-cine	Oil	Tannin /Gum	Timber	Others	Edible	TIV%
Spilanthes oleracea	-	-	-	-	2	-	-	-	-	-	10
Spondias pinnata	-	-	1	-	-	-	-	-	-	2	15
Sporobolus indicus	-	-	2	-	-	-	-	-	2	-	20
Stellaria media	-	-	2	-	-	-	-	-	-	2	20
Stephania glabra	-	-	-	-	2	-	-	-	-	-	10
Syzygium cumini	-	-	-	-	2	-	-	2	2	2	40
Syzygium operculatum	-	-	1	2	-	-	-	-	2	2	35
Tagetes minuta	-	-	-	-	2	-	-	-	1	-	15
Terminalia arjuna	-	-	1	-	2	-	-	1	2	1	35
Terminalia bellirica	-	-	2	2	2	-	-	-	2	2	50
Terminalia chebula	-	-	1	2	2	-	-	-	-	1	30
Terminalia tomentosa	-	-	2	2	-	-	2	2	2	-	50
Thuja orientalis	-	-	-	-	1	-	-	-	2	-	15
Thysanolaena maxima	-	-	2	-	-	-	-	-	2	-	20
Tinospora cordifolia	-	-	2	-	2	-	-	-	2	-	30
Toona ciliata	-	-	-	-	2	-	-	2	2	-	30
Trichosanthes cucumerina	-	-	-	-	1	-	-	-	-	2	15
Trifolium resupinatum	-	-	2	-	-	-	-	-	-	-	10
Trigonella foenum-graecum	-	-	-	-	2	-	-	-	-	2	20
Triticum aestivum	-	-	-	-	2	-	-	-	-	-	10
Tulipa stellata	-	-	-	-	-	-	-	-	1	2	15

Contd...

Table XLII Contd...

Botanical Name	Dye	Fibre	Fodder	Fuel	Medi-cine	Oil	Tannin /Gum	Timber	Others	Edible	TIV%
Urtica ardens	-	-	-	-	-	-	-	-	-	2	10
Urtica dioica	-	-	-	-	2	-	-	-	-	2	20
Vanda tessellata	-	-	-	-	2	-	-	-	-	-	10
Verbascum thapsus	-	-	-	-	2	-	-	-	-	-	10
Vicia faba	-	-	1	-	-	-	-	-	-	2	15
Vicia hirsuta	-	-	2	-	-	-	-	-	-	2	20
Vicia sativa	-	-	2	-	-	-	-	-	-	2	20
Vigna radiata	-	-	2	-	-	-	-	-	1	2	25
Viola patrinii	-	-	-	-	2	-	-	-	-	-	10
Viola pilosa	-	-	-	-	2	-	-	-	-	-	10
Vitex negundo	-	-	-	-	2	-	-	-	2	-	20
Vitis repanda	-	-	-	-	1	-	-	-	-	2	15
Vitis vinifera	-	-	-	-	1	-	-	-	-	2	15
Woodfordia fruticosa	-	-	1	2	2	-	-	-	1	-	30
Wrightia arborea	-	-	1	2	1	-	-	-	-	-	20
Zanthoxylum armatum	-	-	-	-	2	-	-	-	2	2	30
Zea mays	-	-	2	2	2	-	-	-	2	-	40
Zephyranthes carinata	-	-	-	-	-	-	-	-	1	1	10
Zizyphus jujuba	-	-	2	-	2	-	-	2	-	2	40
Zizyphus nummularia	-	-	1	-	2	-	-	-	2	2	35

(60%); followed by *Albizia chinensis, Grewia disperma*, *Grewia optiva*, *Terminalia bellirica*, *Terminalia tomentosa* (50%) each, *Grewia elastica*, *Musa sapientum*, *Albizia julibrissin*, *Albizia lebbeck, Bauhinia vahlii*, *Bauhinia variegata, Bombax ceiba*, *Brassica tournefortii*, *Carissa congesta*, *Carya illinoensis, Celtis australis, Cordia dichotoma, Dalbergia sissoo, Ehretia laevis*, *Emblica officinalis, Mangifera indica*, *Moringa oleifera*, *Murraya koenigii*, *Rhododendron arboretum, Sesamum orientale*, *Syzygium cumini*, *Zea mays* and *Zizyphus jujuba* (40%) each; *Butea monosperma*, *Ficus glomerata*, *Flacourtia indica*, *Phoenix acaulis*, *P. sylvestris, Pyrus pashia*, *Syzygium operculatum*, *Terminalia arjuna* and *Zizyphus nummularia* (35%) each, etc. All these plants have high economic potential values and can be judiciously exploited by using modern technologies for improving life and economy of the inhabitants of this region of N.W. Himalayas in general and the state of H.P. in particular.

CONCLUSION

It is imperative from the forgoing discussion that the region holds good diversity of plants and its rural inhabitants possess rich knowledge of the multifarious usages of the plant resources around them in a sustainable manner since time immemorial and are continuing with the age-old traditional practices. The potential value of these species needs to be popularised in the local, national and international markets. Hopefully, the present data would serve as a useful tool for policy makers to prepare development and action plan for improving and uplifting the life and economy of the hill communities of the state of Himachal Pradesh in particular.

Literature Cited

Abraham, Z. 1981. Ethnobotany of the Todas, the Kotas and the Irulas of the Nilgiris, 308-320. *In:* Jain, S.K. (*ed.*) *Glimpses of Indian Ethnobotany.* Oxford & IBH, New Delhi.

Agarwal, V.S. 2003. *Directory of Indian Economic Plants.* B.S.M.P.S., Dehradun.

Aggarwal, A., Prakash, V. & Sharma, S. 2004. Ethnobotanical studies on some wild plants of lower hills. *Appl. Bot. Abs.* **24**(2): 136.

Agrawal, S.R. 1981. Trees, flowers and fruits in Indian folk songs, folk proverbs and folk tales, 3-12. *In:* Jain, S.K. (*ed.*) *Glimpses of Indian Ethnobotany.* Oxford & IBH, New Delhi.

Ahluwalia, K.S. 1952. Medicinal plants of Kangra valley. *Indian For.* **78**(4): 181-194.

Ahuja, B.S. 1965. *Medicinal Plants of Saharanpur.* Survey of Medicinal Plants, Central Council of Ayurvedic Research Gurukul Kangri Vishwavidalya, Hardwar.

Alcorn, J. B. 1984. *Huastee Mayan Ethnobotany.* Univ. of Texas Press, Austin.

Alcorn, J.B. 1996. Is biodiversity conserved by indigenous people? 233-238. *In*: Jain, S.K. (*ed.*) *Ethnobiology in Human Welfare.* Deep Publ., New Delhi.

Ambasta, S.P. (*ed.*) 1986. *The Useful Plants of India.* C.S.I.R., New Delhi.

Aminuddin & Girach, R.D. 1991. Ethnobotanical studies on Bondo tribe of district Karaput (Orissa), India. *Ethnobotany* **3**: 15-19.

Aminuddin & Girach, R.D. 1993. Observations on ethnobotany of the Bhujia-a tribe of Sonabera plateau. *Ethnobotany* **5**: 83-86.

Aminuddin & Girach, R.D. 1996. Native phytotherapy among the Paudi Bhuinya of Bonai hills. *Ethnobotany* **8**: 66-70.

Anonymous, 1895. Some new ideas. The plants cultivated by aboriginal people and how used in primitive commerce. The Daily Evening Telegraph, Philadelphia, vol. **64**, no 134.

Anonymous, 1948. *Report of the Committee on Indigenous Systems of Medicine,* Vol.–I. Ministry of Health, Govt. of India, New Delhi.

Anonymous, (1948-1976). *The Wealth of India, Raw Materials* Vol.I-XI. P.I.D., C.S.I.R., New Delhi.

Anonymous, 1984. *Ethnobiology in India – A Status Report.* Ministry of Environment and Forest, Govt. of India, New Delhi.

Anonymous, 1985. *The Wealth of India, Raw Materials.* Revised & enlarged Vol. IA., P.I.D., C.S.I.R., New Delhi.

Anonymous, 1986. *The Useful Plants of India.* P.I.D., C.S.I.R., New Delhi.

Anonymous, 1990. *Phytochemical Investigation of Certain Medicinal Plants Used in Ayurveda.* Central Council for Research in Ayurveda & Siddha, New Delhi.

Anonymous, 1994. *Ethnobiology in India-A Status Report.* All India Coordinated Research Project in Ethnobotany. Ministry of Environment and Forests, Govt. of India, New Delhi.

Ansari, A.A. 1991. Ethnobotanical notes on some plants of Khirsu, Pauri Garhwal, U.P. *Ethnobotany* **3**: 105-106.

Arinathan, V., Mohan, V.R. & Britto, J.D. 2003. Ethnomedicinal survey among Palliyar tribals of Srivilliputhur grizzled giant squirrel wildlife sanctuary, Tamil Nadu, 707-710. *In:* Singh, V. & Jain, A.P. *(eds) Ethnobotany and Medicinal Plants of India and Nepal.* Sci. Publ., Jodhpur.

Arora, R.K. 1981. Native food plants of northeastern tribals, 91-106. *In*: Jain, S.K. (*ed.*) *Glimpses of Indian Ethnobotany.* Oxford & IBH Publ. Co., New Delhi.

Arora, R.K. 1985a. Life support species of Himalayan region, 42-45. *In*: Paroda, R.S., Kapoor, P., Arora, R.K. & Bhagmal *Life Support Plants Species, Diversity and Conservation,* N.B.P.G.R., New Delhi.

Arora, R.K. 1985b. *Genetic Resources of Less Known Cultivated Food Plants.* N.B.P.G.R., New Delhi.

Arora, R.K. 1991. Native food plants of tribals of northeast India, 137-152. *In*: Jain, S.K. (*ed.*) *Contributions to Indian Ethnobotany.* Sci. Publ., Jodhpur, India.

Arora, R.K. 1995. Ethnobotany and its role in domestication and conservation of native plant genetic resources, 76-80. *In:* Jain, S.K. (*ed.*) *A Manual of Ethnobotany* (2nd ed) Sci. Publ., Jodhpur.

Arora, R.K. 1997. Ethnobotany and its role in the conservation and use of plant genetic resources in India. *Ethnobotany* **9**: 6-15.

Arora, R.K. & Pandey, A. 1996. *Wild Edible Plants of India- Diversity, Conservation and Use.* N.B.P.G.R., New Delhi.

Arora, R.K., Maheswari, M.L., Chandel, K.P.S. & Gupta, R. 1980. Mano (*Inula racemosa*): little known aromatic plant of Lahaul valley, India. *Eco. Bot.* **34**(2): 174-180.

Asolkar, L.V., Kakkar, K.K. & Chakre, O.J. 1992. *Second Supplement to Glossary of Indian Medicinal Plants with Active Principles,* Part–I (A-K). C.S.I.R., New Delhi.

Aswal, B.S. 1996. Conservation of ethno-medicinal plants diversity of Garhwal Himalaya in India, 133-135. *In*: Jain, S.K. (*ed.*) *Ethnobiology in Human Welfare.* Deep Publ., New Delhi.

Aswal, B.S. & Mehrotra, B.N. 1994. *Flora of Lahaul Spiti* (*A Cold Desert in North West Himalaya*). B.S.M.P.S., Dehradun.

Atkinson, E.T. 1882. *Economic Botany of the Himalayan Region.* Cosmo Publ., New Delhi.

Augustine, J. & Sivadasan, M. 2005. Ethnomedicinal plants of Periyar tiger reserve, Kerela, India. *Ethnobotany* **16**: 44-49.

Badola, H.K. 2001. Medicinal plant diversity in Himachal Pradesh, 407-430. *In*: Samant, S.S., Dhar, U. & Palani, L.M.S. (*eds*) *Himalayan Medicinal Plants: Potential and Prospects.* Gyanodya Prakashan, Nainital, India.

Badola, H.K. & Pal, M. 2002. Endangered medicinal plant species in Himachal Pradesh. *Curr. Sci.* **83**: 797-798.

Bajpayee, K.K. & Dixit, G. 1996. Ethnobotanical studies on food stuffs of tribals of Tarai-Region, Uttar Pradesh. *J. Econ. Taxon. Bot.* Addl. Ser. **12**: 128-132.

Balasubramaniam, P. & Prasad, N. 1996. Medicinal plants among the Irulas of Attapady and Bolumampatti forests in the Nilgiri Biosphere Reserve. *J. Econ. Taxon. Bot.* Addl. Ser. **12**: 253-259.

Ballabh, B. & Chaurasia, O.P. 2006. Ethnobotanical studies on Boto tribe in Ladakh. *Ethnobotany* **18**: 87-95.

Banerjee, A. 1996. Medicinal uses of some flowers by Santhals of West Bengal. *J. Econ. Taxon. Bot.* Addl. Ser. **12**: 314-317.

Banerjee, D.K. & Pal, D.C. 1996. Plants used by the tribals of North Indian plains for hair and scalp protection, 187-188. *In*: Jain, S.K. (*ed.*) *Ethnobiology in Human Welfare*. Deep Publ., New Delhi.

Banerjee, R.N. & Ghora C. 1996. On the domestic use of some unreported plants of West Dinajpur District (West Bengal). *J. Econ. Taxon. Bot.* Addl. Ser. **12**: 325-328.

Baruah, P. & Sarma, G.C. 1984. Studies on the medicinal uses of plants by the Bodo tribals of Assam II. *J. Econ. Taxon. Bot.* **5**: 599-604.

Baruah, P. & Sarma, G.C. 1987. Studies on the medicinal uses of plants by the Northeast tribes-III. *J. Econ. Taxon. Bot.* **11**: 71-76.

Beg, M.J., Beg, M.Z. & Ali, S.J. 2005. Folklore treatment of some common ailments in Eastern U.P., 147-177. *In:* Khan, I.A. & Khanum, A. (*eds*) *Role of Biotechnology in Medicinal and Aromatic Plants.* Ukaaz Publ., Hyderabad.

Beigh, Y.S., Nawchoo, I.A. & Muhammad, I. 2003. Plants in traditional medicine of Kashmir Himalaya, 99-104. *In:* Singh, V. & Jain, A.P. *(eds) Ethnobotany and Medicinal Plants of India and Nepal.* Sci. Publ., Jodhpur.

Belal, A.E. & Springuel, I. 1996. Economic value of plant diversity in arid environments. *Nature & Resources* **32**(1): 33-39.

Bennet, S.S.R. 1983. Ethnobotanical studies in Sikkim. *Indian For.* **109**: 477.

Bennet, S.S.R. 1986. *Name Changes in Flowering Plants of India and Adjacent Regions*. Triseas Publ., Dehradun, India.

Bennet, S.S.R., Biswas, S., Chandra, V., Singh, P.J., Chandra, S. & Dayal, R. 1991. *Food From Forests.* I.C.F.R.E., Dehradun.

Bhalla, N.P., Sahu, T.R., Mishra, G.P. & Dakwale, R.N. 1982. Traditional plant medicines of Sagar district Madhya Pradesh, India. *J. Econ. Taxon. Bot.* **3**: 23-32.

Bhalla, S. & Bhalla, T.C. 2006. Traditional food and beverages of Himachal Pradesh. *Indian J. Trad. Knowledge* **6**(1): 17-24.

Bhandary, M. J. & Chandrasekhar, K.R. 2002. Glimpses of ethnic herbal medicines of Coastal Karnataka. *Ethnobotany* **14**: 1-12.

Bhandary, M.J., Chandrasekhar, K.R. & Kaveriappa, K.M. 1996. Ethnobotany of Gowlis of Uttar Kannada district, Karnataka. *J. Econ. Taxon. Bot.* Addl. Ser. **12**: 244-249.

Bhargava, K.S. 1959. Unusual and supplementary food plants of Kumaon. *J. Bombay Nat. Hist. Soc.* **56**: 26-31.

Bhatt, D.C., Jadeja, B.A., Odedra, N.K. & Baxi, B.S. 2003a. Enumeration of wild plants used as anti-diabetic in Barda hills of Gujarat, India, 897-903. *In:* Singh, V. & Jain, A.P. *(eds) Ethnobotany and Medicinal Plants of India and Nepal.* Sci. Publ., Jodhpur.

Bhatt, D.C., Mehta, S.K. & Mitaliya, K. D. 1999. Ethnomedicinal plants of Shetrunjaya hills of Palitana, Gujarat. *Ethnobotany* **11**: 22-25.

Bhatt, D.C., Mitaliya, K.D. & Babaria, A.V. 2003b. Studies on some ethnobotanical plants of Barda Hills from Saurashtra, 221-227. *In:* Singh, V. & Jain, A.P. *(eds) Ethnobotany and Medicinal Plants of India and Nepal.* Sci. Publ., Jodhpur.

Bhatt, D.C., Mitaliya, K.D. & Mehta, S.K. 2001. Some sacred plants of Gujarat and their medicinal uses. *Ethnobotany* **13**: 146-149.

Bhatt, D.C., Patel, A.M., Parekh, P.P. & Parmar, R.P. 2003c. Plants used as shelter in Aravalli Hills Gujarat (India), 943-946. *In:* Singh, V. & Jain, A.P. (*eds*) *Ethnobotany and Medicinal Plants of India and Nepal.* Sci. Publ., Jodhpur.

Bhattacharjee, S., Tiwari, K.C., Majumdar, R. & Misra, A.K. 1980. Folk-lore medicines from district Kamrup (Assam). *Bull. Medico-Ethno-Bot. Res.* **1:** 441-460.

Bhattacharya, A.K., Tiwari, P., Patra, K. & Sinha, V.K. 2005. Seasonal availability and consumption of non-wood forest products (NWFP) as food among Baiga tribe of Dindori district of Madhya Pradesh, India. *Ethnobotany* **17**: 64-68.

Bhattacharya, G. 1996. Medico-ethno-botanical value of Saurashtra weeds. *J. Econ. Taxon. Bot.* Addl. Ser. **12**: 166-168.

Bhuyan, L.N. 2003. Some plants used as medicine by the Nishi tribe of Arunachal Pradesh: a preliminary study, 447-480. *In:* Singh, V. & Jain, A.P. (*eds*) *Ethnobotany and Medicinal Plants of India and Nepal.* Sci. Publ., Jodhpur.

Billore, K.V., Joseph, T.G. & Dave, S. K. 1998. Interesting folk remedies by the 'Lok Vaidyas' of Rajasthan for 'Swas Roga'. *Ethnobotany* **10**: 42-45.

Binoj, K.M.S. & Balakrishnan, N.P. 1996. Ethnobotanical studies of one genus Euphorbial (Euphorbiaceae). *J. Econ. Taxon. Bot.* Addl. Ser. **12**: 46-49.

Binu, S., Shanavaskhan, A.E., Santhoshkumar, E.S. & Pushpangadan, P. 2003. Plants used as medicine by the Irulas of Palghat district, Kerela, India, 808-814. *In:* Singh, V. & Jain, A.P. (*eds*) *Ethnobotany and Medicinal Plants of India and Nepal.* Sci. Publ., Jodhpur.

Bisht, A. & Jain, S.P. 2006. Review of ethnobotanical studies of genus *Rubus* (Rosaceae) from Western Himalaya. *Ethnobotany* **18**: 127-130.

Bondya, S.L., Khanna, K.K. & Singh, K.P. 2006. Ethnomedicinal uses of leafy vegetable from the tribal folk-lore of Achanakmar- Amarkantak biosphere reserve (Madhya Pradesh and Chhattisgarh). *Ethnobotany* **18**: 145-148.

Bondya, S.L. & Sharma, H.P. 2004. Ethnobotanical studies on plants used in diabetes under the Baharagova block of Jharkhand. *Ethnobotany* **16**: 139-140.

Borthakur, S.K. 1993. Native phytotherapy for child and women diseases from Assam. *Ethnobotany* **5**: 87-91.

Borthakur, S.K. 1996a. Postnatal care of women in traditional system in Assam. *Ethnobotany* **8**: 51-55.

Borthakur, S.K. 1996b. Wild edible plants in markets of Assam, India- an ethnobotanical investigation, 31-34. *In*: Jain, S.K. (*ed.*) *Ethnobiology in Human Welfare*. Deep. Publ., New Delhi.

Borthakur, S.K., Choudhary, B.T. & Gogoi, R. 2004. Folklore hepato-protective herbal recipes from Assam in Northeast India. *Ethnobotany* **16**: 76-82.

Borthakur, S.K. & Gogoi, P. 1994. Indigenous technology of making writing materials among the Tai Khamtis. *Ethnobotany* **6**: 5-8.

Borthakur, S.K. & Sarma, U.K. 1996. Ethnoveterinary medicine with special reference to cattle prevalent among the Nepalis of Assam, India, 197-199. *In*: Jain, S.K. (*ed.*) *Ethnobiology in Human Welfare.* Deep Publ., New Delhi.

Bose, K.C. 1984. *Pharmacopoeia Indica*. B.S.M.P.S., Dehradun.

Brahmam, M., Dhal, N.K. & Saxena, H.O. 1996. Ethnobotanical studies among the Tanla of Malayagiri hills in Dhenkanal district, Orissa, India, 393-396. *In*: Jain, S.K. (*ed.*) *Ethnobiology in Human Welfare.* Deep Publ., New Delhi.

Brahmam, M. & Saxena, H.O. 1990. Ethnobotany of Gardamardan Hills–some noteworthy folk-medicinal uses. *Ethnobotany* **2**: 71-79.

Caius, J.F. 2003. *The Medicinal and Poisonous Plants of India.* Sci. Publ. (India), Jodhpur.

Chakrabarty, T. & Balakrishnan, N.P. 2003. Ethnobotany of the Andaman & Nicobar Islands, India, 859-893. *In:* Singh, V. & Jain, A.P. (*eds*) *Ethnobotany and Medicinal Plants of India and Nepal.* Sci. Publ., Jodhpur, India.

Chakraborty, M.K. & Bhattacharjee, A. 2003. Plants used as masticatories by the ethnic communities of Purulia district, West Bengal, India, 571-572. *In:* Singh, V. & Jain, A.P. (*eds*) *Ethnobotany and Medicinal Plants of India and Nepal.* Sci. Publ., Jodhpur.

Chakraborty, P. 2003. Wild edible plants sold in the daily market in and around of Imphal, Manipur, 481-484. *In:* Singh, V. & Jain, A.P. (*eds*) *Ethnobotany and Medicinal Plants of India and Nepal.* Sci. Publ., Jodhpur.

Chakraborty, P., Pal, G.D. & Parmar, P.J. 2003. Some wild edible plants sold in the daily market of Arunachal Pradesh, 489-495. *In:* Singh, V. & Jain, A.P. (*eds*) *Ethnobotany and Medicinal Plants of India and Nepal.* Sci. Publ., Jodhpur.

Chandra, V. 1997. *Edible Plants of Forestry Origin.* I.C.F.R.E., Dehradun.

Chandrasekar, K. & Srivastava, S.K. 2003. Ethnomedicinal studies in Pin valley national Park, Lahaul-Spiti, Himachal Pradesh. *Ethnobotany* **15**: 44-47.

Chandrasekar, K. & Srivastava, S.K. 2005. New Reports on aphrodisiac plants from Pin Valley National Park, Himachal Pradesh. *Ethnobotany* **17**: 189-190.

Chatterjee, A. & Pakrashi, S.C. (*eds*) 1991. *The Treatise on Indian Medicinal Plants.* Vol.1-5. P.I.D., New Delhi, India.

Chatterjee, A. & Pakrashi, S.C. (*eds*) 1997. *The Treatise on Indian Medicinal Plants.* Vol.1-5. C.S.I.R., New Delhi.

Chaudhuri, R.H.N., Molla, H.A., Pal, D.C & Roy, B. 1982. Plants used in traditional medicines by some tribals of Jalpaiguri district, West Bengal. *Bull. Bot. Surv. India* **24**: 87-90.

Chauhan, N.S. 1974. Medicinal plants of Una forest division, Una district, H.P. lying in Shivalik range. *Nagarjun* **27**: 31-39.

Chauhan, N.S. 1983. Some rare and interesting medicinal plants of H.P. All India Summer Inst. on Agroforestry in the Western Himalaya, July 22-August 7, 1983.

Chauhan, N.S. 1984. *Medicinal Wealth of Pabbar Valley in Himachal Pradesh*, M.Sc. Thesis, H.P.K.V.V., Palampur, H.P.

Chauhan, N.S. 1988. Ethnobotanical study of medicinal plants of Himachal Pradesh, 187-198. *In*: Kaushik, P. (*ed.*) *Indigenous Medicinal Plants.* Today's and Tomorrow's Printers & Publ., New Delhi.

Chauhan, N.S. 1989a. Herbal wealth of cold desert area of Himachal Pradesh, 65-68. *Proc., Nat. Seminar-Cum-Workshop on Cold Desert Area in India.* Govt. Press, Shimla, H.P.

Chauhan, N.S. 1989b. Potential of aromatic plants flora in Himachal Pradesh. *Indian Perf.* **33**(2): 118-122.

Chauhan, N.S. 1989c. Endangered ayurvedic pharmacopoeial plant resources of H.P., 199-205. *In*: Kaushik, P. (*ed.*) *Indiginous Medicinal Plants Including Microbes and Fungi.* Today's and Tommorow's Printers & Publ., New Delhi.

Chauhan, N.S. 1990. Medicinal orchids of Himachal Pradesh. *J. Orchid Soc. India* **4**(1, 2): 99-105.

Chauhan, N.S. 1995. *Plant Resources of Economic Use in Himachal Pradesh.* Dept. of Forest Products, Dr. Y.S. Parmar Univ. Hort. Forestry, Solan (H.P.), India.

Chauhan, N.S. 1996. Ethno-medicobotany of Pabbar valley in district Shimla, Himachal Pradesh (India), 30. *In*: *UHF-IUFRO Intl. Workshop in Prospects of Medicinal Plants,* Solan (H.P.), India.

Chauhan, N.S. 1999. *Medicinal and Aromatic Plants of Himachal Pradesh.* Indus Publ. Co., New Delhi.

Chauhan, N.S. 2003. Important medicinal and aromatic plants of Himachal Pradesh. *Indian For.* **129**(8): 979-998.

Chauhan, N.S. & Lakhanpal, T.N. 2000. Ecological and ethnobotanical studies on the flora of Spiti, Himachal Pradesh- a cold desert in western Himalaya, 293-300. In: Kohli, R.K., Singh, H.P., Vij, S.P., Dhir, K.K., Batish, D.R. & Khurana, D.R. (*eds*) *Man and Forests.* DNAES, ISTS, Bot. Dept., & C.N.S., Punjab Univ., Chandigarh.

Chaurasia, O.P., Ahmed, Z. & Ballabh, B. 2007. *Ethnobotany and Plants of Trans-Himalaya.* Satish Serial Publ., New Delhi.

Chaurasia, O.P., Singh, B. & Ballabh, B. 2001. Potential aromatic floras of Himalayan cold district: Ladakh & Lahul-Spiti. *J. Econ. Taxon. Bot.* **25**(1): 91-97.

Chevallier, A. 1996. *The Encyclopedia of Medicinal Plants.* Dorling Kindersley, London.

Chhetri, D.R. 2005. Ethnomedicinal plants of the Khangchendzonga National Park, Sikkim, India. *Ethnobotany* **17**: 96-103.

Chhetri, R.B. 1994. Further observations on ethnomedicobotany of Khasi Hills in Meghalaya, India. *Ethnobotany* **6**: 33-36.

Chhetri, R.B. & Shrestha, R. 2004. Ethnobotany of some weeds of winter crops in Dhulikhel, Nepal. *Ethnobotany* **16**:108-112.

Chopra, R.N., Badhwar, R.L. & Gosh, S. 1965. *Poisonous Plants of India.* 2 Vols., I.C.A.R., New Delhi.

Chopra, R.N., Chopra, I.C. & Verma, B.S. 1969. *Supplement to Glossary of Indian Medicinal Plants.* C.S.I.R., New Delhi.

Chopra, R.N., Nayar, S.L. & Chopra, I.C. 1956. *Glossary of Indian Medicinal Plants.* C.S.I.R., New Delhi.

Choudhary, K. K. 1996. Studies on medicinal plants of Chikhali (Buldhana- district) with special reference to some aspects of ethnobotany, 152-159. *In:* Rajak, R.C. & Rai, M.K. (*eds*) *Herbal Medicines Biodiversity and Conservation Strategies.* Intl. Book Distributors, Dehradun.

Chowdhery, H.J. 1996. Arunachal Pradesh, its people, ethnobotanical diversity and conservation, 323-328. *In*: Jain, S.K. (*ed.*) *Ethnobiology in Human Welfare.* Deep Publ., New Delhi.

Chowdhery, H.J. & Wadhwa, B.M. 1984. *Flora of Himachal Pradesh.* Vol. 1-3. Bot. Surv. India, Calcutta.

Chowdhury, J.U., Alam, M.K. & Hassan, M.A. 1996. Some traditional folk formularies against dysentery and diarrhoea in Bangladesh. *J. Econ. Taxon. Bot.* Addl. Ser. **12**: 20-23.

Chun-Lin, L. & Jieru, W. 1995. Ethnobotany of Jinuo nationality in Xishuangbanna, Southwest China: edible plants. *Ethnobotany* **7**: 39-49.

Collett, H. 1902. *Flora Simlensis*. Thacker Spink and Co., Calcutta & Shimla.

Cronk, Q.C.B. & Fuller, J.L. 1995. *Plant Invasion: The Threat to Natural Ecosystem.* World Wide Fund for Nature. Chapman & Hall, London.

D'Rozario, A., Bera, S. & Mukherji, D. 1999. *A Handbook of Ethnobotany*. Kalyani Publ., New Delhi.

Dagar, H.S. 2003. Ethnobotanical evaluation of some endemic plants in Andaman and Nicobar Islands, 134-147. *In*: Khan, S.S. (*ed.*) *Vistas in Ethnobotany*, Vol.-I, *Indian Journal of Pure and Applied Biology*. Naval Kishore Press, Lucknow.

Dagar, H.S. & Dagar, J.C. 1996. Ethnobotanical studies of the Nicobarese of Chowra island of Nicobar group of islands. *J. Econ. Taxon. Bot.* Addl. Ser. **12**: 381-388.

Dagar, J.C. & Dagar, H.S. 1999. *Ethnobotany of Aborigines of Andaman-Nicobar Islands*. Survey of India, Port Blair, Andaman-Nicobar Islands, India.

Dar, G.H., Vir J., Kachroo, P. & Buth, G.M. 1984. Ethnobotany of Kashmir-1. Sind valley. *J. Econ. Taxon. Bot.* **5**: 668-675.

Das, A.K. 1997. Less known uses of plant among the Adis of Arunachal Pradesh. *Ethnobotany* **9**: 90-93.

Das , A.K. & Saikia, D. C. 2001. Indigenous practices of treating human liver disorders in Assam. *Ethnobotany* **13**: 87-90.

Das, A.K. & Sharma, G.D. 2003. Ethno-medicinal uses of plants by Manipuri and Barmab communities of Cachar district, Assam, 421-429. *In:* Singh, V. & Jain, A.P. (*eds*) *Ethnobotany and Medicinal Plants of India and Nepal.* Sci. Publ., Jodhpur.

Das, D. 2000. Wild food plants of Midnapur, West Bengal during drought and flood, 535-538. *In:* Maheshwari, J.K. (*ed.*) *Ethnobotany and Medicinal Plants of Indian Subcontinent.* Sci. Publ., Jodhpur.

Das, D. & Agarwal, V.S. 1991. *Fruit Drug Plants of India.* Kalyani Publ., New Delhi.

Das, P.S. & Choudhary, M.D. 2003. A survey on non-conventional food plants of Southern Assam, 416-420. *In:* Singh, V. & Jain, A.P. (*eds*) *Ethnobotany and Medicinal Plants of India and Nepal.* Sci. Publ., Jodhpur.

Dash, S.S., Maiti, A. & Rai, S.K. 2003. Traditional uses of plants among the urban population of Gangtok-Sikkim, 317-324. *In:* Singh, V. & Jain, A.P. (*eds*) *Ethnobotany and Medicinal Plants of India and Nepal.* Sci. Publ., Jodhpur.

Dastur, J.F. 1970. *Medicinal Plants of India and Pakistan.* 2nd *ed.* Taraporevala Sons & Co., Bombay.

Datt, B. 1996. Ethnobotanical resources of Chhatarpur district (Madhya Pradesh), 400-402. *In*: Jain, S.K. (*ed.*) *Ethnobiology in Human Welfare.* Deep Publ., New Delhi.

Deshmukh, V.R. & Rothe, S.P. 2003. Ethno-medicinal study from Malghat tribal region of Amravati district, Maharashtra, 585-591. *In:* Singh, V. & Jain, A.P. (*eds*) *Ethnobotany and Medicinal Plants of India and Nepal.* Sci. Publ., Jodhpur.

Devi, M. 2003. Wild edible plants of Sonipat district, Assam, 396-409. *In:* Singh, V. & Jain, A.P. (*eds*) *Ethnobotany and Medicinal Plants of India and Nepal.* Sci. Publ., Jodhpur.

Dey, A.C. 1980. *Indian Medicinal Plants Used in Ayurvedic Preparations.* B.S.M.P.S., Dehradun.

Dey, K.L.1973. *The Indigenous Drugs of India.* Pama Primlane Publ., New Delhi.

Dey, K. L. & Bahadur, R. 1973. *The Indigenous Drugs of India.* Pama Primlane Publ., New Delhi.

Dhaliwal, D.S. & Sharma, M. 1999. *Flora of Kullu District (Himachal Pradesh).* B.S.M.P.S., Dehradun.

Dhiman, D.R. 1976. *Himachal Pradesh Ki Vanoshdhiya Sampada.* Imperial Printing Press, Dharamsala, H.P.

Diwakar, P.G. & Ansari, A.A. 2003. Economic importance of plant resources of Buldhana district of Maharashtra, 592-600. *In:* Singh, V. & Jain, A.P. (*eds*) *Ethnobotany and Medicinal Plants of India and Nepal.* Sci. Publ., Jodhpur.

Dixit, R.D. & Kumar, R. 2003. Plants used by local people in human welfare, 53-59. *In:* Singh, V. & Jain, A.P. (*eds*) *Ethnobotany and Medicinal Plants of India and Nepal.* Sci. Publ., Jodhpur.

Dixit, R.D. & Pal, D.C. 1996. Role of household remedies in primary healthcare, 108-109. *In*: Jain, S.K. (*ed.*) *Ethnobiology in Human Welfare*, Deep Publ., New Delhi.

Dobriyal, R.M., Singh, G.S., Rao, K.S. & Saxena, K.G. 1997. Medicinal plant resources in Chhakinal watershed in north-western Himalaya. *J. Herbs Spices & Med. Plants* **5**:15-27.

Drake, J.A., Mooney, H.A., Castri, F., Groves, R.H., Kruger, F.J., Rejmanek, M. & Williamson, M. (*eds*) 1989. *Biological Invasions: A Global Perspective.* John Wiley & Sons, New York.

Drury, H. 1978. *Useful Plants of India.* International Books Distributor, Dehradun.

Duhoon, S.S., Kopper, M.N. & Chandra, U. 1996. Seabuckthorn (*Hippophae spp*) - A less known wonder plant of ethno-medico-botanical importance in cold desert of India. *J. Econ. Taxon. Bot.* Addl. Ser. **12**: 43-45.

Duhoon, S.S. & Tripathi, A.K. 2003. Medicinal and curative properties of sesame (*Sesamum indicum* L.), 20-28. *In:* Singh, V. & Jain, A,P. (*eds*) *Ethnobotany and Medicinal Plants of India and Nepal.* Sci. Publ., Jodhpur.

Duke, J.A., Bogenschutz-Godwi, M.J., Judi duCellier & Duke, P.K. 2002. *Handbook of Medicinal Herbs.* C.R.C. Press, New York.

Dwarakan, P. & Ansari, A.A. 1996. Less known uses of plants of Kollimalai (Salem Dt., Tamil Nadu) in South India. *J. Econ. Taxon. Bot.* Addl. Ser. **12**: 284-286.

Dwivedi, A.P. 1993. *Forests– The Non–Wood Resources.* Intl. Book Distributors, Dehradun.

Dwivedi, S.N. 2003. Ethnobotanical studies and conservational strategies of wild and natural resources of Rewa district of Madhya Pradesh, 233-244. *In:* Singh, V. & Jain, A.P. (*eds*) *Ethnobotany and Medicinal Plants of India and Nepal.* Sci. Publ., Jodhpur.

Ganai, K.A. & Nawachoo, I.A. 2003. Traditional treatment of toothache by the Gujjar and Bakerwal tribes of Kashmir in India, 105-107. *In:* Singh, V. & Jain, A.P. *(eds) Ethnobotany and Medicinal Plants of India and Nepal.* Sci. Publ., Jodhpur.

Ganesan, S. & Kesavan, L. 2003. Ethnomedicinal plants used by the ethnic groups Valaiyans of Vellimalai hills (Reserved Forests), Tamilnadu, India, 754-760. *In:* Singh,V. & Jain, A.P. (*eds*) *Ethnobotany and Medicinal Plants of India and Nepal.* Sci. Publ., Jodhpur.

Gangwar, A.K. & Ramkrishnan, P.S. 1990. Ethnobiological notes on some tribes of Arunachal Pradesh, Northeastern India. *Econ. Bot.* **44**: 94-105.

Gaur, R.D., Semwal, J.K. & Tiwari, J.K.1983. A survey of high altitude medicinal plants of Garhwal Himalaya. *Bull. Medico-Ethnobot. Res.* **4**(3-4): 102-116.

Gaur, R.D. & Singh, P.B. 1993. Ethno-medicinal plants of Mandi district, Himachal Pradesh. *Bull. Medico-Ethnobot. Res.* **14**: 1-11.

Geetha, S., Lakshmi, G. & Ranjithakani, P. 1996a. Ethnobotanical review: wild fibre yielding plants of Kolli Hills, Tamil Nadu. *J. Econ. Taxon. Bot.* Addl. Ser. **12**: 250-252.

Geetha, S., Lakshmi, G. & Ranjithkani, P. 1996b. Ethnoveterinary medicinal plants of Kolli Hills, Tamil Nadu. *J. Econ. Taxon. Bot.* Addl. Ser. **12**: 289-191.

Ghosh, A., Maity, S. & Maity, M. 1996. Ethnomedicine in Bankura district, West Bengal. *J. Econ. Taxon. Bot.* Addl. Ser. **12**: 318-320.

Ghosh, S. 1997. Ethnomedicine to modern medicine: an observational study in some villages of West Bengal. *Ethnobotany* **9**: 80-84.

Gill, L.S., Idu, M. & Ogbor, D.N. 1997. Folk medicinal plants; practices and beliefs of the Bini People in Nigeria. *Ethnobotany* **9**: 1-5.

Gill, L.S. & Nyawuame, H.G.K. 1994. Leguminosae in ethnomedicinal practices of Nigeria. *Ethnobotany* **6**: 51-64.

Gill, L.S., Nyawuame, H.G.K., Esezobor, E.I., & Osagie, I.S. 1993. Nigerian folk Medicine : practices and beliefs of Esan people. *Ethnobotany* **5**: 129-142.

Girach, R.D. 1992. Medicinal plants used by Kondh tribe of district Phulbani (Orissa) in Eastern India. *Ethnobotany* **4**: 53-66.

Girach, R.D. 2001. Ethnobotanical notes on some plants of Mahendragiri hills, Orissa. *Ethnobotany* **13**: 80-83.

Girach, R.D. & Aminuddin, 1995. Ethnomedicinal uses of plants among the tribals of Singhbhum district, Bihar, India. *Ethnobotany* **7**: 103-107.

Girach, R.D., Aminuddin, Ahmed, M., Brahmam, M. & Misra, M.K. 1996. Native phytotherapy among rural population of district Bhadrak, Orissa, 162-164. *In*: Jain, S.K. (*ed.*) *Ethnobiology in Human Welfare*. Deep Publ., New Delhi.

Girach, R.D., Aminuddin, Brahmam, M. & Mishra, M.K. 1997. Observations on ethnomedicinal plants of Bhadrak district, Orissa, India. *Ethnobotany* **9**: 44-46.

Goel, A.K., Sahoo, A.K. & Mudgal, V. 1984. A contribution to the ethnobotany of Santal Paragana. *Bot. Surv. India*, Howrah.

Goel, A.K. & Tripathi, S. 2009. Ethnobotanical spectrum of Indian flora: an overview during the past 20 years in the journal Ethnobotany. *Ethnobotany* **21**: 131-134.

Gogoi, J.C., Sharma, D.K. & Borthakur, S.K. 2003a. Folk-lore hepatoprotective herbal recipes for Assam in Northeastern India, 1-21. *In*: Khan, S.S. (*ed.*) *Vistas in Ethnobotany* Vol.-1. *Indian Journal of Applied and Pure Biology*. Naval Kishore Press, Lucknow.

Gogoi, P. & Boissya, C.L. 1984. Information about a few herbal medicines used by the people of Assam (India) against Jaundice. *Himal. Res. Dev.* **2**: 41-44.

Gogoi, R., Bokolial, D. & Hazarika, D. 2003b. Preliminary observations on the medicinal plants of Chandrapur area of Kamrup district, Assam, 410-415. *In:* Singh, V. & Jain, A.P. (*eds*) *Ethnobotany and Medicinal Plants of India and Nepal.* Sci. Publ., Jodhpur.

Gogoi, P. & Borthakur, S.K. 1991. Plants in religio-cultural beliefs of Tai Khamtis of Assam (India). *Ethnobotany* **3**: 89-95.

Gogoi, P. & Borthakur, S.K. 2001. Notes on herbal recipes of Bodo tribe in Kamrup district, Assam. *Ethnobotany* **13**: 15-23.

Gopal, G.V. & Chandra, J.K.S. 2003. Ethnobotany of Soligas Biligirirangana hills, Karnatka, India, 640-647. *In:* Singh, V. & Jain, A.P. (*eds*) *Ethnobotany and Medicinal Plants of India and Nepal.* Sci. Publ., Jodhpur.

Gogoi, R. & Das, M.K. 2003. Observation on some weed of medicinal importance in the Brahmaputra valley of Assam, 434-441. *In:* Singh, V. & Jain, A.P. (*eds*) *Ethnobotany and Medicinal Plants of India and Nepal.* Sci. Publ., Jodhpur.

Goud, P.S. & Pullaiah, T. 1996a. Ethnobotany of Kurnool district– some wild plants used as a food. *J. Econ. Taxon. Bot.* Addl. Ser. **12**: 224-227.

Goud, P.S. & Pullaiah, T. 1996b. Folk veterinary of Kurnool district, Andhra Pradesh. *Ethnobotany* **8**: 71-74.

Gupta, A.K. & Ura, K. 1990. Blending cultural values, indigenous technology and environment: experiences in Bhutan. Paper presented at the Intl. Symp. on Strategies for Sustainable Mountain Development, Kathmandu, Nepal.

Gupta, M.P. 1988. Present status and future prospects for the cultivation and collection of medicinal plants of Himachal Pradesh. *Indian For.* **114**(1): 19-25.

Gupta, R. 1961. Flora of Lam Dal. *Indian For.* **87**(5): 316-324.

Gupta, R. 1962. Some unusual and interesting food plants of Garhwal Himalaya. *J.D. Agric.Trop. Bot. Appl.* **9**(11-12): 532-535.

Gupta, R. 1964. Survey record of medicinal and aromatic plants of Chamba forest division, Himachal Pradesh. *Indian For.* **90**: 454-468.

Gupta, R. 1971. Medicinal and aromatic plants of Bhandal ranges, Churah forest division, Chamba district, Himachal Pradseh. *J. Bomb. Nat. His. Soc.* **68**: 791-803.

Gupta, R. 1981. Plants in folk medicine of the Himalaya, 83-90. In: Jain, S.K. (*ed.*) *Glimpses of Indian Ethnobotany*. Oxford & IBH Publ. Co., New Delhi.

Hajra, P.K. 1987. Plants in magico-religious beliefs and in Sanskrit literature, 98-103. *In:* Jain, S.K. (*ed.*) *A Manual of Ethnobotany.* Sci. Publ., Jodhpur.

Hajra, P.K. & Das, B.K. 1982. Vegetation of Gangtok with special reference to exotic plants. *Indian For.* **107**: 554-566.

Harborne, F.R.S.J.B. & Baxter, H. 2001. *Chemical Dictionary of Economic Plants.* John Wiley & Sons, England.

Hebber, S.S., Harsha, V.H., Shripathi, V. & Hedge, G.R. 2003. Wild edible fruits of Dharwad, Karnatka, 982-988. *In:* Singh, V. & Jain, A.P. (*eds*) *Ethnobotany and Medicinal Plants of India and Nepal.* Sci. Publ., Jodhpur.

Hema, E.S., Sivadasan, M. & Anilkumar, N. 2006. Studies on edible species of Amaranthaceae and Araceae used by Kuruma and Paniya tribes in Wayanad district, Kerela, India. *Ethnobotany* **18**: 122-126.

Hembrom, P.P. 1991. Tribal medicine in Chotanagpur and Santhap Parganas of Bihar, India. *Ethnobotany* **3**: 97-99.

Hembrom, P.P. 1996. Contact therapy practised by Mundas of Chotanagpur (Bihar). *Ethnobotany* **8**: 36-39.

Hembrom, P.P. & Goel, A.K. 2005. Horopathy : ethnomedicine of Mundas. *Ethnobotany* **17**: 89-55.

Henry, A.N., Hosagoudar, V.B. & Kumar, R.K. 1996. Ethno-medico-botany of the Southern Western Ghats of India, 173-180. *In*: Jain, S.K. (*ed.*) *Ethnobiology in Human Welfare*. Deep Publ., New Delhi.

Hooker, J.D. 1872-1897. *The Flora of British India,* Vol. I-VIII. Lalit Mohan Basu, Allahabad.

Hosagoudar, V.B. & Henry, A.N. 1996a. Ethnobotany of Soligas in Biligiri Rangana Betta, Karnataka, Southern India. *J. Econ. Taxon. Bot.* Addl. Ser. **12**: 228-243.

Hosagoudar, V.B. & Henry, A.N. 1996b. Ethnobotany of Kadars, Malasars and Kuthuvans of the Anamalais in Coimbatore district, Tamil Nadu, India. *J. Econ. Taxon. Bot.* Addl. Ser. **12**: 260-267.

Hosagoudar, V.B. & Henry, A.N. 1996c. Ethnobotany of tribes Irular, Kurumban and Paniyan of Nilgiris in Tamil Nadu, Southern India. *J. Econ. Taxon. Bot.* Addl. Ser. **12**: 364-366.

Huidrom, S.B.K. 1996. Plants used in medico-sexual purposes by Meitein community in Manipur state, India. *J. Econ. Taxon. Bot.* Addl. Ser. **12**: 364-366.

Islam, M. 1996. Ethnobotany of certain underground parts of plants of north-eastern region, India. *J. Econ. Taxon. Bot.* Addl. Ser. **12**: 338-343.

Islam, M.A. & Jha, R.M. 2003. Ethnopharmacology of tribals of Ranchi district in Jharkhand, 300-310. *In:* Singh, V. & Jain, A.P. (*eds*) *Ethnobotany and Medicinal Plants of India and Nepal.* Sci. Publ., Jodhpur.

Jadeja, B.A. 1999. Plants used by the tribe Rabari in Barda hills of Gujarat. *Ethnobotany* **11**: 42-46.

Jain A.K. 1992. Clans of Sor tribals of Madhya Pradesh and their role in conservation. *Ethnobotany* **4**: 67-69.

Jain, A.K. & Patole, S.N. 2001. Less known medicinal uses of plants among some tribal and rural communities of Panchmarhi forests (M.P.). *Ethnobotany* **13**: 96-100.

Jain, A.K. & Sharma, H.O. 1996. Ethnobiological studies of Shariya tribe of Central India, 397-399. *In*: Jain, S.K. (*ed.*) *Ethnobiology in Human Welfare*. Deep Publ., New Delhi.

Jain, S.K. 1968. *Medicinal Plants*. National Book Trust, New Delhi.

Jain, S.K. 1981. Observations on ethnobotany of the tribals of Central India, 193-198. *In*: Jain, S.K. (*ed.*) *Glimpses of Indian Ethnobotany*. Oxford & IBH, Publ., New Delhi.

Jain, S.K. 1987a. Ethnobotany its concepts and relevance. pp. 1-12. *Pres. Add. X Bot. Conf.* Patna.

Jain, S.K. 1987b. *A Manual of Ethnobotany*. Sci. Publ., Jodhpur.

Jain, S.K. (*ed.*) 1989. *Methods and Approaches in Ethnobotany*. Deep Publ., New Dehli, India.

Jain, S.K. 1991. *Dictionary of Indian Folk Medicine and Ethnobotany*. Deep Publ., New Delhi.

Jain, S.K. 1995. Ethnobotanical diversity in Zingibers of India. *Ethnobotany* **7**: 83-88.

Jain, S.K. 2002. *Bibliography of Indian Ethnobotany*. Sci. Publ., Jodhpur, India.

Jain, S.K. 2003. Indian plants in Tibetan medicine. *Ethnobotany* **15**: 236.

Jain, S.K. 2006. Ethnobotany in the New Millennium– some thoughts on future direction in Indian ethnobotany. *Ethnobotany* **18**: 1-3.

Jain, S.K. & De, J.N. 1966. Obervations on ethnobotany of Purulia district, West Bengal. *Bull. Bot. Surv. India* **8:** 237-251.

Jain, S.K. & Goel, A.K. 2005. Some Indian plants in Tibetan traditional medicine -1. *Ethnobotany* **17**: 127 -136.

Jain, S.K., Fernandes, V.F., Lata, S. & Ayub, A. 1995. Indo-Amazonian ethnobotanic connections-similar uses of some common plants. *Ethnobotany* **7**: 29-37.

Jain, S.K., Hajra, P.K. & Shanpru, R. 1977. A survey of wild edible plants in bazaars of Meghalaya. *Bull. Meghalaya Sci. Soc.* **2:** 29-34.

Jain, S.K., Mudgal, V., Banerjee, D.K., Guha, A., Pal, D.C. & Das. D. 1984. *Bibliography of Ethnobotany.* B.S.I., Howrah.

Jain, S.K., Ranjan, V., Sikarwar, R.K.S., & Saklani, A. 1994. Botanical distribution of psychoactive plants of India. *Ethnobotany* **6**: 65-75.

Jain, S.K. & Rao, R.R. (*eds*) 1977. *A Handbook of Field and Harbarium Methods.* Today's & Tomorrow's Printers & Publ., New Delhi.

Jain, S.K. & Saklani, A. 1992. Cross-cultural ethnobotanical studies in North East India. *Ethnobotany* **4**: 25-38.

Jain, S.K. & Sharma, P.P. 2000. A review of the parallel or unique indigenous uses of twenty-five plants in India and some African countries. *Ethnobotany* **12**: 51-55.

Jain, S.K., Sinha, B.K. & Gupta, R.C. 1991. *Notable Plants in Ethnomedicine of India.* Deep Publ., New Delhi.

Jain, S.K., Sikarwar, R.L.S. & Pathak, V. 1997. Ethnobotanical aspects of some plants in Latin America. *Ethnobotany* **9**: 16-23.

Jain, S.K. & Srivastava, S. 2001. Indian ethnobotanical literature in last two decades - a graphic review and future directions. *Ethnobotany* **13**: 1-8.

Jain, S.K. & Tarafder, C.R. 1970. Medicinal plant lore of the Santals. A revival of P. O. Boddings's work. *Econ. Bot.* **24:** 241-278.

Jain, S.P. 1984. Ethnobotany of Morni & Kalesar (Ambala, Harayana). *J. Econ. Taxon. Bot.* **5**: 809-813.

Jain, S.P. 1996. Ethno-medico-botanical survey of Chaibasa, Singhbhum district, Bihar. *J. Econ. Taxon. Bot.* Addl. Ser. **12**: 403-407.

Jain, S.P. 2003a. An inventory of threatened medicinal & aromatic plants of North-Western India, 908-913. *In:* Singh, V. & Jain, A.P. (*eds*) *Ethnobotany and Medicinal Plants of India and Nepal.* Sci. Publ., Jodhpur.

Jain, S.P. 2003b. Comparative ethno-medico-botanical studies of tribes of Neterhat Plateau, Chhotanagpur and Singhbhum districts of Jharkhand, India, 295-299. *In:* Singh, V. & Jain, A.P. (*eds*) *Ethnobotany and Medicinal Plants of India and Nepal.* Sci. Publ., Jodhpur.

Jain, S.P. & Puri, H.S. 1994. An ethno-medico-botanical survey of Parvati valley in Himachal Pradesh (India). *J. Econ. Taxon. Bot.* **10**: 231-222.

Jain, S.P., Singh, S.C. & Kumar, S. 2003. Ethno-medico-botanical survey of Raipur district Chhatisgarh state, 266-271. *In:* Singh, V. & Jain, A.P. *(eds) Ethnobotany and Medicinal Plants of India and Nepal.* Sci. Publ., Jodhpur.

Jamir, N.S. 1997. Ethnobiology of Naga tribe in Nagaland: 1- medicinal herbs. *Ethnobotany* **9**: 101-104.

Jamir, N.S. & Rao, R.R. 1990. Fifty new or interesting medicinal plants used by the Zelings of Nagaland (India). *Ethnobotany* **2**: 11-18.

Janaki Ammal, E.K. 1955. An introduction to the subsistence economy of India, 16-22. Background paper No.10. Winner Gref Foundation Intl. Symp. on Man's Role in Changing the Face of the Earth. Princeton Inn. Princeton, New Jeresy.

Janaki Ammal, E.K. 1956. Introduction to the subsistence economy of India, 324-335. *In*: Thomas, W.L. Jr. (*ed.*) *Man's Role in Changing the Face of the Earth.* Univ. of Chicago Press, Illinosis.

Jha, B.J. & Kumar, M. 2003. Herbal remedies for various diseases, 78-87. *In:* Singh, V. & Jain, A.P. (*eds*) *Ethnobotany and Medicinal Plants of India and Nepal.* Sci. Publ., Jodhpur.

Jha, R.R. & Varma, S.K. 1996. Ethnobotany of Sauria Paharias of Santhal Pargana, Bihar: medicinal plants. *Ethnobotany* **8**: 31-35.

Jha, V. & Basak, P.K. 1994. Ethnobotanical aspects of Mithila painting. *Ethnobotany* **6**: 9-18.

Jha, V., Choudhary, U.N. & Saraswati, K.C. 1991. Botanical aspects of an ethnoveterinary prescription in Mithila, North Bihar (India). *Ethnobotany* **3**: 101-104.

Jha, V. & Goel, A.K. 2004. Botanical rosaries and non-flower garlands in folklore life of Mithila region (North Bihar). *Ethnobotany* **16**: 27-32.

Jha, V., Mishra, S., Kargupta, A.N. & Jha, A.1996. Leaves and flowers utilised as supplementary vegetables in Darbhanga (North Bihar) and their ethnobotanical significance. *J. Econ. Taxon. Bot.* Addl. Ser. **12**: 395-402.

Joshi, A. R. & Joshi, K. 2005. Piscicidal plants of Bagmati wasteland, Nepal : Indigenous practices. *Ethnobotany* **17**: 184-186.

Joshi, P. 1995. *Ethnobotany of the Primitive Tribes in Rajasthan.* Rupa Books Pvt. Ltd., Jaipur.

Kala, C.P. 2003. *Medicinal Plants of Indian Trans-Himalaya (Focus on Tibetan Use of Medicinal Resources).* BSMPS, Dehradun.

Kala, C.P. & Manjrekar, N. 2000. Ethno-medicobotany of Indian Trans-Himalaya: a case study from Spiti, 177-183. *In*: Maheshwari, J.K. (*ed.*) *Ethnobotany and Medicinal Plants of Indian Subcontinent.* Sci. Publ., Jodhpur, India.

Kala, C.P. & Rawat, G. S. 2001. Human use and conservation status of wild medicinal herbs in the Bhyundar Valley, Western Himalaya, 547-559. *In*: Pandey, P.C. & Samant, S.S. (*eds*) *Plant Diversity of The Himalaya.* Gyanodaya Prakashan, Nainital, India.

Kalakoti, B.S. 2001. Nainital hills- a transtitional zone for tropical, temperate and alpine plants and medicinal value for selected plants in pharmaceuticals, 583-591. *In*: Pandey, P.C. & Samant, S.S. (*eds*) *Plant Diversity of The Himalaya.* Gyanodaya Prakashan, Nainital, India.

Kalpesh, I., Rao, V.R., Parabia, F.M., Kothari, I.L. & Parabia, M.H. 2004. Ethnobotany of Little Rann of Kachh, Gujarat. *Ethnobotany* **16**: 40-43.

Kant, S. & Dutt, H.C. 2005. Dye yielding plants of Neeru catchment in Doda, Jammu & Kashmir, India. *Ethnobotany* **17**: 197-199.

Kapur, S.K. 1993. Ethnomedico plants of Kangra valley (Himachal Pradesh). *J. Econ. Taxon. Bot.* **17**(2): 395-408.

Kapur, S.K. 1996a. Traditionally important medicinal Plants of Bhaderwah Hills – Jammu Province–III. *J. Econ. Taxon. Bot.* Addl. Ser. **12**: 62-69.

Kapur, S.K. 1996b. Traditionally important medicinal Plants of Bhaderwah Hills – Jammu Province–IV. *J. Econ. Taxon. Bot.* Addl. Ser. **12**: 70-74.

Kapur, S.K. & Nanda, S. 1996. Traditionally important medicinal plants of Bhaderwah Hills (Jammu Province) Part–II. *J. Econ. Taxon. Bot.* Addl. Ser. **12**: 56-61.

Kapur, S.K., Nanda, S. & Srivastava, T.N. 1996. Ethnobotanical uses of PRL-Herbarium – III. *J. Econ. Taxon. Bot.* Addl. Ser. **12**: 50-55.

Kapur, S.K. & Singh, P. 1996. Traditionally important medicinal plants of Udhampur district (Jammu Province) Part–I. *J. Econ. Taxon. Bot.* Addl. Ser. **12**: 75-81.

Kapur, S.K. & Srivastava, T.N. 1996. Traditionally important medicinal plants of Udampur district (Jammu Province) Part–II. *J. Econ. Taxon. Bot.* Addl. Ser. **12**: 82-88.

Karnick, C.R. 1994. *Pharmacopoeial Standards of Herbal Plants.* Sri Satguru Publ., Delhi.

Katewa, S.S. & Arora, A. 1997. Some plants in folk medicine of Udaipur district (Rajasthan). *Ethnobotany* **9**: 48-51.

Katewa, S.S., Choudhary, B.L., Jain, A. & Takhar, H.K. 2001. Some plants in folk medicine of Rajsamand district (Rajasthan). *Ethnobotany* **13**: 129-134.

Katewa, S.S., Nag, A. & Guria, B. 2000. Ethnobotanical studies on wild plants for food from the Aravalli hills of South-East Rajasthan, 259-264. *In:* Maheshwari, J.K. (*ed.*) *Ethnobotany and Medicinal Plants of Indian Subcontinent.* Sci. Publ., Jodhpur.

Kaul, M.K. 1997. *Medicinal Plants of Kashmir and Ladakh Temperate and Cold Arid Himalaya.* Indus Publ. Co., New Delhi.

Kaur, H. & Sharma, M. 2004. *Flora of Sirmaur* (*Himachal Pradesh*). B.S.M.P.S. Dehradun.

Kaushik, P. 1988. *Indigenous Medicinal Plants* (*Including Microbes and Fungi*). Today's & Tomorrow's Printers & Publ., New Delhi.

Kayal, R.N., Saha, S. & Ghosal, P.P. 2003. A floristic account of indian medicinal plants (originated, cultivated, naturalized or introduced) used in Homoeopathic drug preparation, 60-67. *In:* Singh, V. & Jain, A.P. (*eds*) *Ethnobotany and Medicinal Plants of India and Nepal.* Sci. Publ., Jodhpur

Kemp, R. 2003. Ethno-medicinal plants used by the Rengma tribe in Dimapur district, Nagaland (India), 485-488. *In:* Singh, V. & Jain, A.P. (*eds*) *Ethnobotany and Medicinal Plants of India and Nepal.* Sci. Publ., Jodhpur.

Khan, A.A. 2001. Need for an index to compute the relative reliability of ethnobotanical claims. *Ethnobotany* **13**: 84-86.

Khan, A.A. 2005. Ethnobotany in twenty first century in India. *Ethnobotany* **17**: 71-78.

Khan, A.A., Verma, A.K. & Singh, M.P. 2003. Certain ethnobotanical informations on food and medicinal plant from Rewa division of Madhya Pradesh with special reference to rare and endangered species, 249-254. *In:* Singh, V. & Jain, A.P. (*eds*) *Ethnobotany and Medicinal Plants of India and Nepal.* Sci. Publ., Jodhpur.

Khanna, K.K. 2002. Unreported ethnomedicinal uses of plants from the tribal and rural folklores of Gonda district, Uttar Pradesh. *Ethnobotany* **14**: 52-56.

Khanna, K.K. & Kumar, R. 2000. Ethnomedicinal plants used by the Gujjar tribe of Saharanpur district, Uttar Pradesh. *Ethnobotany* **12**:17-22.

Khanna, K.K., Mudgal, V., Shukla, G. & Srivastava, P.K. 1996a. Unreported ethnomedicinal uses of plants from Mirzapur district, Uttar Pradesh. *J. Econ. Taxon. Bot.* Addl. Ser. **12**: 112- 117.

Khanna, K.K., Shukla, G. & Mudgal, V. 1996b. New traditional medicinal uses of plants from Jalaun district, Uttar Pradesh. *J. Econ. Taxon. Bot.* Addl. Ser. **12**: 108-111.

Khanna, K.K., Srivastava, P.K. & Mudgal, V. 1996c. Noteworthy medicinal plants uses from rural folklore of Raebareli district, Uttar Pradesh. *J. Econ. Taxon. Bot.* Addl. Ser. **12**: 118-122.

Khare, C.P. 2004. *Encyclopedia of Indian Medicinal Plants, Rational Western Therapy, Ayurvedic and Other Traditional Usage, Botany.* Springer-Verlag, Berlin Heidelberg, New York.

Kharkongor, P. & Joseph, J. 1981. Folklore medicobotany of rural Khasi and Jaintia tribes in Meghalaya, 124-136. *In*: Jain, S.K. (*ed.*) *Glimpses of Indian Ethnobotany.* Oxford & IBH Publ. Co., New Delhi.

Kirtikar, K.R. & Basu, B.D. 1984. *Indian Medicinal Plants.* Vol. I-IV. Lalit Mohan Basu, Allahabad, India.

Koelz, W.N. 1979. Notes on the ethnobotany of Lahaul, a province of the Punjab. *Quar. J. Crude Drug. Res.* **17**: 1-56.

Kothari, M.J. & Moorthy, S. 1996. Ethnobotany in human welfare of Raigad district–in Maharashtra State, India, 403-407. *In*: Jain, S.K. (*ed.*) *Ethnobiology in Human Welfare.* Deep Publ., New Delhi.

Kothari, M.J., Jadhav, C.R. & Singh, N.P. 2003. Ethnobotanical wealth of two sacred groves in Junnar Taluka, Pune, Maharashtra, 585-591. *In:* Singh, V. & Jain, A.P. (*eds*) *Ethnobotany and Medicinal Plants of India and Nepal.* Sci. Publ., Jodhpur.

Koul, A.K., Karihaloo, J.L. & Hamal, J.A. 1982. Wild edible plants of Kashmir-some less known vegetable substitutes and beverages. *Bull. Bot. Surv. India* **24:** 67-69.

Kowarik, I. 2003. Human agency in biological invasions: secondary releases foster naturalisation and population expansion of alien plant species. *Biol. Invasions* **5**: 293-312.

Krishna, B. & Singh, S. 1987. Ethnobotanical observations in Sikkim. *J. Econ. Taxon. Bot.* **9:** 1-7.

Kshirsagar, R.D. & Singh, N.P. 2000. Less known ethnomedicinal uses of plants in Coorg district of Karnataka state, Sourthern India. *Ethnobotany* **12**: 12-16.

Kshirsagar, S.R. 2003. Notes on miscellaneous uses of plants from tribal areas of Nashik district of Maharashtra, 894-895. *In:* Singh, V. & Jain, A.P. (*eds*) *Ethnobotany and Medicinal Plants of India and Nepal.* Sci. Publ., Jodhpur.

Kshirsagar, S.R., Parabia, M. H. & Reddy, M.H. 2003. Ethnobotany of coastal areas in South Gujarat. *Ethnobotany* **15**: 60-63.

Kulkarni, D.K. & Kumbhojkar, M.S. 1993. Kitchen garden plants of Mahadeokali tribe in Maharashtra. *Ethnobotany* **5**: 119-127.

Kulkarni, D.K. & Kumbhojkar, M.S. 1996. Pest control in tribal areas of western Maharashtra - an ethnobotanical approach. *Ethnobotany* **8**: 56-59.

Kumar, A. 1996. Some ethnomedicinal plants of the Murias of the Indravati tiger reserve, Bastas (Madhya Pradesh). *J. Econ. Taxon. Bot.* Addl. Ser. **12**: 201-205.

Kumar, A. & Lal, J. 2003. Medicinal plant resources of the Kanha tiger reserve, Madhya Pradesh, 255-265. *In*: Singh, V. & Jain, A.P. (*eds*) *Ethnobotany and Medicinal Plants of India and Nepal.* Sci. Publ., Jodhpur, India.

Kumar, A., Tewari, D.D. & Pandey, Y.N. 2003. Ethno-phytotherapeutics among Tharus of Beerpur-Semra forest range of Balrampur district (U.P.), 839-844. *In:* Singh, V. & Jain, A.P. (*eds*) *Ethnobotany and Medicinal Plants of India and Nepal.* Sci. Publ., Jodhpur.

Kumar, K. & Goel, A.K. 1998. Little known ethno-medical plants of Santhal and Paharia tribes in Santhal Paragana, Bihar, India. *Ethnobotany* **10**: 66-69.

Kumar, R.V. & Pullaiah, T. 2003. A survey of wild edible plants of Chenchu tribes of Prakasam district (Andhra Pradesh), 22-31. *In*: Khan, S.S. (*ed.*) *Vistas in Ethnobotany,* Vol.–1. *Indian Journal of Applied and Pure Biology*. Naval Kishore Press, Lucknow.

Kumar, S. 2002. *The Medicinal Plants of North-East India*. Sci. Publ., Jodhpur, India.

Kumar, S., Goyal, S. & Parveen, F. 2004. Ethnoveterinary plants in Indian arid zone. *Ethnobotany* **16**: 91-95.

Kumar, V. 2003. Wild edible plants of Surguja district of Chhattisgarh state, India, 272-282. *In:* Singh, V. & Jain, A.P. (*eds*) *Ethnobotany and Medicinal Plants of India and Nepal.* Sci. Publ., Jodhpur.

Kumar, V. & Rao, R.R. 2001. Some plant beverages used in traditional medicine. *Ethnobotany* **12**: 36-39.

Kumar, V.S. & Rao, P.S.N. 2003. Dye yielding plants of Andaman & Nicobar Islands, 827-838. *In:* Singh, V. & Jain, A.P. (*eds*) *Ethnobotany and Medicinal Plants of India and Nepal.* Sci. Publ., Jodhpur.

Kumar, Y., Fancy, S. & Rao, R.R. 1987. Further contribution to the ethnobotany of Meghalaya: plants used by War Jaintia of Jaintia Hill district. *J. Econ. Taxon. Bot.* **11**: 65-70.

Kurian, J.C. (*ed.*) 1995. *Plants that Heal*. Oriental Watchman Publ. House, Pune, India.

Lal, B., Vats, S.K., Singh, R.D. & Gupta, A. K. 1996. Plants used as ethnomedicine and supplementary food by Gaddis of Himachal Pradesh, India, 384-387. *In*: Jain, S.K. (*ed.*) *Ethnobiology in Human Welfare*. Deep Publ., New Delhi.

Lalramnghinglova, H. 2000. Ethnobotanical and agroecological studies on genetic resources of food plants in Mizoram state, 637-644. *In:* Maheshwari, J.K. (*ed.*) *Ethnobotany and Medicinal Plants of Indian Subcontinent.* Sci. Publ., Jodhpur.

Lalramnghinglova, H. 2002. Ethnobotanical study on the edible plants of Mizoram. *Ethnobotany* **14**: 23-33.

Lalramnghinglova, J.H. 1996. Ethnobotany of Mizoram- a preliminary survey. *J. Econ. Taxon. Bot.* Addl. Ser. **12**: 439-459.

Lata, S. 1996. *Grewia optiva* Drumm. 'Bhimal' : a boon for rural people in Kumaun Hills, India, 471-471. *In*: Jain, S.K. (*ed.*) *Ethnobiology in Human Welfare,* Deep Publ., New Delhi.

Lata, S. 1997. Indigenous knowledge about *Grewia optiva* in Indo-Nepal Himalaya. *Ethnobotany* **9**: 112-116.

Mahanta, D. & Tiwari, S.C. 2005. Natural dye-yielding plants and indigenous knowledge on dye preparation in Arunachal Pradesh, Northeast India. *Curr. Sci.* **88**: 1474-1480.

Mahato, A.K. & Mahato, P. 1996. Ethnobotanical wealth of Chhotanagpur Plateau–IV. Some medicinal plants used against intestinal worms. *J. Econ. Taxon. Bot.* Addl. Ser. **12**: 389-391.

Mahato, R.B. & Chaudhary, R.P. 2005. Ethnomedicinal plants of Palpa district, Nepal. *Ethnobotany* **17**: 152-163.

Maheshwari, J.K. 1983. Development in ethnobotany. *J. Econ. Taxon. Bot.* **4**(1): 1-4.

Maheshwari, J.K. 1987. Interdisciplinary approaches in ethnobotany, 19-34. *In*: Jain, S.K. (*ed.*) *A Manual of Ethnobotany*, Sci. Publ., Jodhpur, India.

Maheshwari, J.K. (*ed.*) 1992. *Ethnobotany in India*. Sci. Publ., Jodhpur.

Maheshwari, J.K. (*ed.*) 1996. *Ethnobotany in South Asia*. Sci. Publ., Jodhpur.

Maheshwari, J.K. & Paul, S.R. 1975. The exotic flora of Ranchi. *J. Bombay Nat. Hist. Soc.* **72**(1): 158-188.

Maheshwari, J. K. & Singh, J. P. 1984. Contribution to the ethnobotany of Bhoxa tribe of Bijnor and Pauri Garhwal districts, U.P. *J. Econ. Taxon. Bot.* **5**: 251-259.

Maheshwari, J.K., Singh, K.K. & Saha, S. 1980. The ethnomedicinal uses of plants of Tharus in Kheri district, Uttar Pradesh. *Bull. Medico-Ethnobot. Res.* **1**: 318-337.

Maheshwari, J.K., Singh, K.K. & Saha, S. 1981. *The Ethnobotany of the Tharus of Kheri District, U.P.* N.B.R.I., Lucknow.

Maheshwari, J.K., Singh, K.K. & Saha, S. 1986. *Ethnobotany of Tribals of Mirzapur District, U.P.* N.B.R.I., Lucknow.

Malhotra, S.K. & Moorthy, S. 1973. Some useful medicinal plants of Chandrapur district (Maharashtra State). *Bull. Bot. Surv. India* **15**: 13-21.

Malik, S. 1996. Belief and healthcare: a study among the Saoras of Koraput district Orissa, 151-152. *In*: Jain, S.K. (*ed.*) *Ethnobiology in Human Welfare*. Deep. Publ., New Delhi.

Maliya, S.D. 2004. Some new or less-known folkmedicine of district Bahraich, Uttar Pradesh, India-2. *Ethnobotany* **16**: 113-115.

Maliya, S.D. 2005. Uses of some Indian plants in folkmedicine of Nepal. *Ethnobotany* **17** : 202-204.

Maliya, S.D. & Singh, K.K. 2003. Some new or less-known folkmedicine of Bahraich district, Uttar Pradesh. *Ethnobotany* **15**: 1332-135.

Manandhar, N.P. 1980a. *Medicinal Plants of Nepal Himalaya*. Ratna Pustak Bhandar, Kathmandu, Nepal.

Manandhar, N.P. 1980b. Some less known medicinal plants of Rasuwa district (Nepal). *Quart. J. Crude Drug Res.* **18**(3): 147-151.

Manandhar, N.P. 1985. Ethnobotanical notes on certain medicinal plants used by Tharus of Dang-deokhuri district, Nepal. *Int. J. Crude Drug Res.* **25**(4): 153-159.

Manandhar, N.P. 1987a. Traditional medicinal plants used by tribals of Lamjung district, Nepal. *Int. J. Crude Drug Res.* **25**(4): 236-240.

Manandhar, N.P. 1987b. An ethnobotanical profile of Manang valley, Nepal. *J. Econ. Taxon. Bot.* **10**(1): 207-213.

Manandhar, N.P. 1989a. Medicinal plants used by Chepang tribes of Makawanpur district, Nepal. *Fitotreapia* **60**: 61-68.

Manandhar, N.P. 1989b. *Useful Wild Plants of Nepal*. Franz Steiner Verlag Wiesbaden GMBH, Stuttgart.

Manandhar, N.P. 1990. Folk-lore medicine of Chitwan district, Nepal. *Ethnobotany* **2**: 31-38.

Manandhar, N.P. 1995. Ethnobotanical notes on unexploited wild food plants of Nepal. *J. Econ. Taxon. Bot.* **7**: 95-101.

Manandhar, N.P. 1996. Traditional practice for oral health care in Nepal. *J. Econ. Taxon. Bot.* Addl. Ser. **12**: 408-413.

Manandhar, N.P. 1997. Unreported wild food plants of Nepal. *Ethnobotany* **9**: 97-100.

Manandhar, N.P. 2000. Botany and ethnobotany of *Urtica dioica* L. (Stinging nettle), 38-40. *In:* Maheshwari, J.K. (*ed.*) *Ethnobotany and Medicinal Plants of Indian Subcontinent.* Sci. Publ., Jodhpur.

Manandhar, N.P. 2002. Ethnobotanical uses of 'Pakhanbed' (*Bergenia ciliata*) in Nepal. *Ethnobotany* **14**: 13-15.

Manandhar, N.P. 2003. Ethnobotanical notes on the family *Leguminosae* in Nepal, 354-357. *In:* Singh, V. & Jain, A.P. (*eds*) *Ethnobotany and Medicinal Plants of India and Nepal.* Sci. Publ., Jodhpur.

Mandal, S.K. & Basu, S.K. 1996. Ethnobotanical studies among some tribals of Niligiri district, Tamil Nadu. *J. Econ. Taxon. Bot.* Addl. Ser. **12**: 268-271.

Manilal, K.S. 1989. Linkages of ethnobotany with other sciences and discipline *Ethnobotany* **1**: 15-24.

Mao, A.A. 1993. A preliminary report on the folklore botany of Mao Nagas of Manipur (India). *Ethnobotany* **5**: 143-147.

Mashelkar, R.A. 2001. Intellectual property rights and the third world. C.S.I.R. *Curr. Sci.* **81**(8): 955.

Masih, S.K. 2003, Diversity of ethnomedicinal wealth in Amar Kantak plateau region (India), 204-214. *In*: Khan, S.S. (*ed.*) *Vistas in Ethnobotany,* Vol.-I, *Indian Journal of Applied and Pure Biology*. Naval Kishore Press, Lucknow.

Matthew, K.M. 1969. Exotic flora of Kodaikanal and Palni hills. *Rec. Bot. Surv. India.* **20**(1): 1-241.

Maya, S., Sarojini M.V. & Sreekandan N.G. 2003. Economic importance of river vegetation of Kerela - a case study, 796-803. . *In:* Singh, V. & Jain, A.P. (*eds*) *Ethnobotany and Medicinal Plants of India and Nepal.* Sci. Publ., Jodhpur.

Mc Neely, J.A. & Pit, D. (*eds*) 1985. *Culture and Conservation, The Human Dimension in Environmental Planning.* Croom Helm, London, New York, Sydney.

Megoneitso & Rao, R.R. 1983. Ethnobotanical studies in Nagaland-4. Sixty two medicinal plants used by the Angami Nagar. *J. Econ. Taxon. Bot.* **4**(1-3): 167-172.

Mehrotra, B.N. 1990. Quality control requirement of medicinal plants used in traditional medicines. *Ethnobotany* **2**: 19-24.

Mehrotra, S. & Mehrotra, B.N. 2005. Role of traditional and folklore herbals in the development of new drugs. *Ethnobotany* **17**: 104-111.

Mehrotra, S., Rawat, A.K.S., Singh, H.K. & Shome, U. 1995. Standardization of popular Ayurvedic adaptogenic preparation 'Chayavanprash' and ethnobotany of its ingredients. *Ethnobotany* **7**: 1-15.

Mehtani, S. & Ingle, P.K. 1999. *Plants Food Flavours*. Nat. Inst. Sci. Comm. New Delhi.

Mishra, D.K., Samanta, G. & Mishra, T.K. 1996a. Ethnomedicine of the tribe Kharia of Midnapore district, West Bengal. *J. Econ. Taxon. Bot.* Addl. Ser. **12**: 329-331.

Mishra, M.K. & Gokhale, S.S. 2003. *Baseline Information on Medicinal Plants Conservation and Sustainable Utilization (Himachal Pradesh).* A project sponsored by UNDP/GEF, MOEF, Govt. of India.

Mishra, S., Jha, V. & Jha, S, 1996b. Plants in ethnoveterinary practices in Darbhanga (North Bihar), 189-193. *In*: Jain, S.K. (*ed.*) *Ethnobiology in Human Welfare*, Deep Publ., New Delhi.

Mishra, T.K., Bhunia, A., Sahoo, T.K. & Roy, S.B. 1997. Ethnobotany vis-à-vis joint forest management movement of Midnapore (West Bengal), India. *Ethnobotany* **9**: 52-55.

Mitaliya, K.D. & Bhatt, D.C. 2003. Ethnobotanical studies on wild edible plants of Saurashtra, Gujarat, India, 228-232. *In:* Singh, V. & Jain, A.P. (*eds*) *Ethnobotany and Medicinal Plants of India and Nepal.* Sci. Publ., Jodhpur.

Mitaliya, K.D., Bhatt, D.C., Patel, J.N. & Patel, S.K. 2003. Observation of plants used for joint disorders by the rural and tribal people of Gujarat, 212-216. *In:* Singh, V. & Jain, A.P. (*eds*) *Ethnobotany and Medicinal Plants of India and Nepal.* Sci. Publ., Jodhpur.

Mohanty, R.B. & Padhy, S.N. 1996. Traditional phytotheraphy for diarrhoeal diseases in Ganjam and Phulbani districts of South Orissa, India. *Ethnobotany* **8**: 60-65.

Mokat, D.N. & Deokule, S.S. 2004. Plants used as veterinary medicine in Ratnagiri district of Maharashtra. *Ethnobotany* **16**: 131-135.

Molla, H.A. & Roy, B. 1996. Some ethnobotanical claims from the Jalpaiguri district of West Bengal. *J. Econ. Taxon. Bot.* Addl. Ser. **12**: 322-324.

Molla, M.A. & Roy, B. 1985. Traditional uses of some medicinal plants by the Robha tribals in Jalpaiguri district, West Bengal. *J. Econ. Taxon. Bot.* **7**(3): 578-580.

Mudgal, V. & Pal, D.C. 1980. Medicinal plants used by tribals of Mayurbhanj (Orissa). *Bull Bot. Surv. India* **22**: 59-62.

Mudgal, V., Pal, D.C., Kayal, R.N. & Saha, S. 1999. *Ethnobotany of Totopora.* B.S.M.P.S., Dehradun, India.

Mukherjee, A. & Namhata, D. 1990. Medicinal plant of tribals of Sundargarh district, Orissa. *Ethnobotany* **2**: 57-60.

Murthy, E.N., Raju, V.S. & Reddy, C.S. 2007. Occurence of exotic *Hyptis suaveolens. Curr. Sci.* **93**(9): 1203.

Murthy, S.R., Rani, S.S. & Pullaiah, T. 2003. Wild edible plants of Andhra Pradesh, India, 613-630. *In:* Singh, V. & Jain, A.P. (*eds*) *Ethnobotany and Medicinal Plants of India and Nepal.* Sci. Publ., Jodhpur.

Nadanakunjidam, S. 2003. Ethnomedicinal observations from Attapadi hills of Western Ghats, 732-740. *In:* Singh, V. & Jain, A.P. (*eds*) *Ethnobotany and Medicinal Plants of India and Nepal.* Sci. Publ., Jodhpur.

Nadanakunjidam, S. & Abirai, S. 2003. Comparative study of traditional medical knowledge of Pondicherry and Karaikal regions in union territory of Pondicherry. *Ethnobotany* **17***:* 112-117.

Nair, N.C. 1964. On a botanical tour to Lahaul and Spiti (Punjab Himalaya). *Bull. Bot. Surv. India.* **6**: 219-235.

Nair, N.C. 1977. *Flora of Bashahr Himalaya.* Int. Bioscience Publ., Hissar.

Nath, A. & Maiti, G.G. 2003. Ethnobotany of Barak valley (Southern Assam) with special reference to folk medicines, 964-974. *In:* Singh, V. & Jain, A.P. (*eds*) *Ethnobotany and Medicinal Plants of India and Nepal.* Sci. Publ., Jodhpur.

Nath, S.C. & Bordoloi, D.N. 1989. Ethnobotanical observation on some medicinal folklore of Tirap district, Arunachal Pradesh. *J. Econ. Taxon. Bot.* **13:** 321-325.

Nautiyal, S. 1981. Some medicinal plant of Garhwal hills- a traditional use. *J. Sci. Res. Pl. Med.* **2:** 12-17.

Nautiyal, S., Maikhuri, R.K., Rao, K.S. & Saxena, K.G. 2003. Ethnobotany of the Thlchha Bhotiya tribe of the buffer zone villages in Nanda Devi biosphere reserve, India, 119-142. *In:* Singh, V. & Jain, A.P. (*eds*) *Ethnobotany and Medicinal Plants of India and Nepal.* Sci. Publ., Jodhpur.

Nayak, P.K., Nayak R.K. & Choudhury, B.P. 2003. A checklist of medicinal plants Kalahandi district in Orissa, 519-532. *In:* Singh, V. & Jain, A.P. (*eds*) *Ethnobotany and Medicinal Plants of India and Nepal.* Sci. Publ., Jodhpur.

Nayar, M.P. 1977. Changing patterns of the Indian flora. *Bull. Bot. Surv. Ind.* **19**: 145-154.

Nayar, M.P. & Sastry, A.R.K. 1987-1988. *Red Data Book on Indian Plants* (Vol. 1-11). Bot. Surv. of India, Calcutta.

Nazarudeen, A., Seeni, S., Koshy, K.C. & Pushpangadan, P. 1996. Folk plants of food, medicine, adornment and repellent used by the Paniyar community in North Kerela. *J. Econ. Taxon. Bot.* Addl. Ser. **12**: 299-305.

Negi, K.S. & Gaur, R.D. 1991. A contribution to the edible wild fruits of Uttar Pradesh hills. *Bull. Bot. Surv. India* **43**: 233-266.

Negi, K.S., Gaur, R.D. & Tiwari, J.K. 1999. Ethnobotanical notes on the flora of Har-Ki-Doon (District Uttarkashi), Garhwal Himalaya, Uttar Pradesh, India. *Ethnobotany* **11**: 9-17.

Negi, K.S. & Pant, K.C. 1990. Notes on ethnobotany of Gangwal- a tribe of Garhwal Himalaya. *Ethnobotany* **2**: 81-85.

Negi, K.S., Tiwari, J.K. & Gaur, R.D. 1985. Economic importance of some common trees in Garhwal Himalaya: an ethnobotanical study. *Indian J. Forestry* **8**: 276-289.

Negi, K.S., Tiwari, J.K., Gaur, R.D. & Pant, K.C. 1993. Notes on ethnobotany of five districts of Garhwal Himalaya, Uttar Pradesh, India. *Ethnobotany* **5**: 73-81.

Negi, P.S. & Hajra, P.K. 2007. Alien flora of Doon valley, North West Himalaya. *Curr. Sci.* **92** (7): 968-978.

Negi, P.S. & Subramani, S.P. 2002. Ethnobotanical study in the village Chhitkul of Sangla valley, district Kinnaur, Himachal Pradesh. *J. Non-Timber Forest Prod.* **9**(3-4): 113-120.

Negi, Y.S. & Bhalla, P. 2002. Collection and marketing of important medicinal and aromatic plants in tribal areas of Himachal Pradesh. *Indian For.* **128**(6): 641-649.

Nunez, D.R. & Castro, C.O. 1995. Medicinal plants and a multipurpose complex sold in the market of Funchal (island of Madeira, Portugal). *Ethnobotany* **7**: 75-82.

Nwosu, M.O. 2001. Ethnobotanical studies on some pteridophytes of Southern Nigeria. *Ethnobotany* **13**: 29-35.

Padhy, S.N. & Dash, S.K. 2003. Ethnobotanical survey of insect repellent plant species in folklore and literature of Orissa State, India, 505-512. *In:* Singh, V. & Jain, A.P. (*eds*) *Ethnobotany and Medicinal Plants of India and Nepal.* Sci. Publ., Jodhpur.

Painuli, R.M. & Maheshwari, J.K. 1996. Some interesting ethnomedicinal plants used by Sahariya tribe of Madhya Pradesh. *J. Econ. Taxon. Bot.* Addl. Ser. **12**: 179-185.

Pal, D.C. 1981. Plants used in treatment of cattle and birds among tribals of Eastern India, 245-257. *In:* Jain, S.K. (*eds*) *Glimpses of Indian Ethnobotany.* Oxford & IBH, Delhi.

Pal, D.C., Guha, A. & Sen, R. 2000. Medicinal plants and plant products used in children diseases among aboriginal in India, 121-130. *In:* Maheshwari, J.K. (*ed.*) *Ethnobotany and Medicinal Plants of Indian Subcontinent.* Sci. Publ., Jodhpur.

Pal, D.C. & Paul, C.R. 2003. Healthful culinary herbal spices of India, 75-77. *In:* Singh, V. & Jain, A.P. (*eds*) *Ethnobotany and Medicinal Plants of India and Nepal.* Sci. Publ., Jodhpur.

Pal, D.C. & Srivastava, J.N. 1976. Preliminary notes of ethnobotany of Singhbhum district, Bihar. *Bull. Bot. Surv. India* **18**: 247-250.

Pal, G.D. 1984. Observations on the ethnobotany of tribals of Subansiri, Arunachal Pradesh. *Bull. Bot. Surv. India* **26**: 26-37.

Panda, S. 1996. Plant use - a recent perspective from Eastern-Himalayas. *J. Econ. Taxon. Bot.* Addl. Ser. **12**: 332-337.

Pandey, A.K., Bora, H.R. & Deka, S.C. 1996. An ethno-medico botanical study of Golghat district, Assam : native plant remedies for jaundice. *J. Econ. Taxon. Bot.* Addl. Ser. **12**: 344-349.

Pandey, D.N. 1998. *Ethnoforestry* (*Local Knowledge for Sustainable Forestry and Livelihood Security*). Himanshu Publ., Udaipur, New Delhi.

Pandey, H.P. & Verma, B.K. 2002. Plants in oral healthcare among the aborigines of Gonda and Balrampur regions, U.P. India. *Ethnobotany* **14**: 81-86.

Pandey, I.B. 2003. Some traditional herbal remedies used in and around Kanpur city (Uttar Pradesh), India. *Ethnobotany* **15**: 129- 131.

Pandey, P.C. & Joshi, J.S. 2001. Cultured plants of Kumaun Himalaya used for medicinal purposes. 117-124. *In* : Samant, S.S., Dhar, U. & Palni, M.S. (*eds*) *Himalayan Medicinal Plants, Potential and Prospects.* Gyanodya Prakashan, Nainital, India.

Pandey, R.P. & Parmar, P.J. 1994. The exotic flora of Rajasthan. *J. Econ. Taxon. Bot.* **18**(1): 105-121.

Panthi, M.P. & Chaudhary, R.M. 2003. Ethnomedicinal plant resources of Arghakhanchi district. *Ethnobotany* **15**: 71-86.

Parmar, C. & Kaushal, M.K. 1982. *Wild Fruits of the Sub-Himalayan Region.* Kalyani Publ., New Delhi- Ludhiana.

Parrotta, J.A. 2001. *Healing Plants of Peninsular India.* CABI Publ., New York.

Patil, D.A. 2000. Sanskrit plant names in an ethnobotanical perspectives. *Ethnobotany* **12**: 60-64

Patil, D.A. 2004. Origin of medicine vis-a vis doctrine of signatures. *Ethnobotany* **16:** 52-58.

Patiole, S.N. & Jain, A.K. 2002. Some wild edible plants of Panchmarhi Biosphere Reserve (M.P.). *Ethnobotany* **14**: 48-51.

Paul, C.R. & Pal, D.C. 2003. Some non-conventional oil sources of Bankura district, West Bengal, 564-567. *In:* Singh, V. & Jain, A.P. (*eds*) *Ethnobotany and Medicinal Plants of India and Nepal.* Sci. Publ., Jodhpur.

Pawar, S. & Patil, D.A. 2004. Herbal folk remedies for urinogenital complaints in Jalgaon district of Maharashtra. *Ethnobotany* **16**: 86-90.

Pearson, R.S. & Brown, H.P. 1932. *Commercial Timber of India*, vol.-I, Central Publication Branch, Govt. of India, Calcutta.

Pei, S.J. 2000. Application of ethnobotany in the studies of ethno-medicine. *Chinese Ethnomed. & Ethnophar. Biol.* **39**: 74-79.

Perrings, C., Dehnen-Schmutz, K., Touza, J. & Williamson, M. 2005. How to manage biological invasions under globalization. *Trends Ecol. Evol.* **20**: 212-215.

Polunin, O. & Stainton, A. 1984. *Flowers of the Himalaya.* Oxford Univ. Press, Delhi.

Polunin, O. & Stainton, A. 1987. *Concise Flowers of Himalaya.* Oxford *Univ. Press,* Delhi.

Prajapati, N.S., Purohit, S.S., Sharma, A.K. & Kumar, T. 2006. *A Handbook of Medicinal Plants- A Complete Source Book.* Agrobios (India).

Prakash, V. 1999. Indian medicinal plants : current status-1. *Ethnobotany* **10**: 112-121.

Prakash, V. 2001. Indian medicinal plants: current status, 45-63. *In*: Samant, S.S., Dhar, U. & Palni, M.S. (*eds*) *Himalayan Medicinal Plants, Potential And Prospects.* Gyanodaya Prakashan, Nainital (U.P.), India.

Prakash, V., Aggarwal, A., Sharma, A. & Mehrotra, R.S. 2003. Ethnobotanical studies on some wild plants of district Hamirpur in Himachal Pradesh, 374. *In:* Prakash, P. (*ed.*) *Proc. Nat. Sem. Biochem. Sci.: Health and Environment Aspects,* Allied Publ., Pvt. Ltd., New Delhi.

Prakash, V. & Mehrotra, B.N. 1987. Ethnobotanical studies on the flora of Khandala, Maharashtra state. *J. Econ. Taxon. Bot.* **9**: 205-208.

Prasad, P., Nagendra, R.S., Narayanan, L.M. & Natarajan, C.R. 1996. Ethnobotany of the Kanikkars of South Tamil Nadu-1. *J. Econ. Taxon. Bot.* Addl. Ser. **12**: 292-298.

Pruthi, J.S.1976. *Spices and Condiments.* National Book Trust, India.

Pullaiah, T. & Kumar, T.D.C. 1996. Herbal plants in Mannanur forest Mahaboobnagar district, Andhra Pradesh. *J. Econ. Taxon. Bot.* Addl. Ser. **12**: 218-220.

Punjani, B.L. 2002. Ethnobotanical aspects of some weeds from Gujarat. *Ethnobotany* **14**: 78-80.

Purohit, K. & Samant, S.S. 1995. *Fodder Trees and Shrubs of Central Himalaya.* Gyanodaya Prakashan, Nainital (U.P.), India.

Pushpangadan, P. 1990. Health status of tribals of India. Key note address, 67-71. Nat. Sem. Health of Tribal people. *Indian Society for Health Administrators (ISHA),* May 21-26, 1990.

Pushpangadan, P. & Kumar, B. 2005. Ethnobotany, CBD,WTD and the biodiversity act of India. *Ethnobotany* **17:** 2-12.

Radhakrishnan, K., Pandurangan, A.G. & Pushpangandan P. 1996. Ethnobotany of wild edible plants of Kerala, India, 48-51. *In*: Jain, S.K. (*ed*) *Ethnobotany in Human Welfare*, Deep Publ., New Delhi.

Ragupathy, S. & Mahadevan, A. 1991. Ethnobotany of Kodiakkarai reserve forest, Tamil Nadu, South India. *Ethnobotany* **3**: 79-82.

Rai, M.K. & Upadhyay, S.K. 1996. Medicinal plants of Chhindwara district: retrospect and prospect, 174-188. *In:* Rajak, R.C. & Rai, M.K. (*eds*) *Herbal Medicines Biodiversity and Conservation Strategies.* International Books Distributor, Dehradun.

Rajan G.B., Khan, I.A. & Khanum, A. 2005. Medicinal plants for urogenital systems, 241-261. *In:* Khan, I.A. & Khanum, A. (*eds*) *Role of Biotechnology in Medicinal and Aromatic Plants.* Ukaaz Publ., Hyderabad.

Rajan, S., Baburaj, S., Sethuraman, M. & Primala, S. 2003. Folk medicine among Nilgiri Irulas : an emerging trend in primary healthcare practice, 791-793. *In:* Singh, V. & Jain, A.P. (*eds*) *Ethnobotany and Medicinal Plants of India and Nepal.* Sci. Publ., Jodhpur.

Rajasekaran, A. & Singh, J. 2009. Ethnobotany of Indian Horse Chestnut (*Aesculus indica*) in Mandi district, Himachal Pradesh. *Indian J. Trad. Knowledge* **8**(2): 285-286.

Rajendran, A. & Henry, A. N. 1994. Plants used by the tribe Kadar in Anaimalai hills of Tamil Nadu. *Ethnobotany* **6**: 19-24.

Rajendran, A., Rama Rao, N. & Henry, A.N. 1996. Hepatic stimulant plants of Andhra Pradesh, India. *J. Econ. Taxon. Bot.* Additional Seires **12**: 221-223.

Rajendran, A. & Sikarwar, R.L.S. 2003. Intra-cultural ethnobotanical studies on the tribe Kadars (Tamilnadu) and Sahariyas (Madhya Pradesh), India, 662-664. *In:*

Singh, V. & Jain, A.P. (*eds*) *Ethnobotany and Medicinal Plants of India and Nepal.* Sci. Publ., Jodhpur.

Rajendran, K., Balakrishnan, R. & Chandrasekaran, S. 2003. Common medicinal plants and their utilization by villagers in east coast districts of Tamilnadu, 715-726. *In:* Singh, V. & Jain, A.P. (*eds*) *Ethnobotany and Medicinal Plants of India and Nepal.* Sci. Publ., Jodhpur.

Rajendran, S.M. 2003. West Sikkim : A treasure house of medicinal plants and their utility by local tribals, 311-316. *In:* Singh, V. & Jain, A.P. (*eds*) *Ethnobotany and Medicinal Plants of India and Nepal.* Sci. Publ., Jodhpur.

Rajore, S. & Batra, A. 2003. *Jatropha curcas* L. : a plant of immense potential value, 36-41. *In:* Singh, V. & Jain, A.P. (*eds*) *Ethnobotany and Medicinal Plants of India and Nepal.* Sci. Publ., Jodhpur.

Rajwar, G.S. 1983. Low altitude medicinal plants of south Garhwal. *Bull. Medico-Ethnobot. Res.* **4:** 14-28.

Ramachandran, V.S. & Nair, V.J. 1981. Ethnobotanical observation on 'Irulars' of Tamil Nadu (India). *J. Econ. Taxon. Bot.* **2**: 183-190.

Rama Rao, N. & Henry, A.N. (*eds*) 1996. *Ethnobotany of Eastern Ghats in Andhra Pradesh, India.* B.S.I., Calcutta.

Ramdas, S.R., Ghotge, N.S., Ashalata, S., Mathur, N.P., Brome, V.G. & Rao, S. 2000. Ethnoveterinary remedies used in common surgical conditions in some districts of Andhra Pradesh and Maharashtra, India. *Ethnobotany* **12**: 100-112.

Rana, J.C. 2004. *Lesser Known Food Plants of Himachal Pradesh.* N.B.P.G.R., Regional Station, Phagli, Shimla (H.P.).

Rana, T.S., Dutt, B. & Rao, R.R. 1996. Strategies for sustainable utilization of plant resources by the Tribals of the Tons Valley, Western Himalaya. *Ethnobotany* **8**: 96-104.

Ranjan, P. 2000. A contribution to some wild edible plants of Indo-Nepal Border area, 661-666. *In:* Maheshwari, J.K. (*ed.*) *Ethnobotany and Medicinal Plants of Indian Subcontinent.* Sci. Publ., Jodhpur.

Ranjan, P. 2003. A contribution to some ethnomedicinally important plants of Nepal, 345-353. *In:* Singh, V. & Jain, A.P. (*eds*) *Ethnobotany and Medicinal Plants of India and Nepal.* Sci. Publ., Jodhpur.

Ranjan, V. 1996. Some ethnomedicinal plants of Lalitpur district, Uttar Pradesh, India, 149-150. *In*: Jain S.K. (*ed.*) *Ethnobiology in Human Welfare*, Deep Publ., New Delhi.

Rao, D.M. & Pullaiah, T. 2001. Ethno-medico-botanical studies in Guntur district of Andhra Pradesh, India. *Ethnobotany* **13**: 40-44.

Rao, N.S. B., Rajasekhar, D., Raju, D. C. & Nagaraju, J. 1996. Ethnomedicinal notes on some plants of Tirumala Hills for dental disorders. *Ethnobotany* **8**: 88-91.

Rao, R.R. 1990. Ethnobiology in the study and conservation of fragile ecosystem: some issue from the Himalayan region. *Ethnobotany* **2**: 45-55.

Rao, R.R. 1996. Traditional knowledge and sustainable development: key role of ethnobiologists. *Ethnobotany* **8**: 14-24.

Rao, R.R. & Jamir, N.S. 1982. Ethnobotanical studies in Nagaland–II; 54 medicinal plants used by Nagas. *J. Econ. Taxon. Bot.* **3**: 11-17.

Rastogi, M.A. 1960. Medicines from the wild. A case study of the Great Himalayan Park. *The Indian Magazine of her People and Culture*, 74-75.

Rastogi, R.P. & Mehrotra, B.N. 1993. *Compendium of Indian Medicinal Plants.* Vol. 3. (1980-1984). C.D.R.I., Lucknow & C.S.I.R., New Delhi.

Rastogi, R.P. & Mehrotra, B.N. 1999. *Compendium of Indian Medicinal Plants.*Vol 1-V CDRI, Lucknow & NISC, New Delhi.

Rau, M.A. 1961. A recent survey of medicinal plant resources of Lahaul valley (Panjab India). *Pakistan J. Soc. & Ind. Res. Spl. Nr. Intl. Symp.* **4**: 201-205.

Ravikumar, K. & Shankar, V.R. 2003. Ethnobotany of Malayali tribe in Melpattu village, Javvadhu hills of Eastern Ghats, Tiruvannamalai district, Tamil Nadu, 715-726. *In:* Singh, V. & Jain, A.P. (*eds*) *Ethnobotany and Medicinal Plants of India and Nepal.* Sci. Publ., Jodhpur, India

Ravishankar, T. 1996. Role of indigenous people in the conservation of plant genetic resources, 310-314. *In*: Jain, S.K. (*ed.*) *Ethnobiology in Human Welfare*. Deep Publ., New Delhi.

Ravishankar, T. & Henry, A.N. 1992. Ethnobotany of Aditlabad district, Andhra Pradesh, India. *Ethnobotany* **4**: 45-52.

Rawat, G.S. & Pangtey, Y.P.S. 1987. A contribution to the ethnobotany of alpine region of Kumaon. *J. Econ. Taxon. Bot.* **11**: 139-148.

Rawat, M.S. & Chowdhary, S. 1998. *Ethno-Medico-Botany of Arunachal Pradesh.* B.S.M.P.S., Dehradun.

Reddy, C.S. 2008. Catalogue of invasive alien flora of India. *Life Sci. Jour.* **5** (2): 84-89.

Reddy, C.S., Bhanja, M.R. & Raju, V.S. 2000. *Cassia uniflora* Miller: a new record for Andhra Pradesh, India. *Indian J. Forestry* **23**(3): 324-325.

Reddy, C.S. & Raju, V.S. 2000. Folklore biomedicines for common veterinary diseases in Nalgonda district, Andhra Pradesh, India. *Ethnobotany* **12**: 113-117.

Reddy, C.S. & Raju, V.S. 2002. Additions to the weed flora of Andhra Pradesh. *J. Econ. Taxon. Bot.* **26:** 195-198.

Reddy, C.S. & Reddy, K.N. 2004. *Cassia rotundifolia* Pers. (Caesalpiniaceae): a new record for India. *J. Econ. Taxon. Bot.* **28:** 73-74.

Reddy, H., Hemambara, V.K. & Venkata Raju, R.R. 1996. Native phytotherapy for snake bite in Nallamalais Eastern Ghats, India. *J. Econ. Taxon. Bot.* Addl. Ser. **12**: 214-217.

Reddy, P.R., Rao, P.P. & Prabhakar, M. 2003a. Ethnomedicinal practices amongst Chenchus of Nagarjunsagar Srisailam Tiger Reserve (NSTR) Andhra Pradesh-plant remedies of cuts, wounds and boils. *Ethnobotany* **15**: 67-70.

Reddy, R.V., Lakshmi, N.V.N. & Venkata Raju, R.R. 1997. Ethnomedicine for ephemeral fevers and Anthrax in cattle from the hills of Cuddapah district, Andhra Pradesh, India. *Ethnobotany* **9**: 94-96.

Reddy, S., Nagesh, K., Reddy, K.N. & Raju, V.S. 2003b. Plants used in ethnoveterinary practices by Gonds of Karimnagar district, Andhra Pradesh, India, 631-634. *In:* Singh, V. & Jain, A.P. (*eds*) *Ethnobotany and Medicinal Plants of India and Nepal.* Sci. Publ., Jodhpur.

Retnam, R.K. & Martin, P. 2006. *Ethnomedicinal Plants.* Agrobios, india.

Robbins, W.W., Herrington, J.P. & Freire-Marreco, B. 1916. *Ethnobotany of the Tewa Indians.* Bureau of American Ethnology Bulletin, no 55. Smithsonian Insts, Washington, D.C.

Roy, B., Halder, A.C. & Pal, D.C. 1998. *Plants for Human Consumption in India.* B.S.I., Calcutta.

Sagwal, S.S. & Gupta, N.K. 1988. Role of forestery extension in exploitation on medicinal and aromatic plants of Himachal Pradesh, 50-54. *In:* Khosla, P.K. & Sehgal, R.N. (*eds*) *Trends in Tree Science.* ISTS Publ., Solan.

Sahni, K.C. 1998. *The Book of Indian Trees.* Bombay Natural History Soci, Oxford Univ. Press, New York.

Sahoo, A.K. & Mudgal, V. 1995. Less known ethnobotanical uses of plants of Phulbani district, Orissa, India. *Ethbobotany* **7**: 63-67.

Sahu, T.R. 1996. Life support promising food plants among aboriginals of Bastar (M.P.), India, 26-30. *In*: Jain, S.K. (*ed.*) *Ethnobiology in Human Welfare.* Deep Publ. New Delhi.

Said, H.M. 1997. *Hamdard Pharmacopoeia of Eastern Medicine.* Sri Satguru Publ., Delhi.

Saikia, N. & Nath, S.C. 2003. Traditional use of some plants in religio-cultural purposes by 'Koch' community in Assam, India, 997-1001. *In:* Singh, V. & Jain, A.P. (*eds*) *Ethnobotany and Medicinal Plants of India and Nepal.* Sci. Publ., Jodhpur.

Saini, D.C. 1996a. Ethnobotany of Tharus of Basti district, Uttar Pradesh. *J. Econ. Taxon. Bot.* Addl. Ser. **12**: 138-153.

Saini, V.K. 1996b. Plants in the welfare of tribal women and children in certain areas of central India, 140-144. *In*: Jain, S.K. (*ed.*) *Ethnobiology in Human Welfare.* Deep. Publ., New Delhi.

Saklani, S. & Jain, S.K. (*eds*) 1994. *Cross-Cultural Ethnobotany of North-East India.* Deep Publ., New Delhi.

Saluja, B. 1996. Commonly available plants tested for lowering blood sugar, 195-198. *In:* Rajak, R.C. & Rai, M.K. (*eds*) *Herbal Medicines Biodiversity and Conservation Strategies.* Intl. Book Distributors, Dehradun.

Samant, S. S., Dhar, U. & Rawal, R.S. 1996. Natural resource use of some natives within Nanda Devi biosphere reserve in Western Himalaya. *Ethnobotany* **8**: 40-50.

Samant, S.S., Dhar, U. & Rawal, R.S. 2001. Diversity, distribution, and indigenous uses of threatened medicinal plants of Askot wildlife sanctuary in Western Himachal : causes and manage - perspectives., 167-184. *In*: Samant, S.S., Dhar, U. & Palni, M.S. (*eds*) *Himalayan Medicinal Plants, Potential and Prospects.* Gyanodaya Prakashan, Nainital, India.

Samwatsar, S. & Diwanji, V.B. 1996a. Plants used for skin diseases, cuts, wounds, and bruises by the tribals of Western M.P. *J. Econ. Taxon. Bot.* Addl. Ser. **12**: 192-195.

Samwatsar, S. & Diwanji, V.B. 1996b. Plants used in snake, scorpion and insect bites/stings by Adibasis of Jhabua (M.P.) India. *J. Econ Taxon. Bot.* Addl. Ser. **12**: 199-200.

Sanyal, D. 1994. *Vegetable Drugs of India.* B.S.M.P.S., Dehradun.

Sarangi, N. & Sahu, R.K. 2005. Ethno-medicinal plants used in venereal and gynaecological disorders in Kalahandi, Orissa. *Ethnobotany* **16**: 16-20.

Sarin, Y.K. 1967. A survey of vegetable raw resources of Lahaul. *Indian For.* **93**(7): 489-499.

Sarin, Y.K. 1990. Some less known but effective folk remedies in North - West Himalayan Region. *Ethnobotany* **2**: 39-43.

Sarin, Y.K. 1996. *Illustrated Manual of Herbal Drugs Used in Ayurveda.* C.S.I.R. & I.C.M.R., New Delhi.

Sarin, Y.K. & Chopra, C.L. 1985. Medicinal and aromatic plant resources of Himachal Pradesh, present status and future prospects. *Workshop Sci. & Tech., Shimla*, Oct. 29-30, 1985.

Sarin, Y.K., Kapoor, L.D. & Chopra, I.C. 1963. *Physochlaina praelta* (Don) Miers: a hyoscyamine yielding plant important in Lahhul's economy. *Indian For.* **89**: 610-611.

Sarkar, N., Rudra, S., Basu, S.K. & Debnath, H.S. 2003. Herbarium- an archine of ethnomedicine, 1-6. *In:* Singh, V. & Jain, A.P. (*eds*) *Ethnobotany and Medicinal Plants of India and Nepal.* Sci. Publ., Jodhpur.

Sarma, S.K., Bhattacharya, D.K. & Devi, B. 2001. Medicinal plants used by Bodo tribe of Nalbari district in Assam. *Ethnobotany* **13**: 135-139.

Satapathy, K.B. & Brahmam, M. 1996. Some medicinal plants used by tribals of Sundargarh district, Orissa, India, 153-155. *In*: Jain, S.K. (*ed.*) *Ethnobiology in Human Welfare*, Deep Publ., New Delhi.

Saxena, A.P. & Vyas, K.M. 1981. Ethnobotanical records on infectious diseases from tribals of Banda district. *J. Econ. Taxon. Bot.* **2**: 191-194.

Saxena, H.O., Brahmam, M. & Dutta, P.K. 1981. Ethnobotanical studies in Orissa, 232-244. *In*: Jain, S.K. (*ed.*) *Glimpses of Indian Ethnobotany.* Oxford & IBH, Publi. Co., New Delhi.

Saxena, K.G. 1991. Biological invasion in the Indian sub-continent: review of invasion by plants, 53-73. *In*: Ramakrishnan, P.S. (*ed.*): *Ecology of Biological Invasion in the Tropics.* Intl. Sci. Publ., New Delhi.

Schultes, R.E. 1960. Tapping our heritage of ethnobotanical lore. *Econ. Bot.* **8**: 224-227.

Schultes, R.E. & Reis, S.V. 1995. *Ethnobotany - Evolution of a Discipline.* Chapman, London.

Schultes, R.E. & Swain, T. 1976. The plant kingdom: a virgin field for new biodynamic constituents, 134-171. *In*: Fina, N.J. (*ed.*) *The Recent Chemistry of Natural Products, including Tobacco.* Proc. *Second Philip Moriss Sci. Symposium,* Richmond.

Seetharam, Y.N., Chalageri, G., Haleshi, C. & Vijay. 1999. Folk medicine and ethnomedicine of North-Eastern Karnataka. *Ethnobotany* **11**:32-37.

Sen, S. & Batra, A. 1997. Ethno-medico-botany of household remedies of Phagi tehsil of Jaipur district (Rajasthan). *Ethnobotany* **9**: 122-128.

Sen, S., Batra, A. & Ali, Z.D. 2002. Some herbal home remedies form Jaipur district, Rajasthan. *Ethnobotany* **14**: 116-122.

Sen, S.K. & Behera, L.M. 2003. Ethnomedicinal plants used against skin diseases in Bargarh district in Orissa. *Ethnobotany* **15**: 90-96.

Sensarma, P. 1991. Herbal veterinary medicines in an ancient-Sanskrit work – The Garuda Purana. *Ethnobotany* **3**: 83-87.

Sethi, G.S. 2003. Potential application biotechnology in medicinal plants of Himachal Pradesh. *In: Proc. Natl. Sem. Emerging Trends in Res. & Business Management of Medicinal & Aromatic Plants,* March 11-13, 2003. CCSHAH, Hissar, Haryana.

Shabnam, S.R. 1964. Medicinal Plants of Chamba. *Indian For.* **90**: 50-63.

Shah, G.L. 1984. Some economically important plants of Salsette Island near Bombay. *J. Econ. Taxon. Bot.* **5:** 753-765.

Shah, G.L., Yadav, S.S. & Badri, N. 1983. Medicinal plants from Dahanu forest division in Maharashtra State. *J. Econ. Taxon. Bot.* **4**: 141-151.

Shah, M.A.U. 2005. Ethnomedicinal plants of Bhakkar tehsil in Punjab, Pakistan. *Ethnobotany* **17**: 171-175.

Shah, N.C. 1997. Ethnobotany of *Cannabis sativa* in Kumaon Region, India. *Ethnobotany* **9**: 117-121.

Shah, N.C. 2008. Prospects of ethnobotany in India. *Ethnobotany* **20**: 16-24.

Shah, N.C. & Joshi, M.C. 1971. An ethnobotanical study of Kumaon region of India. *Econ. Bot.* **25**: 412-422.

Shah, N.C. & Singh, S.C. 1990. Hitherto unreported phytotherapeutical uses from tribal pockets of Madhya Pradesh (India). *Ethnobotany* **2**: 91-95.

Shanavashan, S.B., Santhoshkumar, E.S. & Pushpangadam, P. 2003. Plants used as medicine by the Irulas of Palaghat district, Kerala, India, 808-814. *In*: Singh, V. & Jain, A.P. *(eds) Ethnobotany and Medicinal Plants of India and Nepal.* Sci. Publ., Jodhpur.

Sharma, B.D. 1984. *Exotic Flora of Allahabad.* B.S.I., Dehradun.

Sharma, B.D. & Chandel, K.P.S.1996. Occurence, distribution and diversity of soft fruits in N.W and W. Himalayas and prospects of their conservation and utilization. *Indian J. Pl. Gen. Res.* **9**(2): 237-240.

Sharma, B.D. & Rana, J.C. 1999. Traditional medicinal uses of plants of Himachal hills. *J. Econ. Taxon. Bot.* **23**(1): 173-176.

Sharma, B.D. & Rana, J.C. 2000. Traditional medicinal uses of plants of Himachal Hills, 173-176. *In:* Maheshwari, J.K. (*ed.*) *Ethnobotany and Medicinal Plants of Indian Subcontinent.* Sci. Publ., Jodhpur.

Sharma, B.D. & Rana, J.C. 2005. *Plant Genetic Resources of Western Himalayas: Status and Prospects.* B.S.M.P.S., Dehradun, India.

Sharma, H.M., Sharma, B.M. & Devi, A.R. 2000. Contributions to the edible fruits of Manipur, 615-623. *In:* Maheshwari, J.K. (*ed.*) *Ethnobotany and Medicinal Plants of Indian Subcontinent.* Sci. Publ., Jodhpur.

Sharma, K.R. & Sood, M. 1997. *Important Medicinal Plants of Himachal Pradesh.* Vol-I & II. Dr. Y.S. Parmar Univ. Hort. & For., Solan, H.P.

Sharma, K.S., Chauhan, N.S. & Lal, B. 2003a. Commercially important medicinal and aromatic plants of Parvati valley, Himachal Pradesh, 937-942. *In*: Singh, V. & Jain, A.P. (*eds*) *Ethnobotany and Medicinal Plants of India and Nepal.* Sci. Publ., Jodhpur.

Sharma, M., Devi, H., Radhapyari, A. & Manihar, S.B. 2003b. Ethnomedicinal uses of monocotyledonous plants by the Meitei of Manipur, 473-480. *In:* Singh, V. & Jain, A.P. (*eds*) *Ethnobotany and Medicinal Plants of India and Nepal.* Sci. Publ., Jodhpur.

Sharma, P.K. & Chauhan, N.S. 2000. Ethnobotanical studies of Gaddies- a tribal community of district Kangra, Himachal Pradesh, 301. *In*: Kohli, R.K., Singh, H. P., Vij, S.P., Dhir, K.K., Batish, D.R. & Khurana, D. K. (*eds*) *Man and Forests.* DNAES, IUFRO, ISTS, Botany Dept. & CVS, Pan. Univ. Chandigarh.

Sharma, P.K., Chauhan, N.S. & Lal, B. 2004. Observation on the traditional phytotherapy among the inhabitants of Parvati valley in Western Himalaya, India. *J. Ethnopharmacol.* **92**: 167-176.

Sharma, P.K., Dhyani, S.K. & Shanker, V. 1979. Some useful and medicinal plants of the district Dehradun and Siwalik. *J. Sci. Res. Pl. Med.* **1**: 17-43.

Sharma, P.P. & Singh, N.P. 2001. Ethnomedicinal uses of some edible plants in Dadra Nagar Haveli and Daman (U.T.). *Ethnobotany* **13**: 121-125.

Sharma, R. 2003. *Medicinal Plants of India: An Encyclopedia.* Daya Publ. House, Delhi.

Sharma, S., Gautam, A.K. & Bhadauria, R. 2009. Some important supplementary food plants and wild edible fungi of upper hilly region of district Shimla (Himachal Pradesh), India. *Ethnobot. Leaflets* **13**: 1020-1028.

Sharma, S.C. 1996. A medicobotanical study in relation to veterinary medicines of Shahjahanpur district (Uttar Pradesh). *J. Econ. Taxon. Bot.* Addl. Ser. **12**: 123–127.

Sharma, T.R. & Borthakur, S.K. 2003. Inquire into old lead-II: Ethnoveterinary medicine for treatment of horse in Assam, 69-85. *In*: Khan, S.S. (*ed.*) *Vistas in Ethnobotany*, Vol. I, *Indian Journal of Pure and Applied Biology*. Naval Kishore Press, Lucknow.

Shekhawat, G.S. & Anand, S. 1984. An ethnobotanical profile of Indian Desert. *J. Econ. Taxon. Bot.* **5:** 591-598.

Shome, U., Rawat, A.K.S. & Mehrotra, S. 1996. Time tested household remedies, 96-100. *In*: Jain, S.K. (*ed.*) *Ethnobiology in Human Welfare*, Deep Publ., New Delhi.

Shukla, G. & Verma, B.K. 1996. Roots - a vital plant part to cure body ailments among tribals/rural folklore of Western Bihar. *J. Econ. Taxon. Bot.* Addl. Ser. **12**: 392-394.

Shukla, G., Verma, B.K. & Pandey, H.P. 2003. Folk-lore herbotherapy of Jalaun district, U.P., India, 143-147. *In:* Singh, V. & Jain, A.P. (*eds*) *Ethnobotany and Medicinal Plants of India and Nepal.* Sci. Publ., Jodhpur.

Sikarwar, R.L.S. 1996. Ethnoveterinary herbal medicines in Morena district of Madhya Pradesh, India, 194-196. *In*: Jain, S.K. (*ed.*) *Ethnobiology in Human Welfare,* Deep Publ., New Delhi.

Sikarwar, R.L.S. 2002. Ethnogynaecological uses of plants new to India. *Ethnobotany* **14**: 112-115.

Sikarwar, R.L.S. & Kaushik, J.P. 1992. Some less known medicinal uses of trees among the Sahariyas of Morena district, M.P. India. *Ethnobotany* **4**: 71-74.

Singh, A.K. 2000. A contribution to the ethnobotany of sub-Himalayan region of eastern Uttar Pradesh, 237-246. *In:* Maheshwari, J.K. (*ed.*) *Ethnobotany and Medicinal Plants of Indian Subcontinent.* Sci. Publ., Jodhpur.

Singh, A.K., Singh, R.N. & Singh, S.K. 1987. Some ethnobotanical plants of Terai region of Gorakhpur district. *J. Econ. Taxon. Bot.* **23**: 185-198.

Singh, B., Chaurasia, O.P. & Ballabh, B. 2001. Edible wild plants of trans-Himalayan cold desert, 483-512. *In*: Pandey, P.C. & Samant, S.S. (*eds*) *Plant Diversity of The Himalaya.* Gyanodaya Prakashan, Nainital, India.

Singh, B., Chaurasia, O.P. & Jadhav, K.L. 1996a. An ethnobotanical study of Indus valley (Ladakh). *J. Econ. Taxon. Bot.* Addl. Ser. **12**: 92-101.

Singh, G., Singh, H.B. & Mukherjee, T.K. (*eds*) 2003a. Ethnomedicine of North-east India. Proc. Natl. Seminar on *Traditional Knowledge Base on Herbal Medicines and Plant Resources of North-East India: Protection, Utilization and Conservation.* N.I.S.C.A.I.R., New Delhi.

Singh, G.S. 1990. Ethnobotanical studies of useful plants of the Kullu district in North-Western Himalaya, India. *J. Econ. Taxon. Bot.* **23**: 185-188.

Singh G.S. 2003. Ethnobotanical inventory of Sariska national park in the Aravalli hills of eastern Rajasthan, India, 181-195. *In:* Singh, V. & Jain, A.P. (*eds*) *Ethnobotany and Medicinal Plants of India and Nepal.* Sci. Publ., Jodhpur.

Singh, G.S. 2004. Indigenous knowledge and conservation practices. tribal society of Westren Himalaya: a case study of Sangla valley. *Stud. Trbes Tribals* **2**(1): 29-35.

Singh, G.S., Rao, K.S. & Saxena, K.G. 1998. Environment, ecological and socio-economic impact of introduced crops in Western Himalaya: a case study of Kullu valley. *J. Human Ecol.* **9**: 63-72.

Singh, H. 1988. Ethnobiological treatment of piles by Bhoxas of Uttar Pradesh. *Anc. Sci. Life* **8**: 167-170.

Singh, H. & Sharma, M. 2006. *Flora of Chamba District (Himachal Pradesh)*. B.S.M.P.S., Dehra Dun.

Singh, H., Kumar, B., Hyniewta, T.M. & Bora, P.J. 1997. Ethno-medico-botanical studies in Tripura, India. *Ethnobotany* **9**: 56-58.

Singh, H.B. & Arora, R.K. 1978. *Wild Edible Plants of India.* New Delhi, ICAR.

Singh, H.B., Hynniewta, T.M. & Bora, P.J. 1999. An ethnobotanical note on wild edible plants of Tripura, India. *Ethnobotany* **11**: 26-28.

Singh, H.B., Singh, R.S. & Sandhu, J.S. (*eds*) 2003b. *Herbal Medicine of Manipur: A Coloured Encyclopedia.* Daya Publ. House, New Delhi.

Singh, H.B. & Viswanathan, M.V.1996. Useful pteridophytes of India- a gift of nature to human beings. *J. Econ. Taxon. Bot.* Addl. Ser. **12**: 24 - 36.

Singh, J., Bhuyan, T.C. & Ahmed, A. 1996b. Ethnobotanical studies on the Mishing tribes of Assam with special reference to food and medicinal plants-I. *J. Econ. Taxon. Bot.* Addl. Ser. **12**: 350-356.

Singh, K.K. 1996a. Ethnobotanical observations on Sonbhadra district of southern Uttar Pradesh, India: utilization and conservation, 145-148. *In*: Jain, S.K. (*ed.*) *Ethnobiology in Human Welfare*, Deep Publ., New Delhi.

Singh, K.K. & Kumar, K. 2000a. *Ethnobotanical Wisdom of Gaddie Tribe of Western Himalayas*. B.S.M.P.S., Dehradun.

Singh, K.K. & Kumar, K. 2000b. Observations in ethnoveterinary medicine among the Gaddie tribe of Kangra valley, Himachal Pradesh. *Ethnobotany* **12**: 42-44.

Singh, K.K. & Maheshwari, J.K. 1983. Traditional phytotherapy amongst the tribals of Varanasi district, U. P. *J. Econ. Taxon. Bot.* **4**: 829-838.

Singh, K.K., & Maheshwari, J.K. 1992. Folk Medicine uses of some plants among the Thaarus of Gorakhpur. *Ethnobotany* **4**: 39-43.

Singh, K.K., Palvi, S.K. & Singh, H.B. 1980. Survey of some medicinal plants of Dharchula block-I Pithoragarh district of U.P. *Bull. Medico-ethnobot. Res.* **1**: 1-7.

Singh, K.K. & Prakash, A. 1994. Indigenous phytotherapy among the Gond tribe of Uttar Pradesh, India. *Ethnobotany* **6**: 37-41.

Singh, K.K. & Prakash, A. 1996. Observations on ethnobotany of the Kol tribe of Varanasi district, Uttar Pradesh. *J. Econ. Taxon. Bot.* Addl. Ser. **12**: 133-137.

Singh, K.K. & Rao, R.R. 2003. Studies on some potential ethnomedicine among the tribals of Uttar Pradesh, India, 42-57. *In*: Khan, S.S. (*ed.*) *Vistas in Ethnobotany*

Vol.-I, *Indian Journal of Applied & Pure Biology*. Naval Kishore Press, Lucknow.

Singh, P.B. 1991. *Illustrated Field Guide to Commercially Important Medicinal and Aromatic Plants of Himachal Pradesh*. Soci. Herbal Medicine and Himalayan Biodiversity, Mandi, Himachal Pradesh.

Singh, P.B. 1993. Medicinal plants of ayurvedic importance from Mandi district of Himachal Pradesh. *Bull. Medico-Ethnobot. Res.* **14**: 126-136.

Singh, P.B. 1996b. Wild edible plants of Mandi district in North-West Himalaya, 22-25. *In:* Jain, S.K (*ed.*) *Ethnobiology in Human Welfare*. Deep Publ., New Delhi.

Singh, P.B. 1999. *Illustrated Field Guide to Commercially Important Medicinal and Aromatic Plants of Himachal Pradesh* (with special reference to Mandi district). Published by Society for Herbal Medicine and Himalayan Biodiversity, Mandi, Himachal Pradesh.

Singh, P.B. 2001. Medicinal plant diversity of Kangra district, Himachal Pradesh: a focus on folk, commercial and threatened medicinal plants. 185-196. *In*: Samant, S.S., Dhar, U. & Palani, L.M.S. (*eds*) *Himalayan Medicinal Plants: Potential and Prospects*. Gyanodya Prakashan, Nainital, India.

Singh, P.B. & Aswal, B.S. 1992. Medicinal plants of Himachal Pradesh used in Indian pharmaceutical industry. *Bull. Medico-Ethnobot. Res.* **13**: 172-208.

Singh, P. K. & Singh, K.I. 2003. First aid remedies: an ethno-medico-botanical study of the Meitei community of Manipur state, India, 466-472. *In:* Singh, V. & Jain, A.P. (*eds*) *Ethnobotany and Medicinal Plants of India and Nepal*. Sci. Publ., Jodhpur.

Singh, V. & Chauhan, N.S. 2005. Traditional practices of herbal medicines in the Lahaul valleys, Himachal Himalayas. *Indian J. Trad. Knowledge* **4**(2): 208-220.

Singh, V. & Pandey, R.P. 1996. Ethno-medicinal plants used for venereal and gynaecological diseases in Rajasthan (India). *J. Econ. Taxon. Bot.* Addl. Ser. **12**: 154-165.

Singh, V. & Pandey, R.P. 1998. *Ethnobotany of Rajasthan, India*. Sci. Publ., Jodhpur.

Singh, V. & Singh, P. 1981. Edible wild plants of eastern Rajesthan. *J. Econ. Taxon Bot.* **2:**197-207.

Singh, V.K. & Zaheer, A.A. 1996. Folk medicinal plants used for family planning in India, 184-186. *In*: Jain, S.K. (*ed.*). *Ethhnobiology in Human Welfare*, Deep Publ., New Delhi.

Sinha, R.K. & Sinha, S. 2001. *Ethnobiology*. Surabhi Publ., Jaipur.

Sinha, S.C. 1986. *Ethnobotanical Study of Manipur*, Ph.D. Thesis, Manipur University, Imphal.

Sinha, S.C. 1987. Ethnobotany-medicinal Plants. *Frontier Bot.* **1**: 123-152.

Sinha, S.C. 1996. Wild edible plants of Manipur, India, 42-47. *In*: Jain, S.K. (*ed.*) *Ethnobiology in Human Welfare*. Deep Publ., New Delhi.

Sivakumar, A., Subramanian, M. & Burkanudun, A. 2003. Ethnobotany of Poliyars of Anaimalai hills, Tamilnadu, 679-685. *In:* Singh, V. & Jain, A.P. (*eds*) *Ethnobotany and Medicinal Plants of India and Nepal.* Sci. Publ., Jodhpur.

Siwakoti, M. & Siwakoti, S. 2000. Ethnomedicinal uses of plants among the Satar tribe of Nepal, 99-108. *In:* Maheshwari, J.K. (*ed.*) *Ethnobotany and Medicinal Plants of Indian Subcontinent.* Sci. Publ., Jodhpur.

Siwakoti, M. & Varma, S.K. 1996. Medicinal plants of the Terai of eastern Nepal. *J. Econ. Taxon. Bot.* Addl. Ser. **12**: 423-438.

Sood, S.K., Gupta, P. & Kumar, S. 2010. *Flavoring and Fragrant Resources of India.* Sci. Publ., Jodhpur.

Sood, S.K. & Kaushal, B. 2008. *Fibre and Fodder Resources of India.* Satish Serial Publ. House, Azadpur, Delhi.

Sood, S.K., Nath, R. & Kalia, D.C. 2001. *Ethnobotany of Cold Desert Tribes of Lahaul-Spiti (N.W. Himalaya).* Deep Publ., New Delhi.

Sood, S.K. & Prakash, V. 2007. *Edible Roots and Underground Stems of Ethnic India.* Satish Serial Publ. House, Delhi.

Sood, S.K., Rawat, S. & Rawat, D. 2009a. *Dye, Masticatory and Fumitory Resources of India.* Intl. Book Distributors, Dehradun.

Sood, S.K. & Sudersna. 2008. *Root Drugs Plants of Ethnic India.* Satish Serial Publ. House, Azadpur, Delhi.

Sood, S.K. & Thakur, S. 2004. *Ethnobotany of Rewalsar Himalaya (Dist. Mandi, Himachal Pradesh, India).* Deep Publ., New Delhi.

Sood, S.K., Sharma, S.K., Kumar. N. & Kumar, H. 2009b. *Ethnobotanical Studies on Trees, Shrubs and Climbers of Himalaya.* Satish Serial Publ. House, Azadpur, Delhi.

Srivastava, R.C., Singh, V.P. & Singh, M.K. 2003a. Medicinal plants of Jaunpur district (U.P.), 148-159. *In*: Singh, V. & Jain, A.P. (*eds*) *Ethnobotany and Medicinal Plants of India and Nepal.* Sci. Publ., Jodhpur.

Srivastava, S. & Jain, S.K. 2000. Cumulative index to Ethnobotany, Vol. 1-12, 1989-2000, with analysis of region and themes covered. *Ethnobotany* **12**: 139-161.

Srivastava, S.C., Mitra, S. & Bandyopadhyay, S. 2003b. Some plants of ethnomedicinal importance from Darjeeling district, West Bengal, 972-977. *In:* Singh, V. & Jain, A.P. (*eds*) *Ethnobotany and Medicinal Plants of India and Nepal.* Sci. Publ., Jodhpur.

Srivastava, S.K. & Chandrasekar, K. 2004. Ethnomedicine of the Pin Valley National Park. Himachal Pradesh: plant used in treatment of dysentery. *Ethnobotany* **16**: 62-63.

Srivastava, S.K., Mehrotra, S., Srivastava, G.K. & Chauhan, D.K. 2003c. Folk uses of herbs from Mahabaleshwar region of Sahyadri Hills, 857-863. *In:* Singh, V. & Jain, A.P. (*eds*) *Ethnobotany and Medicinal Plants of India and Nepal.* Sci. Publ., Jodhpur.

Stainton, A. 1988. *Flowers of Himalaya, A Supplement.* Oxford Univ. Press, Delhi.

Subramaniam, A. 2000. A survey of medicinal plants from Chitheri hills in Dharmapuri district, Tamilnadu, 395-416. *In:* Maheshwari, J.K. (*ed.*) *Ethnobotany and Medicinal Plants of Indian Subcontinent.* Sci. Publ., Jodhpur.

Subramani, S.P. & Goraya, G.S. 2003. Some folklores medicinal plants of Kolli hills: record of a Natti Vaidyas Sammelan, 665-678. *In:* Singh, V. & Jain, A.P. (*eds*) *Ethnobotany and Medicinal Plants of India and Nepal.* Sci. Publ., Jodhpur.

Sudhakar, A. & Vedavathy, S. 2000. Wild edible plants used by the tribals of Chotoor district (Andhra Pradesh), India, 321-329. *In*: Maheshwari, J.K. (*ed.*) *Ethnobotany and Medicinal Plants of Indian Subcontinent.* Sci. Publ., Jodhpur.

Sundaravelu, S., Selvaraju, M. & Sundaramoorthy, P. 2003. Medicinal weed flora of South Arcot, Cuddalore district in Tamil Nadu, 87-94. *In*: Khan, S.S. (*ed.*) *Vistas in Ethnobotany* Vol.-I, *Indian Journal of Pure & Applied Biology*. Naval Kishore Press, Lucknow.

Swami, A. & Gupta, B.K. 1996. A note on some commonly occurring medicinal weeds of Udhampur district (J & K). *J. Econ. Taxon. Bot.* Addl. Ser. **12**: 89-91.

Thakur, V. 2005. *Ethnic Indian Plants in Religious Ceremonies*, M.Phil. Thesis, H.P. Univ., Shimla.

Thomas, J. & Britto, A. John De. 2003. Ethnobotanical study in Wynad district in Kerela, 815-824. *In:* Singh, V. & Jain, A.P. (*eds*) *Ethnobotany and Medicinal Plants of India and Nepal.* Sci. Publ., Jodhpur.

Tigga, M.., Lakra, G, S. & Rao, P.S.N. 2003. Medicinal value of latex yielding plants from Andaman & Nicobar Islands, 765-772. *In:* Singh, V. & Jain, A.P. (*eds*) *Ethnobotany and Medicinal Plants of India and Nepal.* Sci. Publ., Jodhpur.

Tirkey, A. 2004. Some ethnomedicinal plant species of Chhattisgarh state. *Ethnobotany* **16**: 118-124.

Tirkey, A., Khan, F. & Khan, S.S. 2003a. Medicinal plants used in the treatment of hair in Raijarh district of Madhya Pradesh (India), 104-110. *In*: Khan, S.S. (*ed.*) *Vistas in Ethnobotany,* Vol.-I, *Indian Journal of Pure and Applied Biology*. Naval Kishore Press, Lucknow.

Tirkey, A., Khan, F. & Khan, S.S. 2003b. Medicinal plants used by the tribals of Raijarh district (M.P.) for Jaundice, 163-170. *In*: Khan, S.S. (*ed.*) *Vistas in Ethnobotany*, Vol. I, *Indian Journal of Pure and Applied Biology*. Naval Kishore Press, Lucknow.

Tiwari, K.C. & Tiwari, V.P. 1996. Some important medicinal plants of the tropical, sub-tropical and temperate region of Siang, Subansiri and Tirap district of Arunachal Pradesh. *J. Econ. Taxon. Bot*. Addl. Ser. **12**: 359-363.

Tosh, J. 1996. Ethnobotanical study of Western Maharastra. *J. Econ. Taxon. Bot.* Addl. Ser. **12**: 169-174.

Trivedi, P.C. 2004. *Medicinal Plants Conservation and Utilization.* Avishkar Publ., Jaipur.

Trivedi, P.C. 2009. *Medicinal Plants Conservation and Utilization* (IInd edition). Avishkar Publ., Jaipur.

Trivedi, P.C. 2010. *Ethnic Tribes and Medicinal Plants.* Pointer Publ., Jaipur.

Uniyal, B. 2003. Utilization of medicinal plants by the rural women of Kullu, Himachal Pradesh. *Indian J. Trad. Knowledge* **2** (4): 366-370.

Uniyal, M.R. 1968. Medicinal plants of Bhagirathi valley lying in Uttarkashi forest division. *Indian For.* **94**: 407-420.

Uniyal, M.R. & Chauhan, N.S. 1971. Medicinal plants of Uhal valley in Kangra Forest Division. H.P. *J. Res. Ind. Med.* **6**(3): 287-299.

Uniyal, M.R. & Chauhan, N.S. 1972. Commercially important medicinal plants of Kullu forest division. H.P. *Nagarjun* **4**(2): 28-32.

Uniyal, M.R. & Chauhan, N.S. 1973a. Traditionally important medicinal plants of Kullu forest Division, H.P. *Jour. Res. Ind. Med.* **7**: 76-85.

Uniyal, M.R. & Chauhan, N.S. 1973b. Traditionally important medicinal plants of Kangra Valley in Dharamsala Forest Circle, Himachal Pradesh. *J. Res. Indian Med.* **8**(1): 76-85.

Uniyal, S.K., Singh, K.N., Jamwal, P. & Lal, B. 2006. Traditional use of medicinal plants among the tribal communities of Chhota Bhangal, Western Himalaya. *J. Ethnobiol. & Ethnomed.* **2**(14): 1-8.

Upadhyay, R. & Chauhan, S.V.S. 2000. Ethnobotanical observations on Koya tribe of Gundala Mandal of Khammam district, Andhra Pradesh. *Ethnobotany* **12**: 93-99.

Upadhyay, R. 2003. Ethnobotanical observations on some rare and endangered herbs of Singrauli in Madhya Pradesh, 989-992. *In:* Singh, V. & Jain, A.P. (*eds*) *Ethnobotany and Medicinal Plants of India and Nepal.* Sci. Publ., Jodhpur.

Upadhyay, R. & Chauhan, S.V.S. 2003. Ethnobotanical uses of plant gums by the tribals. 601-602. *In:* Singh, V. & Jain, A.P. (*eds*) *Ethnobotany and Medicinal Plants of India and Nepal.* Sci. Publ., Jodhpur.

Upadhyay, R. & Singh, J. 2005. Ethno-medical uses of plants from Tikri forest of Gonda district (U.P). *Ethnobotany* **17**: 167-170.

Upadhye, A.S. & Kulkarni, D.K. 2004. Traditional phytotherapy for insect bites among the tribals in Western Maharashtra. *Ethnobotany* **16**: 59-61.

Upadhye, A.S. & Kumbhojkar, M.S. 2003. Ethnobotany of *Butea monosperma* (Lam.) Taub, 94-98. *In:* Singh, V. & Jain, A.P. (*eds*) *Ethnobotany and Medicinal Plants of India and Nepal.* Sci. Publ., Jodhpur.

Upadhye, A.S., Vartak, V.D. & Kumbhojkar, M.S. 1994. Ethno-medico-botanical studies in Western Maharastra, India. *Ethnobotany* **6**: 25-31.

Uphof, J.C. Th. 1968. *Dictionary of Economic Plants.* H.R. Engelmann (J. Cramer), Weinheim.

Vaidya, R.R. & Dhumal, K.N. 2000-2001. Studies on nutritive value in some vegetables used by tribals of Mahabaleshwar. *J. Med. Aromat. Pl. Sci.* **22** (4A), **23** (1A): 312-315.

Varma, S.K. 1997. Comparative studies on folk drugs of tribals of Chotanagpur and Santhal Pargana of Bihar, India. *Ethnobotany* **9**: 70-76.

Vartak, V.D. 1981. Observations on wild-edible plants form hilly regions of Maharashtra and Goa: Resume and future prospects, 261-273. *In*: Jain, S.K. (*ed.*) *Glimpses of Indian Ethnobotany*, Oxford & IBH, New Delhi.

Ved, D.K., Kinchal, G.A., Ravikumar, K., Pravharan, V., Uttakarsh, G., Vijaya Sankar, R. & Indresha, J.H. 2003. *Conservation Assessment and Management Prioritisation for the medicinal plants of Jammu and Kashmir, Himachal Pradesh and Uttranchal.* FRLHT, Banglore, India.

Verma, P., Khan, A.A. & Singh, K.K. 1995. Traditional phyotherapy among the Baiga tribe of Shahdol district of Madhya Pradesh, India. *Ethnobotany* **7**: 69-73.

Verma, S. & Chauhan, N.S. 2006. Studies on ethno-medico-botany of Kunihar forest division, district Solan (H.P.). *Ethnobotany* **18**: 160-165.

Verma, S. & Chauhan, N.S. 2007. Indigenous medicinal plants knowledge of Kunihar Forest Division, district Solan. *Indian J. Trad. Knowledge* **6**(3): 494-497.

Vihari, V. 1995. Ethnobotany of cosmetics of Indo-Nepal border. *Ethnobotany* **7**: 89-94.

Virjee, G., Dar, H., Kachroo, P. & Bhat, G.M. 1984. Taxo-ethnobotanical studies of the rural areas in district Rajouri (Jammu). *J. Econ. Taxon. Bot.* **5**(4): 831-838.

Vishwanathan, M.B. 1997. Ethnobotany of the Malyalis in North Arcot district, Tamil Nadu, India. *Ethnobotany* **9**: 77-79.

Viswanathan, M.B., Kumar, E.H.P. & Ramesh, N. 2001. Ethnomedicines of Kanis in Kalakkad - Mundanthurai tiger reserve, Tamil Nadu. *Ethnobotany* **13**: 60-66.

Viswanathan, M.V. 2000. Edible and medicinal plants of Ladakh (J & K). 151-154. *In*: Maheshwari, J.K. (*ed.*) *Ethnobotany and Medicinal Plants of Indian Subcontinent*. Sci. Publ., Jodhpur.

Vishwanathan, M.V. & Singh, H.B. 1996. Plants used as household remedies In India. 105-107. *In*: Jain, S.K. (*ed.*) *Ethnobiology in Human Welfare*. Deep Publ., New Delhi.

Watt, G. 1889-1896. *A Dictionary of the Economic Products of India.* Vol. 1-VI. Periodical Experts, New Delhi.

Watt, G. 1908. *The Commercial Products of India.* John Murray, London. 249.

Watt, G. 1972. *A Dictionary of Economic Products of India* Vol.-1-VI. Periodical Experts, Delhi (India).

Wickens, G.E.1990. What is economic botany? *Econ. Bot.* **44**: 12-28.

Wielgorskaya, T. 1995. *Dictionary of Generic Names of Seed Plants.* B.S.M.P.S., Dehradun.

Yanzone, G.S., Yonzone, A. & Bhujel, R.B. 1996. Contribution to the ethnobotany of Darjeeling district, India, 388-389. *In*: Jain, S.K. (*ed.*) *Ethnobiology in Human Welfare,* Deep Publ., New Delhi.

Plates

Plate 1

A. *Abelmoschus esculentus*

B. *Abrus precatorius*

C. *Acacia catechu*

D. *Achyranthes aspera*

E. *Achyranthes bidentata*

F. *Acorus calamus*

Plate 2

A. *Adiantum capillus veneris*

B. *Aegle marmelos*

C. Fruit of *Aegle marmelos*

D. *Aerva lanata*

E. *Agave cantula*

F. *Agave wightii*

Plate 3

A. *Ageratum conyzoides*

B. *Ajuga parviflora*

C. *Albizia chinensis*

D. *Albizia julibrissin*

E. *Albizia lebbeck*

F. *Allium cepa*

Plate 4

A. *Allium sativum*

B. *Aloe vera*

C. *Alternanthera sessilis*

D. *Amaranthus paniculatus*

E. *Amaranthus virids*

F. *Araucaria columnaris*

Plate 5

A. *Artemisia capillaris*

B. *Artemisia japonica*

C. *Artocarpus lacucha*

D. *Asparagus adscendens*

E. *Avena sativa*

F. *Bambusa nutans*

Plate 6

A. *Basella alba*

B. *Bauhinia vahlii*

C. *Bauhinia variegata*

D. *Berberis lycium*

E. *Bergenia ciliata*

F. *Bidens biternata*

Plate 7

A. *Boehmeria macrophylla*

B. *Boerhaavia diffusa*

C. *Bombax ceiba*

D. Same; flower enlarged

E. *Brassica nigra*

F. *Brassica tournefortii*

Plate 8

A. *Butea monosperma*

B. *Caesalpinia decapetala*

C. *Cajanus cajan*

D. *Callicarpa macrophylla*

E. *Calotropis procera*

F. *Camellia sinensis*

Plate 9

A. *Cannabis sativa*

B. *Capillipedium assimile*

C. *Capsicum annuum*

D. *Carica papaya*

E. *Carissa congesta*

F. *Carya illinoensis*

Plate 10

A. *Caryopteris bicolor*

B. *Cassia fistula*

C. *Cassia floribunda*

D. *Cassia obtusifolia*

E. *Cassia occidentalis*

F. *Catharanthus roseus*

Plate 11

A. *Cedrus deodara*

B. *Celastrus paniculatus*

C. *Celtis australis*

D. Same; close-up view

E. *Centella asiatica*

F. *Centipeda minima*

Plate 12

A. *Centratherum anthelminticum*

B. *Cestrum nocturnum*

C. *Cheilanthes farinosa*

D. *Chenopodium album*

E. *Chenopodium ambrosioides*

F. *Cicer arietinum*

Plate 13

A. *Cinnamomum camphora*

B. *Cinnamomum tamala*

C. *Cirsium arvense*

D. *Cissampelos parriera*

E. *Citrus grandis*

F. *Citrus jhambiri*

Plate 14

A. *Citrus limon*

B. *Citrus medica*

C. *Citrus medica* var. *medica*

D. *Clematis gauriana*

E. *Coccinia grandis*

F. *Cocculus laurifolius*

Plate 15

A. *Colebrookea oppositifolia*

B. *Colocasia esculenta*

C. *Commelina paludosa*

D. *Cordia dichotoma*

E. *Coriandrum sativum*

F. *Coronopus didymus*

Plate 16

A. *Costus speciosus*

B. Same; rhizome enlarged

C. *Cryptolepis buchanani*

D. *Cucumis melo* var. *agrestis*

E. *Cucumis sativus*

F. *Cucurbita hispida*

Plate 17

A. *Cucurbita pepo*

B. *Curcuma longa*

C. Same; close-up view

D. *Cuscuta reflexa*

E. *Cyclanthera pedata*

F. *Cymbopogon citratus*

Plate 18

A. *Cynodon dactylon*

B. *Cyphomandra betacea*

C. *Dalbergia sissoo*

D. *Datura innoxia*

E. *Datura stramonium*

F. *Debregeasia hypoleuca*

Plate 19

A. *Deeringia amaranthoides*

B. *Dendrocalamus hamiltonii*

C. *Didymocarpus pedicellata*

D. Same; closer view of basal portion

E. *Dioscorea bellophylla*

F. *Dioscorea bulbifera*

Plate 20

A. *Dioscorea oppositifolia*

B. *Diospyros kaki*

C. *Diplazium esculentum*

D. *Diplocyclos palmatus*

E. *Dodonaea viscosa*

F. *Dolichos biflorus*

Plate 21

A. *Duranta repens*

B. *Ehretia acuminata*

C. *Ehretia laevis*

D. *Eleusine coracana*

E. *Emblica officinalis*

F. *Engelhardia colebrookeana*

Plate 22

A. *Equisetum debile*

B. *Eriobotrya japonica*

C. *Eriophorum comosum*

D. *Erythrina suberosa*

E. *Eucalyptus tereticornis*

F. *Eulaliopsis binata*

Plate 23

A. *Euonymus hamiltonianus*

B. *Euphorbia royleana*

C. *Fagopyrum cymosum*

D. *Fagopyrum esculentum*

E. *Ficus benghalensis*

F. Same; closer view

Plate 24

A. *Ficus carica*

B. *Ficus foveolata*

C. *Ficus glomerata*

D. *Ficus hispida*

E. *Ficus palmata*

F. *Ficus religiosa*

Plate 25

A. *Ficus roxburghii*

B. *Ficus rumphii*

C. *Ficus semicordata*

D. *Flacourtia indica*

E. *Foeniculum vulgare*

F. *Fragaria indica*

Plate 26

A. *Fumaria indica*

B. *Galium aparine*

C. *Geranium nepalense*

D. *Glycine max*

E. *Gnaphalium affine*

F. *Gonatanthus pumilus*

Plate 27

A. *Gossypium arboreum*

B. *Grevillea robusta*

C. *Grewia disperma*

D. *Grewia elastica*

E. *Grewia optiva*

F. *Hamiltonia suaveolens*

Plate 28

A. *Hibiscus rosa sinensis*

B. *Holarrhena antidysentrica*

C. *Hypericum oblongifolium*

D. *Impatiens balsamina*

E. *Indigofera cassioides*

F. *Ipomoea batatas*

Plate 29

A. *Ipomoea turbinata*

B. *Jasminum dispermum*

C. *Jasminum grandiflorum*

D. *Jasminum humile*

E. *Jasminum multiflorum*

F. *Jatropha curcas*

Plate 30

A. *Juglans regia*

B. Same; closer view of twig

C. *Justicia adhatoda*

D. *Lannea coromandelica*

E. *Lantana camara*

F. *Lathyrus aphaca*

Plate 31

A. *Lathyrus sativus*

B. *Lawsonia inermis*

C. *Lepidagathis cuspidata*

D. *Lepidium sativum*

E. *Leucaena glauca*

F. *Leucas cephalotes*

Plate 32

A. *Linum usitatissimum*

B. Same; closer view

C. *Litchi chinensis*

D. *Litsea monopetala*

E. *Litsea sebifera*

F. *Litsea semicarpifolia*

Plate 33

A. *Loranthus falcatus*

B. *Lotus corniculatus*

C. *Luffa acutangula*

D. *Mallotus philippinensis*

E. *Mangifera indica*

F. Same; fruit

Plate 34

A. *Medicago polymorpha*

B. *Melia azedarach*

C. *Mentha longifolia*

D. *Mentha piperita*

E. *Michelia champaca*

F. *Micromeria biflora*

Plate 35

A. *Milletia extensa*

B. *Mirabilis jalapa*

C. *Momordica charantia*

D. *Moringa oleifera*

E. *Morus alba*

F. *Mucuna pruriens*

Plate 36

A. *Murraya koenigii*

B. *Murraya paniculata*

C. *Musa sapientum*

D. *Neolitsea umbrosa*

E. *Nerium indicum*

F. *Ocimum basilicum*

Plate 37

A. *Ocimum gratissimum*

B. *Ocimum kilimandscharicum*

C. *Ocimum sanctum*

D. *Opuntia dillenii*

E. *Origanum vulgare*

F. *Oroxylum indicum*

Plate 38

A. *Ougeinia dalbergioides*

B. *Oxalis corniculata*

C. *Panicum antidontale*

D. *Pennisetum americanum*

E. *Pentanema indicum*

F. *Perilla frutescens*

Plate 39

A. *Persea odoratissima*

B. *Phoenix acaulis*

C. *Phoenix sylvestris*

D. *Phyllanthus urinaria*

E. *Pinus roxburghii*

F. *Pistacia khinjuk*

Plate 40

A. *Plantago lanceolata*

B. *Plumbago zeylanica*

C. *Pogostemon benghalense*

D. *Polygala chinensis*

E. *Polygonum barbatum*

F. *Populus ciliata*

Plate 41

A. *Porana paniculata*

B. *Portulaca grandiflora*

C. *Prinsepia utilis*

D. *Prunus cerasoides*

E. *Prunus domestica*

F. *Prunus persica*

Plate 42

A. *Psidium guajava*

B. *Punica granatum*

C. *Pyrus communis*

D. *Pyrus pashia*

E. *Pyrus serotina*

F. *Quercus leucotrichophora*

Plate 43

A. *Randia dumetorum*

B. *Ranunculus arvensis*

C. *Ranunculus laetus*

D. *Ranunculus muricatus*

E. *Raphanus sativus*

F. *Reinwardtia indica*

Plate 44

A. *Rhamnus triquetra*

B. *Rhododendron arboreum*

C. *Ricinus communis*

D. *Rorippa nasturtium - aquaticum*

E. *Rosa brunonii*

F. *Rosa chinensis*

Plate 45

A. *Rosa damascena*

B. *Roylea cinerea*

C. *Rubus ellipticus*

D. *Rubus niveus*

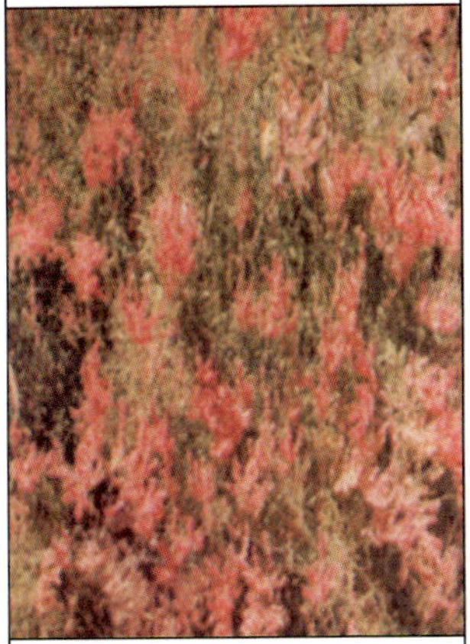

E. *Rumex hastatus*

F. *Rumex nepalensis*

Plate 46

A. *Saccharum spontaneum*

B. *Salix karelini*

C. *Salix tetrasperma*

D. *Salvia coccinea*

E. *Santalum album*

F. *Sapindus mukorosii*

Plate 47

A. *Sapium insigne*

B. *Sapium sebiferum*

C. *Scandix pecten veneris*

D. *Sechium edule*

E. *Selaginella chrysocaulos*

F. *Sesamum orientale*

Plate 48

A. *Sida cordifolia*

B. *Silene conoidea*

C. *Sisymbrium irio*

D. *Solanum erianthum*

E. *Solanum melongena*

F. *Solanum nigrum*

Plate 49

A. *Solanum torvum*

B. *Solanum viarum*

C. *Sorghum bicolor*

D. *Spilanthes oleracea*

E. *Spondias pinnata*

F. *Stellaria media*

Plate 50

A. *Stephania glabra*

B. *Syzygium cumini*

C. *Syzygium operculatum*

D. *Tagetes minuta*

E. *Terminalia arjuna*

F. *Terminalia bellirica*

Plate 51

A. *Terminalia chebula*

B. *Terminalia tomentosa*

C. *Thuja orientalis*

D. *Thysanolaena maxima*

E. *Tinospora cordifolia*

F. *Toona ciliata*

Plate 52

A. *Trichosanthes cucumerina*

B. *Trifolium resupinatum*

C. *Trigonella foenum-graecum*

D. *Triticum aestivum*

E. *Tulipa stellata*

F. *Urtica dioica*

Plate 53

A. *Vanda tessellata*

B. *Verbascum thapsus*

C. *Vicia faba*

D. *Vicia hirsuta*

E. *Vicia sativa*

F. *Vigna radiata*

Plate 54

A. *Viola patrinii*

B. *Viola pilosa*

C. *Vitex negundo*

D. *Vitis repanda*

E. *Vitis vinifera*

F. *Woodfordia fruticosa*

Plate 55

A. *Wrightia arborea*

B. *Zanthoxylum armatum*

C. *Zea mays*

D. *Zephyranthes carinata*

E. *Zizyphus jujuba*

F. *Zizyphus nummularia*

Appendices

Appendix I

INDEX OF FAMILIES

Name of the Family	*Total Number of Genera and Spp*	*Total Number of Spp in the Genera*
Acanthaceae	*Justicia* (1), *Lepidagathis* (1).	2/2
Amaranthaceae	*Achyranthes* (2), *Aerva* (1), *Alternanthera* (1), *Amaranthus* (2), *Deeringia* (1).	5/7
Amaryllidaceae	*Agave* (2), *Zephyranthes* (1).	2/3
Anacardiaceae	*Lannea* (1), *Mangifera* (1), *Pistacia* (1), *Spondias* (1).	4/4
Apiaceae	*Apium* (1), *Centella* (1), *Coriandrum* (1), *Foeniculum* (1), *Scandix* (1).	5/5
Apocynaceae	*Carissa* (1), *Catharanthus* (1), *Holarrhena* (1), *Nerium* (1), *Wrightia* (1).	5/5
Araceae	*Acorus* (1), *Colocasia* (1), *Gonatanthus* (1).	3/3
Araucariaceae	*Araucaria* (1).	1/1
Arecaceae	*Phoenix* (2).	1/2
Asclepiadaceae	*Calotropis* (1), *Cryptolepis* (1).	2/2
Asteraceae	*Ageratum* (1), *Artemisia* (2), *Bidens* (1), *Centipeda* (1), *Centratherum* (1), *Cirsium* (1), *Gnaphalium* (1), *Pentanema* (1), *Spilanthes* (1), *Tagetes* (1).	10/11
Athyriaceae	*Diplazium* (1).	1/1
Berberidaceae	*Berberis* (1).	1/1
Bignoniaceae	*Oroxylum* (1).	1/1
Boraginaceae	*Cordia* (1), *Ehretia* (2).	2/3
Brassicaceae	*Brassica* (2), *Coronopus* (1), *Lepidium* (1), *Rorripa* (1), *Raphanus* (1), *Sisymbrium* (1).	6/7
Cactaceae	*Opuntia* (1).	1/1
Caricaceae	*Carica* (1).	1/1
Caryophyllaceae	*Silene* (1), *Stellaria* (1).	2/2
Celastraceae	*Celastrus* (1), *Euonymus* (1).	2/2
Cheilanthaceae	*Cheilanthes* (1)	1/1
Chenopodiaceae	*Chenopodium* (2), *Basella* (1).	2/3
Combretaceae	*Terminalia* (4).	1/4

Contd...

Table Contd...

Name of the Family	*Total Number of Genera and Spp*	*Total Number of Spp in the Genera*
Commelinaceae	*Commelina* (1).	1/1
Convolvulaceae	*Cuscuta* (1), *Ipomoea* (2), *Porana* (1).	3/4
Crassulaceae	*Kalanchoe* (1).	1/1
Cucurbitaceae	*Coccinia* (1), *Cucumis* (2), *Cyclanthera* (1), *Cucurbita* (2), *Diplocyclos* (1), *Luffa* (2), *Momordica* (2), *Sechium* (1), *Trichosanthes* (1).	9/13
Cupressaceae	*Thuja* (1).	1/1
Cyperaceae	*Eriophorum* (1).	1/1
Dioscoreaceae	*Dioscorea* (3).	1/3
Ebenaceae	*Diospyros* (1).	1/1
Equisetaceae	*Equisetum* (1).	1/1
Ericaceae	*Rhododendron* (1).	1/1
Euphorbiaceae	*Emblica* (1), *Euphorbia* (1), *Jatropha* (1), *Mallotus* (1), *Phyllanthus* (1), *Ricinus* (1), *Sapium* (2).	7/8
Fabaceae	*Abrus* (1), *Acacia* (1), *Albizia* (3), *Bauhinia* (2), *Butea* (1), *Caesalpinia* (1), *Cajanus* (1), *Cassia* (4), *Cicer* (1), *Dalbergia* (1), *Dolichos* (1), *Erythrina* (1), *Glycine* (1), *Indigofera* (1), *Lathyrus* (2), *Leucaena* (1), *Lotus* (1), *Medicago* (1), *Milletia* (1), *Mucuna* (1), *Ougeinia* (1), *Trifolium* (1), *Trigonella* (1), *Vicia* (3), *Vigna* (1).	25/34
Fagaceae	*Quercus* (1).	1/1
Flacourtiaceae	*Flacourtia* (1).	1/1
Fumariaceae	*Fumaria* (1).	1/1
Gesneriaceae	*Didymocarpus* (1).	1/1
Geraniaceae	*Geranium* (1), *Impatiens* (1), *Oxalis* (1).	3/3
Hypericaceae	*Hypericum* (1).	1/1
Juglandaceae	*Carya* (1), *Engelhardia* (1), *Juglans* (1).	3/3
Lamiaceae	*Ajuga* (1), *Colebrookea* (1), *Leucas* (1), *Mentha* (2), *Micromeria* (1), *Ocimum* (4), *Origanum* (1), *Perilla* (1), *Pogostemon* (1), *Roylea* (1), *Salvia* (1).	11/15
Lauraceae	*Cinnamomum* (2), *Litsea* (3), *Neolitsea* (1), *Persea* (1).	4/7
Liliaceae	*Allium* (2), *Aloe* (1), *Asparagus* (1), *Drimia* (1), *Tulipa* (1).	5/6
Linaceae	*Linum* (1), *Reinwardtia* (1).	2/2
Loranthaceae	*Loranthus* (1).	1/1
Lythraceae	*Lawsonia* (1), *Woodfordia* (1).	2/2

Contd...

Table Contd...

Name of the Family	*Total Number of Genera and Spp*	*Total Number of Spp in the Genera*
Malvaceae	*Abelmoschus* (1), *Bombax* (1), *Gossypium* (1), *Hibiscus* (1), *Sida* (1).	5/5
Magnoliaceae	*Michelia* (1).	1/1
Meliaceae	*Melia* (1), *Toona* (1).	2/2
Menispermaceae	*Cissampelos* (1), *Cocculus* (1), *Stephania* (1), *Tinospora* (1).	4/4
Moringaceae	*Moringa* (1).	1/1
Myrtaceae	*Eucalyptus* (1), *Psidium* (1), *Syzygium* (2).	3/4
Musaceae	*Musa* (1).	1/1
Nyctaginaceae	*Boerhaavia* (1), *Mirabilis* (1).	2/2
Oleaceae	*Jasminum* (4).	1/4
Orchidaceae	*Vanda* (1).	1/1
Pedaliaceae	*Sesamum* (1).	1/1
Pinaceae	*Cedrus* (1), *Pinus* (1).	2/2
Plantaginaceae	*Plantago* (1).	1/1
Plumbaginaceae	*Plumbago* (1).	1/1
Poaceae	*Aristida* (1), *Arundinaria* (1), *Avena* (1), *Bambusa* (1), *Capillipedium* (1), *Cymbopogon* (1), *Cynodon* (1), *Dendrocalamus* (1), *Eleusine* (1), *Eulaliopsis* (1), *Heteropogon* (1), *Panicum* (1), *Pennisetum* (1), *Saccharum* (2), *Sorghum* (1), *Sporobolus* (1), *Thysanolaena* (1), *Triticum* (1), *Zea* (1).	19/20
Polygonaceae	*Fagopyrum* (2), *Polygala* (1), *Polygonum* (1), *Rumex* (2).	4/6
Polypodiaceae	*Adiantum* (1).	1/1
Portulacaceae	*Portulaca* (1).	1/1
Proteaceae	*Grevillea* (1).	1/1
Punicaceae	*Punica* (1).	1/1
Ranunculaceae	*Clematis* (1), *Ranunculus* (3).	2/4
Rhamnaceae	*Rhamnus* (1), *Zizyphus* (2).	2/3
Rosaceae	*Eriobotrya* (1), *Fragaria* (1), *Prinsepia* (1), *Prunus* (3), *Pyrus* (3), *Rosa* (3), *Rubus* (2).	7/14
Rubiaceae	*Galium* (1), *Hamiltonia* (1), *Randia* (1).	3/3
Rutaceae	*Aegle* (1), *Citrus* (5), *Murraya* (2), *Zanthoxylum* (1).	4/9
Salicaceae	*Populus* (2), *Salix* (2).	2/4
Santalaceae	*Santalum* (1).	1/1
Sapindaceae	*Dodonaea* (1), *Litchi* (1), *Sapindus* (1).	3/3

Contd...

Table Contd...

Name of the Family	*Total Number of Genera and Spp*	*Total Number of Spp in the Genera*
Saxifragaceae	*Bergenia* (1).	1/1
Scrophulariaceae	*Verbascum* (1).	1/1
Selaginellaceae	*Selaginella* (1).	1/1
Solanaceae	*Capsicum* (1), *Cestrum* (1), *Cyphomandra* (1), *Datura* (2), *Solanum* (5).	5/10
Sterculiaceae	*Helicteres* (1).	1/1
Theaceae	*Camellia* (1).	1/1
Tiliaceae	*Grewia* (3).	1/3
Urticaceae	*Artocarpus* (1), *Boehmeria* (1), *Cannabis* (1), *Celtis* (1), *Debregeasia* (1), *Morus* (1), *Ficus* (12), *Urtica* (2).	8/20
Verbenaceae	*Callicarpa* (1), *Caryopteris* (1), *Duranta* (1), *Lantana* (1), *Vitex* (1).	5/5
Violaceae	*Viola* (2).	1/2
Vitaceae	*Vitis* (2).	1/2
Zingiberaceae	*Costus* (1), *Curcuma* (1).	2/2

Appendix II

INDEX OF USES

Anaemia

Carissa congesta, Saccharum officinarum

Anthelmintic

*Butea monosperma, Centratherum anthelminticum, Chenopodium album, Cordia dichotoma, Momordica charantia, Rumex nepalensis**

Antiemetic

Allium cepa, Citrus grandis, Mentha longifolia

Antidote

Achyranthes bidentata, Cannabis sativa, Capsicum annuum*, Catharanthus roseus, Rumex nepalensis, Sapindus mukorosii**

Antiseptic

Agave cantula, Berberis lycium, Chenopodium album, Cymbopogon citratus*, Solanum viarum**

Astringent

Aegle marmelos, Asparagus adscendens, Bauhinia variegata, Cordia dichotoma, Cymbopogon citratus, Diospyros kaki, Foeniculum vulgare, Litchi chinensis*, Opuntia dillenii, Pyrus communis*, Rhododendron arboreum, Rosa brunonii**

Aesthetic

Araucaria columnaris, Catharanthus roseus, Dodonaea viscosa, Grevillea robusta, Kalanchoe pinnata, Michelia champaca, Nerium indicum, Pinus roxburghii, Populus ciliata, Populus nigra, Rosa chinensis, Thuja orientalis

Bed Wetting

Carya illinoensis†, *Sesamum orientale*

Bedding

Cassia obtusifolia, Justicia adhatoda, Murraya paniculata

Blood Purifier

Berberis lycium, Brassica nigra, Brassica tournefortii, Carissa congesta*, Centella asiatica, Plantago lanceolata, Moringa oleifera*, Roylea cinerea, Stephania glabra, Urtica ardens, Urtica dioica*, Zanthoxylum armatum*

Body-Tones

Coccinia grandis, Celastrus paniculatus, Dendrocalamus hamiltonii, Dioscorea bellophylla, Dioscorea oppositifolia, Ficus glomerata, Geranium nepalense*, Gossypium arboreum, Linum usitatissimum, Mucuna pruriens, Oxalis corniculata, Sida cordifolia, Terminalia bellirica*

Boils

Ajuga parviflora†, *Albizia chinensis, Allium cepa, Aloe vera, Alternanthera sessilis*, Berberis lycium, Bombax ceiba, Butea monosperma, Centella asiatica, Costus speciosus, Curcuma longa*, Cymbopogon citratus*, Dalbergia sissoo, Dodonaea viscosa*, Euphorbia royleana, Ficus hispida*, Ficus roxburghii*, Hibiscus rosa sinensis, Jasminum multiflorum*, Kalanchoe pinnata, Mirabilis jalapa, Plantago lanceolata, Plumbago zeylanica*, Prinsepia utilis*, Ranunculus arvensis*, Ranunculus laetus*†, *Ricinus communis, Roylea cinerea*, Stephania glabra, Syzygium cumini, Triticum aestivum, Verbascum thapsus*, Zizyphus nummularia*

Bone Fracture

Artemisia japonica, Cissampelos parriera, Dodonaea viscosa, Litsea sebifera, Oroxylum indicum, Pentanema indicum, Prunus cerasoides

Burns

Aloe vera, Caesalpinia decapetala, Chenopodium album, Dodonaea viscosa, Impatiens balsamina, Lawsonia inermis, Tagetes minuta**

Condiments

Artemisia capillaris, Coriandrum sativum, Cinnamomum tamala, Foeniculum vulgare, Murraya koenigii, Murraya paniculata, Ocimum basilicum, Ocimum kiliman-dscharicum, Origanum vulgare

Constipation

*Aegle marmelos, Chenopodium album, Drimia indica, Eleusine coracana**, *Euphorbia royleana**, *Ficus palmata, Mallotus philippinensis**, *Opuntia dillenii**, *Oroxylum indicum, Psidium guajava, Pyrus communis**, *Ricinus communis**, *Rosa damascena**, *Rubus ellipticus**, *Sorghum bicolor, Terminalia bellirica, Verbascum thapsus*

Dental Problems

*Albizia lebbeck, Boehmeria macrophylla**, *Butea monosperma**, *Calotropis procera, Carya illinoensis*†, *Cassia occidentalis**, *Centipeda minima, Cinnamomum tamala**, *Citrus limon**, *Citrus medica, Clematis gauriana**, *Engelhardia colebrookeana**, *Ficus benghalensis, Heteropogon contortus**, *Jasminum grandiflorum, Juglans regia, Lantana camara, Lepidagathis cuspidata**, *Mangifera indica, Murraya koenigii**, *Murraya paniculata**, *Ocimum basilicum**, *Plumbago zeylanica**, *Prunus cerasoides**, *Psidium guajava, Ricinus communis**, *Sesamum orientale**, *Solanum viarum, Spilanthes oleracea, Terminalia chebula, Vitex negundo**, *Zanthoxylum armatum*

Diabetes

Ajuga parviflora†, *Carica papaya, Cassia obtusifolia**, *Catharanthus roseus, Clematis gauriana**, *Dodonaea viscosa**, *Eleusine coracana, Ficus palmata**, *Glycine max**, *Kalanchoe pinnata**, *Mangifera indica, Momordica charantia, Psidium guajava, Sechium edule**, *Syzygium cumini**, *Terminalia chebula, Thuja orientalis**, *Tinospora cordifolia, Trigonella foenum-graecum*

Diuretic

*Achyranthes aspera, Aerva lanata**, *Amaranthus viridis, Butea monosperma, Cassia occidentalis, Chenopodium album, Cynodon dactylon, Didymocarpus pedicellata**, *Drimia indica, Emblica officinalis, Grewia disperma**, *Grewia optiva**, *Heteropogon contortus, Phyllanthus urinaria, Portulaca grandiflora, Zea mays*

Dye

Bauhinia variegata, Camellia sinensis, Carya illinoensis, Emblica officinalis, Impatiens balsamina, Juglans regia, Terminalia tomentosa

Ear Complaints

*Allium cepa, Allium sativum, Cheilanthes farinosa, Costus speciosus, Erythrina suberosa**, *Panicum antidontale**, *Tagetes minuta*

Edible

Abelmoschus esculentus, Aegle marmelos, Agave cantula, Agave wightii, Allium cepa, Allium sativum, Aloe vera, Alternanthera sessilis, Amaranthus paniculatus, Amaranthus viridis, Apium leptophyllum, Artocarpus lacucha, Basella alba, Bauhinia vahlii, Bauhinia variegata, Berberis lycium, Bergenia ciliata, Boerhaavia diffusa, Bombax ceiba, Brassica nigra, Brassica tournefortii, Butea monosperma, Cajanus cajan, Callicarpa macrophylla, Capsicum annuum, Carica papaya, Carissa congesta, Carya illinoensis, Cassia floribunda, Cassia obtusifolia, Cassia occidentalis, Celastrus paniculatus, Chenopodium ambrosioides, Chenopodium album, Cicer arietinum, Citrus grandis, Citrus jambhiri, Citrus limon, Citrus medica, Citrus medica var *medica, Coccinia grandis, Colocasia esculenta, Commelina paludosa, Cordia dichotoma, Coriandrum sativum, Coronopus didymus, Cucumis melo* var *agrestis, Cucumis sativus, Cucurbita hispida, Cucurbita pepo, Cyclanthera pedata, Cyphomandra betacea, Debregeasia hypoleuca, Deeringia amaranthoides, Dendrocalamus hamiltonii, Dioscorea bellophylla, Dioscorea bulbifera, Dioscorea oppositifolia, Diospyros kaki, Diplazium esculentum, Dolichos biflorus, Eleusine coracana, Emblica officinalis, Eriobotrya japonica, Euphorbia royleana, Fagopyrum cymosum, Fagopyrum esculentum, Ficus benghalensis, Ficus carica, Ficus clavata, Ficus foveolata, Ficus hispida, Ficus palmata, Ficus roxburghii, Ficus semicordata, Flacourtia indica, Foeniculum vulgare, Fragaria indica, Galium aparine, Gonatanthus pumilus, Glycine max, Grewia disperma, Grewia elastica, Grewia optiva, Indigofera cassioides, Ipomoea batatas, Ipomoea turbinata, Juglans regia, Lathyrus aphaca, Lathyrus sativus, Lepidium sativum, Linum usitatissimum, Litchi chinensis, Litsea monopetala, Luffa acutangula, Luffa cylindrica, Mangifera indica, Medicago polymorpha, Mentha longifolia, Mentha piperita, Momordica charantia, Momordica dioica, Moringa oleifera, Morus alba, Murraya koenigii, Murraya paniculata, Musa sapientum, Neolitsea umbrosa, Opuntia dillenii, Oroxylum indicum, Ougeinia dalbergioides, Oxalis corniculata, Pennisetum americanum, Phoenix acaulis, Phoenix sylvestris, Pinus roxburghii, Portulaca grandiflora, Prunus cerasoides, Prunus domestica, Prunus persica, Psidium guajava, Punica granatum, Pyrus communis, Pyrus pashia, Pyrus serotina, Randia dumetorum, Ranunculus muricatus, Raphanus sativus, Reinwardtia indica, Rhododendron arboreum, Rorippa nasturtium-aquaticum, Rubus ellipticus, Rubus niveus, Rumex hastatus, Rumex nepalensis, Saccharum officinarum, Scandix pecten veneris, Sechium edule, Sesamum orientale, Silene conoidea, Solanum melongena, Solanum nigrum, Solanum torvum, Spondias pinnata, Sorghum bicolor, Stellaria media, Syzygium cumini, Syzygium operculatum, Terminalia arjuna, Terminalia bellirica, Terminalia chebula, Trichosanthes cucumerina, Trigonella foenum-graecum, Tulipa stellata, Urtica ardens, Urtica dioica, Vicia faba, Vicia hirsuta, Vicia sativa, Vigna radiata, Vitis repanda, Vitis vinifera, Zanthoxylum armatum, Zephyranthes carinata, Zizyphus jujuba, Zizyphus nummularia*

Eye Diseases

Abrus precatorius, Albizia lebbeck, Berberis lycium, Boerhaavia diffusa, Butea monosperma, Coriandrum sativum, Lepidium sativum, Linum usitatissimum*, Litsea sebifera, Mangifera indica, Oxalis corniculata, Rorippa nasturtium-aquaticum*, Solanum erianthum*

Febrifuge

Apium leptophyllum, Dendrocalamus hamiltonii, Drimia indica, Fumaria indica, Jasminum grandiflorum, Momordica charantia*, Pogostemon benghalense, Rumex hastatus, Tinospora cordifolia*

Fertility

Curcuma longa, Diplocyclos palmatus, Euphorbia royleana, Mucuna pruriens

Fibre

Abelmoschus esculentus, Agave cantula, Bauhinia vahlii, Bombax ceiba, Eulaliopsis binata, Grewia disperma, Grewia elastica, Grewia optiva, Milletia extensa, Porana paniculata

Fish Poison

Euphorbia royleana, Sapium insigne

Fodder

Albizia chinensis, Albizia julibrissin, Albizia lebbeck, Alternanthera sessilis, Artocarpus lacucha, Arundinaria falcata, Avena sativa, Bambusa nutans, Bauhinia vahlii, Bauhinia variegata, Berberis lycium, Butea monosperma, Caesalpinia decapetala, Capillipedium assimile, Carissa congesta, Celtis australis, Cordia dichotoma, Cryptolepis buchanani, Cymbopogon citratus, Cynodon dactylon, Debregeasia hypoleuca, Dendrocalamus hamiltonii, Ehretia acuminata, Ehretia laevis, Eriophorum comosum, Euonymus hamiltonianus, Ficus clavata, Ficus foveolata, Ficus glomerata, Ficus hispida, Ficus infectoria, Ficus roxburghii, Ficus rumphii, Ficus semicordata, Flacourtia indica, Fumaria indica, Galium aparine, Grewia elastica, Grewia disperma, Grewia optiva, Helicteres isora, Heteropogon contortus, Hypericum oblongifolium, Indigofera cassioides, Lannea coromandelica, Lathyrus aphaca, Leucaena glauca, Litsea monopetala, Litsea semicarpifolia, Loranthus falcatus, Mallotus philippinensis, Medicago polymorpha, Milletia extensa, Moringa oleifera, Morus alba, Musa sapientum, Ougeinia dalbergioides, Panicum antidontale, Pennisetum americanum, Persea odoratissima, Phoenix acaulis, Phoenix sylvestris, Pistacia khinjuk, Polygala chinensis, Polygonum barbatum, Populus ciliata, Populus nigra, Porana paniculata, Prunus cerasoides, Pyrus pashia, Quercus leucotrichophora, Randia dumetorum, Rhamnus triquetra, Rubus ellipticus, Rubus niveus, Salix karelini, Sapium sebiferum, Scandix pecten veneris, Silene conoidea, Sorghum bicolor, Spondias pinnata, Sporobolus indicus, Stellaria media, Syzygium operculatum, Terminalia arjuna, Terminalia bellirica, Terminalia chebula, Terminalia tomentosa, Thysanolaena maxima, Tinospora cordifolia, Trifolium resupinatum, Vicia faba, Vicia hirsuta, Vicia sativa, Vigna radiata, Woodfordia fruticosa, Wrightia arborea, Zea mays, Zizyphus jujuba, Zizyphus nummularia

Fuel

Albizia chinensis, Albizia julibrissin, Albizia lebbeck, Bauhinia variegata, Cajanus cajan, Carissa congesta, Caryopteris bicolor, Cassia fistula, Celtis australis, Colebrookea oppositifolia, Cordia dichotoma, Dalbergia sissoo, Debregeasia hypoleuca,

Dendrocalamus hamiltonii, Dodonaea viscosa, Duranta repens, Ehretia laevis, Erythrina suberosa, Euonymus hamiltonianus, Ficus glomerata, Flacourtia indica, Grewia disperma, Grewia elastica, Grewia optiva, Hamiltonia suaveolens, Helicteres isora, Lannea coromandelica, Leucaena glauca, Mallotus philippinensis, Moringa oleifera, Murraya koenigii, Oroxylum indicum, Pinus roxburghii, Populus nigra, Prunus domestica, Randia dumetorum, Rhamnus triquetra, Rhododendron arboreum, Salix karelini, Salix tetrasperma, Sapium sebiferum, Syzygium operculatum, Terminalia bellirica, Terminalia chebula, Terminalia tomentosa, Woodfordia fruticosa, Wrightia arborea, Zea mays

Gastro-Intestinal Disorders

Aegle marmelos, Albizia lebbeck, Allium sativum, Aloe vera, Apium leptophyllum*, Artemisia capillaris*, Asparagus adscendens, Bauhinia variegata, Bergenia ciliata, Boerhaavia diffusa, Bombax ceiba, Cajanus cajan, Cannabis sativa, Carica papaya, Cassia fistula, Centella asiatica, Centratherum anthelminticum, Cinnamomum tamala, Cissampelos parriera, Didymocarpus pedicellata*, Euphorbia royleana, Ficus benghalensis, Ficus glomerata, Flacourtia indica, Foeniculum vulgare, Indigofera cassioides*, Leucas cephalotes, Mallotus philippinensis, Mentha longifolia, Mentha piperita, Micromeria biflora, Moringa oleifera, Ocimum basilicum, Ocimum gratissimum, Ocimum sanctum*, Origanum vulgare*, Pentanema indicum, Perilla frutescens*, Rosa chinensis*, Rosa damascena, Roylea cinerea*, Selaginella chrysocaulos†, Solanum viarum*, Terminalia bellirica, Terminalia chebula, Tinospora cordifolia, Toona ciliata, Trigonella foenum-graecum, Vitex negundo, Woodfordia fruticosa, Wrightia arborea, Zizyphus jujuba*

Hair Care (Baldness, Hair Loss)

Adiantum capillus veneris, Ajuga parviflora†, Brassica tournefortii, Carya illinoensis, Emblica officinalis, Grewia disperma*, Grewia optiva, Hibiscus rosa sinensis*, Mangifera indica*, Melia azedarach, Pyrus pashia*, Sapindus mukorosii*

Headache

Abrus precatorius, Centipeda minima, Justicia adhatoda, Lawsonia inermis, Melia azadarach, Michelia champaca, Randia dumetorum, Santalum album, Toona ciliata, Vitex negundo**

Hedges

Agave cantula, Carissa congesta, Dendrocalamus hamiltonii, Duranta repens, Lantana camara, Opuntia dillenii, Phoenix acaulis, Rubus niveus, Zizyphus nummularia

Heart Complaints

Allium sativum, Camellia sinensis, Cinnamomum tamala, Eleusine coracana, Momordica charantia, Pennisetum americanum*, Portulaca grandiflora, Reinwardtia indica*, Rhododendron arboreum, Solanum nigrum, Sorghum bicolor*, Terminalia arjuna, Terminalia chebula*

Herpes

Aerva lanata, Ficus benghalensis**

Household Articles

Aristida cyanantha, Bambusa nutans, Bauhinia vahlii, Capillipedium assimile, Cucurbita pepo, Dendrocalamus hamiltonii, Eriophorum comosum, Eulaliopsis binata, Ficus roxburghii, Phoenix acaulis, Phoenix sylvestris, Pinus roxburghii, Sporobolus indicus, Thysanolaena maxima, Zanthoxylum armatum, Zea mays

Injuries

Cannabis sativa, Colebrookea oppositifolia, Curcuma longa, Cynodon dactylon, Dodonaea viscosa, Justicia adhatoda*, Reinwardtia indica*, Solanum erianthum*

Insecticidal

Acorus calamus, Ageratum conyzoides, Ajuga parviflora†*, Roylea cinerea*

Insect Repellent

Artemisia japonica, Lantana camara, Ocimum sanctum

Internal Injuries

Curcuma longa, Eulaliopsis binata*, Oroxylum indicum, Prunus cerasoides, Vanda tessellata*, Vitex negundo*

Kidney Disorders

*Achyranthes aspera, Achyranthes bidentata**, *Aerva lanata**, *Bergenia ciliata, Butea monosperma, Catharanthus roseus**, *Cucumis sativus, Debregeasia hypoleuca**, *Dolichos biflorus, Kalanchoe pinnata, Lawsonia inermis, Portulaca grandiflora, Sesamum orientale** *Sorghum bicolor, Urtica dioica*

Lactation

Acacia catechu, Asparagus adscendens, Cirsium arvense, Cryptolepis buchanani, Grewia optiva, Helicteres isora, Phoenix acaulis, Phoenix sylvestris, Quercus leucotrichophora, Tinospora cordifolia

Leucorrhoea

*Capillipedium assimile**, *Cissampelos parriera, Cymbopogon citratus**, *Dalbergia sissoo**, *Hibiscus rosa sinensis**, *Opuntia dillenii**, *Rosa damascena**, *Woodfordia fruticosa*

Liver Complaints

*Aerva lanata, Aloe vera, Cajanus cajan, Carica papaya, Centella asiatica, Cuscuta reflexa, Emblica officinalis, Melia azedarach**, *Ocimum sanctum**, *Phyllanthus urinaria**, *Portulaca grandiflora, Raphanus sativus, Saccharum officinarum, Terminalia bellirica, Tinospora cordifolia, Urtica dioica*

Lung Diseases

*Acacia catechu, Achyranthes aspera, Acorus calamus, Bombax ceiba, Cassia obtusifolia**, *Cassia occidentalis, Cinnamomum camphora**, *Cinnamomum tamala, Citrus limon, Cordia dichotoma, Curcuma longa, Datura stramonium, Drimia indica, Ficus palmata, Ficus religiosa**, *Justicia adhatoda, Lawsonia inermis, Lepidium sativum, Leucas cephalotes, Micromeria biflora, Momordica charantia**, *Ocimum kilimandscharicum**,

*Ocimum sanctum, Pistacia khinjuk, Pogostemon benghalense, Punica granatum, Randia dumetorum, Solanum viarum**, *Terminalia bellirica, Terminalia chebula, Vanda tessellata**, *Viola patrinii, Viola pilosa*

Magico-Religious

Abrus precatorius, Achyranthes aspera, Aegle marmelos, Aerva lanata, Asparagus adscendens, Berberis lycium, Bombax ceiba, Butea monosperma, Calotropis procera, Capsicum annuum, Cassia fistula, Cestrum nocturnum, Cinnamomum camphora, Citrus limon, Citrus medica var *medica, Cucurbita hispida, Curcuma longa, Cynodon dactylon, Datura innoxia, Datura stramonium, Emblica officinalis, Erythrina suberosa, Euphorbia royleana, Ficus benghalensis, Ficus glomerata, Ficus infectoria, Ficus religiosa, Gnaphalium affine, Gossypium arboreum, Grewia optiva, Hibiscus rosa sinensis, Holarrhena antidysentrica, Jasminum dispermum, Jasminum grandiflorum, Jasminum humile, Jasminum multiflorum, Lotus corniculatus, Mallotus philippinensis, Mangifera indica, Mirabilis jalapa, Musa sapientum, Ocimum kilimandscharicum, Ocimum sanctum, Pistacia khinjuk, Prinsepia utilis, Punica granatum, Reinwardtia indica, Rosa chinensis, Rosa damascena, Saccharum officinarum, Saccharum spontaneum, Salvia coccinea, Santalum album, Sesamum orientale, Sisymbrium irio, Tagetes minuta, Tinospora cordifolia, Tulipa stellata, Vigna radiata, Zephyranthes carinata*

Massages

Acorus calamus, Brassica tournefortii, Prunus persica

Measals

*Aerva lanata**, *Musa sapientum**, *Sisymbrium irio**, *Verbascum thapsus**

Menstruation Complaints

*Aloe vera, Amaranthus viridis, Boehmeria macrophylla**, *Butea monosperma, Holarrhena antidysentrica**, *Terminalia arjuna**

Mouth Odour (Foul-Breath)

*Camellia sinensis, Citrus limon**, *Citrus medica, Zanthoxylum armatum*

Mouth Sores/Ulcers

*Acacia catechu, Bidens biternata, Cajanus cajan, Callicarpa macrophylla**, *Cassia occidentalis**, *Musa sapientum**, *Plantago lanceolata**, *Psidium guajava, Pyrus pashia, Solanum nigrum, Spilanthes oleracea, Syzygium cumini, Vitex negundo**, *Zanthoxylum armatum*

Nasal Problems

Cynodon dactylon, Rhododendron arboreum

Nectar

Caryopteris bicolor, Grevillea robusta, Woodfordia fruticosa

Pains And Aches

*Albizia julibrissin**, *Butea monosperma, Carissa congesta**, *Celastrus paniculatus**, *Cuscuta reflexa**, *Eucalyptus tereticornis**, *Justicia adhatoda**, *Moringa oleifera, Sida cordifolia**, *Tinospora cordifolia**, *Vitex negundo**

Paralysis

*Abrus precatorius, Justicia adhatoda**

Piles

Aegle marmelos, Butea monosperma, Dioscorea bellophylla, Dioscorea bulbifera, Holarrhena antidysentrica, Jatropha curcas, Momordica charantia, Pennisetum americanum*, Punica granatum*, Tagetes minuta**

Purgative

Amaranthus viridis, Cassia floribunda, Cassia occidentalis, Chenopodium album, Emblica officinalis, Ficus carica, Jatropha curcas, Mirabilis jalapa, Terminalia bellirica, Trichosanthes cucumerina*

Refrigerant

Aegle marmelos, Aloe vera, Caesalpinia decapetala, Centella asiatica, Cirsium arvense*, Coriandrum sativum, Cucumis sativus, Cymbopogon citratus, Cynodon dactylon, Emblica officinalis, Euphorbia royleana*, Holarrhena antidysentrica, Lawsonia inermis, Lepidagathis cuspidata*, Morus alba, Musa sapientum, Phoenix acaulis, Phoenix sylvestris*, Portulaca grandiflora, Rhododendron arboreum*, Rubus niveus, Rumex hastatus, Saccharum officinarum, Santalum album, Tinospora cordifolia, Triticum aestivum, Vitis repanda, Vitis vinifera, Zea mays, Zizyphus jujuba*, Zizyphus nummularia*

Rheumetic Pains

Allium sativum, Artemisia capillaris, Basella alba*, Boerhaavia diffusa, Calotropis procera, Cannabis sativa*, Cinnamomum camphora*, Citrus limon, Cuscuta reflexa*, Datura innoxia, Eucalyptus tereticornis*, Litsea sebifera, Michelia champaca, Murraya koenigii, Oroxylum indicum, Prinsepia utilis, Ricinus communis, Sesamum orientale, Spilanthes oleracea*, Trigonella foenum-graecum, Urtica dioica, Vitex negundo*

Rumination

Aerva lanata, Carica papaya, Equisetum debile, Grewia optiva, Leucas cephalotes, Pentanema indicum, Pogostemon benghalense, Syzygium cumini

Skin Diseases

Abrus precatorius, Aerva lanata, Ajuga parviflora†*, Allium sativum, Aloe vera, Butea monosperma, Carica papaya, Carissa congesta, Cassia fistula, Centratherum* *anthelminticum, Coriandrum sativum*, Cucumis sativus, Ehretia laevis*, Ficus benghalensis, Jatropha curcas*, Mallotus philippinensis, Michelia champaca*, Nerium indicum, Pinus roxburghii, Polygonum barbatum*, Ranunculus laetus*†, *Ricinus communis, Santalum album, Sapium insigne*, Stephania glabra, Tagetes minuta, Toona ciliata**

Sore Toes

Carya illinoensis†, *Cassia fistula*, Dalbergia sissoo*, Impatiens balsamina**

Sprain

Artemisia japonica, Cassia floribunda, Dodonaea viscosa, Lantana camara, Litsea sebifera, Murraya koenigii, Prunus cerasoides, Sesamum orientale*, Vitex negundo*

'Sur' (Local Drinks)

Achyranthes bidentata, Albizia chinensis, Callicarpa macrophylla, Celastrus paniculatus, Cymbopogon citratus, Justicia adhatoda, Panicum antidontale, Woodfordia fruticosa, Zizyphus nummularia

Swellings

*Artemisia capillaris**

Thatching

Cocculus laurifolius, Dodonaea viscosa, Phoenix acaulis, Phoenix sylvestris

Vermillion

Citrus jambhiri, Curcuma longa, Santalum album, Selaginella chrysocaulos

Walking Sticks

Arundinaria falcata, Bambusa nutans, Dodonaea viscosa, Prunus cerasoides, Pyrus pashia, Zanthoxylum armatum

Wood

Acacia catechu, Albizia chinensis, Albizia lebbeck, Arundinaria falcata, Araucaria columnaris, Bambusa nutans, Bombax ceiba, Cedrus deodara, Celtis australis, Dalbergia sissoo, Dendrocalamus hamiltonii, Ehretia acuminata, Ehretia laevis, Engelhardia colebrookeana, Erythrina suberosa, Eucalyptus tereticornis, Juglans regia, Litsea monopetala, Litsea sebifera, Litsea semicarpifolia, Mangifera indica, Melia azedarach, Ougeinia dalbergioides, Persea odoratissima, Pinus roxburghii,

Populus ciliata, Populus nigra, Quercus leucotrichophora, Sapindus mukorosii, Sapium sebiferum, Syzygium cumini, Terminalia arjuna, Terminalia tomentosa, Toona ciliata, Zizyphus jujuba

Wounds

*Agave cantula, Callicarpa macrophylla, Cannabis sativa, Cassia fistula**, *Commelina paludosa, Emblica officinalis, Jasminum dispermum**, *Jatropha curcas, Lannea coromandelica, Solanum erianthum**

Misscelaneous

Abrus precatorius, Achyranthes bidentata, Agave cantula, Aegle marmelos, Ageratum conyzoides, Albizia chinensis, Aloe vera, Arundinaria falcata, Bauhinia vahlii, Bombax ceiba, Cajanus cajan, Camellia sinensis, Carissa congesta, Caryopteris bicolor, Cassia fistula, Celtis australis, Centella asiatica, Cestrum nocturnum, Cheilanthes farinosa, Cinnamomum tamala, Citrus grandis, Cocculus laurifolius, Colebrookea oppositifolia, Cryptolepis buchanani, Curcuma longa, Cuscuta reflexa, Cynodon dactylon, Dalbergia sissoo, Dendrocalamus hamiltonii, Diplocyclos palmatus, Dodonaea viscosa, Emblica officinalis, Eriophorum comosum, Eulaliopsis binata, Ficus benghalensis, Ficus hispida, Ficus glomerata, Ficus palmata, Ficus semicordata, Flacourtia indica, Gossypium arboreum, Grewia disperma, Grewia elastica, Grewia optiva, Grevillea robusta, Helicteres isora, Holarrhena antidysentrica, Jatropha curcas, Justicia adhatoda, Lepidium sativum, Litsea monopetala, Loranthus falcatus, Luffa acutangula, Luffa cylindrica, Melia azedarach, Michelia champaca, Milletia extensa, Momordica charantia, Morus alba, Mucuna pruriens,

Musa sapientum, Ocimum basilicum, Ocimum kilimandscharicum, Oxalis corniculata, Pennisetum americanum, Phoenix acaulis, Phoenix sylvestris, Pinus roxburghii, Polygala chinensis, Porana paniculata, Prunus cerasoides, Prunus persica, Pyrus pashia, Randia dumetorum, Rhamnus triquetra, Rhododendron arboreum, Roylea cinerea, Saccharum spontaneum, Salix karelini, Salix tetrasperma, Santalum album, Sapindus mukorosii, Spilanthes oleracea, Syzygium operculatum, Terminalia arjuna, Terminalia bellirica, Terminalia tomentosa, Zanthoxylum armatum

Appendix III

INDEX OF PLANTS USED FOR VARIOUS EDIBLE PREPARATIONS

Fruits

Abelmoschus esculentus, Aegle marmelos, Artocarpus lacucha, Berberis lycium, Bombax ceiba, Callicarpa macrophylla, Capsicum annuum, Carica papaya, Carissa congesta, Carya illinoensis, Cassia floribunda, Cassia occidentalis, Celastrus paniculatus, Citrus grandis, Citrus jambhiri, Citrus limon, Citrus medica, Citrus medica var. *medica, Coccinia grandis, Cordia dichotoma, Coriandrum sativum, Cucumis melo* var. *agrestis, Cucumis sativus, Cucurbita hispida, Cucurbita pepo, Cyclanthera pedata, Cyphomandra betacea, Debregeasia hypoleuca, Diospyros kaki, Eleusine coracana, Emblica officinalis, Eriobotrya japonica, Ficus benghalensis, Ficus carica, Ficus clavata, Ficus foveolata, Ficus hispida, Ficus palmata, Ficus roxburghii, Ficus semicordata, Flacourtia indica, Foeniculum vulgare, Fragaria indica, Grewia disperma, Grewia elastica, Grewia optiva, Juglans regia, Litchi chinensis, Litsea monopetala, Luffa acutangula, Luffa cylindrica, Mangifera indica, Momordica charantia, Momordica dioica, Moringa oleifera, Morus alba, Murraya koenigii, Murraya paniculata, Musa sapientum, Neolitsea umbrosa, Opuntia dillenii, Oroxylum indicum, Pennisetum americanum, Phoenix acaulis, Phoenix sylvestris, Prunus cerasoides, Prunus domestica, Prunus persica, Psidium guajava, Punica granatum, Pyrus communis, Pyrus pashia, Pyrus serotina, Randia dumetorum, Raphanus sativus, Rubus ellipticus, Rubus niveus, Sechium edule, Solanum melongena, Solanum nigrum, Solanum torvum, Spondias pinnata, Sorghum bicolor, Syzygium cumini, Syzygium operculatum, Terminalia arjuna, Terminalia bellirica, Terminalia chebula, Trichosanthes cucumerina, Vicia faba, Vitis repanda, Vitis vinifera, Zizyphus jujuba, Zizyphus nummularia*

Pot-herbs & Vegetable

Abelmoschus esculentus, Agave cantula, Agave wightii, Allium cepa, Allium sativum, Aloe vera, Alternanthera sessilis, Amaranthus paniculatus, Amaranthus viridis, Artocarpus lacucha, Basella alba, Bauhinia variegata, Boerhaavia diffusa, Bombax ceiba, Brassica nigra, Brassica tournefortii, Cajanus cajan, Carica papaya, Cassia floribunda, Cassia occidentalis, Chenopodium album, Chenopodium ambrosioides, Cicer arietinum, Coccinia grandis, Colocasia esculenta, Commelina paludosa, Cordia dichotoma, Coronopus didymus, Cyclanthera pedata, Cucurbita hispida, Cucurbita hispida, Cucurbita pepo, Cyphomandra betacea, Debregeasia hypoleuca, Dendrocalamus hamiltonii, Dioscorea bellophylla, Dioscorea bulbifera, Dioscorea oppositifolia, Diplazium esculentum, Dolichos biflorus, Euphorbia royleana, Fagopyrum cymosum, Fagopyrum esculentum, Ficus hispida, Ficus palmata, Ficus roxburghii, Ficus semicordata, Foeniculum vulgare, Galium aparine, Gonatanthus pumilus, Glycine max, Ipomoea turbinata, Lathyrus aphaca, Lathyrus sativus, Lepidium sativum, Luffa acutangula, Luffa cylindrica, Mangifera indica, Medicago polymorpha, Momordica charantia, Momordica dioica, Moringa oleifera, Musa sapientum, Oroxylum indicum, Oxalis corniculata, Portulaca grandiflora, Ranunculus muricatus, Raphanus sativus, Rorippa nasturtium-aquaticum, Rumex hastatus, Rumex nepalensis, Scandix pecten veneris, Sechium edule, Silene conoidea, Solanum melongena, Solanum torvum, Stellaria media, Trichosanthes cucumerina, Trigonella foenum-graecum, Urtica ardens, Urtica dioica, Vicia faba, Vicia hirsuta, Vigna radiata

Pickle

Abelmoschus esculentus, Agave cantula, Agave wightii, Allium sativum, Artocarpus lacucha, Bauhinia variegata, Capsicum annuum, Cassia occidentalis, Citrus limon, Citrus medica, Citrus medica var. *medica, Colocasia esculenta, Cordia dichotoma, Dendrocalamus hamiltonii, Dioscorea bellophylla, Diplazium esculentum, Emblica officinalis, Gonatanthus pumilus, Mangifera indica, Momordica charantia, Prunus domestica, Prunus persica, Pyrus communis, Pyrus serotina, Solanum melongena, Spondias pinnata, Terminalia arjuna, Terminalia chebula, Vicia sativa*

Flavouring

Allium sativum, Artemisia capillaris, Capsicum annuum, Cinnamomum tamala, Coriandrum sativum, Foeniculum vulgare, Murraya koenigii, Murraya paniculata, Ocimum basilicum, Ocimum kilimandscharicum, Origanum vulgare

Culinary

Aloe vera, Artocarpus lacucha, Basella alba, Bergenia ciliata, Brassica nigra, Brassica tournefortii, Carica papaya, Coriandrum sativum, Gonatanthus pumilus, Indigofera cassioides, Lathyrus aphaca, Lathyrus sativus, Mangifera indica, Morus alba, Ougeinia dalbergioides, Reinwardtia indica, Rhododendron arboreum, Vicia hirsuta, Vigna radiata

Chutney

Allium cepa, Allium sativum, Cassia obtusifolia, Coriandrum sativum, Mentha longifolia, Mentha piperita, Murraya koenigii, Murraya paniculata, Musa sapientum, Punica granatum, Rhododendron arboreum, Sesamum orientale, Spondias pinnata, Zanthoxylum armatum

Soup

Abelmoschus esculentus, Raphanus sativus

Salad

Allium cepa, Raphanus sativus

Sweet-Pudding, Jams & Jellies

Celastrus paniculatus, Cucurbita hispida, Eriobotrya japonica, Foeniculum vulgare, Prunus domestica, Pyrus communis, Pyrus serotina

Chapaati

Eleusine coracana, Pennisetum americanum, Sorghum bicolor

Appendix-IV

INDEX OF PLANTS WITH ADDITIONAL USES

Human-ailments

Anthelmintic	*Rumex nepalensis*
Antidote	*Cannabis sativa, Capsicum annuum, Sapindus mukorosii*
Antiseptic	*Agave cantula, Cymbopogon citratus, Solanum viarum*
Astringent	*Cymbopogon citratus, Litchi chinensis, Pyrus communis, Rosa brunonii*
Blood Purifier	*Brassica tournefortii, Carissa congesta, Moringa oleifera, Urtica dioica*
Body Tone	*Dendrocalamus hamiltonii, Geranium nepalense*
Boils & Sores	*Alternanthera sessilis, Curcuma longa, Cymbopogon citratus, Dodonaea viscosa, Ficus hispida, Ficus roxburghii, Jasminum multiflorum, Plumbago zeylanica, Prinsepia utilis, Ranunculus arvensis, Roylea cinerea, Verbascum thapsus*
Burns	*Chenopodium album, Tagetes minuta*
Cardiac Disorders	*Eleusine coracana, Pennisetum americanum, Reinwardtia indica, Sorghum bicolor*
Constipation	*Eleusine coracana, Euphorbia royleana, Mallotus philippinensis, Opuntia dillenii, Pyrus communis, Ricinus communis, Rosa damascena, Rubus ellipticus*
Dental Problems	*Boehmeria macrophylla, Butea monosperma, Cassia occidentalis, Cinnamomum tamala, Citrus limon, Clematis gauriana, Engelhardia colebrookeana, Heteropogon contortus, Lepidagathis cuspidata, Murraya koenigii, Murraya paniculata, Ocimum basilicum, Plumbago zeylanica, Prunus cerasoides, Ricinus communis, Sesamum orientale, Vitex negundo*
Diabetes	*Cassia obtusifolia, Clematis gauriana, Dodonaea viscosa, Ficus palmata, Glycine max, Kalanchoe pinnata, Sechium edule, Syzygium cumini, Thuja orientalis,*
Diuretic	*Aerva lanata, Didymocarpus pedicellata, Grewia disperma, Grewia optiva,*
Ear Complaints	*Erythrina suberosa, Panicum antidontale*
Eye Diseases	*Lepidium sativum, Linum usitatissimum, Rorippa nasturtium-aquaticum*
Febrifuge	*Dendrocalamus hamiltonii, Momordica charantia*

Contd...

Contd...

Gastrointestinal Problems	*Albizia lebbeck, Apium leptophyllum, Artemisia capillaris, Didymocarpus pedicellata, Indigofera cassioides, Ocimum sanctum, Origanum vulgare, Perilla frutescens, Rosa chinensis, Solanum viarum*
Hair Care	*Adiantum capillus veneris, Grewia disperma, Hibiscus rosa-sinensis, Mangifera indica, Pyrus pashia*
Headache	*Toona ciliata, Vitex negundo*
Herpes	*Aerva lanata, Ficus benghalensis*
Injuries	*Curcuma longa, Justicia adhatoda, Reinwardtia indica*
Internal Injuries	*Curcuma longa, Eulaliopsis binata, Justicia adhatoda, Vanda tessellata, Verbascum thapsus*
Kidney Ailments	*Achyranthes bidentata, Aerva lanata, Catharanthus roseus, Debregeasia hypoleuca, Didymocarpus pedicellata, Grewia disperma, Grewia optiva, Sesamum orientale*
Leucorrhoea	*Capillipedium assimile, Cymbopogon citratus, Dalbergia sissoo, Hibiscus rosa-sinensis, Opuntia dillenii, Rosa damascena*
Liver Complaints	*Melia azedarach, Ocimum sanctum, Phyllanthus urinaria*
Lungs Diseases	*Cassia obtusifolia, Cinnamomum camphora, Ficus religiosa, Momordica charantia, Ocimum kilimandscharicum, Solanum viarum,Vanda tessellata*
Measles	*Aerva lanata, Musa sapientum, Sisymbrium irio, Verbascum thapsus*
Menstruation Complaints	*Boehmeria macrophylla, Holarrhena antidysentrica, Terminalia arjuna*
Mouth Odour	*Citrus limon*
Mouth Sores/Ulcers	*Callicarpa macrophylla, Cassia occidentalis, Musa sapientum, Plantago lanceolata, Vitex negundo*
Paralysis	*Justicia adhatoda*
Pains & Aches	*Albizia julibrissin, Carissa congesta, Celastrus paniculatus, Cuscuta reflexa, Eucalyptus tereticornis, Justicia adhatoda, Sida cordifolia, Tinospora cordifolia, Vitex negundo*
Piles	*Momordica charantia, Pennisetum americanum, Punica granatum, Tagetes minuta*
Purgative	*Cassia floribunda, Eluusine coracana, Euphorbia royleana, Mallotus philippinensis, Opuntia dillenii, Pyrus communis, Ricinus communis, Rosa damascena, Rubus ellipticus*
Refrigerant	*Aloe vera, Cirsium arvense, Euphorbia royleana, Lepidagathis cuspidata, Phoenix sylvestris, Rhododendron arboreum, Zizyphus jujuba*
Rheumatism	*Artemisia capillaris, Basella alba, Cannabis sativa, Cinnamomum camphora, Cuscuta reflexa, Eucalyptus tereticornis, Spilanthes oleracea*

Contd...

Contd...

Skin problems	*Coriandrum sativum, Ehretia laevis, Jatropha curcas, Michelia champaca, Polygonum barbatum, Sapium insigne, Toona ciliata*
Sore-Toes	*Cassia fistula, Dalbergia sissoo, Impatiens balsamina*
Sprain & Fracture	*Artemisia japonica, Sesamum orientale*
Swellings	*Artemisia capillaris*
Wounds	*Cassia fistula, Jasminum dispermum, Solanum erianthum*
Fodder	*Bauhinia vahlii, Cryptolepis buchanani, Cymbopogon citratus, Dendrocalamus hamiltonii, Eriophorum comosum, Fumaria indica, Galium aparine, Grewia disperma, Grewia elastica, Heteropogon contortus, Hypericum oblogifolium, Loranthus falcatus, Polygala chinensis, Porana paniculata, Sapium sebiferum, Silene conoidea, Spondias pinnata, Stellaria media, Syzygium operculatum, Vicia faba, Zizyphus jujuba*
Fuel-wood	*Bauhinia variegata, Cajanus cajan, Caryopteris bicolor, Cassia fistula, Celtis australis, Cordia dichotoma, Debregeasia hypoleuca, Duranta repens, Ehretia laevis, Erythrina suberosa, Euonymus hamiltonianus, Flacourtia indica, Grewia disperma, Grewia elastica, Lannea coromandelica, Leucaena glauca, Moringa oleifera, Murraya koenigii, Oroxylum indicum, Prunus domestica, Randia dumetorum, Rhamnus triquetra, Syzygium operculatum, Terminalia bellirica, Terminalia chebula, Wrightia arborea, Zea mays*
Magico-Religious	*Achyranthes aspera, Aegle marmelos, Aerva lanata, Berberis lycium, Calotropis procera, Cassia fistula, Cestrum nocturnum, Cinnamomum camphora, Citrus medica var. medica, Cucurbita hispida, Emblica officinalis, Erythrina suberosa, Euphorbia royleana, Ficus glomerata, Ficus infectoria, Holarrhena antidysentrica, Jasminum dispermum, Jasminum humile, Lotus corniculatus, Musa sapientum, Ocimum kilimandscharicum, Pistacia khinjuk, Prinsepia utilis, Saccharum spontaneum, Sisymbrium irio, Tagetes minuta, Zephyranthes carinata*
Edible	*Agave cantula, Aloe vera, Apium leptophyllum, Artocarpus lacucha, Cucurbita hispida, Cucurbita pepo, Gonatanthus pumilus, Indigofera cassioides, Morus alba, Murraya paniculata*
Wood	*Engelhardia colebrookeana, Erythrina suberosa, Rhamnus triquetra, Sapindus mukorosii*
House-Hold Articles	*Aristida cyanantha, Bambusa nutans, Cucurbita pepo, Eriophorum comosum, Eulaliopsis binata, Sporobolus indicus, Zea mays*
Condiments	*Artemisia capillaris, Cinnamomum tamala, Ocimum kilimandscharicum*
Miscellaneous Uses	
Wine	*Achyranthes bidentata, Albizia chinensis, Callicarpa macrophylla, Panicum antidontale, Woodfordia fruticosa*

Contd...

Contd...

Hedges	*Agave cantula, Dendrocalamus hamiltonii, Duranta repens, Lantana camara, Rubus niveus*
Dye	*Impatiens balsamina, Terminalia tomentosa*
Nectar	*Caryopteris bicolor, Grevillea robusta, Woodfordia fruticosa*
Killing Lice	*Ageratum conyzoides, Roylea cinerea*
Asthetic	*Grevillea robusta, Michelia champaca*
Bath-Sponge	*Luffa acutangula, Luffa cylindrica*
Binder in mud-flooring	*Eulaliopsis binata, Pinus roxburghii*
Dip-Pens	*Prunus cerasoides, Saccharum spontaneum*
Thatching	*Dodonaea viscosa, Cocculus laurifolius*
Bow-making	*Grewia elastica*
Cheecking lactation	*Ficus benghalensis*
Children Amusements	*Polygala chinensis*
Cleaning Utensils	*Oxalis corniculata*
Cholesterol plaques	*Oxalis corniculata*
Dust	*Abrus precatorius*
Fibres	*Porana paniculata*
Grafting	*Salix tetrasperma*
Hockey Sticks	*Dendrocalamus hamiltonii*
Sound-Sleep (kids)	*Ficus glomerata*
Lubricant	*Ficus glomerata*
Nose-Pin	*Cheilanthes farinosa*
Agricultural Implements	*Cassia fistula*
Pestle	*Acacia catechu*
Killing Leeches	*Bauhinia vahlii*
Smoothening Mud Floor	*Bombax ceiba*
Vermillion	*Curcuma longa*
Walking Sticks	*Arundinaria falcata*

Appendix-V

INDEX OF LOCAL NAMES

Kapah	186
Kapur	103
Karaalein	60
Karale	60
Karela	239
Karkara	330
Karmarua	39
Kashmal	61
Kasmalein	61
Kathmanh	336
Keeoon	357
Kela	249
Keon	357
Keor	194
Khair	24
Khajoor	268
Khajoori	267
Khaloa	367
Khanurein	177
Kharbuja	81
Khasarkhurein	177
Khasra	173
Khatirali	110
Khaukhua	332
Kheera	125
Khidak	94
Khinnu	212
Khirak	94
Khod	205
Khukhani	231
Kuaarein	44
Kulath	148
Kulfa	280
Kungu	319
Kyor	35
Laah	323
Lahsun	43
Laljari	183
Lambia Ghaa	192
Lassan	43
Lasura	119
Lasyada	119
Litchi	220

Lonku	318
Luanhe	56
Lumb	63
Lungadu	145
Maah	359
Maethi	351
Maggar	139
Maindi	214
Makhan	317
Makoda	131
Makoi	326
Malori	262
Malti	201
Mandal	153
Maraingu	363
Mater-phali	212
Meenaki	95
Meetha karela	131
Meetha Karela	240
Meethi-Saunf	180
Mehendadu	147
Mehendu	147
Meis	349
Minmayaaoon	353
Mirach	80
Mithpatra	104
Mooli	296
Morpankh	344
Muara	331
Musakani	140
Nakk-Chhikannu	97
Nashpati	289
Neel-Kanta	150
Neelkanthi	37
Nimbu	110
Oyee	38
Paaja, Pajja	283
Padaar	190
Pahadi ajwain	49
Pahadi-Beeuins	312
Pahadi-Sannanh	223
Pahadi-Taali	317
Palah	70

Appendix-VI

INDEX OF BOTANICAL NAMES